Advances in Processing and Pattern Analysis of Biological Signals

Advances in Processing and Pattern Analysis of Biological Signals

Edited by

Isak Gath and Gideon F. Inbar

Technion-Israel Institute of Technology
Haifa, Israel

Plenum Press • New York and London

Library of Congress Cataloging-in-Publication Data

Advances in processing and pattern analysis of biological signals /
  edited by Isak Gath and Gideon F. Inbar.
       p.   cm.
    Includes bibliographical references and index.
    ISBN 0-306-45215-4
    1. Signal processing--Digital techniques--Congresses.
  2. Electroencephalography--Congresses. 3. Electromyography-
  -Congresses. 4. Electrocardiography--Congresses.   I. Gath, Isak.
  II. Inbar, Gideon F.
  R857.D47A29 1996
  616.07'547--dc20   96-13448
                                                          CIP

Proceedings of the Bat-Sheva De Rotschild Seminar, held March 19 – 25, 1995, at Technion – Israel
Institute of Technology, Haifa, Israel

ISBN 0-306-45215-4

PREFACE

In recent years there has been rapid progress in the development of signal processing in general, and more specifically in the application of signal processing and pattern analysis to biological signals. Techniques, such as parametric and nonparametric spectral estimation, higher order spectral estimation, time-frequency methods, wavelet transform, and identification of nonlinear systems using chaos theory, have been successfully used to elucidate basic mechanisms of physiological and mental processes. Similarly, biological signals recorded during daily medical practice for clinical diagnostic procedures, such as electroencephalograms (EEG), evoked potentials (EP), electromyograms (EMG) and electrocardiograms (ECG), have greatly benefitted from advances in signal processing.

In order to update researchers, graduate students, and clinicians, on the latest developments in the field, an International Symposium on Processing and Pattern Analysis of Biological Signals was held at the Technion-Israel Institute of Technology, during March 1995. This book contains 27 papers delivered during the symposium.

The book follows the five sessions of the symposium. The first section, Processing and Pattern Analysis of Normal and Pathological EEG, accounts for some of the latest developments in the area of EEG processing, namely: time varying parametric modeling; non-linear dynamic modeling of the EEG using chaos theory; Markov analysis; delay estimation using adaptive least-squares filtering; and applications to the analysis of epileptic EEG, EEG recorded from psychiatric patients, and sleep EEG.

In the second section, Investigation of Psychophysiological Phenomena by Processing and Pattern Analysis of Evoked Potentials, methods for single trial-evoked potentials, are described. Techniques, such as parametric modeling, adaptive delay and coherence estimation, wavelet-type decomposition, and higher-order spectral estimation are discussed.

The third section is dedicated to Processing and Pattern Analysis of Neural Cell Activity. Correlation and coherence analysis, as well as system identification methods have proved to be valuable for the understanding of firing patterns of neural populations.

In the fourth section, Processing and Pattern Analysis of ECG in Health and Disease, several authors consider advanced techniques for the analysis of the heart rate variability signal. Applications include analysis of ventricular fibrillation signals, recording and analysis of fetal ECG, and analysis of body surface potential maps.

In the last section, Processing and Pattern Analysis of EMG and Human Movement, basic problems of EMG modeling, such as source characteristics, are discussed. Decomposition of needle EMG recordings, as well as processing of multichannel EMG using signal processing, are treated. Modeling and analysis of human movements for the purpose of understanding motor organization, as well as processing of handwriting movements, conclude this section.

The symposium and this volume would have never emerged without The Bat-Sheva De Rotschild Foundation, whose generous support made the International Workshop on "Advances in Processing and Pattern Analysis of Biological Signals" a reality. Thanks are also due to the Technion, IIT, to the Israel Academy of Sciences and Humanities, and to the Ministry of Science and the Arts for their support. Special thanks are due to Amir Geva for designing the front cover of the book. We are especially indebted to Miss. Deborah E. Shapiro for her meticulous typing and organization of the book and to Mrs. Cecily Hyams for language editing of several chapters.

Isak Gath   Gideon F. Inbar
Department of Biomedical Engineering   Department of Electrical Engineering
Technion-IIT, Haifa   Technion-IIT, Haifa

CONTENTS

Processing and Pattern Analysis of Neuronal Cell Activity

Processing and Pattern Analysis of ECG in Health and Disease

Processing and Pattern Analysis of EMG and Human Movement

Processing and Pattern Analysis of Normal and Pathological EEG

SOME NEW TOOLS FOR EEG MODELING AND ANALYSIS

Will Gersch

Department of Information and Computer Sciences
University of Hawaii
Honolulu, Hawaii 96822

ABSTRACT

Several tools for the modeling and analysis of scalar and multivariate EEG data that are not necessarily stationary (and may in fact be subject to abrupt changes in covariance structure), and not necessarily Gaussian distributed are introduced.

First we show a "smoothness priors" quasi Bayesian method of time series analysis that is applicable to the modeling of scalar non stationary covariance EEG data. That methodology is exploited here to realize the power spectral density of both slowly varying nonstationary covariance EEGs as well as EEGs whose covariance structure changes abruptly. (The latter is achieved via a state space non-Gaussian smoothness priors analysis, that does not require data segmentation.) Secondly we introduce a one channel at-a-time" paradigm which yields the autoregressive (AR) modeling of multivariate stationary and nonstationary covariance EEG time series by successive scalar autoregressive time series modeling. An application of that paradigm to the parsimonious modeling of multivariate stationary EEGs is shown. Such relatively statistically efficient parsimonious modeling can potentially enhance stationary multivariate EEG classification performance. Finally, the smoothness priors modeling of scalar nonstationary covariance time series and the one channel at-a-time paradigms are both exploited to achieve the modeling and analysis of multivariate nonstationary covariance data. An application of the latter methodology to the identification of the epileptic focus in a human epileptic event is shown.

INTRODUCTION

The modeling and analysis of EEG data must contend with scalar and multivariate data that is not necessarily stationary (and may in fact be subject to abrupt changes in covariance structure), and not necessarily Gaussian distributed. In this paper we introduce some methodology that is of potential use for the modeling and analysis of such scalar and multivariate EEGs. The new methods are associated with the application of two concepts.

Advances in Processing and Pattern Analysis of Biological Signals, Edited by Isak Gath and Gideon F. Inbar
Plenum Press, New York, 1996

One is a quasi-Bayesian "smoothness priors" method of time series analysis (reviewed in Gersch & Kitagawa, 1988; Gersch, 1992; Kitagawa & Gersch, 1995), which we use for the modeling of nonstationary covariance time series. The other concept, which we refer to as a "one-channel at-a-time" paradigm is one by which multivariate stationary and nonstationary covariance time series may be modeled one (stationary or nonstationary) autoregressive scalar time series at-a-time.

The first problem, treated in Section 2, is that of modeling and spectral analysis of scalar EEG time series whose spectrum is not constant over time and in fact may change abruptly. A general state space quasi Bayesian-smoothness priors method of time series analysis, which admits not necessarily linear-not necessarily Gaussian, modeling is exploited. Priors on the distribution of the partial autocorrelations (PARCORS), of a time varying coefficient autoregressive model expressed in a lattice structure achieve the realization.

The second problem addressed here in Section 3, is the reduction of the tendency to overparametrize in the parametric modeling of multivariate time series. That reduction is achieved using the one-channel at-a-time paradigm. Multivariate autoregressive models time series are realized one autoregressive channel at-a-time. Subsequently subset selection-regression that potentially reduces the number of parameters in each data channel is used. The reduced parametrization method (with its attendant smaller prediction error property), may be applied to enhance EEG classification performance. An EEG time series example illustrates the enhanced mean square tracking error performance of the procedure.

Finally in Section 4, the one channel at-a-time and smoothness priors methods introduced in the first two problems are combined in the fitting of multivariate time varying autoregressive models and are applied in the analysis of an episode of a multichannel nonstationary covariance human epileptic event EEG.

MODELING NONSTATIONARY COVARIANCE EEGS

Here, scalar nonstationary covariance EEG time series are modeled using time varying autoregressive (TV-AR) coefficient models. Our goal is to model the evolution with time of the power spectrum. The fitted TV-AR model yields an "instantaneous power spectral density". Using this approach we are able to model EEG time series with either slowly varying or abruptly changing spectral structure without resorting to data segmentation.

The generic scalar TV-AR coefficient model of the observed data $y = y_1, ..., y_N$ is given by

$$y_n = \sum_{i=1}^{M} \alpha_{i,n} y_{n-i} + w_n, \quad w_n \sim \text{dist}(0, \sigma_n^2) \tag{1}$$

In general, in (Eq. (1)), the innovations w_n, $n = 1, ..., N$, are constrained to be independent but not necessarily Gaussian distributed or necessarily with constant variance. In such a model of N observations, if the order of the AR model is M there will be $N \times M$ AR model parameters and as many as N innovations variance parameters. Fitting TV-AR model with smoothness priors constraints permits those parameters to be estimated implicitly in terms of only a small number of explicitly estimated "hyperparameters".

Our approach to this important topic has evolved over several years (with Genshiro Kitagawa). Several different smoothness priors constraints methods for TV-AR modeling have been investigated. These are reviewed in Kitagawa and Gersch (1996). The particular TV-AR modeling approach discussed here, which is well suited to modeling time series with either smoothly changing or abruptly changing covariance structure can be achieved by placing not necessarily Gaussian smoothness priors constraints on the partial correlation coefficients (PARCORs), in a lattice structure AR model representation and modeling using a general state space model.

In Section 2.1 the smoothness priors concept is briefly introduced. The PARCORS TV-AR modeling is shown in Section 2.2. Examples of the modeling of EEG time series data with abruptly changing spectrum are shown in Section 2.3.

Smoothness Priors

The conceptual predecessors of our quasi-Bayesian work on smoothness priors is in papers by Whittaker (1923), Shiller (1973) and Akaike(1980). Our own work in smoothness priors, including our earlier approaches to TV-AR modeling, is reviewed in Gersch and Kitagawa (1988), Gersch (1992) and Kitagawa and Gersch (1995). Whittle (1965), Kozin (1977), Grenier (1982) are some earlier non-Bayesian papers on TV-AR modeling.

Perhaps the simplest smoothness priors modeling problem is that of modeling a nonstationary mean (or stochastic trend) time series. In the context of the stochastic trend model, the expression $\nabla^k t_n = w_n$ imposes a prior distribution on the unknown trend parameters $t = (t_1, t_2, ..., t_n)'$. For example if $w_n \sim N(0, \tau^2)$ that distribution may be expressed as $\pi(t|\tau^2) = (2\pi\tau^2)^{-N/2} \exp\{-t't/2\tau^2\}$. In Bayesian terminology τ^2, is known as a hyperparameter (Berger, 1985). Similarly if the unobserved observation noise vector $\varepsilon = (\varepsilon_1, \varepsilon_2, ..., \varepsilon_n)'$ were normally distributed with zero mean and variance σ^2, we might express the conditional data distribution of the observation vector $y = (y_1, y_2, ..., y_n)'$ in the form $p(y|t,\tau^2,\sigma^2) = (2\pi\sigma^2)^{-N/2} \exp\{-(y-t)'(y-t)/2\sigma^2\}$. In that case, $\pi(t|y,\tau^2,\sigma^2)$ the posterior distribution of the trend parameters is (Berger, 1985),

$$\tag{2}$$

The integration of the right hand side of Eq. (2) yields $L(\tau^2,\sigma^2)$ the likelihood for the unknown parameters τ^2 and σ^2

$$\tag{3}$$

A constrained least squares closed form computational procedure that realizes a closed form solution for the maximization of the likelihood of the hyperparameters is due to Akaike (1980). An equivalent state space computational approach for nonstationary mean time series (Gersch & Kitagawa, 1988), is shown in the next section. The critical issue in smoothness priors is to identify appropriate priors. Using the smoothness priors approach to time series analysis, quite complex time series can be modeled that require the maximization of the likelihood for only a small number of hyperparameters.

Parcor Time Varying AR Modeling

In this section we demonstrate the realization of a TV-AR model via the stochastic trend modeling of the PARCORS in a lattice structure AR model. First we show the state space representation of nonstationary in the mean or stochastic trend modeling. That is followed by an introduction to general state space modeling. That method applied to the stochastic trend modeling of the PARCORS with non-Gaussian disturbances yields a TV-AR model which permits the modeling of time series whose covariance changes abruptly.

Stochastic State Space Trend Modeling. Nonstationarity in the mean time series can be expressed as

$$y_n = t_n + \varepsilon_n \tag{4}$$

Here ε_n is a stationary white noise process and t_n is a trend component with

$$\nabla^k t_n = w_n \tag{5}$$

As before, ∇ is the difference operator defined by $\nabla t_n = t_n - t_{n-1}$ and w_n is a white noise with mean zero and dispersion parameter (hyperparameter), τ^2. In what follows for convenience, we permit the distribution of the w_n sequence to be either Gaussian or Cauchy. Then, Eq. (5), together with the specification of its distribution is the prior distribution for the trend.

The generic linear state space model (Anderson & Moore, 1978), is

$$x_n = F_n x_{n-1} + G_n w_n$$

$$y_n = H_n x_n + \varepsilon_n \tag{6}$$

where in general, the joint distribution of the system noise and observation noise is

$$\begin{bmatrix} w_n \\ \varepsilon_n \end{bmatrix} \sim P\left(\begin{bmatrix} 0 \\ 0 \end{bmatrix}, \begin{bmatrix} Q_n & 0 \\ 0 & R_n \end{bmatrix} \right). \tag{7}$$

In Eq. (7) the notation P, denotes an arbitrary zero mean and uncorrelated joint distribution for w_n, ε_n with potentially time dependent dispersion parameters.

The stochastic trend state space model matrix components are then simply given by

$$\text{for } k = 1: \quad F^{(1)} = G^{(1)} = H^{(1)} = 1, \quad x_n = t_n \tag{8}$$

$$\text{for } k = 2: \quad F^{(2)} = \begin{bmatrix} 2 & -1 \\ 1 & 0 \end{bmatrix}, \ G^{(2)} = \begin{bmatrix} 1 \\ 0 \end{bmatrix}, \ H^{(2)} = \begin{bmatrix} 1 & 0 \end{bmatrix}, \ x_n = \begin{bmatrix} t_n \\ t_{n-1} \end{bmatrix} \tag{9}$$

This model is used for the estimation of the trend of nonstationary mean PARCORS.

General State Space Modeling. Consider a system described by a general state space model

$$X_n \sim q(x_n | x_{n-1})$$
$$Y_n \sim r(y_n | x_n) \tag{10}$$

where y_n is the time series and x_n is the unknown state vector. q and r are conditional distributions of x_n given x_{n-1} and of y_n given x_n, respectively. The initial state vector x_0 is distributed according to the distribution $p(x_0 | Y_0)$. This general state space model includes the linear state space model with non-Gaussian white system and observation noises.

General Filtering and Smoothing. For the state estimation of the general state space model, we need to evaluate $p(x_n | Y_m)$, the conditional distribution of x_n given observations $Y_m = \{y_1, y_2, ..., y_m\}$. It can be shown that for the general state space model, the recursive formulas for obtaining the one step ahead predictor, filter and smoother are given as follows (Kitagawa, 1987):

One step ahead prediction:

$$p(x_n|Y_{n-1}) = \int_{-\infty}^{\infty} q(x_n|x_{n-1})p(x_{n-1}|Y_{n-1})dx_{n-1} \tag{11}$$

Filtering:

$$p(x_n|Y_n) = \frac{r(y_n|x_n)p(x_n|Y_{n-1})}{p(y_n|Y_{n-1})} \tag{12}$$

where $p(y_n|Y_{n-1})$ is obtained by $\int r(y_n|x_n)p(x_n|Y_{n-1})dx_n)$.

Smoothing:

$$p(x_n|Y_N) = p(x_n|Y_n) \int_{-\infty}^{\infty} \frac{p(x_{n+1}|Y_N)q(x_{n+1}|x_n)}{p(x_{n+1}|Y_n)} dx_{n+1} \quad . \tag{13}$$

These formulas, Eqs. (11)-(13) show recursive relations between state distributions. The conditional distribution of the state $p(x_n|Y_m)$ is in general non-Gaussian. For the general state space model the log likelihood is obtained by

$$\ell(\theta) = \sum_{n=1}^{N} \log p(y_n|Y_{n-1}) \quad . \tag{14}$$

It should be noted here that $p(y_n|Y_{n-1})$ is the denominator of Eq. (12). Therefore the log-likelihood is obtained as the by-product of the non-Gaussian filter.

Numerical implementation of the general recursive formulas were communicated in Kitagawa (1987). Additional details and numerous examples involving the general state space modeling are in Kitagawa and Gersch (1995).

Realizing the PARCOR TV-AR Model. The coefficients of the TV-AR model in Eq. (1) can be estimated recursively via a lattice structure AR, a successively increasing model order and time order modeling, with stochastic trend varying PARCORS. (Haykin, 1991 for example treats stationary time series PARCOR lattice structure AR modeling.) Here 2 x M hyperparameters are fitted, one at-a-time for each of the forward and backward PARCORS. Let f_n^m and b_n^m represent forward and backward prediction errors of an m-th order AR model at time instant n, and let $\alpha_{mn}^{(m)}$ and $\gamma_{mn}^{(m)}$ denote the forward and backward PARCORS of the m-th order AR model at time n. The zero-th order forward and backward prediction errors are

$$f_n^{(0)} = y_n, \quad b_n^{(0)} = y_n \quad . \tag{15}$$

Then, the relationship between the order updated forward and backward prediction errors and the instantaneous PARCORS are given by

$$f_n^{(m-1)} = \alpha_{mn}^{(m)} b_{n-m}^{(m-1)} + f_n^{(m)}$$

$$\nabla\alpha_{mn}^{(m)} = v_n \sim G_{\pi_{1m}^2}$$

$$b_{n-m}^{(m-1)} = \gamma_{mn}^{(m)} f_m^{(m-1)} + b_n^{(m)}$$

$$\nabla\gamma_{mn}^{(m)} = u_n \sim G_{\tau_{2m}^2} \tag{16}$$

with $f_n^{(0)} = b_n^{(0)} = y_n$, $f_n^{(m)}$ and $b_n^{(m)}$ respectively the forward and backward prediction errors of the autoregressive model of order m. In the stationary case ($v_n = u_n = 0$) and $\alpha_{mn}^{(m)} = \gamma_{mn}^{(m)}$ are identical to the partial autocorrelation coefficients.

The instantaneous updated forward and backward AR model parameters for model order $i = 1, \ldots, m-1$, then become,

$$\alpha_{in}^{(m)} = \alpha_{in}^{(m-1)} + \alpha_{m,n}^{(m)} \gamma_{m-i,n}^{(m-1)}$$

$$\gamma_{in}^{(m)} = \gamma_{in}^{(m-1)} + \gamma_{mn}^{(m)} \alpha_{m-i,n}^{(m)}$$

(17)

The innovation in Eq. (16) is the stochastic difference equation representation for the order updated forward and backward instantaneous PARCORs. $f_n^{(m-1)} = \alpha_{mn}^{(m)} b_{n-m}^{(m-1)} + f_m^{(m)}$ is a simple time varying linear regression. The $b_{n-m}^{(m-1)}, f_n^{(m-1)}, f_n^m$ respectively represent the independent and dependent variables and the residual for n = 1, ..., N and m = 1, ..., M. The expression $\nabla \alpha_{mn}^{m} = v_n$ is a smoothness priors (random walk) constraint on the evolution of the forward PARCOR of the m-th order model and the notation G identifies an arbitrary distribution. The τ_{1m}^2, $m = 1,..., M$ are the hyperparameters of the distribution of the forward PARCORS. The maximization of the likelihood of a single hyperparameter τ_{1m}^2 (or τ_{2m}^2), is required at a time. State space computations for the likelihood of the hyperparameter models are used (for convenience we let $\tau_{1m}^2 = \tau_{2m}^2 = \tau_m^2$). Akaike's AIC is used to determine the AIC best PARCOR order model. In the examples considered here, a zero mean Gaussian distribution is used in the modeling of slowly varying EEG covariance and a zero mean Cauchy distribution is used in the modeling of abruptly changing covariance structure.

After filtering and smoothing the time-varying AR coefficients, the instantaneous spectrum of the nonstationary process is estimated by

$$S_n(f) = \frac{\sigma_n^2}{|1 + \sum_{j=1}^{M} \alpha_{jn} \exp(2\pi i j f)|^2}$$

(18)

Examples

Examples of the simulation nonstationary covariance time series and their modeling by the PARCORS TV-AR coefficient model are shown in Kitagawa and Gersch (1995). Here we show EEG data examples corresponding to segments of seizure episodes in humans modeled using the stochastic PARCOR model. Figure 1A shows a segment of an abruptly varying human epileptic event EEG (data provided by I. Gath). Figures 1B and 1C respectively show the corresponding instantaneous power spectral density and instantaneous PARCORS for Gaussian distributed and Cauchy distributed PARCOR system noise. The Gaussian PARCORS can only realize slowly varying spectra. It is clear that the Cauchy system noise modeled changing PARCORS model more successfully captures the abrupt change in the power spectral density than does the Gaussian system noise modeling. In each case, the computational issue is to determine an optimum hyperparameter for each individual model order PARCOR stochastic trend function (Gaussian distributed or Cauchy distributed). Compared to the Gaussian system noise computations, the Cauchy noise modeling is computationally intensive.

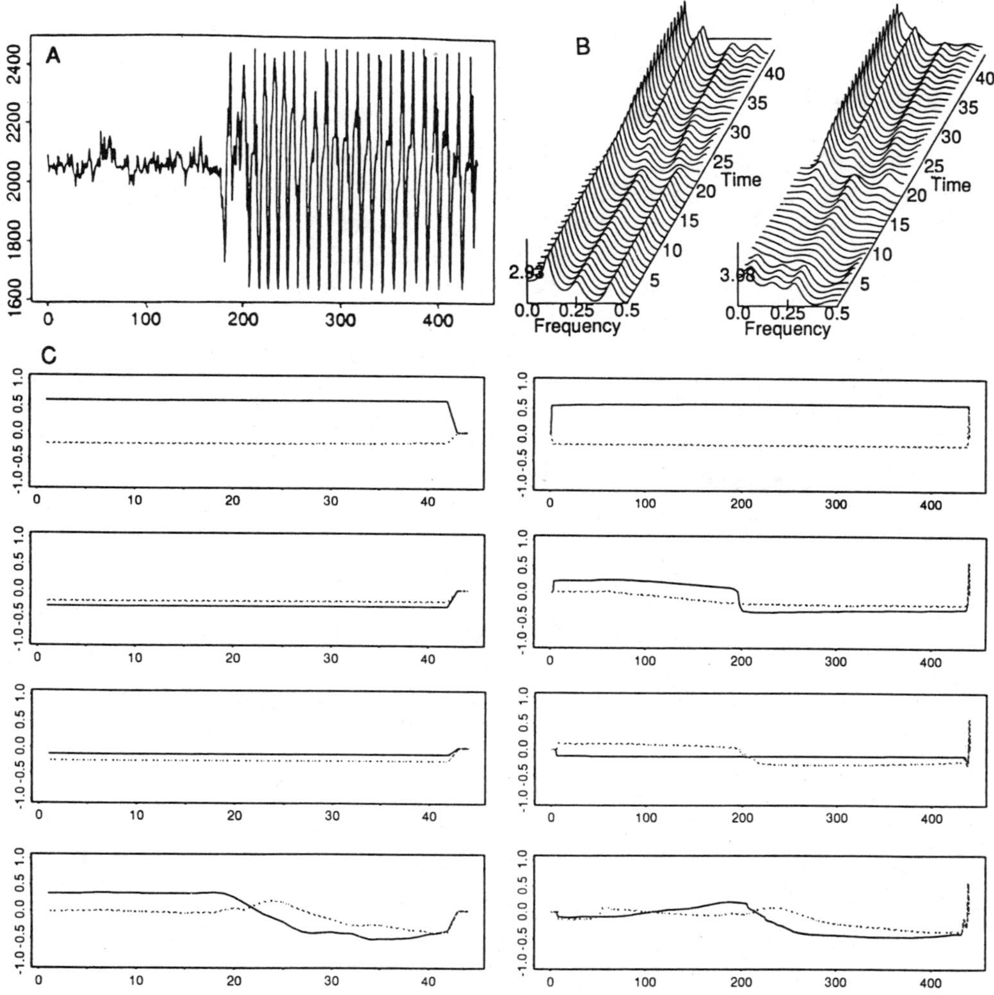

Figure 1. A) Human epileptic event. B) Gaussian and Cauchy modeled changing spectrum. C) Gaussian and Cauchy modeled changing PARCORS.

THE INSTANTANEOUS RESPONSE-ORTHOGONAL INNOVATIONS MODEL

In this section an instantaneous response-orthogonal innovations (IR-OI), representation of multivariate or vector AR (VAR) time series is introduced and exploited to accomplish the fitting of constant coefficient stationary VAR models to multivariate time series, one scalar AR time series at-a-time. We refer to this idea as the "one channel at-a-time" paradigm. Using a linear regression model AR modeling and subsequent subset selection, yields a parsimonious representation of the individual scalar AR time series models. The resulting parsimonious VAR modeling can potentially enhance the Kullback-Leibler metric-nearest neighbor rule classification performance of VAR modeled multivariate EEG time series. Also, In Section 4, the IR-OI representation is generalized to the time varying VAR model modeling of multivariate nonstationary covariance EEGs and applied to the analysis of a human epileptic event EEG.

Background

Starting with a d channel time series and using the instantaneous response-orthogonal innovations representation, these data are modeled as d uncorrelated scalar autoregressive time series as a linear combination of the data in each channel. The order of each of the scalar AR models is determined using Akaike's AIC. The individual orthogonal scalar AR models are combined to yield the multichannel IR-OI variance model. That model is related to the more conventional VAR model with a general innovations variance by invertible algebraic transformations.

The classical multivariate Yule-Walker equation (Whittle, 1963; Lutkepohl, 1991) fitting of VAR models to stationary multivariate time series data solve linear equations in the estimated data covariances. For relatively short duration data sets, covariance estimation and subsequent AR model fitting, can be quite poor in a statistical sense. The IR-OI representation VAR model that we use, is based on computations on the instantaneous data thereby bypassing covariance computations.

The VAR model of order M for a d-channel time series is (Whittle, 1993)

$$\sum_{m=0}^{M} A_m X(n-m) = \varepsilon(n), \quad E[\varepsilon(n)] = 0, \quad Var[\varepsilon(n)] = V\delta_{n,o} \tag{19}$$

In Eq. (19), A_m, $m = 0, ..., M$ are $d \times d$ AR coefficient matrices, A_0 is the $d \times d$ identity matrix and $A_m \neq 0$ for $m = M$. The $d \times d$ matrix V is the innovations matrix. In fitting AR models to data, the VAR order M can be determined with the Akaike's AIC criterion (Akaike, 1974) or the (preferred for shorter data length series) AIC_C criterion (Hurvich & Tsai, 1989, 1992). For the d dimensional VAR model in Eq. (19) there are $d \times d \times M$ AR coefficient parameters and $(d \times (d + 1))/2$ innovations matrix parameters to be estimated. Thus the number of VAR model parameters to be estimated increases as the square of the dimension d while the number of observations increases linearly in proportion to d. VAR models have an inherent tendency to overparameterize.

The prediction of a future value of a time series uses the fitted VAR model. Each of the VAR model parameters is necessarily estimated with an accompanying error which in turn contributes to the prediction error variance. Hence modeling with a statistically excessive number of estimated parameters is costly in prediction error variance. Similar considerations apply to the Kullback Leibler nearest neighbor rule classification of VAR modeled time series. Therefore, we are motivated to achieve a parsimonious VAR modeling.

Our own approach to an instantaneous response-orthogonal innovations VAR parameterization was motivated by Sakai (1982) who in turn was motivated by Pagano (1978). In fact an IR-OI VAR model was probably first introduced in Akaike (1968). Applications of the OI-IR VAR type modeling method include Kitagawa and Akaike (1981), Newton (1982) and Takanami and Kitagawa (1991). Each of those papers dealt with the VAR modeling of stationary time series. Gersch and Stone (1990), Gersch (1992), and Jiang and Kitagawa (1993) exploit the one channel at-a-time paradigm IR-OI modeling in different ways to achieve the time varying, TV-VAR, modeling of multivariate nonstationary covariance data. Stone (1993) is a Ph.D thesis in which the one channel at-a-time paradigm is exploited for parsimonious VAR modeling and enhanced one-step-ahead prediction of stationary time series and several other applications.

Constant Coefficient IR-OI VAR Time Series Modeling

A parsimonious VAR model is realized by expressing each of the IR-OI representation individual scalar AR models in the form of a linear regression model and subsequently employing a subset selection procedure. The one channel at-a-time IR-OI model is related to the usual VAR model with general innovations variance by an invertible algebraic transformation. In the one channel at-a-time VAR modeling of stationary time series, the order of each of the scalar AR models is determined using the Akaike AIC or the Hurvich and Tsai AIC_C. With each scalar AR model represented in a linear regression model, subset selection is achieved via an all-subsets search also using the AIC.

Some Details. Let $y_n = (y_{1,n}, y_{2,n}, ..., y_{d,n})^t$, $(n = 1, 2, \cdots, N)$ be a d-variate time series. Here, let the stationary time series vector autoregressive (VAR), model of that series be given by

$$y_n = \sum_{l=1}^{p} A_l y_{n-\ell} + w_n \, , \tag{20}$$

where p is the autoregressive order, A_l is the $d \times d$ autoregressive coefficient matrix and w_n is assumed to be a d-variate Gaussian white noise sequence with mean zero and the innovations matrix W.

Premultiplying both sides of Eq. (20) by the unique unit diagonal lower triangular Cholesky factor of W so that the covariance matrix of w_n is reduced to a diagonal matrix D, yields the instantaneous response-orthogonal innovations model,

$$y_n = B_0 y_n + \sum_{l=1}^{p} B_l y_{n-\ell} + \varepsilon_n \, . \tag{21}$$

Here the coefficient matrix of the instantaneous response, B_0, is in the form:

$$B_0 = \begin{bmatrix} 0 & 0 & \cdots & 0 \\ b_0(2,1) & 0 & \cdots & 0 \\ \vdots & \ddots & \ddots & \vdots \\ b_0(d,1) & \cdots & b_0(d,d-1) & 0 \end{bmatrix} , \tag{22}$$

and ε_n is a d-variate white noise sequence with mean zero and diagonal covariance matrix $D = \text{diag} \, (\sigma_1^2 , \sigma_2^2 ,..., \sigma_d^2)$. Using simple algebraic operations, Eq. (21) can be re-expressed as

$$y_n = (I - B_0)^{-1} \sum_{l=1}^{p} B_l y_{n-l} + (I - B_0)^{-1} \varepsilon_n \, . \tag{23}$$

Thus the IR-OI model in Eq. (21) is related by an algebraic transformation to the usual VAR model in Eq. (19) by

$$A_l = (I - B_0)^{-1} B_l, \quad l = 1, 2, \cdots, m,$$
$$W = (I - B_0)^{-1} D (I - B_0)^{-1t} \tag{24}$$

To summarize, a motivating reason to use the instantaneous response-orthogonal innovations VAR modeling is that it can be realized by successive scalar model computations.

Some Additional Details. Another useful way of interpreting the relationship between the conventional TV-VAR representation in Eq. (20) and the instantaneous response model in Eq. (21) is that the latter can be thought of an interlacing of the data in the former. That is, first consider interlacing the d-channel time series $y(t)$ into a scalar series $z(t)$ using,

$$z(j + d(t - 1)) = y_j(t), \quad j = 1, ..., d \quad . \tag{25}$$

Then, re-express the $z(t)$ sequence into a d-vector AR model according to

$$z(k + d(t - 1)) + \sum_{j=1}^{p_k} \alpha_k(j) z(k + d(t - 1) - j) = \varepsilon(k + d(t - 1)), k = 1, ..., d \quad . \tag{26}$$

with $\varepsilon(k + d(t - 1)) \sim N(0, \sigma_k^2)$ i.i.d. and with p_k the order of the k-th channel AR model. For example for $d = 3$, Eq. (26) can be written,

$$\begin{bmatrix} 1 & 0 & 0 \\ \alpha_2(1) & 1 & 9 \\ \alpha_3(2) & \alpha_3(1) & 1 \end{bmatrix} \begin{bmatrix} z(1 + 3(t-1)) \\ z(2 + 3(t-1)) \\ z(3 + 3(t-1)) \end{bmatrix} + \begin{bmatrix} \alpha_1(3) & \alpha_1(2) & \alpha_1(1) \\ \alpha_2(4) & \alpha_2(3) & \alpha_2(2) \\ a_3(5) & \alpha_3(4) & \alpha_3(3) \end{bmatrix} \begin{bmatrix} z(1 + 3(t-2)) \\ z(2 + 3(t-2)) \\ z(3 + 3(t-2)) \end{bmatrix} \cdots = \begin{bmatrix} \varepsilon(1 + 3(t-1)) \\ \varepsilon(2 + 3(t-1)) \\ \varepsilon(3 + 3(t-1)) \end{bmatrix} \tag{27}$$

Using Eq. (26), and $(t - 1) = n$ the $y(t)$ sequence can be written in terms of the $z(t)$ of Eq. (27) in the matrix form

$$\sum_{m=0}^{M} B_m z(n - m) = \varepsilon(n) \tag{28}$$

where the B_0 is lower unit triangular and the d vector sequence $\varepsilon(n)$ has diagonal covariance matrix $D = \mathrm{diag}(\sigma_1^2, ..., \sigma_d^2)$. That is, the model in Eq. (28) has precisely the same structure of the model in Eq. (21) with $z(n) = y_n$ and $\varepsilon(n) = \varepsilon_n$.

Parsimonious Multichannel Autoregressive Modeling

Here the one channel at-a-time paradigm is exploited to achieve parsimonious VAR modeling. The d channel multivariate time series data are modeled as d independent scalar autoregressive time series in a linear regression model form. Subset selection is achieved within this scalar AR linear regression model framework and the resulting subsetted scalar models are recombined by a linear transformation to realize the parsimonious multichannel autoregressive model. The VAR model is used to evaluate the tracking performance of the full and subset selection models. Mean square error tracking performance may be used as a heuristic to compare the relative performance of the IR-VAR subset selection model and the non-subset selection VAR models.

An Example. Here for illustrative purposes we compare the attempts at achieving parsimonious time series modeling in a segment of a $d = 3$ channel EEG times series. (Stone (1993) showed successful examples comparing parsimonious IR-VAR modeling and tracking of simulated multivariate stationary time series data and of real stationary and nonstationary mean time series data.)

A $d = 3$, $n = 70$ segment of a human EEG epileptic event was considered for analysis. VAR models were fitted to this data using the Whittle-AIC modeling paradigm and by the

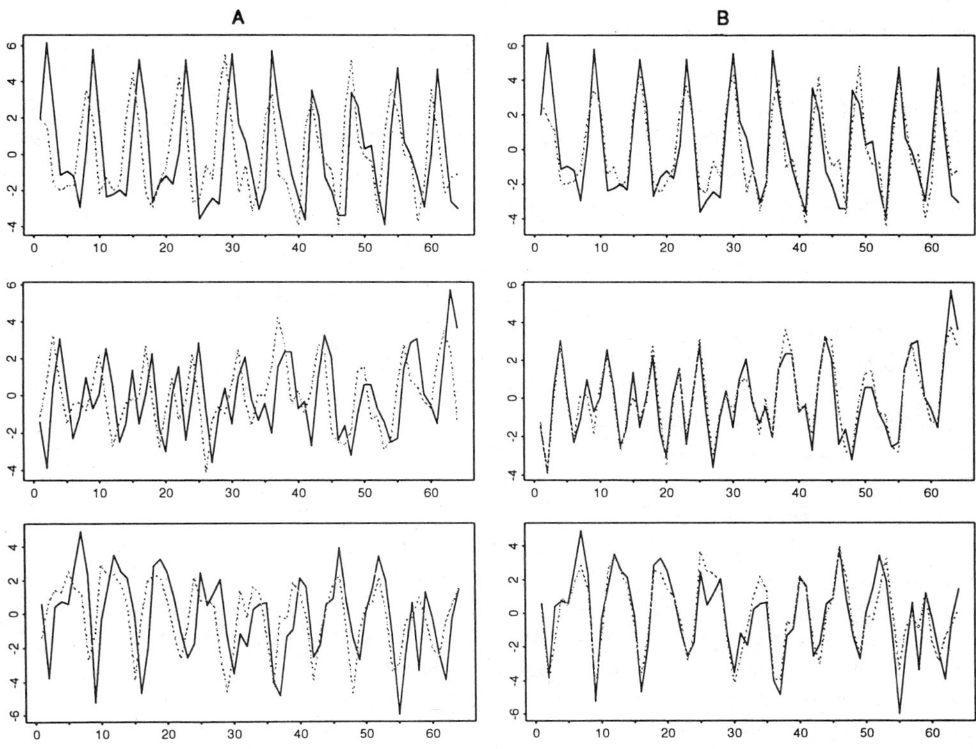

Figure 2. A) Nonsubset model and B) subset model tracking.

IR-OI subset selection parsimonious VAR method. Whittle's algorithm yields successively increasing order multivariate autoregressive models recursively. The model order was selected using Akaike's AIC. A VAR_7 model was selected as the Whittle-AIC best model fitted with a total of 9 x 7 VAR plus six innovations (or 69), model parameters. The subset-selection parsimonious VAR model corresponding to a IR-OI VAR_7 model consisted of 24 IR-OI and three innovations (or 27) model parameters. The tracking performance of the two models is illustrated in Fig. 2. The dark line being the original data and the dotted line being the fitted model tracking behavior. The corresponding mean-square tracking errors for the Whittle-AIC and parsimonious IR-OI VAR models is as shown in Table 1.

The tabulated results are quite understandable. The Whittle-AIC model of the each of the three channels is effectively computed on 7 x 3 (21) data variables. The corresponding IR-OI models for the 1st, 2nd and 3rd channels were computed on subsets of 5, 10 and 9 variables respectively. The subsetted IR-OI first channel model, having a smaller number of explanatory variables has a smaller R^2, hence a poorer

Table 1. Mean-square tracking errors, Whittle-AIC and subset
IR-OI models

	Channel 1	Channel 2	Channel 3
Whittle-AIC	1.82874	0.86667	1.94101
Subset IR-OI	2.11089	0.50363	0.96060

fit and a larger mean square tracking error. On the other hand, referring to (12), we see that there is an instantaneous or zero-lag feed through from the first channel into the 2nd and 3rd channels. As a result even though the corresponding 2nd and 3rd subsetted IR-OI channels are regressed on fewer explanatory variables than the non-subsetted Whittle-AIC modeled channels, their tracking performance is superior. Over-all in this example, the mean square tracking error of the subsetted or parsimonious IR-OI model is 77% of that of the Whittle-AIC model.

Parsimonious VAR EEG Modeling and EEG Classification. For both exploratory EEG population screening and normalized baseline EEG classification, we advocate the use of Kullback-Leibler nearest neighbor (KL-NN), classification rules (Gersch, 1979, 1981). The Kullback-Leibler number is a measure of dissimilarity between distributions. In the KL-NN rule classification a measure of dissimilarity between a new to-be-classified time series and each of a set of categorically labeled sample is computed. The new time series is classified with the label of the particular time series which is least dissimilar. The dissimilarity measure between the time series is computed as an estimate of the KL number as if the time series were normally distributed.

VAR models of the labeled time series samples can be interpreted as templates of those time series. In the KL-NN rule classification, the new time series is compared against the templates (filtered through the labeled time series VAR models), and the label of the template (VAR model) for which the tracking error is smallest, is the label of the new time series. In that context, parsimonious VAR models have a smaller uncertainty region around them than less parsimonious models, enhancing the separability of the categorically labeled time series and consequently, potentially enhancing EEG classification performance.

The motivation for advocating a KL-NN classification rule, in at least the exploratory stages of the population screening problem, is based on the fact that it is not certain at the outset, that there is sufficient information in the time series to achieve statistically reliably classification. The credibility of such a conjecture is strained by evidence of broad intersubject categorical EEG variability. In the population screening problem the KL-NN classification rule approach yields the Cover and Hart (1967)-Rogers and Wagner (1978) asymptotic and finite sample properties of formal NN rules. These properties allow a test of the conjecture implicit in the population screening problem, i.e. that there is sufficient information in the time series to achieve an acceptable level of misclassification error performance with only a modest-size labeled sample data set. Details and statistical results of non-trivial KL-NN rule normalized baseline EEG classification are reported in Gersch *et al.* (1979) and Gersch (1981). Gevins (1980), indicates that the normalized baseline classification results are among the strongest ever obtained.

MODELING MULTIVARIATE NONSTATIONARY COVARIANCE EEGS

A commonly used approach to the modeling of multivariate nonstationary covariance time series, is to segment the series into approximately stationary time series segments and to model each segment as a multivariate stationary time series model (Gath *et al.*, 1992, for example). The problem with this approach is that it may not adequately capture an appropriate balance between the local and global statistical properties of the time series or have sufficiently satisfactory time-frequency resolution properties. The direct fitting of a time varying coefficient multivariate autoregressive (TV-VAR) model (Gersch & Kitagawa, 1983, and subsequently in the modeling of nonstationary covariance EEGs, Gersch, 1987), appeared to be a solution to those problems. Because is modeled using instantaneous data rather than covariance data, the IR-OI TV-VAR model shown here has even finer time-frequency resolution properties than the earlier Gersch and Kitagawa (1983) or Gersch (1987) model.

The TV-VAR model is

$$y_n = \sum_{m=1}^{p} A_{mn} y_{n-m} + w_n, \quad n = 1, \cdots, N \ . \tag{29}$$

to nonstationary covariance economic data. In Eq. (29), the y_n's are d-dimensional vectors and w_n, 1, $\cdots$, N is a zero-mean uncorrelated normally distributed d-dimensional vectors sequence with instantaneous covariance matrix, W_n.

In the model above, for a d channel N observation time series there are N x d observations and d x d x P x N autoregressive model parameters and d x (d + 1) x N/2 innovations matrix parameters. In order to fit such a model with so many more parameters than observations, it is necessary to efficiently parametrize the time evolution of the parameters. The method used in Gersch and Kitagawa (1983) to characterize the evolution of each of the d x d x P autoregressive parameters in terms of a discrete Laguerre polynomial in a least squares computational solution. The order of the multivariate AR model and the degree of the Laguerre polynomial were determined using the AIC. The results obtained in the aforementioned papers seemed to be quite reasonable. On the other hand in that procedure, the Laguerre polynomial coefficients were determined by global statistical considerations. In retrospect it was clear that it was necessary to develop a multivariate time varying AR parameter modeling procedure that was more sensitive to local time variations. Two different TV-VAR modeling methods were developed that have the property of being more sensitive to local time variations of the covariances.

In one method (Kitagawa & Gersch, 1985), smoothness priors constraints are placed directly on the evolution of the individual AR parameters of the time varying AR model. Jiang and Kitagawa (1993), is an extension of that approach to multivariate nonstationary covariance modeling using TV-VAR modeling. In the other approach (Gersch & Stone, 1990), smoothness priors constraints are placed on the evolution of the partial correlation coefficients (PARCORs) of scalar AR models in an application of the instantaneous response orthogonal innovations modeling to time varying coefficient multivariate AR models.

The IR-OI VAR model idea introduced earlier, is generalized here to the TV-VAR model, The generalization is achieved via the smoothness priors treatment of the PARCORs of the forward and backward innovations for each scalar IR-OI AR model. Our primary objective for achieving this modeling is in the evolution with time of the power spectral density matrix. The concept of an instantaneous power spectral density matrix introduced in Gersch and Kitagawa (1983), yields that quantity.

A matrix algebra representation of the IR-OI TV-VAR model is shown below. Also, some details of the realization of IR-OI stochastic PARCOR TV-VAR modeling are discussed. An example of the TV-VAR modeling of a human epileptic event EEG is shown.

A Instantaneous TV-VAR Time Series Model

A time varying coefficient d-variate IR-OI TV-VAR model is,

$$y_n = B_{0n}y_n + \sum_{l=1}^{p} B_{l,n}y_{n-l} + \varepsilon_n \tag{30}$$

where $\varepsilon_n = (\varepsilon_{1,n}, \varepsilon_{2,n},..., \varepsilon_{d,n})' \sim N(0, D)$ with $D = \text{diag}(\sigma_1^2, \sigma_2^2,...,\sigma_d^2)$. The time varying coefficient matrices are expressed by

$$B_{0n} = \begin{bmatrix} 0 & 0 & \cdots & 0 \\ b_{0n}(2,1) & 0 & \cdots & 0 \\ \vdots & \ddots & \ddots & \vdots \\ b_{0n}(d,1) & \cdots & b_{0n}(d,d-1) & 0 \end{bmatrix}, \quad B_{ln} = \begin{bmatrix} b_{ln}(1,1) & \cdots & b_{ln}(1,d) \\ \vdots & \ddots & \vdots \\ b_{ln}(d,1) & \cdots & b_{ln}(d,d) \end{bmatrix}, \quad l = 1, 2, \cdots, p. \tag{31}$$

The instantaneous matrix power spectral density matrix corresponding to the TV-VAR model in Eq. 29 is,

$$P_{f,n} = A_{f,n}^{-1} W_n A_{f,n}^{-1*}, \quad -\frac{1}{2} \le f \le \frac{1}{2}, \tag{32}$$

with $A_{f,n}$ the instantaneous whitening filter given by

$$A_{f,n} = I - \sum_{l=1}^{p} A_{l,n} \exp(-2\pi i f l).$$

Time Varying PARCOR VAR Modeling

Background. Whittle (1963) was a singular development in the fitting of multichannel autoregressive (AR) models to stationary time series data. In extending the Levinson algorithm for fitting AR models to scalar stationary time series to the multichannel case, Whittle found it necessary to employ both forward and backward multichannel AR models. The latter was critical in the implementation of a multichannel lattice-form algorithm, Wiggins and Robinson (1965). That realization and several subsequent multichannel realizations involved matrix computations. Scalar implementations of multichannel recursions, which required no matrix processing, were realized in equivalent structures by Sakai (1982), Lev-Ari (1987), and several others. Sakai's implementation (motivated by Pagano, 1978), transforms multichannel AR models into scalar AR models, by converting the multichannel Levinson algorithm into a scalar equivalent which Sakai calls the "circular lattice filter".

Sakai's algorithm is based on sample correlation function computations and in fact is a one channel at-a-time structure. However Sakai only deals with stationary time series modeling and Lev-Ari concern was with a conceptual theoretical structure, not with implementation. Our approach to the TV-VAR modeling algorithm shown here was directly motivated by Sakai's paper. Our algorithm is computed on the instantaneous data and we use a random walk smoothness priors trend model in the regression equation for the PARCORs.

Instantaneous PARCOR TV-VAR Modeling. The k-th channel zero-th order forward and backward innovations are

$$@EQ = \varepsilon_{k+nd}^{(0)} = y_{k+nd}, \quad \eta_{k+nd}^{(0)} = y_{k-1+nd}. \tag{33}$$

Imposing the nonstationarity of the PARCORs, the relations between the order update innovations and the instantaneous PARCORs are given by

$$\varepsilon^{(j+1)}_{k+nd} = \varepsilon^{(j)}_{k+nd} + \alpha^{(j+1)}_{j+1,n}(k)\eta^{(j)}_{k-1+nd}$$

$$\nabla\alpha^{(j+1)}_{j+1,n}(k) = v_n(k) \sim G(\tau^2_1)$$

@EQ =

$$\eta^{(j+1)}_{k+nd} = \eta^{(j)}_{k+nd} + \beta^{(j+1)}_{j+1,n}(k)\varepsilon^{(j)}_{k+nd}$$

$$\nabla\beta^{(j+1)}_{j+1,n} = u_n(k) \sim G(\tau^2_2) \tag{34}$$

In Eq. (34), the notation G identifies an arbitrary distribution. In what follows we consider only $G(\tau) = N(0,\tau^2)$, which is satisfactory for modeling relatively slowly changing covariances. However, as mentioned earlier, the distribution G can be heavy tailed for the modeling of abruptly changing covariances.

Following from Eq. (34), the instantaneous order updated forward and backward model parameters then become,

$$\alpha^{(j+1)}_{i,n}(k) = \alpha^{(j)}_{i,n}(k) + \alpha^{(j+1)}_{j+1,n}(k)\beta^{(j)}_{j+1-i,n}(k - 1)$$

$$\beta^{(j+1)}_{i,n}(k) = \beta^{(j)}_{in}(k - 1) + \beta^{(j+1)}_{j+1,n}(k)\alpha^{(j)}_{j+1-i,n}(k-1) \tag{35}$$

From Eq. (34) the order updated instantaneous PARCORs, $\alpha^{(j+1)}_{j+1}(k)$, $\beta^{(j+1)}_{j+1,n}(k)$ are scalar regression coefficients and the updated forward and backward innovations are the residuals of the regressions. The key innovation in Eq. (34) is the stochastic difference equation representation for the order updated forward and backward instantaneous PARCORs. Kalman filter computations for the likelihood of the τ^2_1 and τ^2_2 hyperparameter models are computed for each successive AR model order. The AIC statistic is used to determine the AIC best order PARCOR model. Ultimately then, a d-channel TV-VAR model with M x N x d^2 AR parameters is based on the explicit computation of the likelihood of 2 x d hyperparameters.

An Example

Examples of the TV-VAR modeling of simulated multivariate nonstationary covariance time series and comparisons of the theoretical instantaneous power spectral density matrix with that estimated by the instantaneous PARCOR TV-VAR model are shown in Stone (1993). Here an example of the data analysis of a three-channel data episode of an "aura" event, a pre-epileptic seizure EEG event by the TV-VAR model is shown. Conceptually, the analysis to identify the causal or focal channel follows from previous work in the analysis of human epileptic EEGs (Gersch & Goddard, 1970; Gersch, 1987).

Figure 3 illustrates data from three sites of a 7-second data record of the electrical activity at the onset of a spontaneous seizure event. (In fact, 16 channels of data were continuously monitored by telemetry.) The recorded data was sampled at the rate of 200 samples per second. Every 4th data point (Nyquist frequency = 25 Hertz), was used for analysis. The objective of the analysis was to determine whether the electrical activity at any one of the observed anatomical recording sites could be interpreted as "driving", or causing the electrical activity at the other sites. The data is complex, and there is no obvious visual clue that might unambiguously identify an initiating or driving site or identify when driving is present.

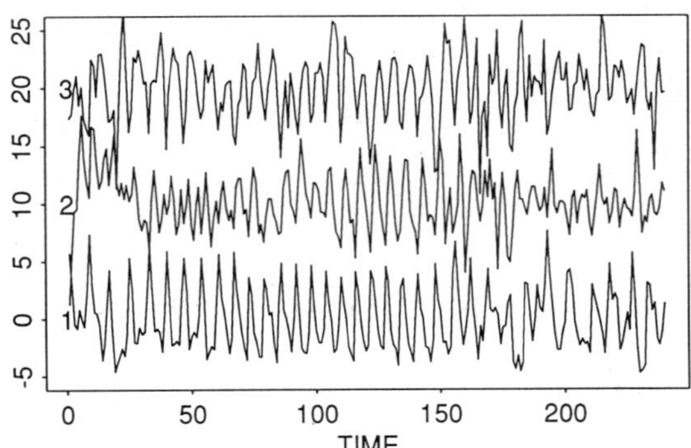

Figure 3. Human epileptic event segment.

The appearance of each individual EEG trace reveals that the local structure of the time series changes with time. Our analysis involved the fitting of a TV-VAR model to the data and a subsequent spectral analysis using the concept of an instantaneous power spectral density matrix (Gersch & Kitagawa, 1983; Gersch, 1987),

$$P_{f,n} = A_{f,n}^{-1} W_n A_{f,n}^{-1\prime} \tag{36}$$

where the multichannel AR model instantaneous system function $A_{f,n}$ is given by $A_{f,n} = I - \sum_{m=1}^{M} A_{m,n} \exp(-2\pi i m f)$.

The top row of Figure 4 shows the instantaneous power spectral density matrix as estimated from the fitted time varying VAR model versus frequency for channels 1, 2 and 3. We identify causality or driving at an instant in time by a frequency domain analysis via the instantaneous power spectral density matrix. The concept of causality that we use was introduced for stationary time series in Gersch and Goddard (1970). Our concept of causality differs from the "feedback free" definition of causality in econometrics (Geweke, 1982), or the time precedence definition more commonly used in the location of epileptic foci from EEG time series (Harris *et al.*, 1994, for example). The concepts of instantaneous coherence and instantaneous partial coherence versus frequency are involved in our concept of instantaneous causality or instantaneous driving. Driving is operationally defined in terms of a statistical test for zero partial coherence. (Coherence and partial coherence are defined for example in Brillinger, 1974.)

Let $W_{f,n}^2(x,y)$ denote the value of the spectral coherence between channels x and y at frequency f at time n. Similarly, let $W_{f,n}^2(x,y|z)$ denote the value of the partial spectral coherence between channels x and y conditioned on channel w at frequency f at time n. Then, our definition of causality is equivalent to the following conditions on spectral coherences and partial coherences

$$W_{f,n}^2(x,y) \ne 0, \quad W_{f,n}^2(x,w) \ne 0, \quad W_{f,n}^2(y,w) \ne 0,$$
$$W_{f,n}^2(x,y|w) \ne 0, \quad W_{f,n}^2(x,w|y) \ne 0, \quad W_{f,n}^2(y,w|x) \equiv 0 \ .$$

That is, the identification of causality at frequency f at time n is determined by the detection of zero partial instantaneous coherence. (Tests for zero partial coherence are in Brillinger, 1974.) The instantaneous coherences and partial coherences are computed from the instantaneous power spectral density matrix. The TV-VAR model yields estimates of the

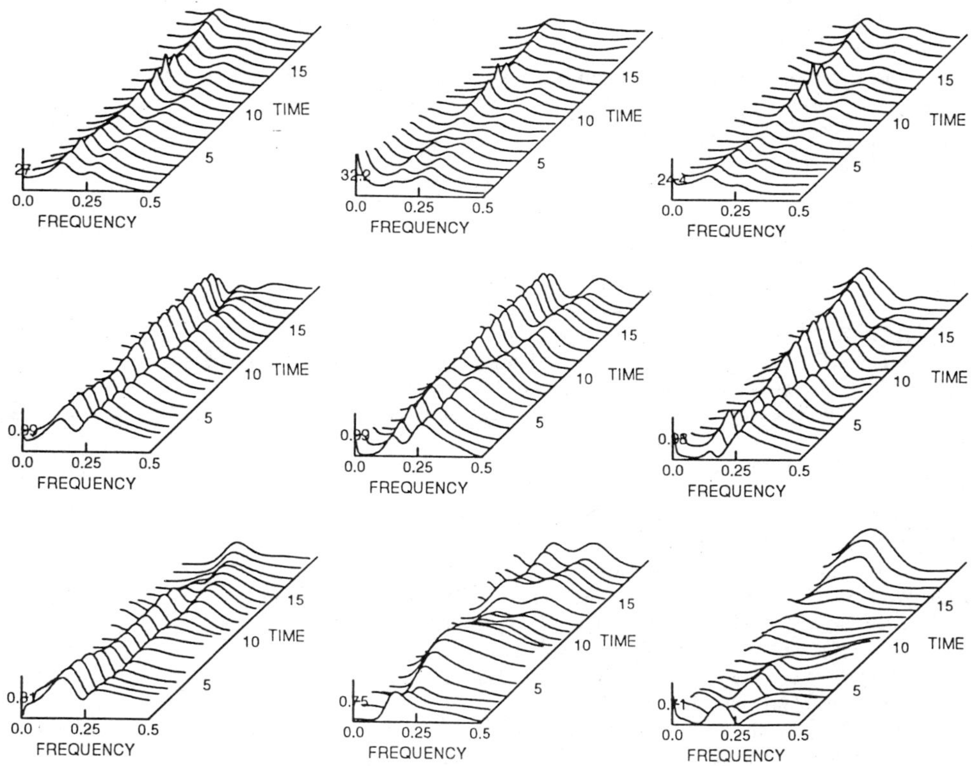

Figure 4. Human EEG spectrum analysis.

spectral coherence and partial spectral coherence at every instant in time. Computational results of both the evolving spectral coherence and evolving partial spectral coherence for each of the three possible pairs of three different data channels are shown in Figure 4 in rows 2 and 3. The computational results are used to determine whether or not and when one channel drives the other two.

The partial coherence between channels 1 and 2 and between channels 1 and 3 with the partialing done on the excluded third channel (respectively the first and second columns in the third row of Figure 4), remain significantly correlated. We conclude that channels 2 and 3 have relatively little explanatory power. On the other hand, the partial coherence between channels 2 and 3 with the influence of channel 1 removed (the third columns in the third row of Figure 4), results in the appearance of a relatively flat partial coherence surface statistically indistinguishable from zero over the frequency interval between 7 and 11 Hertz for the first 2.5 seconds of the record. We conclude that the pairwise coherence between the three channels is a consequence of the fact that channel 1 was driving channels 2 and 3 in the vicinity of 7.25 Hertz.

The observed result that driving persists for only a short time in the human seizure event is consistent with experience communicated personally by J. Gotman. The implication of this finding is that the mechanism that initiates a seizure is not the mechanism that sustains a seizure. Clearly, both mechanisms are worthy of study.

Because it is computed on instantaneous data, the TV-VAR model shown here yields the best time-frequency resolution properties achieved to date. The results derived from the estimated instantaneous power spectral density matrix yield a useful diagnostic to aid in the identification of candidate epileptic seizure driving channels. Finally, we note that the instantaneous TV-VAR model can be used as a research tool to study the evolution of the delays and transfer functions between brain sites during epileptic seizures.

REFERENCES

Akaike, H., 1968, On the use of a linear model for the identification of feedback systems, *Annals Inst. of Stat. Math.* 20:425-439.

Akaike, H., 1973, Information theory and an extension of the maximum likelihood principle, *2nd Int. Sym. on Information Theory*, Petrov, B. N., and Csaki, F., eds., Akademiai Kiado, Budapest, pp. 267-281.

Akaike, H., 1980, Likelihood and the Bayes procedure, In: *Bayesian Statistics*. Bernado, J. J., De Groot, M.H., Lindley, D. V., and Smith, A. F. M., eds., University Press, Valencia, Spain, PP. 143-166.

Anderson, B. D. O., and Moore, J. B., 1979, *Optimal Filtering*, Prentice Hall, New York.

Berger, J. O., 1985, *Statistical Decision Theory and Bayesian Analysis 2nd Edition*, Springer-Verlag, New York.

Brillinger, D. R., 1975, *Time Series Data Analysis and Theory*, Holt, Rinehart and Winston Inc., New York.

Cover, T. M., Hart, P. G., 1967, Nearest neighbor pattern classification, *IEEE Trans. Information Theory* IT-19:827-833.

Gath, I., Feuerstein, C., Pham, C. T., and Rondouin, G., 1992, On the tracking of rapid dynamic changes in seizure EEG, *IEEE Trans. Biomed. Eng.* 39:952-958.

Gersch, W., 1981, Nearest neighbor rule classification of stationary and nonstationary time series, In: *Applied Time Series Analysis II*, Findley, D. F. ed, Academic Press, New York, pp. 221-270.

Gersch, W., 1987, Non-stationary multichannel time series analysis. In: *EEG Handbook vol. II: Methods of Analysis of Brain Electrical and Magnetic Signals*, Gevins, A. S., and Remond, A., Eds. Academic Press, New York, pp. 261-296.

Gersch, W., 1992, Smoothness Priors. In: *New Directions in Time Series Part II*, Brillinger, D., Caines, P., Geweke, J., Parzen, E., Rosenblatt, M., and Taqqu, M. S., eds., *The IMA Volumes in Math, and its Applications*, volume 46, Springer-Verlag, pp. 113-146.

Gersch, W., and Goddard, G. V., 1970, Epileptic focus location, spectral analysis method, *Science* 150:701-702.

Gersch, W., and Kitagawa, G., 1983, A multivariate time varying autoregressive modeling of nonstationary econometric time series, *Proc. Bus. and Econ. Stat. Section, Am. Stat. Assoc.*, Toronto, Canada, August 1983.

Gersch, W. and Kitagawa, G., 1988, Smoothness priors in time series, In: *Bayesian Analysis of Time Series and Dynamic Systems*, Spall, J. C., ed., Marcel Dekker, New York, pp. 431-476.

Gersch, W., Martinelli, F., Yonemoto, J., Low, M. D., and McEwen, J. A., 1979, Automatic classification of electroencephalograms: Kullback Leibler nearest neighbor rules, *Science* 205:193-195.

Gersch, W., and Stone, D., 1990, Multichannel time varying autoregressive modeling: A circular lattice-smoothness priors realization, *Proc. 29th IEEE Conf. on Decision and Control*, Honolulu, Hawaii.

Gevins, A. S., 1980, Pattern recognition of human brain electrical potentials, *IEEE Trans. Pattern Analysis and Machine Intelligence* PAMI-2:383-404.

Geweke, J., 1982, The measurement of linear independence and feedback between multiple time series, *J. Am. Stat. Assoc.* 77:304-324.

Grenier, Y., 1982, Time varying lattices and autoregressive models: Parameter estimation, *Proc. of ICASSP 1982*, Paris, France.

Harris, B., Gath, I., Rondouin, G., and Feuerstein, C., 1994, On time delay estimation of epileptic EEG, *IEEE Trans. Biomed. Eng.* 41:820-829.

Haykin, S., 1991, *Adaptive Filter Theory, 2nd Edition*, Prentice Hall, New Jersey.

Hurvich, C. M., and Tsai, C. L., 1989, Regression and time series model selection in small samples, *Biometrika* 76:297-307.

Hurvich, C. M., and Tsai, C. L., 1992, A corrected Akaike information criterion for vector autoregressive model selection, *Report SOR-92-10*, New York University, L. N. Stern School of Business.

Jiang, X.-Q., and Kitagawa, G., 1993, A time varying vector autoregressive modeling of nonstationary time series, *Signal Processing* 33:315-331.

Kitagawa, G., 1987, Non-Gaussian state space modeling of nonstationary time series, *J. Am. Stat. Assoc.* 82:1032-1063.

Kitagawa, G. and Akaike, H., 1981, On TIMSAC-78, In: *Applied Time Series Analysis II*, Findley, D. F., ed., Academic Press, New York, pp. 499-547.

Kitagawa, G., and Gersch, W., 1985, A smoothness priors time varying AR coefficient modeling of nonstationary time series, *IEEE Trans. Automatic Control* AC-30:48-56.

Kitagawa, G., and Gersch, W., 1995, *Smoothness Priors Analysis of Time Series*, Springer-Verlag, New York, in press.

Kozin, F., 1977, Estimation and modeling of nonstationary time series, *Proc. Symp. Appl. Comp. Methods in Eng.*, Univ. Southern California, Los Angeles, pp. 603-612.

Lev-Ari, H., 1987, Modular architectures for adaptive multichannel lattice algorithms, *IEEE Trans. Acoust. Speech and Signal Process.* ASSP-35:543-552.

Lutkepohl, H., 1991, *Introduction To Multiple Time Series Analysis*, Springer-Verlag, Berlin.

Newton, J. J., 1982, Using periodic autoregressions for multiple spectral estimation, *Technometrics* 24:109-116.

Pagano, M., 1978, Periodic and multiple autoregressions, *Annals of Statistics* 6:1310-1317.

Rogers, W. H., and Wagner, T. J., 1978, A finite sample free performance bound for local discrimination, *Annals of Statistics* 6:506-514.

Sakai, H., 1982, Circular lattice filtering using Pagano's method, *IEEE Trans. Acoustics, Speech and Signal Processing* ASSP-30:279-281.

Shiller, R., 1973, A distributed lag estimator derived from smoothness priors, *Econometrica* 41:775-778.

Stone, D., 1993, One channel at-a-time multichannel autoregressive modeling: Applications to stationary and nonstationary covariance time series, *Ph. D. Thesis* in Communication and Information Sciences, University of Hawaii.

Takanami, T., and Kitagawa, G., 1991, Estimation of the arrival times of seismic waves by multivariate time series model, *Annals Inst. of Stat. Math.* 43:407-433.

Whittaker, E. T., 1923, On a new method of graduation, *Proc. Edinborough Math. Assoc.* 78:81-89.

Whittle P., 1963, On the fitting of multivariable autoregressions and the approximate canonical factorization of a spectral density matrix, *Biometrika* 50:129-134.

Whittle P., 1965, Recursive relations for predictors of non-stationary processes, *J. Royal Stat. Soc.* 27B:523-532.

Wiggins, R. A., and Robinson, E. A., 1965, Recursive solution to the multichannel filtering problem, *J. of Geophys. Res.* 70:1885-1891.

2

SIGNAL PROCESSING OF EEG: EVIDENCE FOR CHAOS OR NOISE. AN APPLICATION TO SEIZURE ACTIVITY IN EPILEPSY

Fernando H. Lopes da Silva,[1,2] Jan-Pieter Pijn,[2] and Demetrios N. Velis[2]

[1] Institute of Neurobiology
University of Amsterdam
[2] Institute of Epilepsy "Meer en Bosch"
Heemstede, The Netherlands

ABSTRACT

The EEG is an important signal for the diagnosis of functional disturbances of the brain, and in particular, of epilepsy. The non-linear dynamical analysis of EEG signals recorded during seizure activity in comparison with on-going signals allowed us to formulate a hypothesis about the generation of epileptic activity. According to this model, epilepsy should be envisaged as a dynamical disease of neuronal networks, that may exhibit different types of attractors, i.e., may present bifurcations. One of these attractors is characterized by the generation of irregular oscillations, typical of epileptic seizures.

INTRODUCTION

The EEG is an important signal for the diagnosis of functional disturbances of the brain, and in particular, of epilepsy. Epilepsy is a pathological state characterized by the occurrence of seizures (i.e., ictal activity), that in most cases occur in an unpredictable way; the seizures consist of sudden changes of neuronal activity, which disturb the normal functioning of neuronal networks. These changes may be expressed as disturbances of sensory and/or motor activity and of the flow of consciousness. During an epileptic seizure, neuronal networks exhibit typical oscillations that appear as more or less sudden modifications of the on-going EEG. These oscillations usually propagate throughout the brain, involving progressively more brain systems. In some cases, it appears as if the seizure activity affects immediately generalized systems of the brain and assumes a global character from the start. In between seizures, the brain may produce what is called interictal activity characterized by paroxysmal discharges of more or less circumscribed populations of neurons that in most cases are not expressed in obvious behavioural symptoms.

Advances in Processing and Pattern Analysis of Biological Signals, Edited by Isak Gath and Gideon F. Inbar
Plenum Press, New York, 1996

Here we approach the question of how to enhance the capacity of analyzing seizure patterns in EEG signals for diagnostic purposes, and also for obtaining more insight into the processes underlying epileptic manifestations. We start from the assumption that it is important to formulate a hypothesis about the generation of EEG signals in order to reach a significant answer to this question.

In general, two basic hypotheses are currently considered to account for the generation of EEG signals. Both are based on the observation that the exact characteristics of EEG signals appear to be essentially unpredictable. (i) One hypothesis states that EEG signals are realizations of random or stochastic processes. According to this hypothesis, EEG signals are considered as forms of filtered noise (for discussion, see review by Lopes da Silva, 1993). (ii) An alternative hypothesis arises from the proposition that EEG signals result from the activity of neuronal networks with non-linear properties that have complex dynamics. Indeed, non-linear dynamic systems with more than two degrees of freedom can display unpredictable behavior over long time scales (Schuster, 1984). In other words, they can display what mathematicians call deterministic chaos. This led to the hypothesis that EEG signals are manifestations of deterministic chaos (Babloyantz & Salazar, 1985; Freeman & Viana Di Prisco, 1986).

We examined this question directly by developing a method for testing which of these hypotheses is the most likely to apply to EEG signals recorded under different circumstances, with special attention to the case of typical EEG seizure activity. First, we describe the general methodology used; second, we present some EEG data, obtained from animals and from patients, to which this computational methods were applied; third, we deduce some general conclusions that can be helpful for understanding the dynamics of neuronal networks both under normal and epileptic conditions.

GENERAL METHODOLOGY TO IDENTIFY THE PRESENCE OF CHAOS IN EEG SIGNALS

The basic elements of the theory of deterministic chaos, which are of interest here, are briefly considered below. In general, the dynamics of systems can be analyzed in phase-space by studying the behavior of the trajectories of representative variables. In general, three types of attractors can be distinguished: point, limit-cycle, and chaotic. The former is the attractor found in systems exhibiting oscillations that die out at a rate according to the magnitude of damping; in phase-space, the trajectories spiral towards a point that defines the equilibrium state. The latter two are found in non-linear dynamical systems. The limit-cycle consists of a periodic oscillation, whereas the chaotic attractor is characterized by a quasi-periodic complex oscillation that has a broadband power spectrum (Thompson & Stewart, 1986). This can be exemplified using a mathematical model: a non-linear differential equation known as the *Duffing equation*, as shown in Fig. 1. This figure illustrates the fact that the same equation may exhibit different solutions (modes of behavior), depending on the choice of parameters and/or on the initial conditions. These different solutions can be characterized in several ways: by phase-plots, power spectra and correlation dimensions. The latter gives in a compact way information about the system's dynamics. Indeed, a convenient way to characterize a non-linear dynamical system is to apply the algorithm developed by Grassberger and Procaccia (1983), by means of which the correlation dimension can be estimated.

Let us consider that the EEG signal consists of a set of samples x_n, with $n = 1,.....,N+k$. We can construct m-dimensional vectors V_m (i) with $i = 1,... N$, i.e., the signal can be embedded in m-dimensional space as follows:

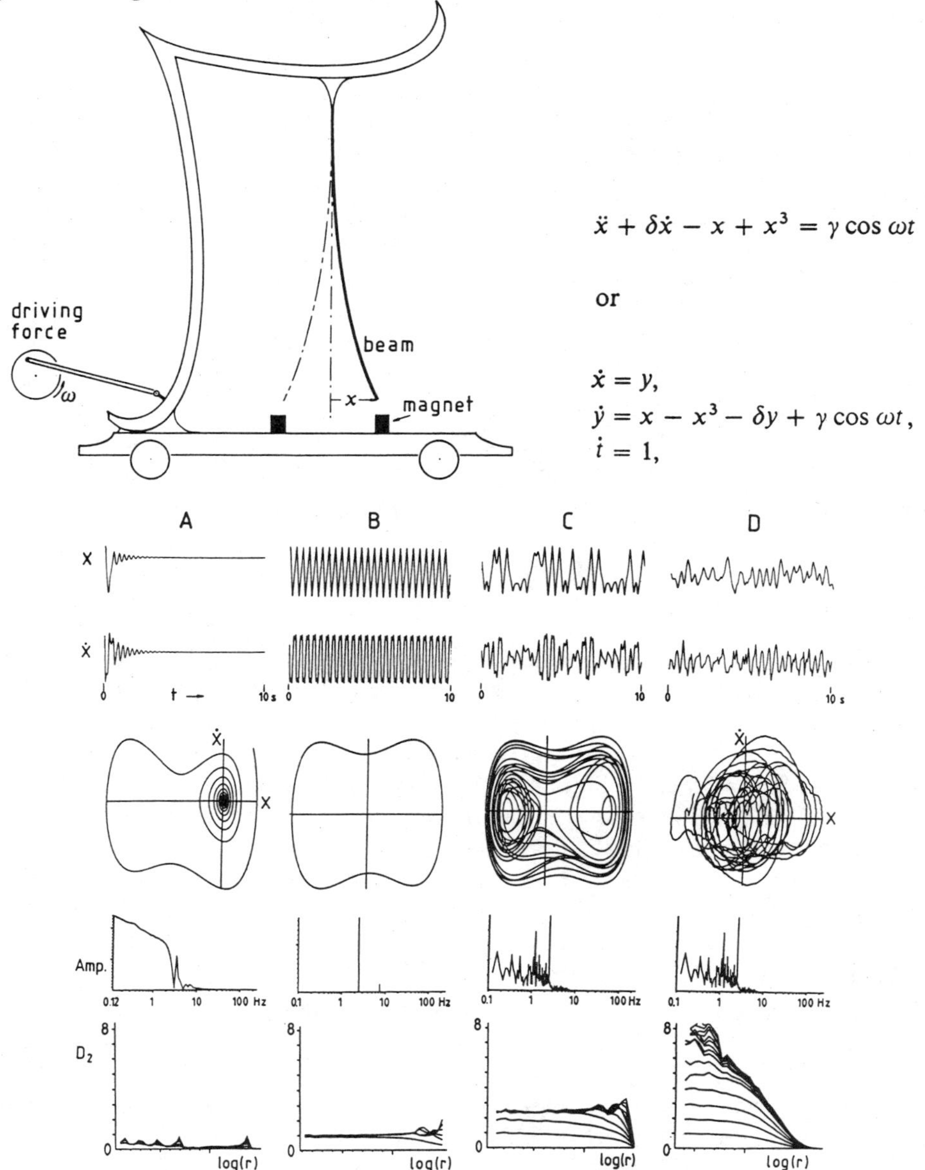

Figure 1. Demonstration of different types of dynamical behavior of the Duffing equation. The Duffing equation (upper right corner) is a good description of the behavior of a metal beam above two magnets. $\delta\dot{x}$ denotes the friction. $- x + x^3$ describes the attractive forces of the magnets on the beam, while $\cos \omega t$ is an external driving force. In A $\gamma = 0$, therefore the beam settles itself close to one of the magnets. In phase space (third row), this is expressed as a point attractor. $D_2 \approx 0$ for such transient behavior (fifth row). In B and C $\gamma = 0.3$, $\omega = 1$ and $\delta = 0.15$. With these values, limit cycle behavior (B) occurs for a starting point outside the region of the magnets, while chaotic behavior (C) occurs for a starting point in-between the magnets. For signals A and B $x = 1.6081$, $\dot{x} = 0.8783$, $t = 0.773$ (precisely on the limit cycle). For C $x = 0.5$, $\dot{x} = 0.5$, and $t = 0$. The limit cycle gives $D2 = 1$ while chaotic behavior gives $2 < D2 < 3$ which corresponds to a set of three first order differential equations. The signals from D were made by randomizing the phase angles of the signals from C. Both signals C and D have the same amplitude spectra (fourth row). The D2 plot in the case of the random surrogate signal (D) does not show the plateau as seen for the chatic signal (C). The spectra and D2 calculations were carried out for the signals represented in the first row. For A, however, the epoch analysed had a length of 2500 samples (half the length shown), while for B, C and D the epochs consisted of 16000 samples (3.2 times the length shown) (for the time axis, we considered the sampling rate to be 500 Hz).

$$V_m(i) = (x_{i+k_0}, x_{i+k_1}, x_{i+k_2}, \ldots\ldots\ldots x_{i+k_{m-1}}) \tag{1}$$

The values of k_l, also called delays, are fixed with $k \geq k_l \geq 0$, ($l = 0, \ldots\ldots$ m-1) and k_l k_p if p l. According to a theorem of Takens (1981), the time evolution of this vector has the same topological properties as the real attractor of the signal generating system. The point is to measure the distances between all $V_m(i)$ vectors and to determine how many of these distances are smaller than a certain value. This is done by computing the correlation integral $C(r,m)$, where r is the radius of a sphere with centre $V_m(i)$ in R^m, as follows:

$$C(r,m) = 1/N^2 \sum_i \sum_j h(r - d(V_m(i), V_m(j))) \tag{2}$$

where h is the Heaviside or step function (h $(x<0) = 0$, h $(x \geq 0) = 1$), and d is the distance between any two vectors, for which we used the largest difference between all corresponding components. If an attractor is present in the time series, the correlation dimension of this attractor D_2, is defined, according to a theorem of Takens (1981) for small r and large m and N, by the following relation:

$$C(r,m) \propto r^{D_2} \tag{3}$$

Thus, D_2 can be estimated by the slope of log $C(r,m)$ versus log r. For the calculations, we modified the standard Grassberger-Procaccia algorithm (1983) by choosing the delays k_l as follows: $k_0 = 0$; $k_1 = WT$, where the window W is a small integer (in our case we choose $W = 3$) and T is the first zero-crossing of the auto-correlation function; $k_2, k_3, \ldots$ were chosen recursively such that for each i the number $k_0, \ldots k_i$ filled the interval [0,WT] as homogeneously as possible. In addition, we used data pre-sorting according to the procedure proposed by Theiler (1986), which allowed to compute $C(r,m)$ for all values of m without repeated distance evaluations. Care was taken that D_2 was not underestimated in case the time series was strongly auto-correlated. To avoid this artefact, we set the initial values of the embedding vectors as follows: $j = i + a$, where a > 1 and i ended at N-a. By taking a of the order T, this artefact can be avoided (Theiler, 1986).

We should note that EEG signals most often present nonstationary properties. This is certainly the case for EEG signals recorded during epileptic seizures. To account for this problem, the EEG epochs were chosen as short as possible. However, to obtain reliable estimates of D_2, we have to use at least 4000 data points. Since we used a sampling rate of 512 Hz, this means that the selected EEG epochs had a duration of 8 sec (4096 samples). In some cases, we explored another novel way of dealing with nonstationary signals of this sort. Namely we computed recurrence plots.

When the behavior of a system changes from one mode to another, exhibiting nonstationarity, the distance between any vector (constructed according to Eq. (1) of the first mode and a vector of the second mode) will become relatively large, in comparison with the distances between any two vectors belonging to one and the same mode. This fact makes it possible to visualize the occurrence of transitions in the behavior of a system and to distinguish different modes of operation. This is done by making a recurrence plot. A recurrence plot is constructed as follows: along the axes, the index i in Eq. (1) is represented, running from the first to the N-th sample. For each pair of indices (i, j), a point is plotted at position (i, j) if the distance in m-dimensional space between the vectors V(i) and V(j) is smaller than a fixed threshold. The final plot depends, of course, on the value of this threshold. In practice, however, the same type of plot is obtained for a wide range of threshold values. An example is given in Fig. 2 for two modes of the Duffing equation.

Figure 2. A recurrence plot used to distinguish different types of dynamics and to detect transitions between different modes of behavior. Limit cycle and chaotic behavior were generated as in B and C of figure 1, respectively. In this case initially the system's behavior was in a chaotic mode. The transition from chaotic to limit cycle behavior was induced by adding to the external oscillating cos ωt force an extra short-lasting pulse. The discriminating power of the recurrence plot between these two types of behavior is striking. It may be noticed that periodic behavior results in stripes parallel to the diagonal at regular distances. The chaotic behavior results in stripes of limited length. The smaller the largest Lyapunov exponent λ, the larger the (average) length of the stripes. Important is that the parts of the recurrence plots that correspond to the two different modes of behavior emerge as disjunct areas in the plot. Note also that a short lasting limit cycle like behavior occurring within the chaotic mode is immediately recognized, and appears as lines on the quadrants representing the correlation between the two modes of activity.

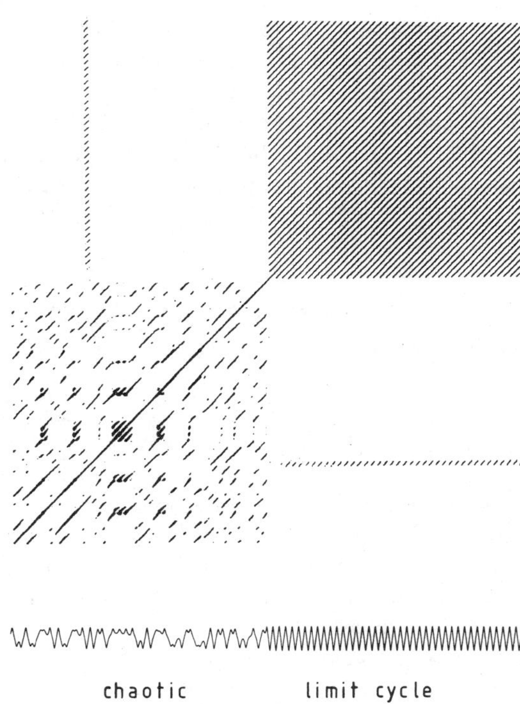

THE CONFIDENCE OF THE ESTIMATES OF D_2: REAL AND SURROGATE SIGNALS

To test whether the estimated value of D_2 really represented the correlation dimension of an attractor or was determined by the limitations of the computational process, we introduced surrogate signals. These were obtained by computing the amplitude and the phase spectrum of a real signal using the Fourier transformation. Then the phase angles of the Fourier components were replaced by random numbers. Applying the inverse Fourier transformation, the corresponding surrogate signals were obtained. The real and the surrogate signals have exactly the same power spectra, but completely different phase spectra. The surrogate signal should have a D_2 value that corresponds to that of time-limited random noise. Thus the values of D_2 of the real and surrogate signals should differ if the former deviates significantly from random noise. The method of surrogate signals was tested extensively on its merits by Theiler *et al.* (1992).

We conclude that the signal is generated by a non-linear process of possible chaotic nature if the curves of D_2 versus $\log(r)$ of the real signal, for increasing embedding dimensions m, converge asymptotically to a plateau for relatively small values of r, whereas those for the corresponding surrogate signal tend to a high value and do not converge to a plateau. The height of the plateau of D_2 versus $\log(r)$ gives an estimate of D_2. This is shown in Fig. 1 for the case of the Duffing equation. A comparison of the plots of D_2 for the real (Fig. 1C) and surrogate signals (Fig. 1D) shows the clear difference between what one expects from a chaotic attractor and an epoch of random noise. In the same figure, we see that in the case of the limit cycle, the value of D_2 is one, as theoretically expected.

We should note that there are other types of dimensions being used to characterize the properties of attractors, besides the correlation dimension, such as the Hausdorff dimension, the Kolmogorov entropy, the Lyapunov exponents, and the information dimension (Farmer *et al.*, 1983). We choose here to use the correlation dimension D_2 because it is the most widely used in similar studies. Using this form of analysis, several investigators have shown that an epileptic seizure is characterized by a low-dimensional process that most likely corresponds to a chaotic attractor (Babloyantz and Destexhe, 1986; Saermark *et al.*, 1990; Pijn *et al.*, 1991). In addition, Iasemidis *et al.* (1990) found that the positive Lyapunov exponent decreases abruptly at the onset of an epileptic seizure for the signals recorded nearest to the focus, and it increases as the seizure progresses.

THE CORRELATION DIMENSION OF ON-GOING AND TYPICAL SEIZURE EEG SIGNALS

A most valuable contribution of dimensional analysis was the demonstration (Fig. 3) that during the development and propagation of an epileptic seizures, the value of correlation dimension D_2 of the corresponding EEG signal decreases sharply whenever the neuronal networks of different brain areas become involved in the seizure activity as demonstrated in an experimental model of epilepsy of the rat, the kindling model of the limbic cortex (Pijn *et al.*, 1991). This indicates that a bifurcation took place between the pre-seizure random state, characterized by a non-saturating high-valued correlation dimension, and the seizure state with a correlation dimension between 2 and 3 (Pijn *et al.*, 1991). The recruitment of successive brain areas into the epileptic seizure activity fits closely with the decrease of the correlation dimension of the corresponding areas. Thus, the epileptic seizure state may be characterized as a low dimensional non-linear dynamical process having a small number of degrees of freedom, in spite of the fact that the underlying neuronal networks contain very large numbers of elements that are described by large number of parameters. It is interesting to add that a multi-neuronal model, developed by Traub and collaborators (Traub & Wong, 1982; Traub & Miles, 1991), which also contains a large number of degrees of freedom (9,900 cells, each with a number of compartments and five non-linear conductances), generates seizure discharges with a correlation dimension of 4.2. Thus, the general tendency, both in the natural case and in the simulations, is that seizure activity corresponds to a relatively low dimensional process. Whether one can speak of the existence of a real chaotic attractor in these cases is still a matter for further investigation.

It is important, of course, to investigate whether the same type of analysis will also yield relevant results when applied to EEG signals recorded from epileptic patients. Figure 4 illustrates the results of the correlation dimension analysis applied to the onset of a typical seizure originating in the hippocampus of a patient suffering from medial temporal lobe epilepsy (TLE). At seizure onset, we found a difference between the plots of D_2 versus log (r) of the real and the surrogate signals only for the EEG signal recorded from the focal area (HCR). The latter shows a D_2 value of approximately 4. At about 20 sec later, the seizure became generalized. Approximately at the same moment, a transition took place at the focal area towards a long lasting large amplitude oscillatory state. The transition is clearly visible in time domain. The recurrence plot of Fig. 5 indicates that it is a transition towards a disjunct state in phase space. This is shown for one of the signals from the hippocampus also illustrated in Fig. 4.

THEORETICAL FRAMEWORK: EPILEPSY AS A DYNAMICAL DISEASE

The correlation analyses presented here show that epileptic seizure activity is characterized by a low value of D_2, usually between 2 and 4, that differs clearly from that of surrogate signals and also from the value found for EEG signals recorded from the same

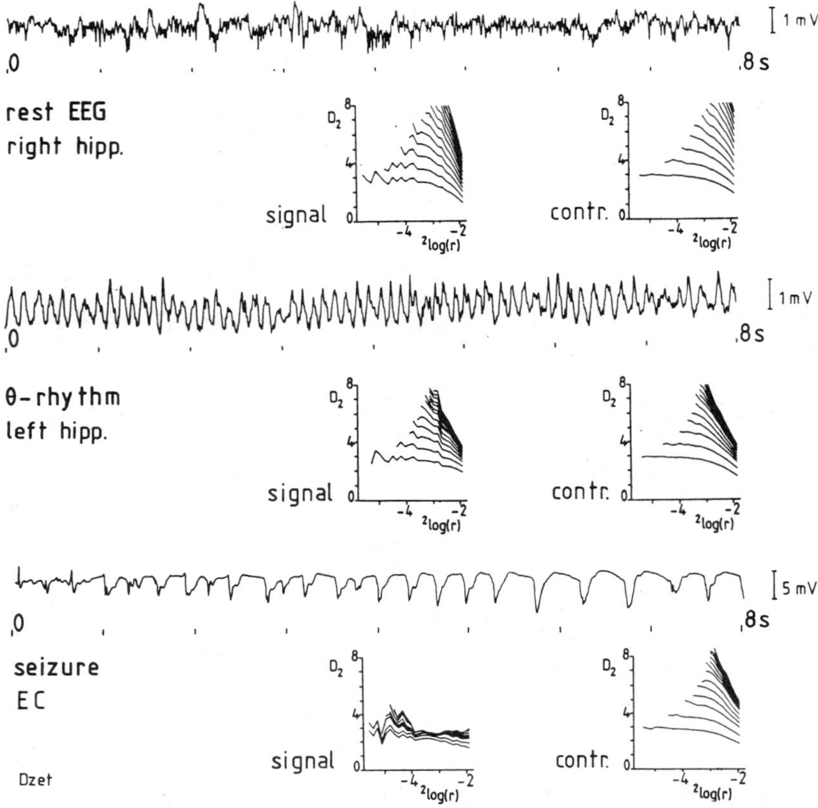

Figure 3. Epochs of EEG signals recorded from the hippocampus of an unanesthetized rat under three different behavioural conditions: restful alertness, locomotion characterized by the presence of a strong theta rhythm and during an epileptiform seizure triggered from the entorhinal cortex. Along with the raw EEG traces, the plots of D_2 against $^2\log r$ for the real signals and for the corresponding surrogate signals are shown. Note that these plots are similar for the two first conditions, but not for the case of the epileptic seizure. The latter shows a relatively low value, between 2 and 4, that may correspond to a chaotic attractor (adapted from Pijn *et al.*, 1991).

brain area before, or after, the seizure. Therefore, we may state that a bifurcation occurs in the dynamics of the underlying neuronal networks as an epileptic seizure occurs. This is a typical feature of complex non-linear systems, such as neuronal networks. Indeed, they can have multiple attractors, presenting bifurcations between different states (Freeman & Skarda, 1985; Elbert *et al.*, 1993). A bifurcation represents a qualitative change and depends on a set of control parameters that define the operating regime of the system.

To understand epilepsy in terms of the dynamics of neuronal networks, it is important to consider that epilepsy is manifested as the sudden occurrence of a typical oscillation of relatively long duration triggered by some initial and/or input conditions. Under normal circumstances, such an input would not cause more than a transient and uneventful change of brain activity; but in the epileptic brain, it can cause a catastrophic massive series of discharges. Why this happens in the epileptic brain may be understood by assuming that some alterations in the operating regime of the networks have occurred, such that their dynamics undergo a qualitative change. The recently developed theory of non-linear dynamics offers the possibility to understand, in formal terms, the changes in state that occur in a neuronal network during epilepsy.

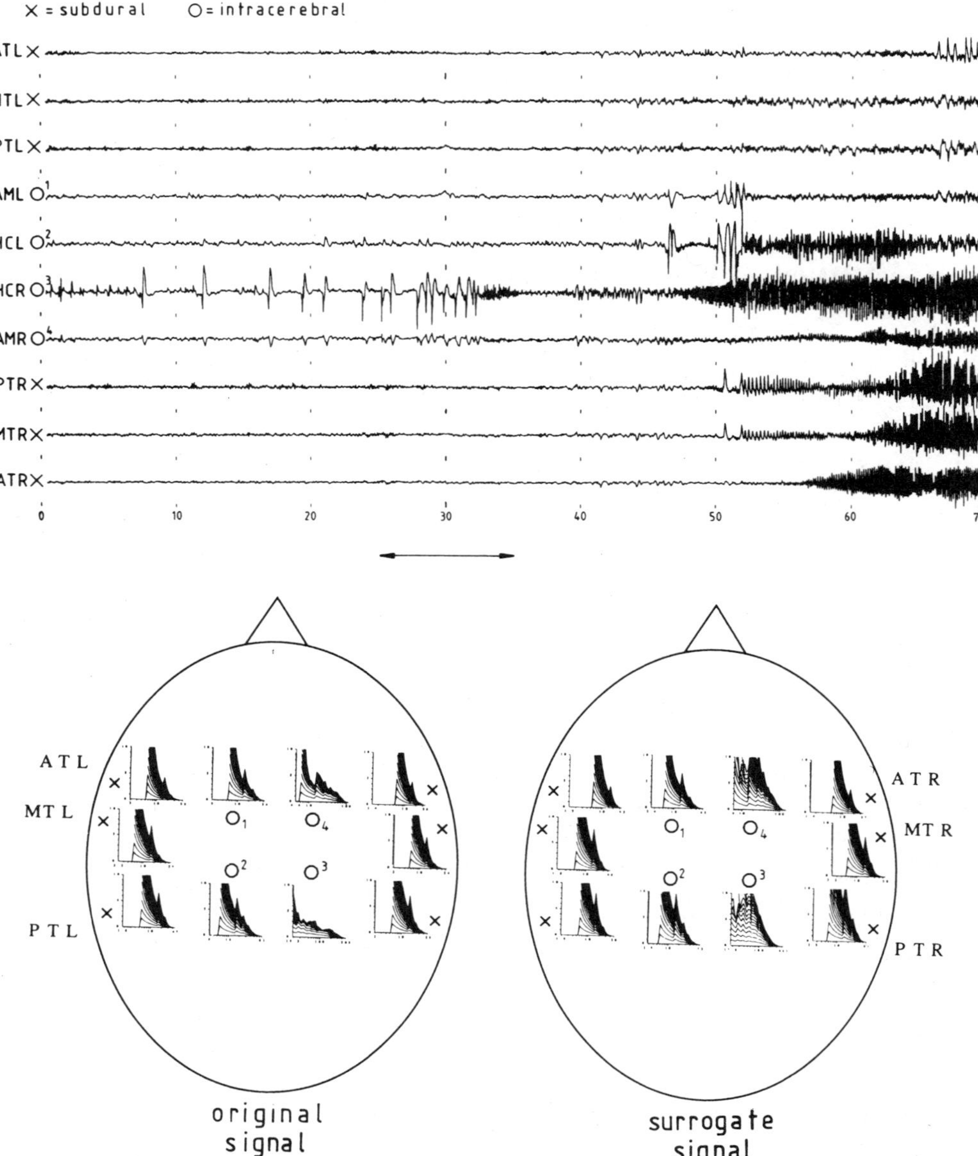

Figure 4. D_2 analysis of an epoch (indicated by means of double-headed arrow) of intracranially recorded EEG signals (ATL = anterior temporal left, MTL = mid temporal left, PTL = posterior temporal left, AML = amygdala left, HCL = hippocampus left, R = analogous sites from the right hemisphere). Epileptiform activity is present in the HCR signal in the shape of individual interictal spikes before the epoch indicated. During this epoch, the HCR activity becomes continuously epileptiform indicating seizure onset. From second 50 onwards, seizure activity becomes evident in other areas as well, roughly coinciding with the time the seizure became clinically manifest. In the D_2 plots, the local slope of $\log(C(r,m))$ versus $^2\log(r)$ is plotted with m varied from 1 to 17. The ordinate of the D_2 plots ranges from 0 to 10. For the abscissa auto scaling was used. The largest radius along the abscissa was equal to the difference between the maximum and the minimum of the signal within the epoch analyzed, while the smallest radius was 32.4 times smaller than the largest one. Only for the HCR and AMR (marked 03 and 04), a very clear difference was found between the D_2 plots of the original and the surrogate signals.

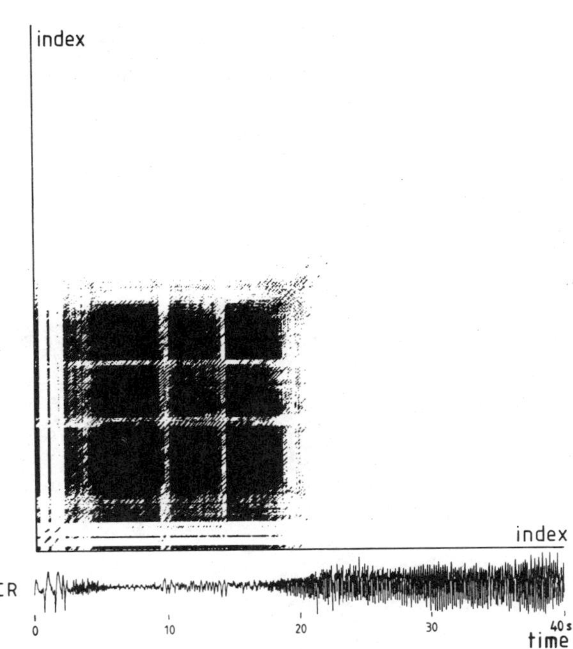

Figure 5. Recurrence plot for the signal recorded shortly after seizure onset in HCR (lasting from second 30 to 70 in Fig. 4). From second 0 to 20 various dark patterns are visible, with lighter intermissions between them. The dark patterns indicate the existence of a large number of points corresponding to small distances between successive vectors; this pattern is interrupted due to the presence of sharp transients. During the first 20 seconds, the signal appears to occupy the same area of the phase space. After second 20, however, only some sparse black points are visible, indicating that the signal has undergone a transition to a different subspace.

We may hypothesize that the main difference between a normal and an epileptic brain is essentially that the operating regime of a given neuronal network of the latter is much closer to a bifurcation point (near to region A2 in Fig. 6) leading to chaos than that of the former. In contrast with the normal case where the operating regime is situated far from such a bifurcation point (region A1 in Fig. 6), in an epileptic neuronal network, the distance between operating and bifurcation points is so small that the system may easily switch from a stable equilibrium to a chaotic attractor even in response to a very weak stimulus.

It should be noted that until recently, studies of the generation of epileptiform seizure activity in neuronal networks have emphasized that these can be described as limit-cycle oscillations (Kaczmarek & Babloyantz, 1977), but in the light of more recent data, as discussed above, it appears that these oscillations are more complex than those produced by a simple periodic system.

Finally, we consider briefly some of the basic physiological and biophysical phenomena that cause the changes in the stability of a neuronal network leading to the generation of an epileptogenic focus.

Currently, we only have fragmentary knowledge of these processes. The relevant information has necessarily to be obtained from experimental studies using animal models of the progressive development of epileptogenesis. In this respect, one of the most useful models has been the kindling model of epilepsy of limbic structures (Goddard, 1967; Kamphuis & Lopes da Silva, 1990; Lothman *et al.*, 1991; McNamara, 1989; Racine, 1972). An interesting feature of the development of a kindled focus in the CA1 area of the hippocampus is the progressive change of the excitatory/inhibitory balance in detriment of the latter as estimated from field potentials that reflect the behavior of local neuronal populations (Kamphuis *et al.*, 1988, 1992). This means that the main control parameter of the neuronal network changes in the sense that the operating regime of the network moves close to a bifurcation point, so that multiple attractors may coexist and chaotic dynamics may occur. The control parameter of the whole neuronal network should be seen as a ratio between all pooled parameters, i.e., those responsible for excitation (glutamatergic synaptic

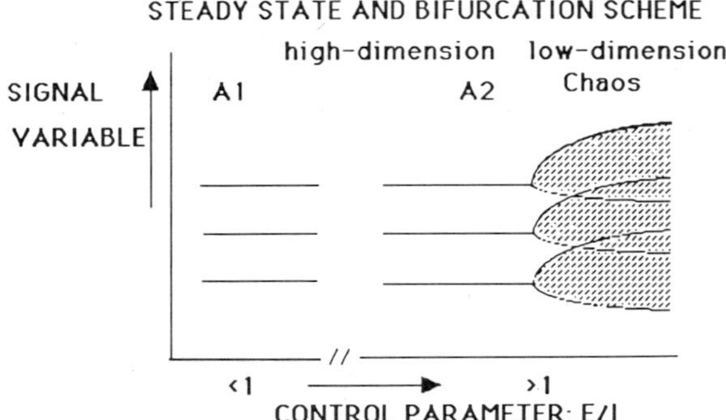

Figure 6. Theoretical representation of the possible regimes of a non-linear neuronal network with complex dynamics, with a bifurcation to a chaotic mode of behavior. This depends on a control parameter represented by the ratio between excitatory (E) and inhibitory (I) processes. The hypothesis is that the regime of the network will be high-dimensional, and not distinct from random noise, for relatively small values of the ratio E/I (region A1 - A2), and it will present a bifurcation to a low-dimensional chaotic regime for large values of E/I.

activity including AMPA, NMDA, metabotropic or other receptors) and for inhibition (GABAergic synaptic activity, both A and B types, after-hyperpolarizations and other K^+ mediated conductances). It is likely that different combinations of changes of these various parameters may lead to the same type of global change in network stability leading to epileptogenesis. Nevertheless, the detailed appearance of the epileptiform discharges as such, is most probably determined by the exact type of parameters that are impaired, as illustrated by the model studies of Traub and collaborators.

In conclusion, the non-linear dynamical analysis of EEG signals recorded during seizure activity in comparison with on-going signals has led to develop a theoretical framework that permits to formulate a clear hypothesis about the generation of epileptic activity. According to this model, epilepsy should be envisaged as a dynamical disease of neuronal networks, that may exhibit different types of attractor, i.e., may present bifurcations. One of these attractors is characterized by the generation of irregular oscillations, typical of epileptic seizures.

ACKNOWLEDGMENTS

We thank Dr. Jaap de Goede and Marcel van der Heyden for the recurrence plots and for their comments. The secretarial assistance of Trúc Ngô-Hà and Cristine Knaap-Cabi is gratefully acknowledged. These investigations were supported by a grant from the National Epilepsy Fund of The Netherlands (Utrecht).

REFERENCES

Babloyantz, A., and Salazar, J. M., 1985, Evidence for chaotic dynamics of brain activity during the sleep cycle, *Phys. Lett.* 111A:152-156.

Babloyantz, A., and Destexhe, A., 1986, Low-dimensional chaos in an instance of epilepsy, *Proc. Nat. Acad. Sci. (U.S.A.)* 83:3513-3517.

Elbert, Th., Ray, W. J., Kowalik, Z., Skinner, J. E., Graf, K. E., and Birbaumer, N., 1994, Chaos and physiology: deterministic chaos in excitable cell assemblies, *Physiol. Rev.* 7:1-7.

Farmer, J. D., Ott, E., and Yorke, J. A., 1983, The dimension of chaotic attractors, *Physica* D7:153.

Freeman, W. J., and Skarda, C. A., 1985, A perspective on brain theory: non-linear dynamics of neural masses, *Brain Res. Rev.* 10:147-175.

Freeman, W. J., and Viana di Prisco, G., 1986, Relation of olfactory EEG to behavior: time series analysis, *Behav. Neurosc.* 100:753-763

Goddard, G., 1967, Development of epileptic seizures through brain stimulation at low intensity, *Nature* 214:1020-1021.

Grassberger, P., and Procaccia, I., 1983, Measuring the strangeness of strange attractors, *Physica* 9:183-208.

Iasemidis, L. D., Sackellares, J., Zaveri, H., and Williams, W., 1990, Phase space topography and the Lyapunov exponent of electrocorticograms in partial seizures, *Brain Topography* 2:187-201.

Kaczmarek, L. K., and Babloyantz, A., 1977, Spatiotemporal patterns in epileptic seizures, *Biol. Cybern.* 26:199-208.

Kamphuis, W., Lopes da Silva, F. H., and Wadman, W. J., 1988, Changes in local evoked potentials in the rat hippocampus (CA1) during kindling epileptogenesis, *Brain Res.* 440:205-215.

Kamphuis, W., and Lopes da Silva, F. H., 1990, The kindling model of epilepsy: The role of GABAergic inhibition, *Neurosci. Res.* Comm. 6:1-9.

Kamphuis, W., Monyer, H., De Rijk, T., and Lopes da Silva, F. H., 1992, Hippocampal kindling increases the expression of glutamate receptor -A Flip and -B Flip mRNA in dendate granule cells, *Neurosci. Lett.* 148:51-54.

Lopes da Silva, F. H., 1993, EEG analysis: Theory and practice. In: Niedermeyer, E., and Lopes da Silva, F. H. (eds.), *Electroencephalography: Basic Principles, Clinical Applications and Related Fields*, 3rd ed. Williams & Wilkins: Baltimore, pp. 1097-1123.

Lothman, E. W., Bertram, E. H. III and Stringer, J. L., 1991, Functional anatomy of hippocampal seizures, *Progress in Neurobiology* 37:1-82.

McNamara, J. O., 1989, Development of new pharmacological agents for epilepsy: lessons from the kindling model, *Epilepsia* 30:S13-S18.

Pijn, J. P., Van Neerven, J., Noest, A., and Lopes da Silva, F. H., 1991, Chaos or noise in EEG signals: dependence on state and brain site, *Electroenceph. clin. Neurophysiol.* 79:371-381.

Racine, R. J., 1972, Modification of seizure activity by electrical stimulation. II. Motor seizure, *Electroenceph. clin. Neurophysiol.* 32:281-294.

Saermaak, K., Labech, J., Bak, C. K., and Sabers, A., 1990, Magnetoencephalography and attractor dimension: normal subjects and epileptic patients. In: Basar, E., and Bullock, H. (eds.), *Brain Dynamics*, Springer-Verlag:Berlin, pp. 149-158.

Schuster, H. G., 1984, *Deterministic Chaos. An Introduction.* Physik, Weinheim, 270 pp.

Takens, F. Detecting strange attractors in turbulence. Dynamical systems and turbulence. In: Rand, D. A., and Youg, L. S. (eds.), *Lecture Notes in Mathematics*, Springer: New York, vol. 898, pp. 365-381.

Theiler, J., 1986, Spurious dimension from correlation algorithms applied to limited time-series data, *Phys. Rev. A* 34:2427-2432.

Theiler, J., Eubank, S., Longtin, A., Galdrikian, B., and Farmer, J. D., 1992, Testing for nonlinearity in time series: the method of surrogate data, *Physica* 58D: 77-94.

Thompson, J. M. T., and Stewart, H. B., 1986, *Non Linear Dynamics and Chaos*, Wiley and Sons: Chichester, p. 376.

Traub, R. D., and Wong, R. K. S., 1982, Cellular mechanism of neuronal synchronization in epilepsy, *Science* 216:745-747.

Traub, R. D., and Miles, R., 1991, *Neuronal Networks of the Hippocampus*, Cambridge University Press: Cambridge, pp. 281.

3

MARKOVIAN ANALYSIS OF EEG SIGNAL DYNAMICS IN OBSESSIVE-COMPULSIVE DISORDER

Alex A. Sergejew[1] and Ah Chung Tsoi[2]

1 Centre for Applied Neurosciences
  School of Biophysical Sciences and Electrical Engineering
  Swinburne University of Technology
  Hawthorn, Victoria 3122, Australia
2 Department of Electrical and Computer Engineering
  University of Queensland
  St. Lucia, Queensland 4072, Australia

ABSTRACT

Electroencephalograph (EEG) signal dynamics of recordings taken during an unstructured "eyes closed" recording session from 13 patients with severe obsessive-compulsive disorder (OCD) and eleven normal control subjects were investigated using Markov modelling of an autoregressive representation of the EEG signal. Limited state transition dynamics were observed in the EEG of all subjects with OCD but in none of the controls. These findings are discussed in relation to hypothesized disturbances in mental state transitions in OCD.

INTRODUCTION

The state/trait dichotomy is used widely in psychology and psychiatry. At the state level the aim is to quantify observed or reported states of mind or behavior over the course of short periods of time, for example in the assessment of response to treatment. Typically, this is attempted by the use of structured interviews to rate symptoms, but even with extreme care to standardize rating criteria the resultant measures are quite coarse and the time resolution is very limited. The trait level attempts to address longer time frames and usually describes habitual or characterological tendencies that underlie the instantaneous state changes.

Carr (1983) in reviewing the concept of clinical state in psychiatry, suggested that *"one task of psychiatric research ... would be to use the state concept as a frame of reference in working from the observed clinical phenomena (i.e. range of states) and inputs to the*

Advances in Processing and Pattern Analysis of Biological Signals, Edited by Isak Gath and Gideon F. Inbar
Plenum Press, New York, 1996

underlying 'configurations of information-processing systems' and the relationships between such systems by measuring outcomes within the context of psychiatric illness." He lamented, however, that *"although certain states may have particular biologic concomitants, in most cases they cannot yet be used as sufficient differentiating criteria in themselves to define particular states."*

The work described here attempts to address this agenda explicitly, while acknowledging that the concept of psychological state or mental state is not yet able to be clearly defined despite prolonged attempts at formalisms such as those developed by Horowitz (1979). It is proposed that, on the basis of phenomenological descriptions, obsessive-compulsive disorder (OCD) may be regarded as a disorder in which mental state transitions are regularly and stereotypically limited. The biological concomitant investigated here is the electroencephalogram (EEG) signal acquired during an unstructured recording session. EEG state is defined by autoregressive (AR) model parameters estimated for fixed model order and fixed consecutive time intervals assumed to be "quasistationary" (Bodenstein & Praetorius, 1977). An objective state transition description, based on a Markov modelling approach, is used to formally quantify EEG state transitions, and the experimental findings used to test a hypothesis which is based on phenomenological descriptions of OCD.

The attempt to associate mental states with EEG states as proposed here, although unusual, has been anticipated in the context of psychiatric disorders by Wright and colleagues (Kydd & Wright, 1986; Wright *et al.*, 1985; Wright, 1990). The association suggested between the two state domains is not intended to be a literal equivalence relationship. A similar association was, nonetheless, implicit in the theory proposed by Remond and Renault (1972), who considered the EEG as being composed of elementary patterns such that a given EEG record of any kind could be produced from consecutive elements taken from a complete pattern collection or "alphabet." Remond postulated that the letter statistics of an EEG recording taken from a given individual could be compared with statistics obtained from a large number of EEGs grouped according to clinical syndromes, and proposed that the given individual could be found to belong to that pathological class which most resembles the individual's EEG letter statistics. Jansen *et al.* (1981) described one of the first of the few attempts to implement this theory, albeit in the context of sleep staging. To the best of our knowledge, the present work is the first reported attempt at an implementation for OCD.

An association between the domains of mental state and EEG signal state is also implicit in the recent explosion of interest in "cognitive" event-related potentials or ERPs (see Giannitrapani (1985) for example). In ERP research, replicatable patterns of stimulus-locked EEG changes in response to specific paradigms or cognitive tasks are identified with or equated to sequential processing by distributed neural structures. The experimental control imposed by the cognitive paradigm enables associations or correlations to be made between electrocortical events and cognitive events, in a way not conceptually possible with the relatively uncontrolled and unstructured ongoing "resting" or "baseline" EEG. However, since it is not possible to control abnormal mental states and the flow over time of thoughts in a thought disorder, ERP paradigms may have less to offer directly in quantifying these neuropsychiatrically significant phenomena. Despite the commonly expressed misgivings (Torello, 1989) regarding the lack of control in the so-called "resting" state, the work described here is based on analysis of the "resting" EEG specifically because "cognitive processes are left free to wander."

Obsessive-Compulsive Disorder

The DSM-IIIR defines obsessive-compulsive disorder as being characterized by "recurrent obsessions or compulsions." Obsessions are defined as *"persistent ideas, thoughts, impulses or images that are experienced ... as intrusive and senseless."* Compulsions are defined as *"repetitive, purposeful and intentional behaviors that are performed in response to an obsession, according to certain rules or in a stereotyped fashion."* The DSM-IV gives a description of the diagnostic features of the disorder without offering definitions of the two terms.

There is some disagreement (Horowitz, 1979) over the phenomenology of OCD. The "experiential" view is that obsessions and/or compulsions occur as intermittent impositions of a stereotyped abnormal mental state over an otherwise unaffected flow of thought. In contrast, the "obsessive personality" view is that the flow of thinking in individuals affected by OCD appears to constantly and almost predictably return to the obsessions despite the distress that these symptoms engender.

There is some electrophysiological and neuropsychological evidence implicating frontal lobe dysfunction in OCD (Malloy *et al.*, 1989; Perros *et al.*, 1992), and it has been suggested that the "frontal lobe" phenomenon of perseveration may be viewed as a not dissimilar and very striking type of disturbance in the flow of mental states. Another striking and possibly related type of disturbance in the flow of thinking (Jenike & Brotman, 1984) which is termed "involuntary forced thinking" has been reported in a group of temporal lobe epileptic patients with demonstrable EEG evidence of temporal spiking and/or slowing.

EEG and OCD

The literature on the EEG in OCD is relatively limited, with some work predating the DSM diagnostic classification system being relatively difficult to evaluate. Earlier studies employed visual inspection of EEG recordings. EEG abnormalities, especially in the theta band, have long been reported (Pacella *et al.*, 1944; Rockwell & Simmons, 1947; Insell *et al.*, 1983; Jenike & Brotman, 1984) in OCD patients, with some of these abnormalities described as being consistent with temporal lobe epilepsy (Epstein & Bailine, 1971). In contrast, however, Shagass (1984) reported no significant differences in the EEG between OCD and normal control groups or groups of other "neurotic" disorders. Flor-Henry *et al.* (1979), in the first study to employ quantitative EEG analysis in OCD, reported finding reduced left temporal beta variability. Malloy *et al.* (1989) employed topographic mapping of frequency domain EEG data in OCD and reported finding a slight tendency to higher-amplitude slow wave activity in the left fronto-parietal regions. Recently, Perros *et al.* (1992) reported finding reduced theta-2 variability and increased theta-2 power in the left posterior frontal to midtemporal region. They also briefly reviewed recent neurological and brain imaging studies of OCD, and concluded that the evidence supports a neurologic substrate for the disorder.

Our work described herein follows on from the study by Perros *et al.* (1992), and is based on the same EEG data from OCD patients and normal control subjects as that reported in their paper.

Hypothesis

The tacit assumptions behind any hypothesis purporting to link EEG states with mental states (Kydd & Wright, 1986; Wright & Kydd, 1986) are that the cerebral cortex maintains intrinsic mechanisms to control mental states, and that there is a linkage between mental or cognitive events and electrocortical events, so that disturbances in cortical homeostatic mechanisms leading to disturbed mental state transitions might be inferred from disturbances in the transition of states in the generated EEG signal.

Our hypothesis is based on these tacit assumptions and follows simply from the phenomenological description of mental state transitions in OCD. We hypothesize that there will be periods in which only a finite number of brain states will be approached by OCD subjects, these periods being quantifiable by Markov modelling of their EEG. The corollary of this hypothesis is that for a normal subject the number of possible brain states, if not infinite, is countably infinite. For the OCD individuals we predict a tendency to approach (if not revisit) a small set of transitions, and further predict that the size of this set will depend on the severity of the obsessional symptoms as measured by standard neuropsychological assessment.

The null hypothesis is that there would be no difference found in the state transitions between normal subjects and individuals suffering from OCD on the basis of Markov modelling of their EEG signal analysis. We would reject the null hypothesis if we found evidence of state transition rules in EEG data from the OCD group which varied quantitatively from those in the normal group.

With regard to the two views of the phenomenology of OCD, the "experiential" view clearly implies that limited EEG state transitions will be found in OCD patients only near the occurrence of their target obsessive and/or compulsive symptoms, whereas the alternative "obsessive personality" view implies that limited EEG states will be found even in the absence of temporal proximity to target symptoms. While chronic ambulant EEG recording technology could acquire EEG data around the time of occurrence of target symptoms, this study sought to test the hypothesis using data acquired during a standard clinical EEG recording session. Experimental results which were unable to falsify the null hypothesis would therefore leave open the possibility of the experimental hypothesis being true but only applicable in the proximity of target symptoms. Because the hypothesis is predicated on estimation of state transition rules in ongoing "resting" or "baseline" EEG, no cognitive "activation" tasks were employed.

In the work described herein, the acquired EEG signal data was preprocessed using an autoregressive (AR) model prior to Markov modelling.

EEG and AR Modelling

The two main signal processing approaches are frequency domain and time domain analysis. With the former approach, the EEG frequency spectrum typically varies over time, indicating that the EEG is a nonstationary signal. In time domain analysis the EEG signal is assumed to be derived from a signal model. There are many possible nonlinear models which, in principle, permit the analysis of a signal like the EEG over longer time durations than a linear model. However, almost all of the EEG literature based on time domain analysis has employed linear models, in which the underlying signal is assumed to be generated by a linear mechanism (within a "window" of time). Two traditional linear models are the autoregressive (AR) and the autoregressive moving average (ARMA) models.

Characterization of the EEG by AR or ARMA modelling techniques has become widely applied (e.g. Cerutti *et al.*, 1986; Crowell *et al.* 1977; Gath *et al.* 1983) since its first reported application by Fenwick *et al.* (1969) and Gersch (1970). In many circumstances these techniques were applied with the intent only to reduce data rather than to obtain direct estimates of physically meaningful parameters, however in the work by Wright *et al.* (1990), AR parameters were directly related to a physical description within a physiologically realistic neural model. By contrast, in an effort to produce operationally meaningful parameters, Isaksson *et al.* (1981) demonstrated an ARMA model which decomposed an EEG signal into a smoothed spectrum divided into the EEG frequency bands traditionally used in clinical neurophysiology. The use of an AR model as a linear prediction filter leads to a means of adaptively segmenting an EEG signal into near-stationary epochs, as was first suggested by Bohlin (1971), and implemented by many investigators (for example Bodenstein *et al.*, 1985; Tsoi, 1987, 1988) since. Lopes da Silva *et al.* (1974) applied a threshold criterion to the forward prediction error estimates to detect the spikes in an EEG, and Bohlin (1977), among others, replicated these results using a Kalman filter implementation of an AR model which was able to estimate the spike waveform.

In the work described here, EEG data was preprocessed using an AR model principally because using a parsimonious AR representation of the signal reduces the computational burden of the subsequent Markov modelling. Our motivation for employing an AR model for EEG signal preprocessing embraces the interpretations in neurophysiology of the WKSL model (Wright *et al.*, 1990a,b, 1994; Wright & Sergejew, 1991), and the potential for implementation of an adaptive segmentation technique (Jensen & Cheng, 1988), together with our success in classifying AR-preprocessed EEG data in psychiatric disorders (unpublished work in progress).

Implementation issues which we have investigated prior include the use of one versus more channels of EEG data, optimum fixed data segment length, and optimum AR model order.

Single Channel EEG Data. Analysis of a single channel of EEG data, if justifiable, leads to a geometric reduction of the computational burden.

The use of the state of a single channel as being representative of the state of the full matrix for each sampling time of EEG recorded "at rest" was tested (Tsoi & Sergejew, 1992; Sergejew & Tsoi, 1992) using the prediction error model

$$x(t) = \sum_{i=1}^{C} w_i y_i(t) + \varepsilon(t) \tag{1}$$

where $x(t)$ is the digitized value at a given EEG channel at time t, and $y_i(t)$, $i = 1, 2, ..., C$ are the instantaneous digitized values of the remaining C channels of recorded EEG. The w_i terms are model coefficients, and $\varepsilon(t)$ is a random residual which is assumed to be Gaussian, with zero mean and an unknown variance σ^2. The standard deviation of the prediction error for all the EEG data reported here was consistently found to be small, of the order of a few percent of the maximum amplitude of the signal.

Also tested was an alternative prediction error model

$$x(t) = \sum_{i=1}^{C} w_i y_i(t) + \sum_{i=1}^{C} \sum_{j=1}^{n} w_{ij} y_i(t - j) + \varepsilon(t) \tag{2}$$

where n is a constant and $n > 0$. Choosing n to be small, and systematically investigating $n = 1, 2, ... 5$, it was consistently found that the prediction error variance did not decrease significantly from that obtained previously.

We therefore chose to use only the data from channel C_z, located at the vertex of the scalp. This site is much less prone to contamination by artefacts such as EOG or muscle artefact, and is also clear of the sites reported to be abnormal in the topographic quantitative EEG studies of OCD cited above.

Optimum Signal Segment Length. In lieu of implementing an adaptive segmentation procedure, some preliminary work was carried out to ascertain the approximate length of time during which the EEG signal can be considered stationary. An autoregressive (AR) model of the form

$$y(t) = \sum_{i=1}^{M} w_i y(t - i) + \varepsilon(t) \tag{3}$$

was applied to the EEG signals (which were sampled at 128 Hz) where $y(t)$ is the digitized value at time t, w_i, $i = 1, 2, ..., M$ are the AR coefficients to model order M, and $\varepsilon(t)$ is a random residual with an variance σ^2 for a time window comprising T samples.

If the signal segment can be approximated by a stationary AR model then the variance of the prediction error will remain approximately constant. Conversely, if the signal segment is not stationary, the prediction error variance will differ from the approximately stationary one. These properties were checked for a range of EEG signal segment lengths, sliding a time window over the signal using time windows T = 128, 256, 384, 512, and estimating the corresponding prediction error variances. The model error found from each window segment was compared to the previous segment of the signal using a likelihood ratio method. The likelihood ratio remains approximately constant if the signal segment in a new window is approximately stationary, whereas the likelihood ratio differs if the segment of signal in the new window differs from the previous window.

In general, the likelihood ratio was found to remain approximately constant for time windows of T = 128 to T = 256, whilst it fluctuated widely for longer time windows. The smaller time window T = 128 was conservatively chosen to ensure the validity of the stationarity (or "quasistationarity") assumption in subsequent analyses.

Following conventions in speech analysis, a 128 point sample of data is termed a "frame" of data. Each EEG recording signal of N digitized samples can be subdivided into $N/128$ frames. For each frame of data the AR coefficients are estimated as above. To a certain extent, the AR coefficients in each frame describes the signal content within that frame, hence the frame-by-frame AR coefficients may be considered as a compressed representation of the signal. Furthermore, the dynamical behavior of the signal can also be represented by these frame-by-frame AR coefficients.

Again, using notations in speech analysis, the AR coefficients in each frame define a "state." The sequence of states parsimoniously represents the dynamical behavior of the EEG signal. The sequential properties of EEG signal states thus defined may be formally quantified using a Markov model.

Optimum AR Model Order. Several approaches to estimating AR model order have been suggested, but Akaike's criterion (Akaike, 1969) appears to be most often used in EEG work, and involves minimizing

$$AIC = T \ln\sigma^2 + 2(M + 1) \tag{4}$$

Optimum model order was found with M = 8 or M = 10 for all the EEG data reported here, and in subsequent analysis a fixed model order with M = 10 was used.

EEG and Markov Modelling

In a Markov model the observed signal is postulated to be generated by an underlying system which undergoes state transitions. There are a number of possible models depending on how the state transitions occur. In the simplest Markov model, the present state depends only on information available at the previous state and not before.

There are two possibilities: the transition of state is either directly observable externally, or it is not. In the event that the state transition is observable directly it is quite easy to compute the state transition matrix, by simply counting the number of transitions to a particular state and dividing this by the total number of state transitions.

If, on the other hand, the state transition is not directly observable externally, the situation is more complicated. This case is called a Hidden Markov Model (HMM) if the unobservable state transition is a Markov process (Rabiner & Juang, 1986). In a more complicated model, called a Semi Markov Model, the present state is dependent on a number of states in the past. The hidden Markov model encompasses the case where the state transitions are directly observable externally.

The problem addressed by the hidden Markov modelling technique is how to infer the state sequence of a "hidden" random process from the values of a related process, the observation sequence. The parameter set of an HMM is usually estimated either by the Baum algorithm (Baum & Petrie, 1966; Forney Jr, 1973) or the so-called segmental K-means algorithm (Juang & Rabiner, 1990). Efficient algorithms have been developed to estimate the state transitions and the unknown parameters of the signal model from the measured signal alone (see Bourlard and Wellekens (1990) for example). Once a Markovian model is obtained from the signal (either from the explicit externally observable state changes or from the hidden Markov model), it is possible to use standard Markov chain techniques (Meyne & Tweedie, 1993) to determine the steady state probability or system "stability," and to quantify the "attractors" of the given model.

Markov models and their applications are described in greater detail elsewhere (Bishop *et al.*, 1975; Plewis, 1985). In psychiatry, Markov models have been used in many contexts, including the course of depression (Dunn & Skuse, 1981), psychotherapy process (Badalamenti, 1992), and the epidemiology of mental illness (Marhall & Goldhamer, 1955). Tuckwell (1989) has recently reviewed the extensive literature in which the theory of stochastic processes has arisen in unit-cellular neuronal studies.

Markov modelling has been only intermittently applied to EEG signal analysis since the pioneering work by Bohlin (1977) who studied the first order Markov process generated by a Kalman filter implementation of a basic AR.model. On the basis of his observations, Bohlin postulated that EEG signals could conveniently be classified into three groups according to the signal state transition matrices: a fixed or unchanging process, a slowly changing process, or a fast changing process. Jensen and Cheng (1988) published a more recent example of a Markov model analysis of EEG, in which the technique was used to adaptively segment the signal.

METHOD

The data analyzed here is the same as that described in Perros *et al.* (1992), in which full details of subject selection and EEG recording methodology are given. Their study comprised of quantitative EEG and spatial topographic mapping, whereas our study focuses on temporal dynamics from a single channel of EEG data using Markov modelling.

Subjects and EEG Recordings

Patient data was taken retrospectively from consecutive outpatients referred for routine neurophysiological investigation. All patients were diagnosed as meeting DSM-IIIR criteria for obsessive-compulsive disorder by their referring psychiatrist, and each had been suffering from severe and intractable OCD for at least one year. Since the data came from a retrospective study, symptom severity rating scale data was not available. Healthy normal subjects were recruited and also underwent routine neurological examinations. Any patients or control subjects with a neurological history were excluded from the study, leaving study populations of 13 patients with severe OCD and 11 normal control subjects.

EEG recordings were performed with eyes open and eyes closed during resting and hyperventilation. EEG signals were acquired at 128 Hz using a 19 channel Bio-Logic Brain Atlas III system from electrodes placed according to the 10-20 system, and signals were referenced to linked earlobes with channel F_{pz} acting as the ground electrode.

All recordings were visually inspected by an experienced neurophysician, who reported definite abnormalities in 6 patients and mild abnormalities in another 5 of the 13 patients in the OCD group. Individual findings are tabulated in Perros *et al.* (1992). The significance of these findings is not clear, but it should be noted that these observations were not made "blind" to diagnosis.

This paper reports only on the data taken at rest with eyes closed, for which the average recording duration was about 4 minutes per patient.

Data Processing

The analysis of data was performed on the Cray YMP system at the Swinburne University of Technology.

All recordings were inspected offline and all EEG segments found to be contaminated by artefact were discarded. Data from channel C_z alone was preprocessed using ordinary least-squares AR model estimation with a fixed model order, with $M = 10$, for consecutive fixed frames or segments of 1 second duration each containing 128 digitized values. Other AR models were also investigated, including maximum-entropy, unconditional least squares, and maximum likelihood parameter estimation methods, but these did not produce significantly different subsequent results and are not reported.

Data Analysis

In order to reduce the dimensionality of the signal state descriptions, the states of all frames for all subjects (as defined by the 10 AR coefficients) were classified into ten classes of states using the unsupervised clustering method described by Mucciardi and Gose (1972), and using the Mahalanobis distance metric. For each subject and each frame of EEG data, each frame state was then represented by its membership in one of the ten classes of states. The simple first-order Markov state transition matrix for each individual was then determined by counting the number of transitions to each state class.

The state transition matrices for all individuals in a diagnostic group (OCD patients or normal controls) were added together, and the summed state transition matrices then compared between groups.

RESULTS

Figure 1 shows the first-order Markov state transition matrices for the two diagnostic groups.

The normal control subjects range freely across many, if not most, state transitions. The transition matrices for each of the normal control subjects strongly resembles the group matrix shown. In contrast, the OCD patients tend to visit a small set of state transitions. It can be seen that large portions of whole rows and columns of the transition matrix are not visited in the OCD group matrix. The transition matrices for each of the OCD individuals resembles the group matrix, with all individual matrices being more sparse than the aggregated matrix to varying degrees.

The results show consistently limited EEG state transition dynamics in the OCD patients as compared with the dynamics from the normal control subjects. These results are in accord with the prediction arising from our hypothesis based on the phenomenological description of mental state transitions in OCD. Because symptom rating scale data were not available, we were unable to test our prediction of the dependency of the size of the set of transitions upon symptom severity.

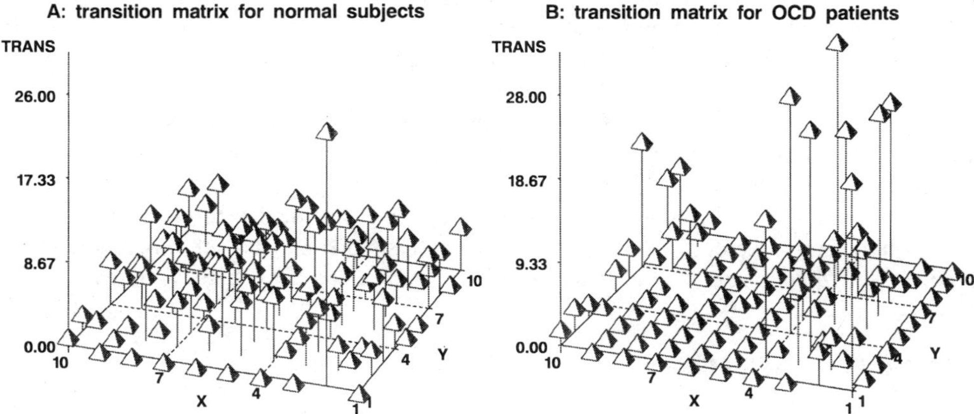

Figure 1. First-order Markov state transitions for EEG signal dynamics estimated for (A) normal control subjects, and (B) obsessive-compulsive disorder (OCD) patients. The vertical axes (labelled "trans") are arbitrarily scaled transition probabilities, the "X" and "Y" axes span 10 arbitrary classes of EEG frame "state." The transition matrices are arranged so as to indicate the probability of visiting state "Y" given state "X." EEG signal dynamics for the normal control subjects can be seen to traverse many state transitions, whereas the signal dynamics from OCD patients traverse a limited set of transitions.

DISCUSSION

While the numbers of subjects in either diagnostic group are small, the results appear to demonstrate a quantitative difference from normal in the state transition probabilities of EEG signals in obsessive-compulsive disorder. EEG signal dynamics in OCD appears to differ from normal, in a way consistent with phenomenological descriptions of the dynamics of mental states in this disorder. These limited EEG state transition dynamics were demonstrated in the OCD patients distal in time from the occurrence of their target symptoms.

While artefactual differences between the two clinical groups must be considered a possibility, the use of data from EEG channel C_z alone, together with the artefact rejection phase of data processing, should minimize the possibility of artefacts contaminating the findings. It should further be noted that, in contrast to patients with some other psychiatric disorders, patients with OCD are not particularly known to be prone to the production of artefact. Work is currently in progress to gauge the performance of the signal analysis methodology described here on a collection of files containing known contaminants such as noise, EOG, EMG, 50 Hz mains interference, and so on.

Further work on the signal processing is also underway. We are investigating other methods of "collapsing" the dimensionality of state feature vectors. We have also successfully implemented two versions of the hidden Markov model, one with a discrete output probability, and another with a semi continuous output probability. These models estimate the transition matrix directly, without the state classification step above. Early indications are that the hidden Markov models successfully quantify the hypothesized limit cycle dynamics of the OCD subjects. Although there is much more that needs to be done to complete the characterization of EEG dynamics from the hidden Markov model, the HMM results do not appear to differ substantially from the simple Markov model results described here and will be described and presented elsewhere.

A prospective study is underway, in which a larger cohort of OCD patients receive neuropsychological Yale-Brown Obsessive Compulsive Scale rating of their symptom severity in addition to neurophysiological investigation. An associated study is also underway, investigating a patient group suffering from frontal lobe syndromes (FLS). On the basis of the phenomenonolgy of perseveration, we hypothesize the occurrence of a fixed point or points in EEG signal dynamics in this syndrome.

We also plan to test the specificity of these findings by comparison with EEG data from patients in other diagnostic categories, and in the presence of secondary symptoms such as depression and anxiety.

We would like to conclude by suggesting that the identification of EEG state transition dynamics, using the methods described in this work, may have future applications in objectively quantifying mental state transitions in other neuropsychiatric contexts.

ACKNOWLEDGMENTS

Supported by a grant from the Australian Research Council. The authors acknowledge the invaluable assistance of Drs G.W. Price, P. Mann, E.S. Young, and P. Hollingworth, and Mr P. Perros of the Clinical Electrophysiology Unit, Wolston Park Hospital, Wacol, Queensland, Australia. The first author (A.A.S.) acknowledges the support of the Swinburne-Technion Academic Cooperative Program.

REFERENCES

Akaike, H., 1969, Fitting autoregressive models for prediction, *Annals Inst. of Statistical Mathematics* 21:243-247.

Badalamenti, A. F., Langs, R. J. and Kessler, M., 1992, Stochastic progression of new states in psychotherapy, *Statistics in Med.* 11:231-242.

Baum, L. E. and Petrie, T., 1966, Statistical inference for probabilistic functions of finite state Markov chains, *Annals of Mathematical Statistics* 37:1554-1563.

Bishop, Y. M. M., Fienberg, S. E. and Holland, P. W., 1975, *Discrete Multivariate Analysis: Theory and Practice.* Cambridge, MA: MIT Press.

Bodenstein, G. and Praetorius, H. M., 1977, Feature extraction from the electroencephalogram by adaptive segmentation, *Proc. IEEE* 65:642-652.

Bodenstein, G., Schneider, W. and Malsburg, C. V. D., 1985, Computerised EEG pattern classification and probability-density-function classification. Description of the method, *Computers in Biol. & Med.* 15:297-313.

Bohlin, T., 1971, Analysis of EEG signals with changing spectra, IBM Systems Development Division, IBM Nordic Laboratory, Sweden, *Technical Paper* 18. 212.

Bohlin, T., 1977, Analysis of EEG signals with changing spectra using a short-word Kalman estimator, *Mathematical Biosciences* 35:221-259.

Bourlard, H. and Wellekens, C. J., 1990, Links between Markov models and multilayer perceptrons, *IEEE Trans. in PAMI* 12:1167-1178.

Carr, V., 1983, The concept of clinical state in psychiatry: a review, *Comprehensive Psychiatry* 24:370-391.

Cerutti, S., Liberati, D., Avanzini, G., Francesetti, S., and Panzica, F., 1986, Classification of the EEG during neurosurgery. Parametric identification and Kalman filtering compared, *J. Biomed. Eng.* 8:244-254.

Crowell, D. H., Jones, R. H., Kapuniai, L. E. and Leung, P., 1977, Autoregressive representation of infant EEG for the purpose of hypothesis testing and classification, *EEG and Clin. Neurophys.* 43:317-324.

Dunn, G. and Skuse, D., 1981, The natural history of depression in general practice: stochastic models, *Psychological Medicine* 11:755-764.

Epstein, A. W. and Bailine, S. H., 1971, Sleep and dream studies in obsessional neurosis with particular reference to epileptic states, *Biological Psychiatry* 4:149.

Fenwick, P. B. C., Mitchie, P., Dollimore, J. and Fenton, G. W., 1969, Application of the autoregressive model to EEG analysis, *Aggressologie* 10:533-564.

Flor-Henry, P., Yeudall, L. T., Koles, Z. J. and Howarth, B. G., 1979, Neuropsychological and power spectral EEG investigations of the obsessive-compulsive syndrome, *Biological Psychiatry* 14:119-130.

Forney, Jr., G. D., 1973, The Viterbi algorithm, *Proc. IEEE* 61:268-278.

Gath, I., Lehmann, D. and Bar-On, E., 1983, Fuzzy clustering of EEG signal and vigilance performance, *Int. J. Neurosci.* 20:303-312.

Gersch, W., 1970, Spectral analysis of EEG's by autoregressive decomposition, *Mathematical Biosciences* 7:205-222.

Giannitrapani, D., 1985, *The Electrophysiology of Intellectual Function.* Basel: Karger.

Horowitz, M. J., 1979, *States of Mind.* NY: Plenum Publishing.

Insell, T. R., Donnelly, E. F., Lalakea, M. L., Alterman, I. S. and Horman, S., 1983, Neurological and neuropsychological studies of patients with obsessive-compulsive disorder, *Biological Psychiatry* 18:741-751.

Isaksson, A., Wennberg, A. and Zetterberg, L. H., 1981, Computer analysis of EEG signals with parameteric models, *Proc. IEEE* 69:451-461.

Jansen, B. H., Hasman, A. and Lenten, R., 1981, Piecewise analysis of EEGs using AR-modelling and clustering, *Comput. & Biomed. Res.* 14:168-178.

Jenike, M. A. and Brotman, A. W., 1984, The EEG in obsessive-compulsive disorder, *J. Clin. Psych.* 45:122-124.

Jensen, B. H. and Cheng, W. K., 1988, Structural EEG analysis: an explorative study, *Int. J. Biomed. Comput.* 23:221-237.

Juang, B. H. and Rabiner, L. R., 1990, The segmental K-means algorithm for estimating parameters of hidden Markov models, *IEEE Trans. in ASSP* 38:1639-1641.

Kydd, R. R. and Wright, J. J., 1986, Mental phenomena viewed as changes of state in a finite-state machine, *Australian and New Zealand J. of Psychiatry* 20:158-165.

Lopes da Silva, F. H., Dijk, A. and Smits, H., 1974, Detection of nonstationarities in EEGs using the autoregression model - An application to the EEGs of epileptics. In: *CEAN - Computerised EEG Analysis* (Dolce, G. and Kunkel, H., eds), Stuttgart: Gustav Fischer Verlag, pp. 180-199.

Malloy, P., Rassmussen, S., Braden, W., and Haier, R. J., 1989, Topographic evoked potential mapping in obsessive-compulsive disorder: evidence of frontal lobe dysfunction, *Psychiatry Res.* 28:63-71.

Marshall, A. W. and Goldhamer, H., 1955, An application of Markov processes to the study of the epidemiology of mental disease, *Am Statistical Assoc. J.* March: 99-129.

Meyne, S. P. and Tweedie, R. L., 1993, *Markov Chains and Stochastic Stability.* Berlin: Springer-Verlag.

Mucciardi, A. N. and Gose, E. E., 1972, An automatic clustering algorithm and its applications to high-dimensional spaces, *IEEE Trans. on SMC* 2:247-254.

Pacella, B. L., Potatin, P. and Nagler, S. H., 1944, Clinical and EEG studies in obsessive-compulsive states, *Am. J. Psych.* 100:830-838.

Perros, P., Young, E. S., Ritson, J. J., Price, G. W. and Mann, P., 1992, Power spectral EEG analysis and EEG variability in obsessive-compulsive disorder, *Brain Topography* 4:187-192.

Plewis, I., 1985, *Analysing Change: Measurement and Explanation Using Longitudinal Data.* Chichester: Wiley.

Remond, A. and Renault, B., 1972, La theorie des objects electrographiques, *Rev EEG Neurophysiol.* 3:241-256.

Rabiner, L. R. and Juang, B. H., 1986, An introduction to hidden Markov models, *IEEE ASSP Mag.* 3(1):4-16.

Rockwell, F. V. and Simmons, D. J., 1947, The electroencephalogram and personality organization in the obsessive-compulsive reactions, *Arch. Neur. and Psych.* 57:71-80.

Sergejew, A. A. and Tsoi, A. C., 1992, How many electrodes are enough? Insights from the application of the zero-delay frequency wavenumber spectral analysis of EEG and AEP. *Pan Pacific Workshop on Brain Electric and Magnetic Topography,* Melbourne, Australia, February 17-18 1992; *Brain Topography* 5:65 (abstract).

Shagass, C., Shagass, C., Roemer, R. A., Straumanis, J. J., and Josiassen, R. C., 1984, Psychiatric diagnostic discriminations with combinations of quantitative EEG variables, *Br. J. Psych.* 144:581-592.

Torello, M. W., 1989, Topographic mapping of EEG and evoked potentials in psychiatry: Delusions, illusions and realities. *Brain Topography* 1:157-174.

Tsoi, A. C., 1987, A new method for analysing electroencephalogram (EEG) signals. *Proc. of the Int. Symposium on Signal Processing and its Applications,* Brisbane, Australia, August 1987, 2:803-807.

Tsoi, A. C., 1988, Parametrisation of electroencephalogram signals, a possible methodology. *Proc. of the IFAC Symp. on Identification and System Parameter Estimation,* Beijing, China, August 27-31 1988, 3:1369-1376.

Tsoi, A. C. and Sergejew, A. A., 1992, An application of artificial neural network technique in EEG signal analysis: Evidence of global disconnection in schizophrenia. *Pan Pacific Workshop on Brain Electric and Magnetic Topography,* Melbourne, Australia, February 17-18 1992; *Brain Topography,* 5:65.

Tuckwell, H. C., 1989, *Stochastic Processes in the Neurosciences*. Philadelphia: SIAM.

Wright, J. J., 1990, Reticular activation and the dynamics of neuronal networks, *Biological Cybernetics* 62:289-298.

Wright, J. J. and Kydd, R. R., 1986, Schizophrenia as a disorder of cerebral state transitions, *Australian and New Zealand J. of Psych.* 20:167-178.

Wright, J. J., Kydd, R. R. and Lees, G. J., 1985, State-changes in the brain viewed as linear steady-states and non-linear transitions between steady-states, *Biological Cybernetics* 53:11-17.

Wright, J. J., Kydd, R. R. and Sergejew, A. A., 1990, Autoregression models of EEG: results compared with expectations for a multilinear near-equilibrium biophysical process, *Biological Cybernetics* 62:201-210.

Wright, J. J. and Sergejew, A. A., 1991. Radial coherence, wave velocity and damping of electrocortical waves, *EEG and Clin. Neurophys.* 79:403-412.

Wright, J. J., Sergejew, A. A. and Liley, D. T. L., 1994, Computer simulation of electrocortical activity at millimetric scale, *EEG and Clin. Neurophys.* 90:365-375.

Wright, J. J., Sergejew, A. A. and Stampfer, H. J., 1990, Inverse filter computation of the neural impulse giving rise to the auditory evoked potential, *Brain Topography* 2:293-302.

4

EEG SLEEP STAGING USING VECTORIAL AUTOREGRESSIVE MODELS

Arnon Cohen, Felix Flomen, and Nir Drori

Electrical and Computer Engineering Department
Biomedical Engineering Program
Ben-Gurion University
Beer-Sheva, Israel

ABSTRACT

Sleep studies require the use of several channels of EEG. The analysis of vector EEG, exhibits significant advantages over scalar analysis. Novel algorithms for segmentation, classification and compression of vector EEG are described. The statistics of the suggested measures for segmentation and classification are discussed. The algorithms were evaluated on four patients, yielding mean correct sleep staging of about 85%.

INTRODUCTION

The study of sleep is of extreme importance in several clinical applications. Sleep studies involve the recordings of various signals such as EEG, eye and body movements, breathing and more. Very large amounts of data are acquired in these studies, hence automatic processing in terms of segmentation, classification and compression is needed. This paper deals with probably the most important signal in sleep study - the EEG. In practice several channels of EEG are simultaneously acquired. Most often the number of channels used is 3-4, but it may be as large as 32. It is well known that multichannel analysis, otherwise known as vector analysis or multivariate analysis, is superior to single channel analysis (scalar processing) since the cross correlation between all channels are taken into account.

A new system is described for the vectorial analysis of sleep EEG signals. The non-stationary vector EEG signal is first automatically segmented into (almost) stationary records, using a vector AR segmentation algorithm. The segmentation algorithm is a

Advances in Processing and Pattern Analysis of Biological Signals, Edited by Isak Gath and Gideon F. Inbar
Plenum Press, New York, 1996

generalization of the scalar spectral error measure (SEM), (Bodenstein & Praetorius, 1977). Each segment is then classified into a sleep state, using a nearest neighbor classifier with Kulback-Liebler based distortion measure.

The multichannel EEG signal is assumed to be an M dimensional vector stochastic process, sampled with sampling interval of T seconds. The discrete vector process $\mathbf{y}(n)$, is given by:

$$\mathbf{y}(n) = [y_1(n), y_2(n), \ldots, y_M(n)]^T \tag{1}$$

where $y_i(n)$ is the ith EEG channel at time nT. The process has a multichannel autocorrelation matrix R(k), defined by:

$$R(k) = E\{\mathbf{y}(n+k)\mathbf{y}^T(n)\} = R^T(-k); \quad -\infty < k < \infty \tag{2}$$

The autocorrelation of the process, at time k, is thus an M*M matrix of the form:

$$R(k) = \begin{bmatrix} r_{11}(k) & r_{12}(k) & \ldots & r_{1M}(k) \\ r_{21}(k) & r_{22}(k) & \ldots & r_{2M}(k) \\ \cdot & \cdot & \ldots & \cdot \\ r_{M1}(k) & r_{M2}(k) & \ldots & r_{MM}(k) \end{bmatrix} \tag{3}$$

where $r_{ij}(k)$ is the cross correlation between the scalar processes, $y_i(n)$, and $y_j(n)$. In general the matrix R(k) is not symmetric, except for the case k=0.

The power spectral density (PSD) function, P(f), (Marple, 1987) of the vector process is related to its autocorrelation matrix by:

$$P(f) = T \sum_{k=-\infty}^{\infty} R(k)\exp(-j2\pi fkT) \tag{4}$$

The PSD matrix is an M*M matrix, given by:

$$P(f) = \begin{bmatrix} p_{11}(f) & p_{12}(f) & \ldots & p_{1M}(f) \\ p_{21}(f) & p_{22}(f) & \ldots & p_{2M}(f) \\ \cdot & \cdot & \ldots & \cdot \\ p_{M1}(f) & p_{M2}(f) & \ldots & p_{MM}(f) \end{bmatrix} \tag{5}$$

where $p_{ij}(f)$ is the cross spectral density function between the ith and jth channels. In general, the spectra matrix is Hermitian, semi positive definite for all f, with real values on the main diagonal and complex values on the off diagonal elements (Jenkins & Watts, 1986).

It is very useful to model the process by a parametric model. The Autoregressive Moving Average, $ARMA^{(M)}(p,q)$, (Hannan, 1970; Priestly, 1981) is such a model. The model is defined by:

$$\mathbf{y}(n) = -\sum_{j=1}^{p} A_j \mathbf{y}(n-j) + \sum_{i=0}^{q} B_i \xi(n-i); \quad B_0 = I_M \tag{6}$$

where I_M is the M*M unity matrix, $\xi(n)$ is an M dimensional white noise excitation process with zero expectation and covariance matrix Σ. The parametric matrices A and B are:

$$A_i = \begin{bmatrix} a_{11}^i & a_{12}^i & \cdots & a_{1M}^i \\ a_{21}^i & a_{22}^i & \cdots & a_{2M}^i \\ \cdot & \cdot & \cdots & \cdot \\ a_{M1}^i & a_{M2}^i & \cdots & a_{MM}^i \end{bmatrix}; 1 \le i \le p \quad B_i = \begin{bmatrix} b_{11}^i & b_{12}^i & \cdots & b_{1M}^i \\ b_{21}^i & b_{22}^i & \cdots & b_{2M}^i \\ \cdot & \cdot & \cdots & \cdot \\ b_{M1}^i & b_{M2}^i & \cdots & b_{MM}^i \end{bmatrix}; 1 \le i \le q \quad (7)$$

The ARMA$^{(M)}$(p,0) is known as the AR$^{(M)}$(p) model, which is given by:

$$y(n) = -\sum_{j=1}^{p} A_j y(n - j) + \xi(n - i) \quad (8)$$

The process y(n) is asymptotically stationary if all the zeroes of det

$$\left(I_M + \sum_{j=1}^{p} A_j z^{-j} \right)$$

are inside the unit circle (Hannan, 1970). It can be shown that with the first p+1 terms of the M dimensional correlation matrix, the matrices A_i and Σ may be calculated by the generalized Yule-Waker equation:

$$\left[I_M : A_1 : \cdots : A_P \right] R = \Sigma \left[I_M : 0_M : \cdots : 0_M \right] \quad (9)$$

where 0_M is an M*M matrix of zeroes, and R is given by:

$$R = \begin{bmatrix} R(0) & R(1) & \cdots & R(p) \\ R(-1) & R(0) & \cdots & R(p-1) \\ \cdot & \cdot & \cdots & \cdot \\ R(-p) & R(-p+1) & \cdots & R(0) \end{bmatrix} \quad (10)$$

The matrix R is a semi positive definite, block Toeplitz, symmetric matrix with dimension of [M(p+1)*M(p+1)].

The values of the correlation matrix for k>p can be calculated by the iterative equation:

$$R(k) = -\sum_{i=1}^{p} A_i R(k - i); \quad k > p \quad (11)$$

The spectra matrix can be calculated by using equation (4), or directly from the model's parameters by:

$$P_{AR}^{(M)}(z) = A^{-1}(z) \Sigma A^{-T}(z^{-1}) \quad (12)$$

with:

$$A(z) = I_M + \sum_{i=1}^{p} A_i z^{-i} \quad (13)$$

The AR model is attractive since every regular vector stochastic process can be represented by a vector AR model with infinite order (Kay, 1988), and since the model's parameters (the matrices A_i and the noise covariance Σ) can relatively easily be estimated using equation (9) by, for example, the Levinson, Wiggins, Robinson (LWR) algorithm (Whittle, 1963; Haykin & Kesler, 1979; Kay, 1988).

Vector AR models have been suggested in the past to represent EEG signals. Franaszczuk has suggested it to enhance spectral resolution (Franaszczuk, *et al..*, 1985) and Gersch and his co-workers have suggested the use of vector AR models for EEG classification (Gersch & Yonemoto, 1977a; Gersch *et al.*, 1977b; Gersch *et al.*, 1979). In this paper we shall discuss distance measures to be used in the segmentation of vector EEG and in its classification.

EEG SEGMENTATION

The EEG signal is not stationary. Spectral analysis of such processes is possible when time-frequency or time-scale transformations (Wiggner-Ville transformation, Wavelets transformation) are employed. However, a much simpler way is to assume the process is "quasi" stationary. The process is assumed to be a concatenation of stationary and ergodic windowed processes. Each segment is processed with the conventional spectral analysis tools. In very highly non stationary signals, such as speech, the signal is segmented into short, fixed duration, segments. The EEG signal may contain long duration "stationary" segments, the duration of which depends on the state of the subject. For such a process, segmentation into fixed duration segments is very ineffective. Adaptive segmentation is desired. EEG processing algorithms have been proposed for both fixed duration (Issakson *et al.*, 1981) and adaptive scalar segmentation (Bodenstein *et al.*, 1977; Sanderson *et al.*, 1980). The adaptive algorithm must obey two basic constraints:

- A segment must be short enough, so that the stationary assumption be kept.
- A segment must be long enough, so that the model's parameters can reliably be estimated.

We adopt the basic segmentation philosophy used with the scalar SEM algorithm (Bodenstein *et al.*, 1977). The spectra matrix of an initial data window is estimated (using VAR model). The spectra of adjacent "running" data windows are compared with that of the initial window. A segment is terminated when the spectra of the running window differ that of the initial window by more than some pre-determined threshold. A distortion measure, to determine the difference between two spectra matrices is required. The knowledge of the statistics of the measure is important for the determination of the classification threshold. We define such a measure and call it the Vector Spectral Error Measure (VSEM).

We shall use the Hilbert-Schmidt definition of the norm of an M*M matrix:

$$\|A\|^2 = \left(M^{-1}\mathrm{tr}\left[AA^H\right]\right) = M^{-1}\sum_{i=0}^{M-1}\sum_{j=0}^{M-1}a_{ij}^{\ 2} \tag{14}$$

where $|\,|*|\,|$ is the norm operator, $[*]^H$ is the Hermitian operator and $\mathrm{tr}[*]$ is the trace of the matrix. For two stationary M dimensional, stochastic processes, with spectra matrices $P(\theta)$ and $S(\theta)$, we define the following distortion measure:

$$D(P,S) = \frac{1}{2\pi}\int_{-\pi}^{\pi}\left\|S^{-1/2}(\theta)P(\theta)S^{-T/2}(\theta) - I_M\right\|^2 d\theta \tag{15}$$

The measure is positive for $P \neq S$ and is zero for $P=S$. It is non symmetric namely $D(P,S) \neq D(S,P)$. The definition (15) results from the analysis of the "whiteness" of the residue error vector. Assume we have a process with spectra matrix $P(\theta)$, we fit a VAR model $A(z)$ to the process. The residue is the process at the output of the inverse filter, it is the estimation of the excitation process $\xi(n)$ (equation 8). The spectra matrix of the residue, $W(\theta)$ is given by:

$$W(\theta) = A(\theta)P(\theta)A^H(\theta) \tag{16}$$

The covariance matrix, Σ, of the residue process is:

$$\Sigma = \frac{1}{2\pi} \int_{-\pi}^{\pi} W(\theta)d\theta \tag{17}$$

If $P(\theta)$ is a spectra matrix of a true M dimensional VAR process of order p, and $A(z)$ is the correct model, then the residue vector is white and:

$$W(\theta) = \Sigma; \quad -\pi \leq \theta \leq \pi \tag{18}$$

For a general vector stochastic process we get:

$$W(\theta) = \Sigma + E(\theta); \quad -\pi \leq \theta \leq \pi \tag{19}$$

The fitness of the model to the process may be evaluated by the distortion between the matrices $W(\theta)$ and Σ, hence:

$$D(W(\theta), \Sigma) = \frac{1}{2\pi} \int_{-\pi}^{\pi} \left\| \Sigma^{-1/2} W(\theta) \Sigma^{-T/2} - I_M \right\|^2 d\theta \tag{20}$$

The last measure may be expressed in the autocorrelation domain:

$$W(\theta) = \sum_{k=-\infty}^{\infty} R(k)\exp(-j2\pi k\theta) \tag{21}$$

$$D(W(\theta), \Sigma) = \frac{1}{2\pi} \int_{-\pi}^{\pi} \left\| \Sigma^{-1/2} [\sum_{k=-\infty}^{\infty} R(k)\exp(-j2\pi k\theta)]\Sigma^{-T/2} - I_M \right\|^2 d\theta \tag{22}$$

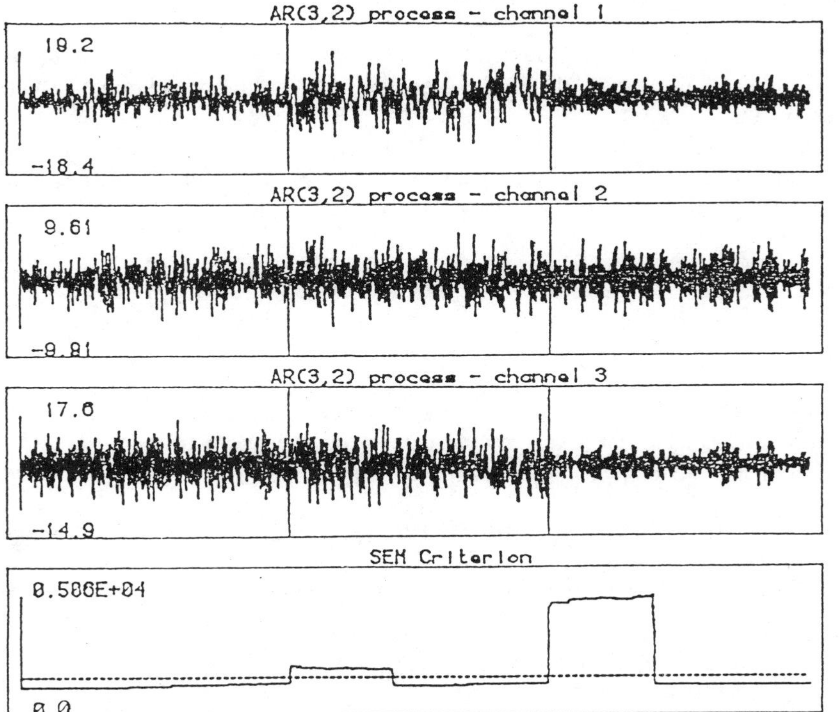

Figure 1. Adaptive segmentation with VSEM (synthesized data; broken line = threshold level).

This can be shown to equal:

$$D(W(\theta), \Sigma) = \left\| \Sigma^{-1/2} R(0) \Sigma^{-T/2} - I_M \right\|^2 + 2 \sum_{k=1}^{\infty} \left\| \Sigma^{-1/2} R(k) \Sigma^{-T/2} \right\|^2 \tag{23}$$

Equation (23) requires the estimation of infinite number of correlation matrices. This is of course impractical. We therefor define the vector spectral error measure (VSEM) as:

$$D_{VSEM}(W(\theta), \Sigma) = \left\| \Sigma^{-1/2} R(0) \Sigma^{-T/2} - I_M \right\|^2 + 2 \sum_{k=1}^{L} \left\| \Sigma^{-1/2} R(k) \Sigma^{-T/2} \right\|^2 \tag{24}$$

Equation (24) is the generalization of the scalar SEM (Bodenstein, 1977). Its first part shows the deviation of the "energy" of the output of the inverse filter from the model's residue's energy, and the second part shows the deviation from whiteness. The parameter L as well as the threshold level for determining segment termination are determined experimentally.

To demonstrate the adaptive segmentation algorithm, a 3 dimensional process was simulated, by a time varying $AR^{(3)}(2)$ model, excited by a 3 dimensional white noise process. The time varying model was simulated by switching between three different $AR^{(3)}(2)$ models, every 500 samples. The reference and sliding windows duration was 200 samples, L=4 was used and the threshold was determined experimentally. Figure 1 shows the adaptive segmentation results. Segmentation was determined at 509 (instead of 500) and 1005 (instead of 1000).

EEG CODING AND CLASSIFICATION

Kulback-Leibler (KL) Distortion Measure

A multichannel vector qauntization (VQ) scheme is suggested for both coding and classification of EEG signals. The VQ is a generalization of the well known LBG (Linde, *et al.*, 1980) VQ.

We shall show that with the use of a Kulback-Leibler (KL) distortion measure (Gray *et al.*, 1980) an optimal VQ for VAR parameters may be designed.

The KL distortion measure is based on the I-Divergence statistical measure:

$$D_{KL}(w_1, w_2) = \int p(x|w_1) P(w_1) \log \left[\frac{p(x|w_1) P(w_1)}{p(x|w_2) P(w_2)} \right] dx \tag{25}$$

where x is a stochastic process and w_1 and w_2 are its parameters. For two M dimensional vector processes, with zero expectation vectors and spectra matrices $S_1(\theta)$ and $S_2(\theta)$ the distortion measure becomes (Flomen, 1990):

$$D_{KL}(S_1, S_2) = \frac{1}{2\pi} \int_{-\pi}^{\pi} \left[tr\{S_1(\theta) S_2^{-1}(\theta) - I_M\} - \log\det\{S_1(\theta) S_2^{-1}(\theta)\} \right] d\theta \tag{26}$$

When the spectra matrices are VAR, introducing (12) into (26) yields (Gersch *et al.*, 1979):

$$D_{KL}(S_1, S_2) = tr\{A_2 R_1 A_2^T \Sigma_2^{-1} - I_M\} - \log\det\{\Sigma_2^{-1} \Sigma_{1\infty}\} \tag{27}$$

where A_2 and Σ_2 are the parameters and gain matrices of the VAR process with spectra matrix $S_2(\theta)$, R_1 is the autocorrelation matrix corresponding to $S_1(\theta)$, and $\Sigma_{1\infty}$ is the covariance

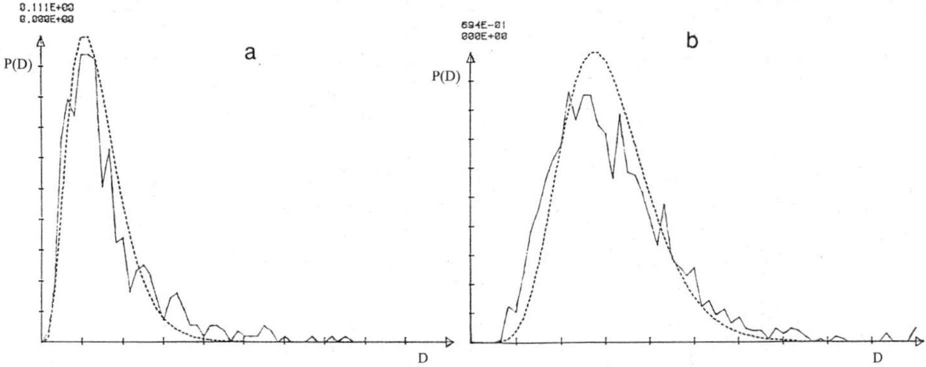

Figure 2. Histograms and theoretic distributions of the D_{KL} distortion measure. (a) $AR^{(2)}(1)$ model. (b) $AR^{(3)}(2)$.

matrix of the residue vector of an infinite order predictor. Statistical analysis of (27) shows (Flomen, 1990) that the distortion measure is asymptotically chi-squared distributed with M^2p degrees of freedom. Figure 2 shows the histograms and the theoretical distributions of the measure, estimated by an $AR^{(2)}(1)$ and $AR^{(3)}(2)$ models.

Vector Quantization of VAR Models

A "conventional" K dimensional vector quantizer (VQ), with N quantization levels, consists of two units: encoder and decoder. The encoder maps a **K** dimensional input vector, $\mathbf{x}(n)$, into a symbol, $\mathbf{u}(n)$, $\mathbf{u}(n) \in U$. Each symbol $\mathbf{u}$ points to a codeword $\mathbf{c}_i$ out of a codebook C of N codewords. The VQ partitions the K dimensional space with N partitions $\mathbf{s}_i$; i=1,2...,N. The VQ is optimal when two conditions are met:

- Given the codebook, C, an input vector $\mathbf{x}(n)$ is encoded such that:

$$\mathbf{x}(n) \in \mathbf{s}_i \quad \text{iff} \quad d(\mathbf{x}(n),\mathbf{c}_i) \leq d(\mathbf{x}(n),\mathbf{c}_j) \quad 1 \leq j \leq N \qquad (28)$$

- The codebook C is determined such that the following criterion is minimized:

$$J = \sum_{i=1}^{N} E\{d(\mathbf{x},\mathbf{c}_i)|\mathbf{x} \in \mathbf{s}_i\}P(\mathbf{x} \in \mathbf{s}_i) \qquad (29)$$

In our case it is desired to code M*M matrices rather than K dimensional vectors. The VQ algorithm is generalized to include this case. We want to encode a time segment of an M dimensional vector stochastic process having the spectra matrix $P(\theta)$ by means of a codebook of M dimensional VAR models of order p, so that the encoding distortion be minimized. We propose a two steps method:

- Encode the spectra matrix $P(\theta)$ by means of $W(\theta)$, using a distortion measure: $d_1(P(\theta),W_i(\theta))$.
- Encode the spectra matrix $W(\theta)$ by means of a matrix codeword $W_i(\theta)$, i=1,2,...,N, minimizing the distortion: $d_2(W(\theta),W(\theta))$.

For the quantizer to be optimal, the general distortion $d(P(\theta),W_i(\theta))$ must be also minimized. We use the KL distortion measure for both d_1 and d_2. The minimization of the first step requires the minimization of $D_{KL}(P(\theta),W(\theta))$ under the constraint $W = A^{-1}\Sigma A^{-T}$. Derivation of equation (27) with respect to the matrices A and Σ, shows that the optimal A and Σ obey equation (9)(Flomen, 1990) and the minimum distortion is given by:

$$D_{KL}(P, A^{-1}\Sigma A^{-T}) = -\log\det\{[ARA^T]^{-1}\Sigma_\infty\} \tag{30}$$

The distortion of the second step is calculated (Flomen, 1990) by introducing the nearest codeword, A_R and S_R, into the distortion measure yielding:

$$D_{KL}(W, A_R^{-1}\Sigma_R A_R^{-T}) = \mathrm{tr}\{A_R RA_R^T\Sigma_R^{-1} - I_M\} - \log\det\{\Sigma_R^{-1}[ARA^T]\} \tag{31}$$

The total distortion is given by adding (30) and (31):

$$D_{KL}(P, A_R^{-1}\Sigma_R A_R^{-T}) = \mathrm{tr}\{A_R RA_R^T\Sigma_R^{-1} - I_M\} - \log\det\{\Sigma_R^{-1}\Sigma_\infty\} \tag{32}$$

The last equation describes the encoding distortion of the process $P(\theta)$ by the VQ encoder.

The codebook is designed by finding N codewords that minimize the mean distortion in the training data. Assume that in the jth partition we have L(j) training matrices. The codeword for representing this partition will be given by minimizing the expression:

$$\frac{1}{L(j)}\sum_{i=1}^{L(j)} D_{KL}(P_i, A_{Rj}^{-1}\Sigma_{Rj}A_{Rj}^{-T}) = \frac{1}{L(j)}\sum_{i=1}^{L(j)}\{\mathrm{tr}[A_{Rj}R_i A_{Rj}^T\Sigma_{Rj}^{-1} - I_M] + \log\det\Sigma_{Rj} - \log\det\Sigma_{i\infty}\} \tag{33}$$

$$= \mathrm{tr}\{A_{Rj}[\frac{1}{L(j)}\sum_{i=1}^{L(j)} R_i]A_{Rj}^T\Sigma_{Rj}^{-1} - I_M\} + \log\det\Sigma_{Rj} - \frac{1}{L(j)}\sum_{i=1}^{L(j)}\log\det\Sigma_{i\infty}\}$$

Note that the last term of (33) does not depend on the codeword, therefor will not take part in the optimization. The minimization of (33) with respect to the codeword A_{Rj} and Σ_{Rj} is given by (Flomen, 1990):

$$A_{Rj}\overline{R}_j = \Sigma_{Rj}[I_M : 0_M : \ldots\ldots : 0_M] \tag{34}$$

with:

$$\overline{R}_j = \frac{1}{L(j)}\sum_{i=1}^{L(j)} R_i \tag{35}$$

The codeword is calculated by solving equation (34) using (for example) the LWR algorithm. From (34) we see that the codeword is achieved by fitting a VAR model to the mean partition autocorrelation matrix. This can also be stated in the frequency domain: the codeword is the VAR model which best fits the mean spectra matrix of the partition. The procedure of the VQ algorithm is the same as LBG, using the KL distortion and equations (34) (35).

To evaluate the algorithm, two tests were performed. Four different $AR^{(2)}(2)$ were used to synthesize data. The codebook size was 4 and initial conditions were randomly chosen. The algorithm converged to a codebook very similar to the original matrices with global distortion close to $M^2p=8$. In the second test a 3 channel EEG signal was used. The signal consisted of 1288 segments of 200 samples duration. Each segment was represented by a $AR^{(3)}(6)$ model. Initial conditions were randomly chosen. Figure 3 describes the convergence of the algorithm. The VQ algorithm functioning as an encoder, was compared with the conventional VQ in a task of EEG

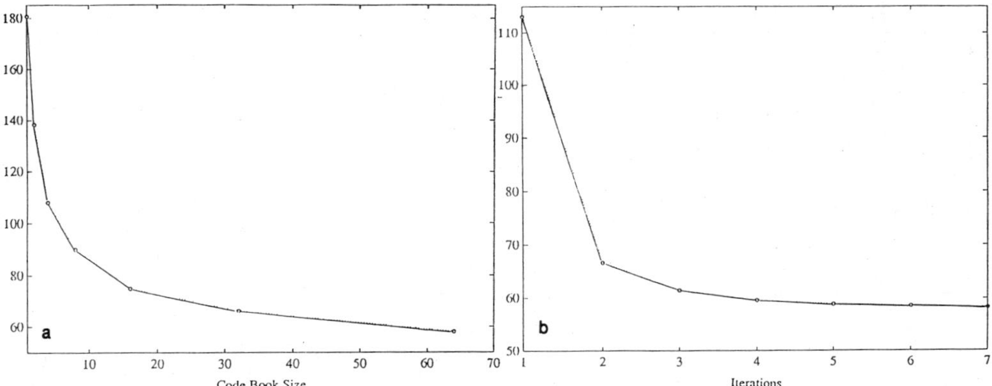

Figure 3. VQ of EEG. (a) Global distortion vs. codebook size. (b) Global distortion (with codebook size = 64) vs. iterations.

compression. Figure 4 shows the reconstruction error vs. bit rate for both the single (conventional) and multichannel VQ algorithms.

SLEEP STAGING

Automatic sleep staging using EEG and other signals has been in use for many years. Most of the algorithms are based on scalar processing (Nhiro *et al.*, 1980). Some, however

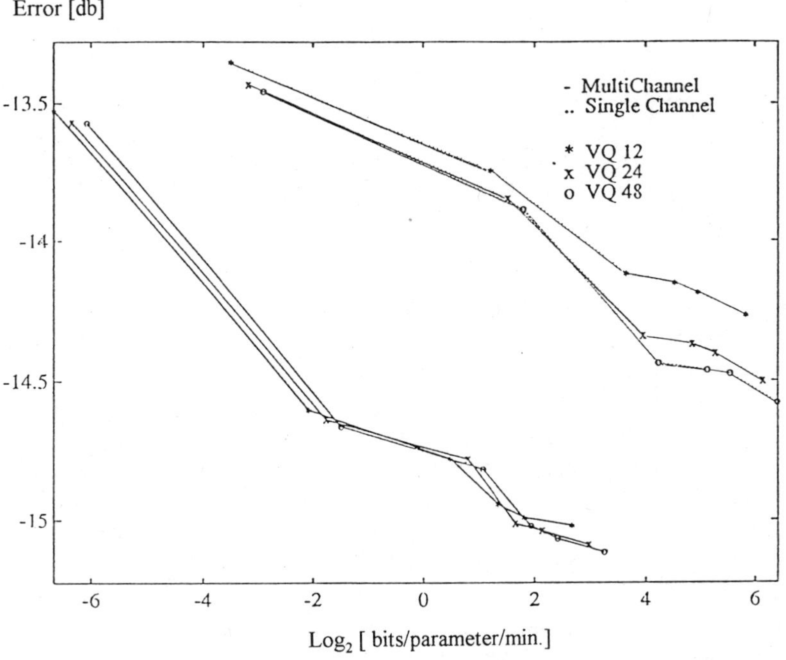

Figure 4. Error vs. bit rate in EEG coding.

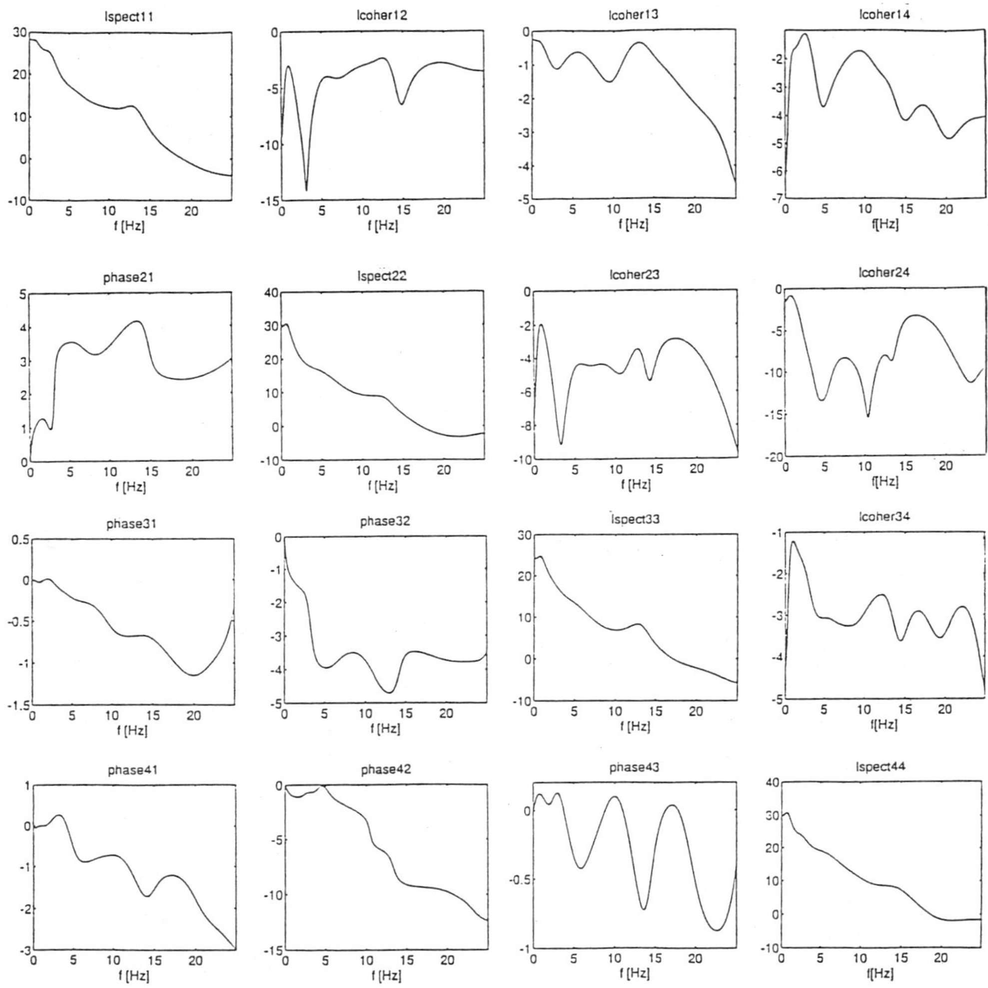

Figure 5. An example of a sleep stage 3 spectra matrix.

are based on vector processing (Gersch, 1977b; Gersch, 1979). The classification system suggested here is based on the adaptive segmentation of the EEG signals, representation by VAR models, VQ and classification by the nearest neighbor rule.

To evaluate the algorithm a small data base was acquired. Seven EEG channels, one EMG and one EOG signals were recorded from four male subject during night sleep of about 5 hours, at the sleep lab. Manual sleep staging was performed by an expert. For the evaluation only four EEG channels were used (C3-A2; FP2-T4; T4-O2; C4-O2). The models used were $AR^{(4)}(6)$, the model's order was determined using the AIC method. Since only EEG signals were used, the classifier was designed to classify among sleep stages number 2, number 3 and REM.

Figure 5 shows an example of a spectra matrix of sleep stage 3. The upper part of the matrix are the spectra while the lower part contains the phase.

A summary of the classification results is shown in table 1.

Table 1. Classification of sleep stages

Sleep Stage	Minutes	Percent of the Time Classified as:		
		State 2	State 3	REM
Stage 2	168.9	83.25	3.70	13.05
Stage 3	89.9	18.50	81.50	0.00
REM	75.4	13.72	0.00	86.28

CONCLUSIONS

Algorithms for multichannel adaptive segmentation, compression and recognition have been presented. The adaptive algorithm is a generalization of the SEM algorithm used in the scalar case. A comparison of the new algorithm with the conventional SEM, on a small EEG data base shows that for equal mean segments rate, the new algorithm provides lower error than the SEM. The multichannel VQ encoder was compared with the single channel encoder. It was shown that for EEG encoding, for a constant distortion, the new algorithm provides 10 times lower bit rate. The VQ encoder can be designed by an algorithm that can be considered a generalization of the LBG VQ algorithm. An EEG sleep staging algorithm was suggested, based on the VQ algorithm and the nearest neighbor rule. The classification results are similar to the ones reported in the literature. A more detailed study, using a larger data base is required in order to compare the classification method with other methods.

REFERENCES

Bodenstein, G., and Praetorius, H. M., 1977, Feature extraction from the electroencephalogram by adaptive segmentation, *Proc. IEEE*, 35:642-652.

Franaszczuk, P. J., Blinowska, K. J. and Kowalczyk, M., 1985, The application of parametric multichannel spectral estimates in the study of electrical brain activity, Biol. Cybern., 51:239-247.

Gersch, W. and Yonemoto, J., 1977a, Parametric time series models for multivariate EEG analysis, Computers in Biomed. Res., 10:113-125.

Gersch, W. , Yonemoto, J., and Naitoh, P., 1977b, Automatic classification of multivariate EEGs using an amount of information and the eigenvalues of parametric time series model features, *Computers in Biomed. Res.*, 17:352-361.

Gersch, W., Martinelli, F., Yonemoto, J., Low, M.D. and McEwan, J. A., 1979, Automatic classification of EEGs: Kullback-Leibler nearest neighbor rules, Science, 205:193-195.

Gray, R. M., Buzo, A., and Gray, A., H., 1980, Distortion measures for speech processing, IEEE Trans. Acoust. Speech & Sig. Proc., ASSP-28:367-376.

Hannan, E. J., 1970, Multiple Time Series, John Wiley & Sons, N.Y.

Haykin, S. and Kesler, S., 1979, Prediction error filtering and maximum entropy spectral estimation, in: Haykin, S., (ed.), Topics in Applied Physics, 34:9-70, Springer Verlag, Berlin.

Isaksson, A., Wennberg, A., and Zetterberg, L. H., 1981, Computer analysis of EEG signals with parametric models, Proc. IEEE, 69:451-461.

Jenkins, G. M. and Watts, D. G., 1986, Spectral Analysis and its Applications, Holden Day, San Francisco, Ca.

Kay, S. M., 1988, Modern Spectral Estimation: Theory and Applications, Prentice Hall, Engelwood Cliffs, N.J.

Linde,Y., Buzo, A., and Gray, R. M., 1980, An algorithm for vector quantizer design, IEEE Trans. Comm., COM-28:84-95.

Marple, S. L., 1987, Digital Spectral Analysis with Applications, Prentice Hall, Engelwood Cliffs, N.J.

Nhiro, I., Hideyuki, S., Akira, I., and Nobus, S., 1980, Computer classification of the EEG time series by Kulback information measure, *Int. J. Syst. Sci.*, 11:677-687.

Priestly, M. B., *Spectral Analysis and Time Series*, 1981, Academic Press, N.Y.

Sanderson, A. C. Segen, J., and Richey, E., 1980, Hierarchical modeling of EEG signals, *IEEE Trans. Patt. Anal. Mach. Intell.*, PAMI-2: 405-414.

Whitle, P., 1963, On the fitting of multivariate autoregression, and the approximate canonical factorization of spectral density matrix, *Biometrika*, 50:129-134.

5

PROCESSING OF EPILEPTIC EEG

Isak Gath,[1] Bernard Harris,[1] Yoram Salant,[1] Claude Feuerstein,[2] Olaf Henriksen,[3] and Gérard Rondouin[4]

[1] Department of Biomedical Engineering
Technion, IIT, Haifa, Israel
[2] Laboratoire de Physiologie, Section Meurophysiologie
INSERM U318 CHU, Grenoble, France
[3] The National Center for Epilepsy
Sandvika, Norway
[4] Laboratoire de Medécine Expérimentale, INSERM U249
University of Monpellier, France

ABSTRACT

Processing of epileptic EEG is required for better understanding of the mechanisms underlying the generation and propagation of seizure electrical activity, as well as for mapping of epileptic foci in patients with intractable focal epilepsy who are candidates for surgery. Most processing methods applied to the EEG require linearity; but, before the question of linearity/non linearity is handled, the non stationary character of the EEG signal has to be tackled.

The nonstationary multichannel epileptic EEG has been submitted to adaptive segmentation, dividing the signal into quasi-stationary segments of uneven length. Next, the linear or nonlinear character of the system has been investigated on these quasi-stationary segments using methods for system identification. Vectorial parametric modeling of the multichannel signal has been undertaken on short, quasi-stationary EEG segments, for which no significant deviations from linearity were detected.

Estimation of the delay between various EEG channels by adaptive least-square filtering (a lattice-ladder type) was carried out on EEG from rats with induced focal epilepsy in order to map the epileptic focus. EEG from patients with primary generalized epilepsy was subject to an efficient method for spectral estimation, based on vectorial parametric modeling. Preictal signal segments were processed in order to detect changes which occur prior to the clinical outbreak of the seizure.

Advances in Processing and Pattern Analysis of Biological Signals, Edited by Isak Gath and Gideon F. Inbar
Plenum Press, New York, 1996

INTRODUCTION

Electroencephalography (EEG) is an important clinical tool in the diagnosis, differential diagnosis and follow up of epileptic patients. Multichannel EEG recording is carried out using surface and/or intracerebral electrodes, and interpretation of the multichannel signal is pertinent to such questions as mechanism of generation of the seizure, location of the source of the seizure activity and mode of propagation.

Most patients suffering from epilepsy, whether of focal origin or primary generalized, respond well to treatment with antiepileptic drugs. However, in a smaller subgroup, seizure frequency and severity can not be properly controlled by medication, and/or the dosage and combination of multiple drugs cause intolerable side effects. Possible alternative therapeutic approaches are surgical removal of the epileptic focus for patients with focal epilepsy (Ojemann & Engel, 1987), and biofeedback and electrical stimulation for patients with primary generalized epilepsy (Kaplan, 1975; Kuhlman & Allison, 1977; Ishijima *et al.*, 1975; Velasco *et al.*, 1987; Penry & Dean, 1990; George & Michael, 1991).

In that case, the following medical and basic problems arise:

- What is the precise location and extent of the epileptic focus?
- What is the mode of propagation of the seizure?
- For primary generalized epilepsy, are there any changes in the *preictal* EEG that precede the occurrence of a clinical seizure?

Analysis of the EEG signal may aid in obtaining answers to these questions. In the daily clinical routine, interpretation of the multichannel EEG signal is very often carried out visually. This has a number of drawbacks. Visual interpretation is time consuming, it is highly subjective and depends on the skill and school of the clinical neurologist, and some characteristics of the preictal or ictal signal may not be evident by visual inspection. The main objectives of computerized analysis of epileptic EEG are to achieve significant reduction in the amount of data treated and to extract clinically (or basic) relevant quantitative parameters. Computer analysis of epileptic EEG data has been primarily applied for spike detection in interictal EEG and for automatic detection of seizure activity as means of data reduction (Panych & Wada, 1990). However, signal processing of the multichannel seizure EEG can provide information on basic mechanisms involved in the *generation and propagation* of the seizure.

Two major problems arise in the analysis of the multichannel EEG signal. The EEG is inherently a non-stationary time series, and the nonstationarity is present both for the mean (baseline shifts) and/or for the covariance (Gersch, 1987). Any modeling of the input-output relationship has to take into consideration that the system is *not time invariant*, and prior to the processing, the signal should be divided into segments which are "piecewise stationary" (Gath & Bar-On, 1980; Gath & Michaeli, 1991).

The second obstacle in the analysis procedure is that most of the common methods for processing of the multichannel EEG assume linearity, and the question arises whether the relations between EEG signals from two channels might be nonlinear (Lopes da Silva *et al.*, 1989; Moddemeijer, 1989; Blinowska & Malinowski, 1991). These two problems are linked together. Before the system is investigated for linearity/nonlinearity, it is imperative that *only* segments which are *"piecewise stationary"* are considered.

In a true focal epilepsy, the epileptic focus triggers the electrical (EEG) and clinical seizure. When a seizure develops into a generalized convulsive disorder, propagation is carried out along neuronal pathways, and the primary focus may also act as a pacemaker. There are, however, factors that may complicate this description (Gersch, 1987):

- Modifications in the configuration of the neuronal nets during the seizure may cause changes in the mode of propagation of the seizure.
- Multiple locations other than the focus may start to fire independently and are decoupled from the pacemaker.
- Feedback may be established between different channels during the seizure.

Thus, time relationships calculated between different EEG channels may become misleading. Therefore, conclusions as to propagation and pacemaking will be drawn from processing of the *first few seconds* of the EEG signal only. It is assumed that during this period there is only one pacemaker and no feedback. In addition, before applying signal processing methods to epileptic EEG recordings from humans, the methods have to be verified in animal experiments. Depth and surface EEG recordings from rats with *induced* focal epilepsy will serve as test material. In these recordings the exact location of the focus is known.

When choosing methods for processing of the multivariate EEG signal it has to be taken into account that the signal exhibits both preictal and ictal dynamic changes that might be of a very short duration (Gath *et al.*, 1992). Thus, for computation of power spectra, for example, a high time/frequency resolution of the events around the onset of the electrical seizure is needed. This cannot be achieved by nonparametric spectral estimation. The reason is that, in order to obtain consistent power spectral estimates, averaging of several periodogram is required to decrease the variability of the estimate, and this averaging technique, inherent in the FFT procedure, will cause "smearing" of short duration dynamic changes. An alternative is to use parametric modeling of the signal (Gersch & Goddard, 1970; Gersch, 1987), but one which can provide a better performance for short record length than that of the commonly used covariance method for autoregressive spectral estimation (Gath *et al.*, 1992).

The second section of the present study will treat the problem of nonstationarity of the data and discuss methods for single and dual channel segmentation of the EEG signal into quasi-stationary segments. In section three, the question of linearity vs. nonlinearity will be investigated. In chapter four a method is described for the detection of short duration preictal dynamic changes, with a high time/frequency resolution. Chapter five contains a method for calculation of small interchannel delays based on adaptive least-squares filtering, in order to determine the exact location and size of an epileptic focus in cases of focal epilepsy. Finally, conclusions are drawn in chapter six.

ADAPTIVE SEGMENTATION OF THE EEG

In order to obtain quasi-stationary segments from the inherently nonstationary EEG signal, adaptive segmentation has been carried out.

Consider the signal as "piecewise stationary" by assuming the underlying mechanism of the time series to remain constant for an arbitrary length of time, followed by a fairly abrupt change in the model parameters. Thus, between each two neighboring signal segments there will be a stepwise change of the parametric structure of the segment to a new parameter set. This calls for detection and estimation of segment boundaries by *adaptive segmentation*. Two cases can be recognized, segmentation carried out separately on each of the various recorded EEG channels, and adaptive segmentation carried out simultaneously on pairs of EEG channels.

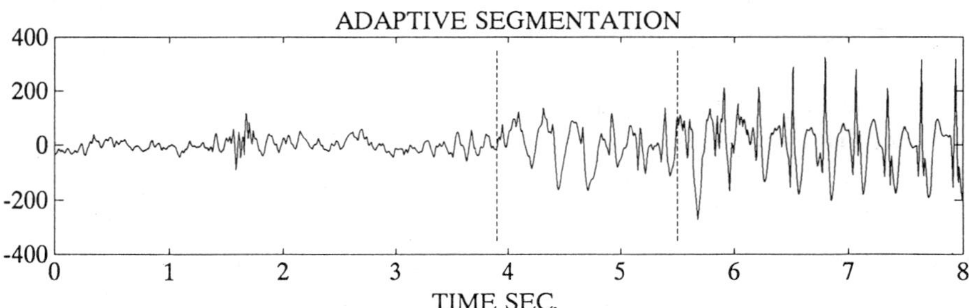

Figure 1. Adaptive segmentation. A reference window (continuous line) and a sliding window (dashed line) are placed on the epileptic EEG signal.

Adaptive Segmentation Carried Separately on Several EEG Channels

The segmentation is based on calculating the ratio between the spectra of a sliding window and a reference window (Bodenstein & Praetorius, 1977; Gath & Bar-On, 1980).

Autoregressive modeling of the EEG is carried out, and a criterion T for a change in the spectral content of the signal is computed from the autocorrelations of the residuals in the reference window and the moving window:

$$T = \sum_{k=0}^{M} \left[\frac{r_f(k) - r_m(k)}{r_f(0)} \right]^2 \tag{1}$$

where $r_f(k)$, $r_m(k)$ are the autocorrelations of the residuals in the reference and moving windows, respectively, and $r_f(0)$ the total power of the error autocorrelation in the reference window. Whenever T exceeds a predetermined value, the segment is closed, the sliding window is moved passed the end of the segment, and the process of segmentation restarted.

Adaptive Segmentation Carried Out Simultaneously on Two EEG Channels

The previous criterion for segmentation of a single EEG channel cannot be extended to two EEG channels simultaneously by simply using the vectorial autoregressive model instead of the scalar case. This is because the ratio between the cross spectra for the sliding and reference windows includes both cross spectra and autospectra terms for each of the two channels.

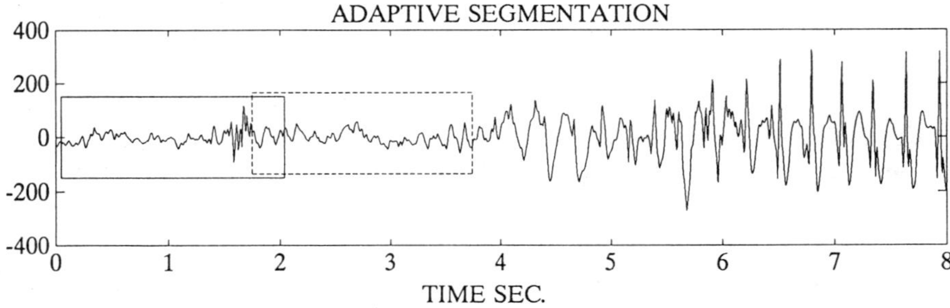

Figure 2. Single channel adaptive segmentation of the EEG signal in Fig. 1. Three EEG segments are detected. The segment boundaries are denoted by the dashed vertical lines.

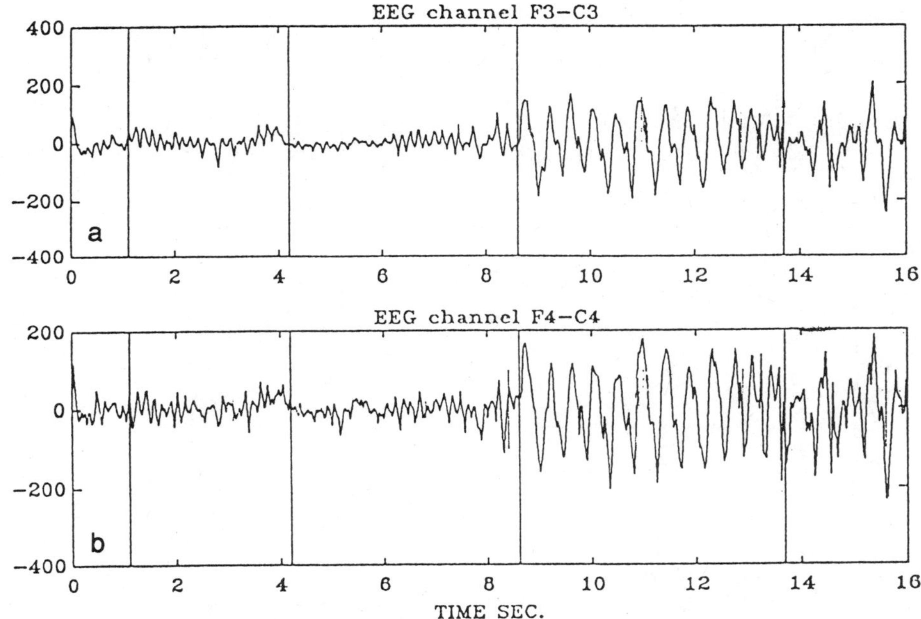

Figure 3. Dual channel segmentation of epileptic EEG signal. The vertical lines show the segment boundaries.

A criterion for dual-channel segmentation is based on the Euclidean distance between the cross spectra of the two channels in the two time windows (a sliding window (m) and a reference window(f)), using the cross-correlations (Gath & Michaeli, 1991):

$$T_2 = \frac{\sum_{k=-M}^{M} \left[r_{y1y2}^{f}(k) - r_{y1y2}^{m}(k) \right]^2}{r_{y1y1}^{f}(0) r_{y2y2}^{f}(0)} \tag{2}$$

where $r_{y1y2}^{f}(k)$ and $r_{y1y2}^{m}(k)$ are the cross spectra in the reference and sliding windows, respectively, and $r_{y1y1}^{f}(0)$ and $r_{y2y2}^{f}(0)$ are the respective autospectra. The segments boundary is defined whenever T exceeds a predetermined value.

INPUT-OUTPUT RELATIONSHIP IN SEIZURE EEG

As has been pointed out in the Introduction, the existence of nonstationarities in the EEG signal can interfere with the investigation, whether the system is linear or nonlinear, and therefore, when testing for linearity, *only* segments which are *"piecewise stationary"* are considered.

The system is assumed to be time invariant and is modeled, using an ARX model as follows:

The present value of the output, $y(t)$, is given as a linear combination of previous outputs, previous and present inputs and the prediction error, $e(t)$

$$y(t) = = \sum_{m=1}^{p} a_m y(t - m) + \sum_{n=1}^{q} b_n u(t - k - n + 1) + e(t) \tag{3}$$

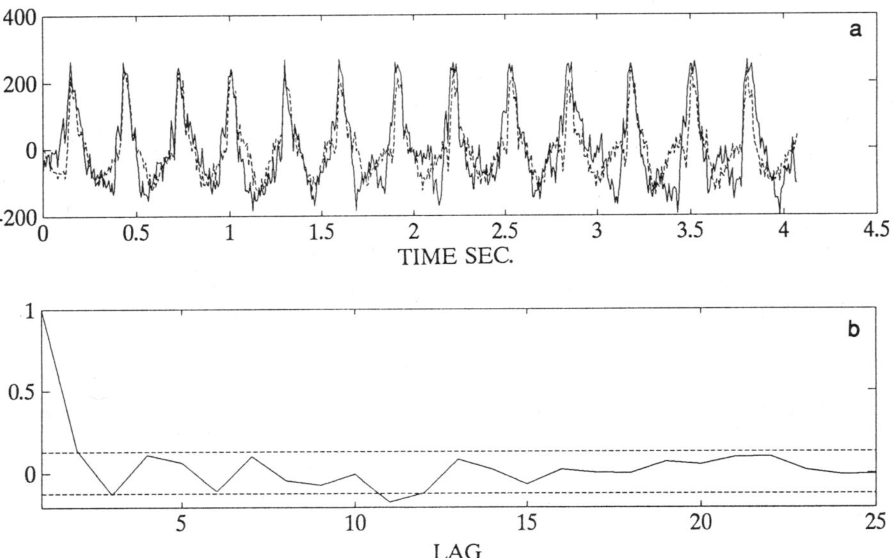

Figure 4. The effect of nonstationarity on testing for linearity. a. ARX modeling of 4.5 sec long epileptic EEG segment (nonstationary). Original signal: continuous line. Modeled signal: dashed line. b. Autocorrelation of residual for the model in a. There is a deviation from the 99% confidence limits and therefore, the hypothesis of linearity is rejected.

where p, q are the orders of the system, k delay between input and output, and a_m, b_n the parameters of the system.

In order for the model to be a good description of the system, the prediction error, $e(t)$, has to be white. Thus, if whiteness analysis on $e(t)$ falls within, for instance, the 99% confidence limits, and

$$\frac{1}{T} \sum_{t=1}^{\tau} e(t)u(t - \tau) = 0, \quad \tau > 0 \tag{4}$$

it can be assumed that the system is not nonlinear, and linear methods for processing can be used.

PREDICTION OF SEIZURES BY MULTIVARIATE PARAMETRIC MODELING OF THE EEG

The aim is the detection of changes in the EEG signal that occur *prior* to the outbreak of the clinical seizure, and which are time locked to the electrical (EEG) epileptic activity. These changes might be of a very short duration and a high time/frequency resolution of the events around the onset of the electrical seizure is needed (Gath *et al.*, 1992). Thus, nonparametric methods for EEG spectral estimation give inaccurate results and should be substituted by multivariate AR modeling of the EEG signal (Berk, 1974).

Regarding multivariate AR modeling, it has been shown that the performance of the commonly used covariance method for AR estimation, using the vectorial version of the Levinson algorithm (Whittle, 1963; Wiggins and Robinson, 1965), is not good enough for

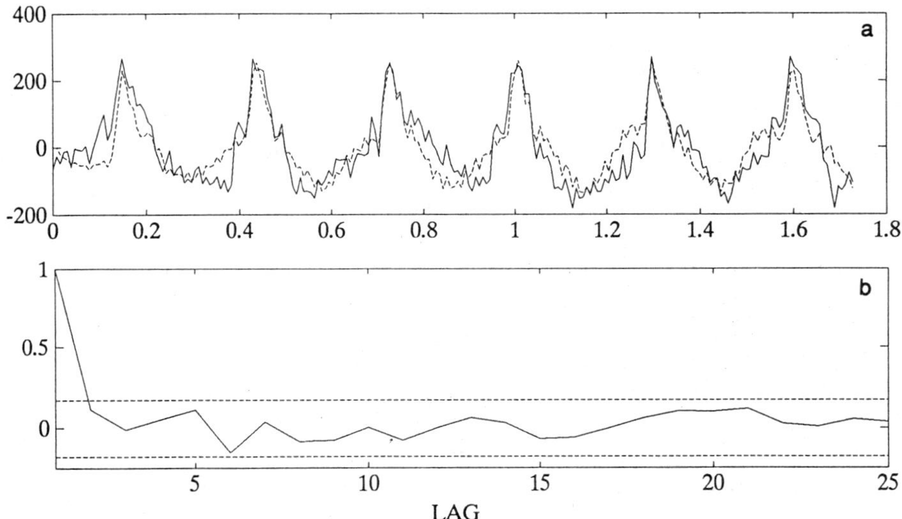

Figure 5. Testing for linearity of a quasi-stationary epileptic EEG segment. a. Original (continuous line) and modeled signal (dashed line). Segment length: 1.75 sec (adaptive segmentation). b. Autocorrelation of residual for the model in a. There is a no deviation from the 99% confidence limits and therefore, the hypothesis of linearity is retained.

short record length (Pham and Tong, 1990). The Dickinson method, based on residual energy ratios, has been shown in these cases to provide better spectral estimates for short record epileptic EEG (Gath *et al.*, 1992).

The Dickinson method is based on a direct estimation of the partial correlation matrices using Cholesky decomposition, and then uses the vectorial version of the Levinson algorithm to compute the AR parameters. Let $y(l)$ be the m-vector valued observed process and denote the forward and backward prediction of $y(l)$ by:

$$\hat{y}_k^f(l) = -\sum_{j=1}^{k} A_{k,j}^f y(l - j) \tag{5}$$

$$\hat{y}_k^b(l) = -\sum_{j=1}^{k} A_{k,j}^b y(l + j) \tag{6}$$

where $A_{k,j}^f$ and $A_{k,j}^b$ are the forward and backward AR coefficient matrices of the k'th order model, respectively. The partial correlation matrix of order k is given by:

$$P_{k+1} = U_k^{-1/2} E\left[e_k^f(l)e_k^b(l)^T\right]\left(V_k^{-1/2}\right)^T \tag{7}$$

where $U_k^{-1/2}$ and $V_k^{-1/2}$ are the inverted Cholesky factors of the covariance matrices U_k and V_k of the forward and backward prediction errors, $e_k^f(l)$ and $e_k^b(l)$, respectively.

From the estimated partial correlation matrices, the forward and backward autoregressive coefficient matrices are computed, using the vectorial version of the Levinson algorithm.

The transfer function $H(z)$ between the innovation $e(l)$ and $y(l)$ is given by:

$$H(z) = \left(I + \sum_{j=1}^{p} A_{p,j}^{f} z^{-j} \right)^{-1} \tag{8}$$

From $H(z)$ and the covariance matrix of $e_{p}^{f}(l)$, the power spectral density can be estimated:

$$\hat{S}_{yy}(f) = \hat{H}(f)\hat{U}\hat{H}^{H}(f) \tag{9}$$

where $H^{H}(f)$ is the hermitian of $H(f)$.

The AR parameters are estimated using a short sliding window with high overlapping. Two methods are used to predict an oncoming epileptic seizure. These are based on the findings that, as the seizure is building up, a tendency towards increased synchronization between various channels is conspicuous, as well as a tendency towards instability (positive feedback).

The first method traces the movement of the poles of the transfer function, $H(z)$, towards the unit circle, indicating a tendency towards instability.

The second method is based on tracking of changes in the coherence function:

$$\gamma^{2}(f) = \frac{S_{12}^{2}(f)}{S_{11}(f)S_{22}(f)} \tag{10}$$

where $S_{11}(f)$ and $S_{22}(f)$ are the respective autospectra, and $S_{12}(f)$ the cospectrum. A sharp rise in coherence indicates increased synchronization.

Using these two methods, preictal changes have been found, indicating increased inter-channel synchronization, and a tendency towards instability for time epochs 1-7.0 sec preceding the epileptic activity.

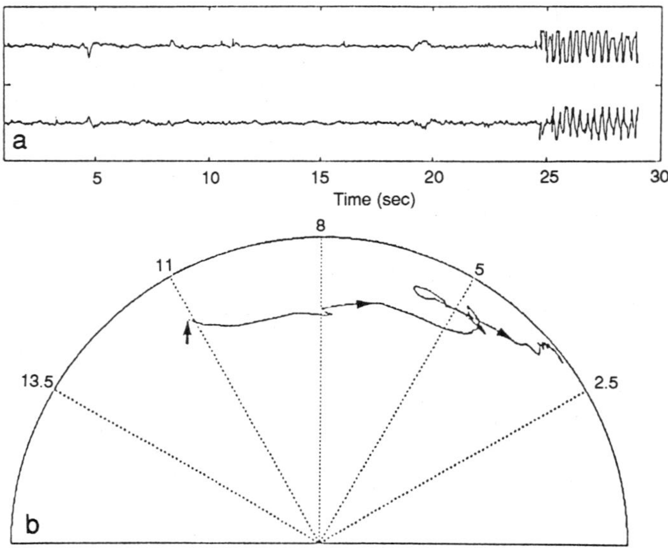

Figure 6. Pole tracking for prediction of seizures. A patient with primary generalized epilepsy. a. Monopolar EEG recording, channels F_4 and O_1. The epileptic EEG activity starts at 24.5 sec. b. Movement of the dominant pole towards the unit circle (instability). The pole trajectory covers the time epoch 22-26 seconds.

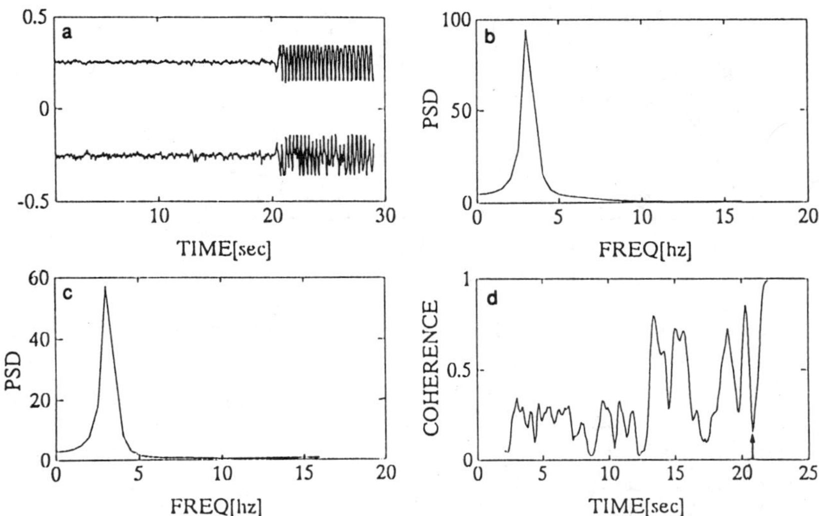

Figure 7. Changes in coherence preictally in a patient with primary generalized epilepsy. a. Monopolar EEG recording, channels F_3 and O_2. The epileptic activity starts at 20 sec. b and c. Power spectrum density for each of the two EEG channels in (a). d. Squared coherence for frequency of 3.5 Hz as a function of time. A rise in coherence can be seen at around 13 seconds.

ESTIMATION OF TIME-DELAYS FOR LOCALIZATION OF EPILEPTIC FOCI

Time delay estimation (TDE) was carried out by adaptive lattice algorithm, as the method of choice for the multichannel epileptic EEG signal. The advantages of this algorithm are:

- Adaptive filtering can track a slowly changing delay very well, even when the dynamics of the input signals are changing rapidly.
- The lattice-ladder filter has fast convergence time and good spectral properties.

The adaptive lattice-ladder filter has a modular lattice upper part, which is the AR predictor of $x(n)$ and a lower ladder structure which is the prediction filter of $y(n)$.

The lattice filter structure estimates $\hat{y}(n)$ as a weighted sum of present and previous samples of $x(n)$

$$\hat{y}(n) = -\sum_{k=0}^{P} W_k(n|P)x(n - k) \tag{11}$$

where a new set of filter coefficients, $W_k(n|P)$, are calculated at each time n. The estimation error:

$$W(n|P) = \arg \min\left\{\sum_j \lambda^{n-j}(y(j) - \hat{y}(j))^2\right\} \tag{12}$$

where λ is a "forgetting factor" such that the distant past affects the solution far less than recent estimates. The delay is computed by a phase approach, calculating the spectrum

$$\mathbf{W}(n|f) = \sum_{k=0}^{P} W_k(n|P)e^{i-2\pi fk} \tag{13}$$

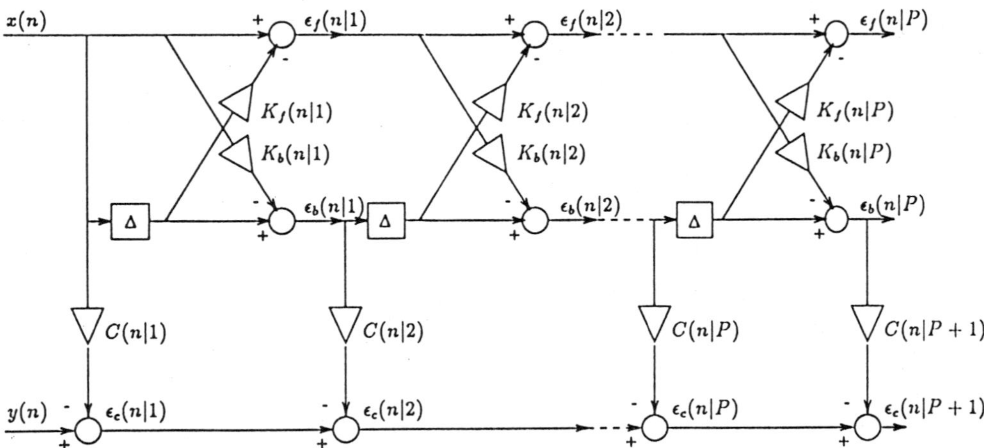

Figure 8. A diagram of the adaptive lattice-ladder filter for TDE (from Harris *et al.*, *IEEE Trans. Biomed. Eng.* 41:820-829, 1994; ©1994 IEEE).

The algorithm for estimation of time delays between the various EEG channels was tested on artificially generated signals and EEG signals with artificially introduced delay, and on real data from rats with focal epilepsy. For the simulation studies, time delays were generated according to the following model:

$$x(t) = s(t) + n_1(t)$$
$$y(t) = N\{\alpha s(t - D)\} + n_2(t) \tag{14}$$

1. $x(t)$ and $y(t)$ are the measured signals
2. $s(t)$ is a deterministic signal
3. $n_1(t)$ and $n_2(t)$ are white, jointly uncorrelated Gaussian processes
4. D is the delay between the channels
5. α is an attenuation or amplification
6. $N\{\cdot\}$ is a non-linear distortion operator.

The nonlinear distortion was used in some of the simulation experiments, in order to investigate the robustness of the TDE algorithm (to examine its ability to deal with simple form for nonlinearities).

The real data was obtained from experiments on rats with induced focal epilepsy. Epileptic foci were generated by intracerebral microinjection of Kainic acid into the right amygdala. Intracerebral EEG electrodes were introduced stereotaxically and EEG recording was carried out from the amygdala and hippocampus on both sides. In addition, surface EEG recording was made from the right parietal cortex.

Using the adaptive lattice-ladder algorithm for TDE in epileptic EEG it could be varified that it was superior to conventional TD estimators, such as the general crosscorrelation method.

From Fig. 9 it can be seen that there is a consistency with the assumption that the right amygdala leads the other channels. The obtained estimate is smooth and constant for a significant period of time. In addition, high coherence (above 0.8) was calculated during this time for most of the frequency band under consideration.

Using the adaptive phase method for time delay estimation, a schematic diagram can be drawn for the pathways of propagation of the seizure (Fig. 10).

CONCLUSIONS

The present study deals with processing of multichannel epileptic EEG data, obtained from epileptic patients and from rat experiments. The two main purposes are definition of the localization and extent of the epileptic focus in cases of focal epilepsy, and detection of preictal changes that characterize an incoming seizure.

The multichannel seizure EEG signal is inherently nonstationary, and therefore, adaptive segmentation of the signal is carried out to generate signal segments which are "piecewise stationary". Linear methods for processing of the multichannel EEG signal can be applied, only if the hypothesis of linear relationship between quasi-stationary EEG segments from various EEG channels is not rejected.

Multivariate parametric modeling of the multichannel EEG, using the method of residual energy ratios, gives spectral estimates with high time/frequency resolution. Using this method, computation of coherence and pole trajectories can be used to reveal changes that occur in the EEG signal *prior* to the outbreak of the clinical seizure and which are linked to the seizure. Thus, early signs of an incoming seizure can be detected from the pre-seizure EEG.

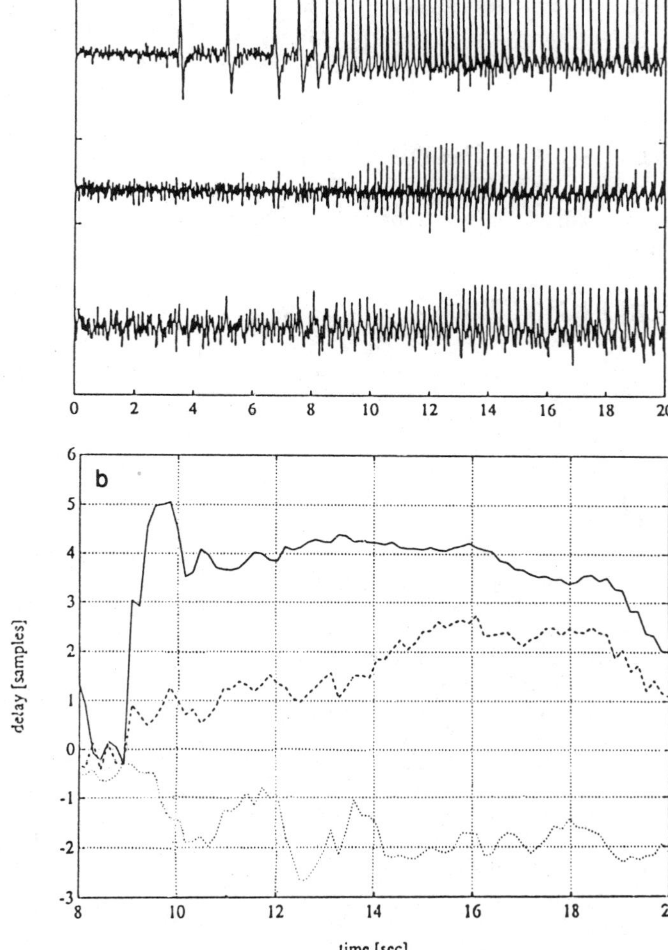

Figure 9. Time delay estimation on real data from a rat with induced focal epilepsy. a. Depth EEG recording from right amygdala (upper trace), right hippocampus (middle trace) and surface EEG recording from the cortex (lower trace). b. Time delays calculated using the adaptive phase method. Solid line: delay between right amygdala and right hippocampus. Dashed line: delay between right amygdala and cortex. Dotted line: delay between right hippocampus and cortex (from Harris *et al., IEEE Trans. Biomed. Eng.* 41:820-829, 1994; ©1994 IEEE).

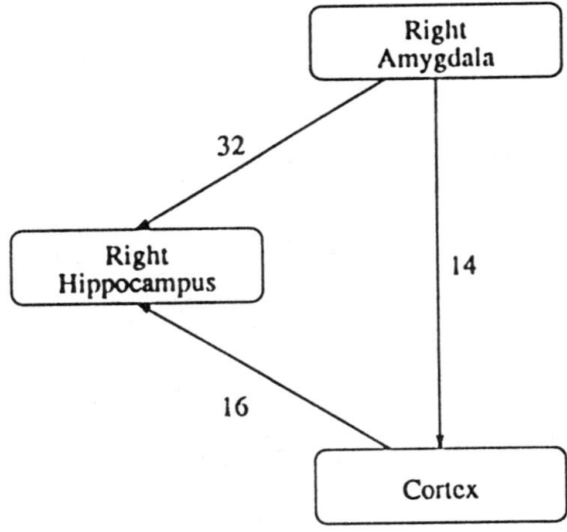

Figure 10. A schematic diagram showing the pathways of propagation of the seizure in Fig. 9. Rat with induced epileptic focus in the right amygdala. The delays, measured in msec, are denoted along the arrows of propagation.

Concerning focal epilepsy, time delay estimation can be applied to the multichannel EEG signal, in order to provide complementary information as to the localization and extent of the epileptic focus. Adaptive least-square filtering (a lattice-ladder type) has been found to give smooth and constant time delay estimates, which in control experiments in rats were consistent with the true location of the epileptic focus.

ACKNOWLEDGMENT

This work was supported by the Loewengart Research Fund.

REFERENCES

1. Berk, K. N., 1974, Consistent autoregressive spectral estimates, *Ann. Stat.* 2:489-502.
2. Blinowska, K. J., and Malinowski, M., 1991, Non-linear and linear forecasting of the EEG time series, *Biol. Cybern.* 66:159-165.
3. Bodenstein, G., and Praetorius, H. M., 1977, Feature extraction from EEG by adaptive segmentation, *Proc. IEEE* 65:642-652.
4. Dickinson, B. W., 1978, Autoregressive estimation using residual energy ratios, *IEEE Trans. Inform Theory.* 24:503-506.
5. Dickinson, B. W., 1979, Estimation of partial correlation matrices using Cholesky decomposition, *IEEE Trans. Automat. Contr.* 24:302-305.
6. Gath, I., and Bar-On, E., 1980, Computerized method for scoring of polygraphic sleep recordings, *Comp. Progr. Biomed.* 11:217-223.
7. Gath, I., Feuerstein, C., Pham, D. T., and Rondouin, G., 1992, On the tracking of rapid dynamic changes in seizure EEG, *IEEE Trans. Biomed. Eng.* 39:52-958.
8. Gath, I., Michaeli, A., and Feuerstein, C., 1991, A model for dual channel segmentation of the EEG signal, *Biol. Cybern.* 64:225-230.
9. George, R. E., and Michael, J. E., 1991, Efficacy of vagal nerve stimulation for the treatment of partial seizures, *Epilepsia* 32:S3-80.
10. Gersch, W., 1987, Non-stationary multichannel time series analysis, in Gevins, A. S., and Remond, A. (eds), *EEG Handbook Revised Series* Elsevier, Amsterdam, vol. II:261-296.
11. Gersch, W., and Goddard, G. V., 1970, Epileptic focus location: spectral analysis method, *Science* 169:701-702.
12. Ishijima, B., Mori, T., and Yoshimasu, N., 1975, Neuronal activities in human epileptic foci and surrounding areas, *Electroencephal. Clin. Neurophysiol.* 39:643-650.

13. Kaplan, B. J., 1975, Biofeedback in epileptic equivocal relationship of reinforced EEG frequency to seizure reduction, *Epilepsia* 15:477-485.
14. Kuhlman, W. N., and Allison, T, 1977, EEG feedback training in the treatment of epilepsy: some questions and some answers, *Pavlovian J. Biol. Sci.* 12:112-122.
15. Lopes da Silva, F., Pijn, J. P., and Boeijinga, P., 1989, Interdependence of EEG signals: linear vs. nonlinear associations and the significance of time delays and phase shifts, *Brain Topography* 2:9-18.
16. Moddemeijer, R., 1989, Delay-Estimation with Application to Electroencephalograms in Epilepsy, *Ph.D Thesis*, University of Twente, Enschede.
17. Ojemann, G. A., Engel, J., 1987, Acute and chronic intracranial recording and stimulation, in J. Engel (ed), *Surgical Treatment of the Epilepsies*, 263-288, Raven Press, New York.
18. Panych, L. P., and Wada, J. A., 1990, Computer applications in data analysis, in Wada, J. A., and Ellingson, R. J., (eds), *Clinical Neurophysiology of Epilepsy, Handbook of Electroencephalography and Clinical Neurophysiology Revised Series*, Elsevier, Amsterdam, Vol. IV, 361-385.
19. Penry, J. K., and Dean, J. C., 1990, Prevention of intractable partial seizures by intermittent vagal stimulation in humans, *Epilepsia* 31:S40-43.
20. Pham, D. T., and Tong, D. Q., 1990, Maximum likelihood estimation for multivariate autoregressive model, *Research report RR 807-M*, Laboratoire LMC/IMAG, Université Joseph Fourier, Grenoble.
21. Velasco, F., Velasco, M., Ogarrio, C., and Fanghanel, G., 1987, Stimulation of the centromedian thalamic nuclei in the treatment of convulsive seizures, a preliminary report, *Epilepsia* 28:421-430.
22. Whittle, P., 1963, On the fitting of multivariate autoregressions and the approximate canonical factorization of a spectral density matrix, *Biometrika* 50:129-134.
23. Wiggins R. A., and Robinson, E. A., 1965, Recursive solution to the multichannel filtering problem, *J. Geophysical Res.* 70:1885-1891.

6

SIMULTANEOUS EEG RECORDINGS FROM OLFACTORY AND LIMBIC BRAIN STRUCTURES: LIMBIC MARKERS DURING OLFACTORY PERCEPTION

Leslie Kay,[1] Walter J. Freeman,[2] and Larry R. Lancaster, Jr.[2]

[1] Graduate Group in Biophysics
[2] Dept. of Molecular and Cell Biology, Neurobiology Division
University of California
Berkeley, California 94720

ABSTRACT

We approach the question of how the brain may expect a known olfactory stimulus. EEG data were recorded from rats simultaneously from the olfactory bulb (OB), prepiriform cortex (PPC), entorhinal cortex (EC), and dentate gyrus (DG) of the hippocampus. The animals were trained in a classical paradigm to identify two odors.

The problem of nonstationarity in analyzing the EEG time series was compounded by the brief duration and only loosely time-locked perceptual event. We used several statistical and dynamical methods to show the relationship of activity among the four structures during odor identification. Tools used were correlation distributions and coherence in different behavioral time segments and a new technique to measure the system's transit in and out of putative chaotic attracted states.

Results show three significant features of the system, which occur reliably during odor identification and unreliably or insignificantly in control periods. We show gamma burst (40-160 Hz) coherence among all four studied structures during odor identification. We have previously identified this burst period as a perceptually defined event, so this ties the hippocampus strongly to olfactory perception. The second feature is the apparent centrifugal transmission of a proposed biasing signal in the beta band (15-35 Hz) from the entorhinal cortex to the olfactory bulb just before the onset of the gamma burst in the olfactory bulb. We have coined the term "preafference" to identify this as an expectation signal, after a generalization of the model of reafference presented by von Holst and Mittelstaedt in 1950. The third feature is a gamma burst apparently initiated in the prepyriform cortex, which is then passed back to the olfactory bulb just after the primary bulb burst. We label this the reafferent or "handshaking" signal, which could serve to adjust the dynamics of the system as it diverges from what was expected before the stimulus arrival.

Advances in Processing and Pattern Analysis of Biological Signals, Edited by Isak Gath and Gideon F. Inbar
Plenum Press, New York, 1996

                                                                     L. Kay et al.

INTRODUCTION

A well-trained animal can identify an odor in one or two sniffs, within a reaction time of 250 msec, especially when it has been trained to expect one of a small set of olfactory stimuli. Expectation is most likely the psychological answer to this problem. If one expects an odor, one can almost perceive it within one's mind before it is delivered. So, the general question becomes "how does the brain expect?" and more specifically, "how does the brain prepare itself to identify a known stimulus?"

This form of expectation has been described by physiologists and other scientists throughout the past 2500 years (Grüsser, 1993), always by a different name. Helmholtz discussed the notion of an "effort of will" within the visual (or other) parts of the brain as a mechanism by which the brain prepares itself for an expected stimulus (Helmholtz, 1910), as exhibited by the optokinetic response. More recently von Holst and Mittelstädt (1950) coined the term "reafference" and gave a mechanistic explanation for the same phenomenon that Helmholtz described. The advantage of setting out the proposed reafferent mechanism was that it suggested the action of specific, but unknown, neural pathways exerting influence over primary sensory and motor "effector" areas. In the same year Roger Sperry (1950) published an account of the optokinetic response in the fish, in which he described the necessity of a "corollary discharge" which would prepare the optic and motor areas of the brain for the expected change of the visual field accompanying movement.

In rats and other non-primate mammals the olfactory areas in the brain are intimately connected with the limbic system. Even in humans, the olfactory areas are more closely tied to the limbic system than the other senses are (Heimer, 1995). Damage to the limbic system drastically affects an animal's ability to learn, to identify odors, and to form intention. It is for this reason that we look to the limbic system for signs of how the brain "expects."

In previous work we showed that there are clear perceptual markers in the olfactory bulb EEG recorded from an array of 64 electrodes placed on the surface of the bulb (Freeman & Viana Di Prisco, 1986). When an animal is in the process of identifying an odor characteristic bursts are seen in the EEG, and these burst segments are coherent across the array. The amplitude patterns over the array of a common waveform extracted from these bursts can be segregated by odor when the odors tested are meaningful to the animal (i. e., paired with a CS+ or CS-). These are distinguished from the patterns of bursts produced during control periods and from the less periodic background activity between bursts. These bursts, occurring approximately 100-300 ms after stimulus arrival, are presumed to be perceptual markers for several reasons. First, the time of occurrence is too late to be due solely to stimulus arrival, and so can be distinguished from "sensation." Second, the patterns are only distinguishable from the control patterns when the odor is meaningful to the animal. Third, these patterns are relatively stable, but not invariant (Freeman, 1992), which distinguishes them from the novelty response produced when an animal smells an unfamiliar odor (to which it quickly habituates, if the odor is not reinforced). Finally, the patterns change as the animal learns new odors, has prolonged experience in its environment (drift with time), and if the meaning of the odor is changed as in contingency reversal (Freeman, 1992). These results were found on EEG band pass filtered in the gamma band (30-100 Hz in rabbits), and thus we label these bursts high frequency perceptual markers.

Bressler (1984, 1987a,b) showed that the gamma bursts present in the olfactory bulb drive those in the prepyriform cortex. Boeijinga and Lopes da Silva (1988, 1989) have shown somewhat lower frequency (below 60 Hz) coherence between the olfactory bulb bursts and those in the prepyriform cortex and entorhinal cortex in the beta band in cats. In the limbic system, gamma activity in the entorhinal cortex is correlated with that in the dentate gyrus (Bragin *et al.*, 1993), although it was neither discussed nor explored in that study, whether it was correlated with sniffing or olfaction. There have been reports of coherent low frequency bursts (20 Hz) in rats in the olfactory bulb and dentate gyrus in response to sniffing toluene or the urine of a predator (Heale & Vanderwolf, 1994; Vanderwolf, 1992). The theta rhythm (3-10 Hz) in the dentate gyrus of rats has been shown to be correlated with the sniffing cycle during olfactory learning, but not at other times (Macrides *et al.*, 1982).

In this report we describe the activity in the full EEG spectrum (3-200 Hz) throughout the olfactory and limbic areas in the rat. Recordings discussed here were not from surface spatial arrays but from electrodes placed stereotaxically into the depth of the brain regions noted. We predict that limbic activity which may control the perceptual state of the bulb should precede the perceptual event (gamma burst) in the bulb. If it is an expectation response, then it may indeed occur during a control period in trained and motivated animals, but would be most predictable and most closely time-locked during odor identification. We choose the entorhinal cortex as the most likely sight of a reafference response because of its reciprocal connectivity with every sensory modality (Witter *et al.*, 1989).

SYSTEM OVERVIEW

The four major parts of the olfactory system are the olfactory receptors in the nasal mucosa, the olfactory bulb, which receives the direct input from the olfactory receptors via the olfactory nerve, the anterior olfactory nucleus, through which activity from each of the two bulbs is transmitted to the other, and the prepyriform or primary olfactory cortex.

The olfactory bulb mitral cells send dendrites into glomeruli, where incoming axons from the receptor neurons synapse directly onto the mitral cells and also onto the surrounding periglomerular cells. The periglomerular cells, although GABAergic, have an excitatory influence on the mitral cells and provide range compression for the olfactory input (Freeman, 1975,1993). The mitral cells are connected with each other, with the tufted cells, and with the granule cell interneurons. They are also the source of projection axons leaving the olfactory bulb to synapse in the anterior olfactory nucleus, which sends axons to the basal dendrites of granule cells in the contralateral olfactory bulb. The mitral cells in the olfactory bulb and nucleus also synapse in the prepyriform cortex and the entorhinal cortex. Other projections go to the olfactory tubercle with synaptic relays to the hypothalamus, the midbrain, and the medial dorsal nucleus of the thalamus. All known actions of the mitral cells upon other cell populations are excitatory. The granule cell inhibitory interneurons form a negative feedback loop with the mitral cell population via reciprocal dendrodendritic synapses (Freeman, 1975). The granule cells have a bipolar geometry, while the mitral cells have radial symmetry, so it is the granule cell population which is responsible for the field potentials recorded from macro electrodes in or on the surface of the olfactory bulb (Freeman, 1972a,b; Rall & Shepherd, 1968). Because of the negative feedback relationship of the excitatory and inhibitory populations, the field potential (from the granule cells) lags the population activity of the mitral cells by one quarter cycle of the dominant frequency, the probability of a mitral cell's firing being greatest one quarter cycle after the peak of negativity in the granule cell population dendritic current (Eeckman & Freeman, 1990; Freeman, 1975). A similar negative feedback relationship of the excitatory and inhibitory cell populations exists in the prepyriform cortex, the difference being that the summed dendritic current (local field potential) is produced by the pyramidal cells (Freeman, 1975), which are also the projection neurons from the prepyriform cortex to the entorhinal cortex, the olfactory bulb and the olfactory nucleus, and thalamus (Heimer, 1968).

Two parts of the limbic system are discussed here. The first is the entorhinal cortex, which is called limbic cortex and sensory integration cortex, because it is intimately connected with both the hippocampus and the sensory cortices. The entorhinal cortex is the major source of input to and output from the hippocampus. It is a six layered cortex, although the central lamina dissecans (between layers 3 and 4) of residual glia and polymorph cells is often labeled as a distinct layer. The layer II stellate cells receive synapses directly from the olfactory bulb, and themselves project via the perforant path to the dentate gyrus and, to a lesser degree, to area CA3 of the hippocampus proper. Layer III pyramidal cells project into area CA1 of the hippocampus. The deeper layers (V and VI) receive projections from the CA1 and CA3 regions of the hippocampus and project to the parahippocampal and perirhinal regions, as well as the olfactory bulb (de Olmos *et al.*, 1978), nucleus, and cortex. The second limbic structure from which we recorded is the dentate gyrus of the hippocampus. The dentate gyrus granule cells receive synapses from the perforant path stellate cells and project throughout the rest of the hippocampus into the CA3 and CA1 regions via the mossy fibers into CA3 and then the Schaeffer collaterals into CA1 (Lopes da Silva *et al.*, 1990; Schwerdtfeger *et al.*, 1990; Witter *et al.*, 1989).

METHODS

Experiments

Adult male Sprague-Dawley rats were chronically implanted with bipolar electrodes in the lateral olfactory tract (LOT; for stimulation), the olfactory bulb, the prepyriform cortex, the entorhinal cortex, and the dentate gyrus on the left side of the brain. The surgical procedure is described in detail elsewhere (Eeckman & Freeman, 1990; Kay & Freeman, 1994). All animal handling and surgical procedures were performed according to the NIH guidelines on the care and use of laboratory animals.

Behavioral experiments must be designed to distinguish perceptual events from sensory events, and to distinguish true recognition of an odor from recognition of a nonspecific change in the background odor. For this reason we chose to train the animals to recognize two odors, only one of which was rewarded, and to test them also on separate trials with air to control for noise, pressure, and temperature cues (Freeman *et al.*, 1983). Animals were trained to wait at an odor port and sniff at incoming odors, remaining in the port for 1-2 seconds, and then to receive a food or water reward with presentation of odor A (mint). With presentation of odor B (almond or coconut) no reward was given, and animals generally left the odor port after one presentation.

Data were collected from five trained rats from the four olfactory and limbic regions simultaneously. Additional recordings were made with three rats using an elastic pneumograph belt placed around the animal's ribcage to record respiration (ribcage movement) simultaneously with local field potentials. Because of the difficulty of maintaining the optimal position of the pneumograph belt on active rats, pneumograph recordings were not made during the odor identification sessions.

Analysis

The problem of nonstationarity in analyzing the EEG time series was compounded by the brief duration and internally initiated perceptual events. Sample data from three different motivational states are shown in Fig. 1. While evoked potentials recorded in each structure from electrical stimulation of the LOT are explicitly time-locked, providing for easy averaging and analysis, a perceptual event is created endogenously, and so is not tightly locked to the original stimulus. This, combined with a relatively slow olfactory cycle making the stimulus arrival ambiguous, smoothes out short-lived events when records are averaged. Even when records are averaged to a respiratory peak, such as the first peak after stimulus

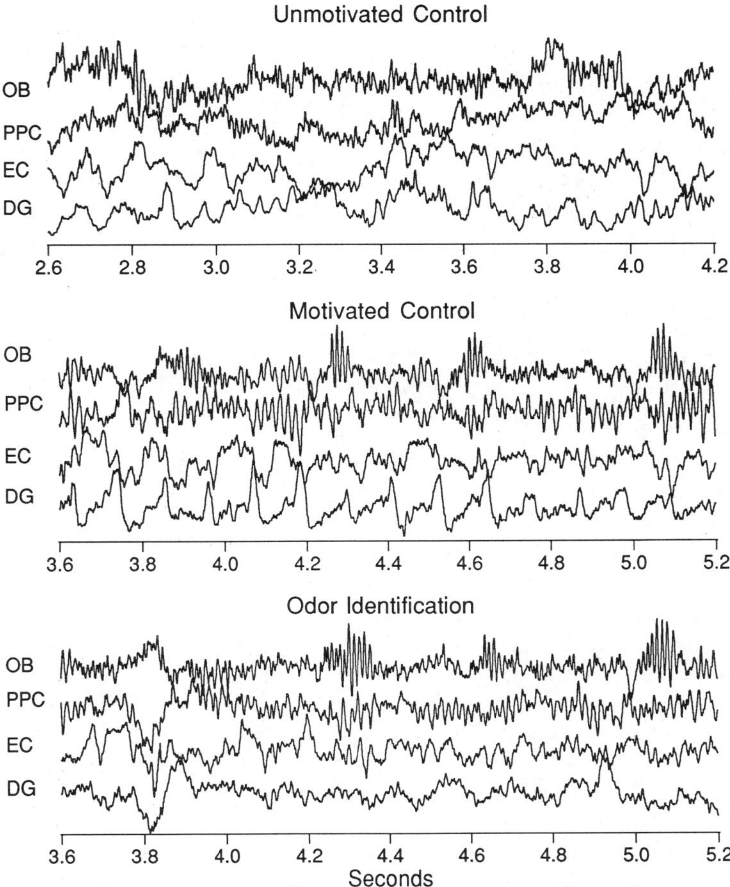

Figure 1. 1. 4 seconds of data recorded during three different motivational states. Unmotivated control: rat is not hungry and is not being presented with an odor. Motivated control: rat is hungry and is being presented with the control odor (plain air) during the odor identification period. Odor identification: rat is hungry and is identifying an odor. Abbreviations: OB - olfactory bulb; PPC - prepyriform cortex; EC- entorhinal cortex; DG - dentate gyrus. Data are recorded at 1000 Hz, 1-300 Hz analog filters.

arrival, there is no apparent structure to the averaged signal, except the single low frequency respiratory wave on which the signals are averaged. The phase and frequency in the gamma range of the EEG are random variables. Thus it is necessary to average or sum not the signal itself but measures made during perceptually relevant periods over many trials.

We used several statistical and dynamical methods to show the relationship of activity among the four structures during odor identification. The first method is a standard coherence measure. A 256 ms window was the largest used for this analysis, as this time length is the maximum amount of time that the system is in one definable state (burst, inhalation, reafferent, etc.). In fact, it is often closer to the 128 ms time frame. When spectra are taken over longer periods, stationarity is often violated and brief events are lost in the often multiple state changes in the window. If an event occurs often enough, a long time window or many successive overlapping time windows are advisable to approximate stationarity, but in this case, many of the phenomena are short-lived and occur infrequently, and they would be lost in the averaging when looking at long records. Thus, we averaged measures not over a long time frame in one record but over the same short time frame in many records.

For coherence during burst periods, the peaks in the theta band from the low pass filtered data (< 11 Hz) were chosen as indicators of the peak of the inhalation cycle in the EEG, and windows from band pass filtered data (10-160 Hz) centered at the time periods indicated by the theta peaks were used to calculate coherence. The same analysis was conducted on the shuffled data (see below) for comparison. The coherence (R) was calculated using the standard equation (Brillinger, 1981):

$$R_{ab}^{(T)}(\lambda) = \frac{f_{ab}^{(T)}(\lambda)}{\sqrt{\left[f_{aa}^{(T)}(\lambda)f_{bb}^{(T)}(\lambda)\right]}}.$$

The power spectra (f) were estimated with a Hamming window in the time domain, and the coherence was estimated from the averaged auto and cross spectra from the EEG.

The second method was an adapted cross-correlation analysis. First, behaviorally significant time periods were chosen, either control or odor identification. Then successive windows were taken for the entire 1 second time period across simultaneous time series from two of the four structures, and correlation coefficients were calculated for each window. The correlation values were collected for the same time period from 20 records, and the distributions of correlations were compared with the results from the same procedure performed on resampled data (see below) using the Kolmogorov-Smirnov test for significance. This procedure was repeated for every relative lag of the two series from -40 ms (series 1 leads series 2 by 40 ms) to +40 ms (series 2 leads series 1 by 40 ms) to look for significant dips in the p values at nonzero lags.

Resampled data was constructed by the following method to retain the most stringent test against spurious correlation. New records were created by choosing non-simultaneously recorded data segments from the same structures as each original data set, so that the OB, PPC, EC, and DG recordings in each shuffled record are from different original records, thus maintaining the same temporal characteristics within each recording site, but effectively randomizing the correlations between structures.

The third method (Lancaster *et al.*, 1995) was inspired by previous hypotheses that the period of identification or perception should, in dynamical terms, be indicative of the system's transit in a basin of attraction (Freeman, 1994; Kay *et al.*, 1995). The "smoothness" of the distribution of a measured variable (voltage in this case) is greatest when the flow has not yet approached an attractor and becomes less, as "spikiness" increases, the longer that the flow is subject to the same attracting set (Lancaster *et al.*, 1995). As the system is altered by parameter changes, old attractors are destroyed and new attractors are created. Thus, the "spikiness" of the distribution of voltage in a small time window (e. g., 500 ms) should decrease when the old attractor is destroyed and the system has not yet been constrained to the new attractor, and then it should increase as the system approaches the new attractor. A method taking advantage of this feature (Lancaster *et al.*, 1995) computed on olfactory bulb and entorhinal cortex time series was used to determine that the appearance of attractor-constrained dynamics in the entorhinal cortex preceded a similar event in the olfactory bulb.

A histogram of the voltage values of the time series is made using j data points from a window of data (length j). The histogram is divided into m pairs of equal-area segments

$$\left\{P_{1,2}, P_{3,4}, \ldots, P_{2m-1,2m}\right\}.$$

The differences between consecutive pairs (D) are then calculated:

$$D = \sum_{k=1}^{m} d_{(2k-1,2k)}, \text{ where } d_{(2k-1,2k)} = P_{(2k,\cdot)} - P_{(\cdot,2k-1)}.$$

Then, the spikiness indicator (I) is given by

$$I = \frac{D}{G}, \text{ where } G = \sqrt{\frac{2\sum_{i=1}^{j}\left(i^2\right)F_{(i,j)}}{n}}$$

and

$$F_{(i,j)} = \left(2n\right)^k \sum_{k=i}^{j}\left[-1^{(k-i)}\left(2m\right)^{-k}\binom{j}{k}\binom{2k}{k+i}\right]$$

OLFACTORY AND LIMBIC SIGNALS

The Low Frequency Respiratory Wave

Simultaneous recording of EEG and the output from a pneumograph belt strapped around the animal's ribcage shows that, like the rabbit and cat, the low frequency component of the olfactory bulb EEG is highly correlated with the respiratory cycle. The classic olfactory gamma bursts ride the crest of this wave, which indicates the end of inhalation (Fig. 2). In a study conducted on rats Macrides *et al., 1*982) showed correlation of the dentate gyrus theta rhythm with respiration during olfactory learning. We did not record respiration during training, so we were unable to verify this finding. However we found no significant coherence between either the dentate gyral or entorhinal cortical theta rhythms and the low frequency respiratory wave in the olfactory bulb during behavior or control periods. In some cases a correlation did appear when the rat's respiratory frequency was in the theta band, but without consistent phase relationships.

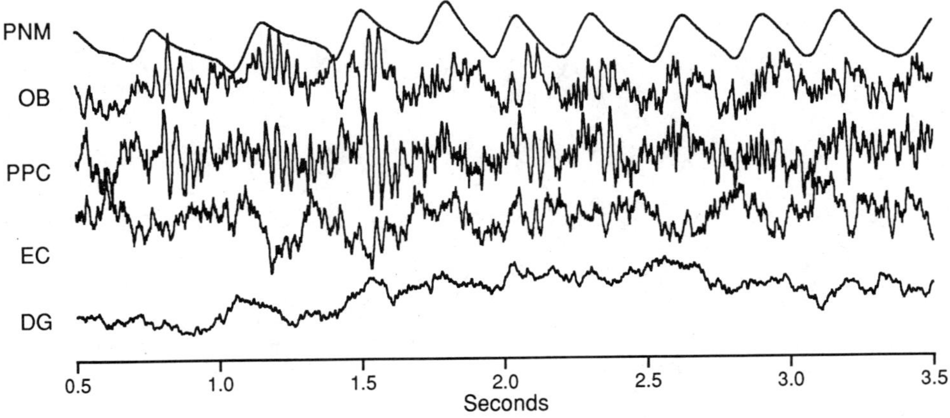

Figure 2. Three seconds of simultaneous pneumograph (PNM) and local field potential recordings. Abbreviations as in Fig. 1.

Inward Transmission of Activity

Visual inspection of data filtered in the gamma band (35-160 Hz) showed that these bursts appear to be transmitted to each of the three remaining structures in temporal succession (Fig. 3). Coherence was calculated between the olfactory bulb EEG and that recorded from each of the other three structures. Significant coherence peaks were found in the gamma band in all animals between the olfactory bulb and each of the other three structures only during the odor identification period and only by selecting windows centered on the peak of the theta rhythm, which brackets the time period in which the olfactory bulb gamma burst occurs. Figure 4 shows coherence spectra for one animal compared with coherence spectra calculated from the same data set shuffled (described above). In some cases coherence was found in the prestimulus period for one set of structures, but this was not consistent across animals.

Although coherence is significantly above the values estimated from resampled data, there was no consistent linear phase gradient through the coherent regions, the slope of which is useful as a lag predictor (Boudreau & Freeman, 1963; Brillinger, 1981). This may be due to several factors: 1) coherence levels are often not high enough to give a reliable phase measurement, 2) phase relationships of burst activity in the pairs of structures are inconsistent but limited to a small range, 3) the wide band nature of the event, combined with a somewhat

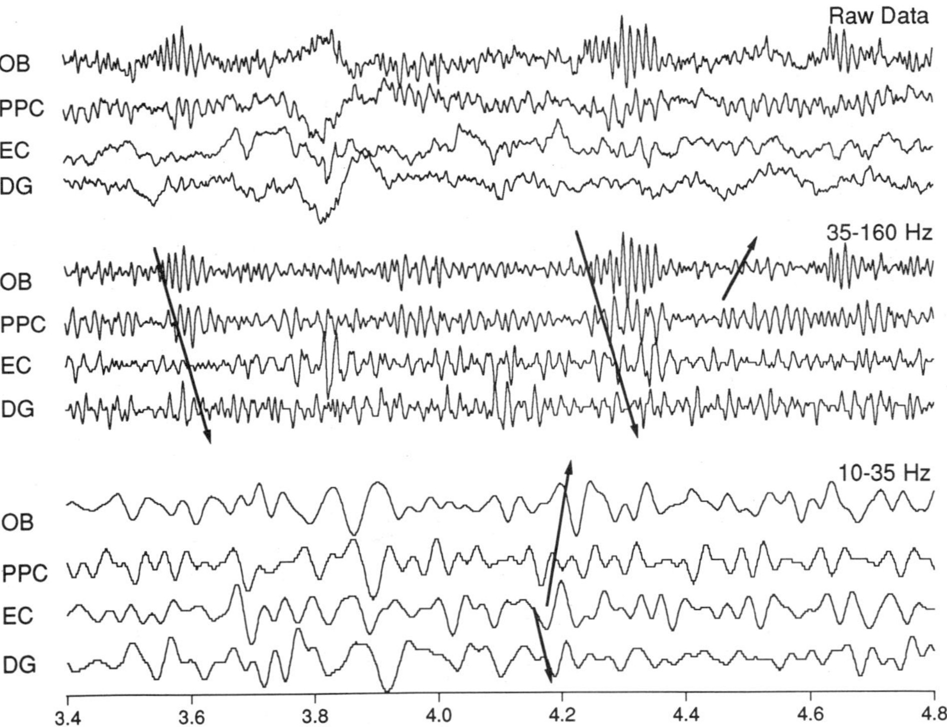

Figure 3. Data (1. 4 seconds) recorded during identification of rewarded odor. Odor arrived at approximately 3. 5 seconds. Raw data in top trace from structures indicated by abbreviations. Lower sets of traces are same data filtered in the gamma (35-160 Hz) and beta (10-35 Hz) bands. Arrows indicate direction of apparent transmission of activity and are not meant to indicate a specific lag. Right and downward pointing arrows indicate the inwardly transmitted "afferent" burst. Right and upward pointing arrow on the gamma data indicates the "reafferent" burst. Right upward and downward arrows on the beta data indicate the "preafferent" signal.

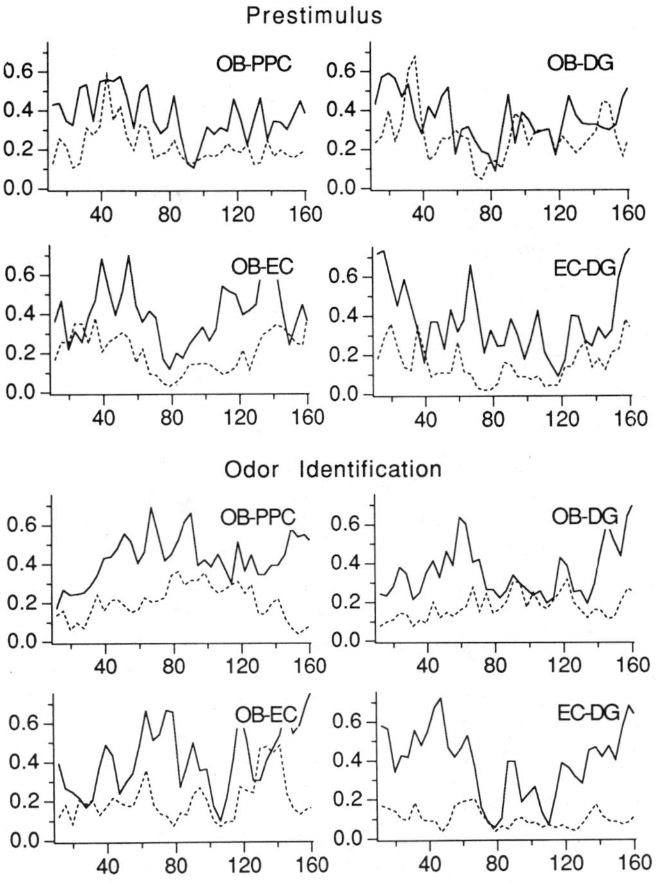

Figure 4. Coherence calculated as described in methods from 10 prestimulus and 10 matching odor identification periods (50 peaks in the theta band for each). Solid line indicates coherence from the data. Dashed line is coherence calculated from the same data resampled (see text). Abbreviations indicate pairs of structures used for the coherence measure.

nonlinear correlation in activity may induce a randomness to the average phase relationships, or 4) the error of measurement of phase is too high to be able to give reliable values. There are indications that all four of these factors contributed to this problem.

On the other hand, the cascade of bursts in Fig. 3 suggests that there may be a consistent lag from the burst in the OB to each of the other sites. By comparing distributions of correlation values across sites during the odor identification periods to the same analysis performed on shuffled data, preferred lags are indicated. Figure 5 shows the distributions estimated from 20 one second odor identification periods from one animal at lags consistent with those predicted by estimated conduction and synaptic delay times along known anatomical pathways. The indicated lags were among those few which were significantly different from the shuffled distributions and from the distributions at zero lag. Distributions were also examined to verify that the skewness was in the positive direction and to verify that the shuffled distributions and those occurring at zero lag were either Gaussian or nearly Gaussian, centered at or near a correlation of zero.

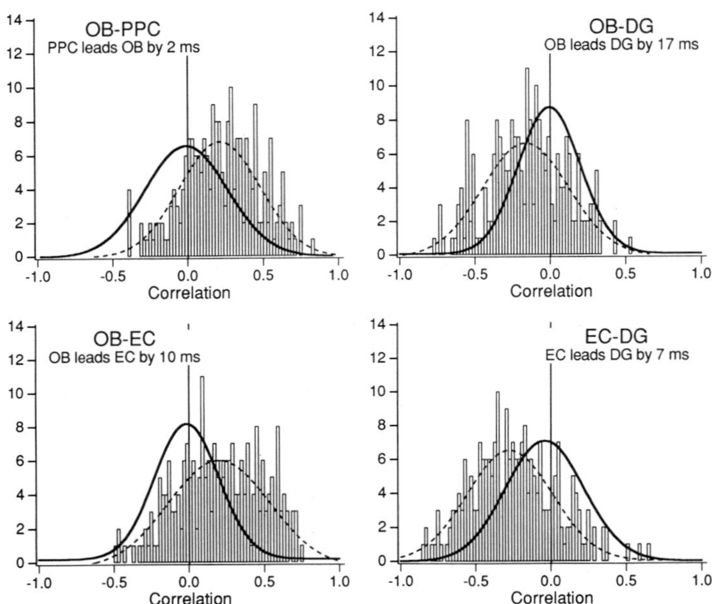

Figure 5. Correlation distributions from one experiment at optimal lags. Vertical axes are numbers in bin. Solid line indicates distribution from shuffled data. Histograms and fitted curves (dashed line) indicate distributions from real data. Skews in OB-DG and EC-DG are negative because the polarity of the signal from the position of dentate recording electrode was reversed relative to the other recordings (deep positive instead of deep negative).

We did not at any time observe the nearly periodic 20 Hz bursts reported by Heale and Vanderwolf (1994). This may be due to the difference in stimuli (food odors vs toluene or urine of a predator).

Centrifugal Signals

Centrifugal signals, originating in the limbic cortex, were present in two distinct frequency bands with highly stereotyped relationships to the inwardly transmitted primary olfactory bulb burst. The first signal is a mid to low frequency signal arising in the entorhinal cortex, in this subset of brain regions, and occurring later (15-25 ms) in the olfactory bulb. Visual inspection of the data filtered in the theta and beta band (3-35 Hz) shows a reliable one cycle of beta band oscillation occurring just before the onset of the primary burst (Fig. 3). In some animals this signal is in the theta range (approximately 10 Hz). We call this the "preafferent" signal because of this relationship to the perceptually relevant primary burst, which is created in response to afferent sensory input.

Again, estimation of correlation and optimal lags was performed using the adapted correlation measure. Significant correlations were evident at two sets of lags in the beta band. The first is the apparent inward transmission of beta band activity at lags similar to those found in the gamma band. The second set were in the opposite direction at slightly longer lags, owing to the probably slower conducting centrifugal pathways. Both distributions were examined and determined to be skewed in the same direction and separated by less than one cycle of the dominant frequency, thus helping to verify that the optimal lags were not due to the positive and negative correlations separated by one half cycle or two positive lags separated by one cycle of the dominant frequency.

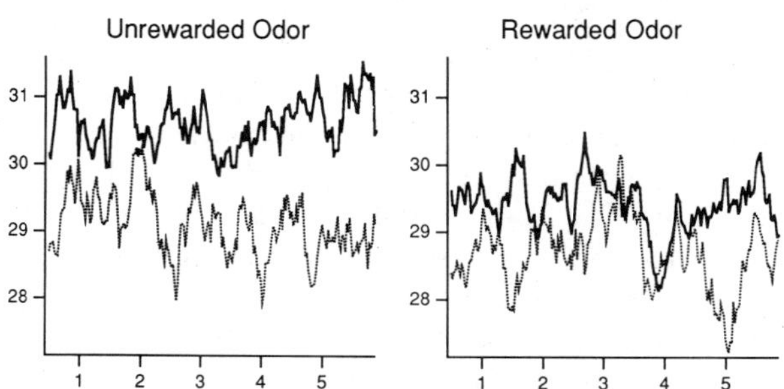

Figure 6. Average measure of distribution smoothness. Solid line is from OB, shaded line is from EC. Note that the dip in "spikiness" indicating a parameter shift from the old putative attractor and the subsequent rise to a new attractor occurs in both structures only at the presumed odor identification point for the rewarded odor. A similar shift does not occur elsewhere. Units are arbitrary.

This signal being of only brief duration and possibly nonlinearly related between the limbic and the olfactory areas, we used a new method of analysis, which does not rely on strict signal coherence. The method relies on the hypothesis that when an odor identification is made, the system may be viewed as subject to a new attracting set. This method should show an decrease in "spikiness", or an increase in smoothness, as the old attractor is destroyed, and a subsequent increase in spikiness when the activity in a recording site approaches a new attractor. Figure 6 shows results from one animal, where the entorhinal cortex approaches a putative attractor before the olfactory bulb does (by 50 - 100 ms), suggesting that the dynamics in the entorhinal cortex set up the highly interconnected system for the stimulus arrival. These indicators are calculated from 20 records each for the unreinforced and reinforced odors, and correlation of the "transits" through domains of smoothness are calculated on a record by record basis. When averaged, these correlation series show a peak at the same time as the drop of the spikiness indicators in the reinforced odor series.

The second centrifugal signal occurs in the gamma band. During odor identification the olfactory bulb primary burst is often followed by a secondary burst at a slightly lower frequency (Fig. 3). We have found that this burst is less reliable than the beta band "preafference" signal, but when it does occur it is most often during odor recognition, always following the primary burst, and at 2/3 to 4/5 the frequency of the primary burst. Preliminary analysis using visual inspection and the adapted correlation technique indicates that this signal originates in the prepyriform cortex and is passed back to the olfactory bulb. We label this the "reafferent" signal, because of its relationship to the primary afferent burst.

DISCUSSION

The highly nonstationary nature of the EEG and interactions between brain regions during perceptual processing makes the results from many standard statistical methods ambiguous and unsatisfactory. In addition, the nonlinearity of relationships between measured signals in brain regions separated by one or even two synapses adds further difficulty to analysis of data recorded during perceptual periods. We have presented here several methods which have proved useful for analyzing these data.

Results show three significant events separated by stereotyped frequency relationships, relative times of occurrence, and direction and origin of apparent transmission.

Coherence of the primary, inwardly transmitted olfactory bulb burst throughout the olfactory tract and into the dentate gyrus, places the limbic system back into the functional rhinencephalon, as these bursts, which occur predominantly during odor identification, have already been shown to be perceptually defined events (Freeman & Baird, 1987). The preafferent signal in the lower frequency band, originating in the entorhinal cortex (or another region, from which we have not yet recorded, possibly the CA1 region of the hippocampus), places the limbic system at the center of intentional processing during olfactory perception. This signal may be seen as an expectation signal, which prepares the olfactory bulb for the desired stimulus.

The reafferent burst seen to follow many, but not all, primary bursts during odor identification may be unreliable precisely because of its function. The notion of the reafferent signal presented by von Holst and Mittelstädt places its function as an error correction mechanism. For this reason, we would only expect to see the signal when further adjustment to the dynamical environment of the olfactory system is necessary for either proper identification or proper behavior associated with the stimulus. Thus, if the animal identifies the odor on the first sniff, then there is no need for the adjustment signal.

We can now propose a "tuning hierarchy" for olfactory bulb activity segregated by the various frequency bands in the EEG. The first level of preparation of the bulb dynamics comes from the respiratory wave present in the olfactory bulb EEG, and verified here for the rat. We have discussed the role of this low frequency event elsewhere as a biasing signal, which raises the background excitation of the system to prepare it to enter a state change associated with the olfactory burst (Freeman, 1992; Freeman & Barrie, 1994; Kay *et al.*, 1995). This biasing is an external force, due to receptor input and the interaction of the periglomerular cells in the olfactory bulb (Freeman, 1992). This is most usefully viewed as generalized biasing (Kay *et al.*, 1995), which enables the system to enter any basin of attraction. An internal and more specific biasing signal could be the proposed preafference signal presented in this report. It is mostly likely driven by intention, by expectation, and arises in the limbic system. It should enable the system to enter the expected odor's attractor basin. In other words, it may prepare the olfactory system for arrival of the expected stimulus. Each of these biasing signals occurs only once or twice (in the case of the beta band signal) for each sampling period (respiratory cycle).

When the system undergoes a change to a learned perceptual and dynamical state, a spatial pattern of activity indicative of a particular odor arises in the olfactory bulb, and this pattern is seen in the high frequency gamma burst evident in the olfactory bulb EEG. We have noted previously (Bressler & Freeman, 1980) that the burst frequency is related to the respiratory frequency, such that about 5 cycles of the gamma burst are evident for each inhalation. This allows for a higher signal to noise ratio of the spatial pattern, giving a more coherent pattern of activity upon arrival in the more inward lying regions of the system. It is only this coherent activity that is seen by the other structures, as the rest is "laundered out" by the dispersion and delay lines connecting the structures (Freeman, 1994). This nearly periodic activity is not used to entrain the internal dynamics of the structure. The coherence within the olfactory bulb arises much more quickly than entrainment would provide and is described by "anomalous dispersion" due to its fast propagation consistent with a state change or a phase transition in a physical system (Freeman, 1994).

All of these phenomena together provide a mechanism for perceptual activity consistent with Tsuda's notion of "chaotic itinerancy," where a system transits through attractor basins and is not drawn almost irretrievably to a static state as the system encodes its own evolution through context and history (Tsuda, 1991). In this theory the transit is represented by a history-dependent, non-static order parameter. Thus the transit of the system is part of the perceptual activity, rather than a simple arrival at a predetermined and static point or limit-cycle attractor. This idea is consistent with the notion of the stream of thought presented by William James (James, 1890/1977) and allows for a flexible, fast, and reliable memory process. These ideas do not require that the system be provably chaotic, in fact they do not rely on chaos per se. They do require that the interconnected brain regions be viewed as interacting dynamical systems and even as parts of one large system, which must by necessity be studied in parts. This system remembers, interacts with its environment, and updates its context via expectation, which provides it with a rich and constantly changing behavioral and experiential repertoire.

ACKNOWLEDGMENTS

The authors thank Brian Burke, Vinod Menon, and Theoden Netoff for invaluable technical assistance and helpful discussions. This work was supported by grants to W. J. F. from the National Institute of Mental Health (MH06686) and Office of Naval Research (N00014-93-1-0938). L. K. was supported by ONR AASERT grant N00014-93-1-1951.

REFERENCES

Boeijinga, P. H. and Lopes da Silva, F. H., 1988, Differential distribution of beta and theta EEG activity in the entorhinal cortex of the cat, *Brain Res.* 448: 272-286.

Boeijinga, P. H. and Lopes da Silva, F. H., 1989, Modulations of EEG activity in the entorhinal cortex and forebrain olfactory areas during odour sampling, *Brain Res.* 478: 257-268.

Boudreau, J. C. and Freeman, W. J., 1963, Spectral analysis of electrical activity in the prepyriform cortex of cat, *Experimental Neur.* 8: 423-430.

Bragin, A., Jando, G., Nadasdy, Z., Hetke, J., Wise, K. and Buzsaki, G., 1993, Beta frequency (40-100 Hz) patterns in the hippocampus: modulation by theta activity, *Soc. for Neuroscience 23rd Annual Meeting*, Washington, D. C.

Bressler, S. L., 1984, Spatial organization of EEGs from olfactory bulb and cortex, *Electroencephalography and Clin. Neurophysiology* 57: 270-276.

Bressler, S. L., 1987a, Relation of olfactory bulb and cortex. I. Spatial variation of bulbocortical interdependence. *Brain Res.* 409: 285-293.

Bressler, S. L., 1987b, Relation of olfactory bulb and cortex. II. Model for driving of cortex by bulb, *Brain Res.* 409: 294-301.

Bressler, S. L. and Freeman, W. J., 1980, Frequency analysis of olfactory system EEG in cat, rabbit and rat, *Electroencephalography and Clin. Neurophysiology* 50: 19-24.

Brillinger, D. R., 1981, *Time Series: Data Analysis and Theory*. San Francisco, Holden-Day, Inc.

de Olmos, J., Hardy, H. and Heimer, L., 1978, The afferent connections of the main and the accessory olfactory bulb formations in the rat: an experimental HRP-study, *J. Comparative Neurology* 181: 213-244.

Eeckman, F. H. and Freeman, W. J., 1990, Correlations between unit firing and EEG in the rat olfactory system, *Brain Res.* 528: 238-244.

Freeman, W. J., 1972a, Measurement of oscillatory responses to electrical stimulation in olfactory bulb of cat, *J. Neurophysiology* 35: 762-779.

Freeman, W. J., 1972b, Depth recording of averaged evoked potential of olfactory bulb, *J. Neurophysiology* 35: 780-796.

Freeman, W. J., 1975, *Mass Action in the Nervous System*. New York, Academic Press.

Freeman, W. J., 1992, Tutorial on neurobiology: from single neurons to brain chaos, *Int. J. Bifurcation and Chaos* 2: 451-482.

Freeman, W. J., 1993, Valium, histamine, and neural networks, *Biological Psychiatry* 34: 1-2.

Freeman, W. J., 1994, Neural mechanisms underlying destabilization of cortex by sensory input, *Physica D* 75: 151-164.

Freeman, W. J. and Baird, B., 1987, Relation of olfactory EEG to behavior: Spatial analysis, *Behavioral Neuroscience* 101: 393-408.

Freeman, W. J. and Barrie, J. M., 1994, Chaotic oscillations and the genesis of meaning in cerebral cortex. In: *Temporal Coding in the Brain* (G. Buzsaki *et al.* eds.) Springer-Verlag, Berlin, pp. 13-37.

Freeman, W. J. and Viana Di Prisco, G., 1986, EEG spatial pattern differences with discriminated odors manifest chaotic and limit cycle attractors in olfactory bulb of rabbits. In: *Brain Theory*, G. A. Palm, ed. Springer-Verlag, Berlin, pp. 99-119.

Freeman, W. J., Viana Di Prisco, G., Davis, G. W. and Whitney, T. M., 1983, Conditioning of relative frequency of sniffing by rabbits to odors, *J. Comparative Psychology* 97: 12-23.

Grüsser, O. -J., (1993) Historical remarks on the ideas involved in the reafference principle. *Abstracts, Soc. for Neuroscience 23rd Annual Meeting*, Washington, D. C.

Heale, V. R. and Vanderwolf, C. H., (1994) Olfactory processing in the dentate gyrus of rats, *Soc. for Neuroscience 24th Annual Meeting*, Miami Beach.

Heimer, L., 1968, Synaptic distribution of centripetal and centrifugal nerve fibres in the olfactory system of the rat. An experimental anatomical study, *J. Anatomy* 103: 413-432.

Heimer, L., 1995, *The Human Brain and Spinal Cord: Functional Neuroanatomy and Dissection Guide*. New York, Springer-Verlag.

Helmholtz, H. V., 1910, *Treatise on Physiological Optics*. New York, Optical Society of America.

James, W. (1890/1977), The stream of thought. In: *The Writings of William James* (J. J. McDermott ed). The University of Chicago Press, Chicago, pp. 21-74.

Kay, L., Shimoide, K. and Freeman, W. J., 1995, Comparison of EEG Time Series from rat olfactory system with model composed of nonlinear coupled oscillators. International, *J. Bifurcation and Chaos,* 3:849-858.

Kay, L. M. and Freeman, W. J., 1994, Coherence of gamma oscillations throughout olfactory and limbic brain structures in rats, *Soc. for Neuroscience 24th Annual Meeting*, Miami Beach.

Lancaster, L. R. and Kay, L., 1995, A method to detect nonstationary commonalities between two time series, *CNS '95*, Monterey, CA.

Lancaster, L. R., Kay, L. and Freeman, W. J., 1995, A New Attractor Analysis Technique applied to Rat Olfactory System EEG, *World Cong. on Neural Networks - 1995*, Washington, D. C. .

Lopes da Silva, F. H., Witter, M. P., Boeijinga, P. H. and Lohman, A. H. M., 1990, Anatomic organization and physiology of the limbic cortex, *Physiological Rev.* 70: 453-511.

Macrides, F., Eichenbaum, H. B. and Forbes, W. B., 1982, Temporal relationship between sniffing and the limbic theta rhythm during odor discrimination reversal learning, *J. Neurosci.* 2: 1705-1717.

Rall, W. and Shepherd, G. M., 1968, Theoretical reconstruction of field potentials and dendrodendritic synaptic interactions in olfactory bulb, *J. Neurophysiology* 31: 884-915.

Schwerdtfeger, W. K., Buhl, E. H. and Germroth, P., 1990, Disynaptic olfactory input to the hippocampus mediated by stellate cells in the entorhinal cortex, *J. Comparative Neurology* 292: 163-177.

Sperry, R. W., 1950, Neural basis of the spontaneous optokinetic response produced by visual inversion, *J. Comp. Physiol.* 43: 482-489.

Tsuda, I., 1991, Chaotic itinerancy as a dynamical basis of hermeneutics in brain and mind, *World Futures* 32: 167-184.

Vanderwolf, C. H., 1992, Hippocampal activity, olfaction, and sniffing: an olfactory input to the dentate gyrus, *Brain Res.* 593: 197-208.

von Holst, E. and Mittelstädt, H., 1950, Das Reafferenzprinzip: Wechselwirkung zwischen Zentralnervensystem und Peripherie, *Naturwissenschaften* 37: 464-476.

Witter, M. P., Groenewegen, H. J., Lopes da Silva, F. H. and Lohman, A. H. M., 1989, Functional organization of the extrinsic and intrinsic circuitry of the parahippocampal region, *Progress in Neurobiology* 33: 161-253.

Investigation of Psychophysiological Phenomena by Processing and Pattern Analysis of Evoked Potentials

7

SINGLE SWEEP ANALYSIS OF EVOKED AND EVENT RELATED POTENTIALS

Sergio Cerutti,[1] Anna M. Bianchi,[2] and Diego Liberati[3]

[1] Department of Biomedical Engineering
Polytechnic University, Milano
[2] Laboratory of Biomedical Engineering
S. Raffaele Foundation, Milano
[3] CNR System Theory Centre
Department of Electronic Engineering
Polytechnic University, Milano

ABSTRACT

Different approaches of single sweep analysis of evoked and event related potentials are presented. The autoregressive with eXogenous input (ARX) model is described with different applications in the study of dynamical changes of the brain responses and in artifact removal. A study of the modifications induced in the EEG by a sensory stimulation via ARX and AR models is also described. Finally, the wavelet transform is employed for the reconstruction of the single evoked response.

INTRODUCTION

Evoked or event related potentials (EP or ERP), that are due to the selective responses of the central nervous system (CNS) to different tasks or stimuli, carry relevant pathophysiological information about the status of the whole complex motor or sensory system. In addition, the signal recording and standard analysis, which require simple instrumentation and processing, and the not invasivity of the procedure itself, make the method a powerful tool for neurophysiological studies and clinical applications.

However, the standard procedure of analysis, based on synchronized averaging, presents some disadvantages for the difficulty in verifying the required hypothesis for the signal to be processed. Among others, we may mention:

- the assumption of zero mean noise, that is mainly constituted by the background EEG signal and may present baseline drifts;
- the strict stationarity of the evoked response that cannot be reasonably maintained for an entire session of a few hundreds of stimuli or more.

Advances in Processing and Pattern Analysis of Biological Signals, Edited by Isak Gath and Gideon F. Inbar
Plenum Press, New York, 1996

Thus, the improvement in signal to noise ratio by a factor of , (where N is the number of averaged sweeps) is only theoretical, and the obtained average is only a general and generic information about "mean" response of the CNS, which do not take into account modification in the system during the recording session (due to modified attention, adaptation to the stimulus, learning of the task, etc.) and to variations in the superimposed noise.

All the information about the dynamics of the responses during the recording session are lost, while it can play a relevant role in understanding physiological or pathological mechanisms in the CNS as well as in the whole motor or sensory paths.

That is the principal reason why numerous research groups studied and proposed various methods for the evaluation of the single sweep ERP or EP.

In this paper, various and different single sweep analysis methods, proposed in literature, are briefly recalled. Then a single sweep filter, based on a parametric model, is described in detail with some examples of application in which an improvement of S/N ratio of the order of 15-20 dB is generally achieved on the single responses, thus allowing to enhance parameters on a single sweep basis. Finally a new approach based on wavelet transformation (WT) is presented, to achieve a decomposition of the single responses into orthonormal functions which have a direct dynamic representation in time-scale and in time frequency domains.

SINGLE-SWEEP ANALYSIS OF EVOKED AND EVENT RELATED POTENTIALS

Event-related brain potentials are of paramount importance in non-invasive investigation of the human behavior in performing a variety of externally-triggered or self-paced tasks. The simultaneous activity of many other cortical neurons less involved in the investigated process acts as an overwhelming source of unavoidable noise, corrupting the recordings in such a way to prohibit a direct analysis of them (Auerbach et al., 1977; Barlow; 1979; Gevins, 1984). The averaging procedure over a proper number of sweeps, conventionally used in order to enhance the signal-to-noise ratio, has been very useful in terms of identifying the mean characters of the brain response to various stimuli, but does not have in itself the capability of tracking the sweep-to-sweep dynamics of those components in the event-related potential that are peculiar of that one special happening of the investigated task. Accordingly, a research path has been followed in the past twenty years, in order to provide simple and efficient tools able to perform a single-sweep analysis, suitable to cope with research objectives focusing on the studies of higher human functions like memory, attention and learning.

McGillem and Aunon (1977) proposed the application of a time-invariant linear minimum-mean-squared-error filter on the basis of the average covariance estimate for signal and noise, in order to estimate brain potentials evoked by single stimulus. A time-varying multidimensional extension was developed by Yu and McGillem (1983). The pre-stimulus information was also taken into account by Westerkamp and Aunon (1987). A Kalman filter was proposed, with different approaches to the state space system modeling, by Bohlin (1977) by von Sprekelsen and Bromm (1988), by Liberati *et al.* (1991a), in order to model and estimate the time variant signal and noise interaction. The increasing in performances is paid in terms of computational complexity, in order to compute the statistical features of the data required to identify the parameters of the model: more simple time-invariant approach, based on a Wiener filter with estimation of EEG noise in the pre-stimulus (Cerutti *et al.* 1986) provides on the other side only partial results which necessitate of further processing.

The improvements in single-sweeps analysis of event related potentials allowed to describe (Mocks *et al.*, 1984) and to classify (Moser and Aunon, 1986) the trial-to-trial variability in the brain response to events of different nature.

A simple but efficient method, based on stochastic identification of the single-sweep variability around the average EEG signal, is the ARX filter whose hints and selected applications are reported in the following of the paper.

METHODS

ARX^n Model

In the upper branch of Fig. 1, the EEG signal e(k) is described as the output of a linear time invariant system driven by stochastic white noise w(k) with null mean value and variance σ^2(Isaksson *et al.*, 1981; Jansen *et al.*, 1981; Cerutti *et al.*, 1985). Least squares identification techniques, provide the transfer function 1/A(z) under the form of an all-pole structure, in the general frame of AR (AutoRegressive) model,

$$A(z) = 1 + \sum_{i=1}^{N} a_i z^{-i} \tag{1}$$

where $\{a_i\}$ are the NAR coefficients.

The model in Fig.1 describes the event-related potential y(k) as the sum of the background EEG activity e(k) and a deterministic component u(k) which represents the information related to the event. u(k) is obtained after filtering a reference signal u_r(k), through a more general recursive filter with a transfer function given by B(z)/A(z), where

$$B(z) = \sum_{j=0}^{M-1} b_j z^{-j} \tag{2}$$

The result of the parametric identification provides the determination of the u(k) signal (i.e. the filtered evoked response) on a sweep-to-sweep basis, by filtering the recorded y(k) signal: u_r(k) may be computed as an averaging of a proper number of single responses (Liberati *et al.*, 1989).

The model is called ARX (AutoRegressive with eXogenous input) being u_r the exogenous signal. The least squares identification algorithm is described in (Cerutti *et al.*, 1987; Cerutti *et al.*, 1988; Liberati *et al.*, 1989), together with the computation of Akaike's FPE and AIC figures of merit (Akaike, 1970) for the determination of the optimal order and the whiteness test on the prediction error (Anderson's test on its autocorrelation function).

An easy extension to ARX^n models can be made by considering up to n exogenous inputs which may account also for:

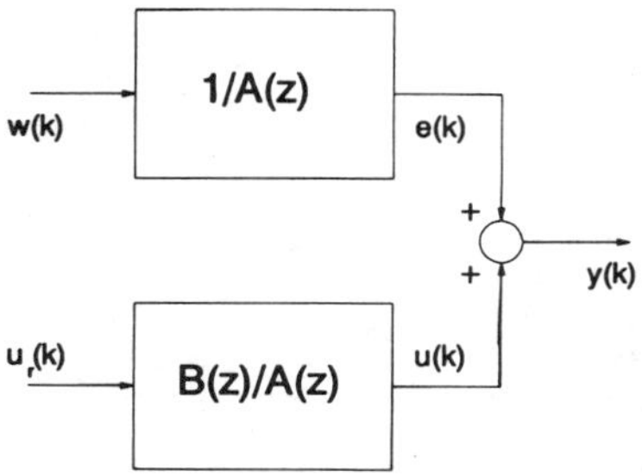

Figure 1. Block diagram of the ARX model for the analysis of evoked or event related potential.

- other source of disturbances and/or
- external noise contributions such as eye movements, blinking, other artifacts, etc.

Ellipses of Confidence

ARX^n model represents an efficient tool in determining the behavior of evoked or event related potentials on a single sweep basis. In fact, it describes the EEG as the output of an AR model, superimposed to the stimulus related activity, independently from the pre-stimulus conditions and provides reference exogenous signals which are supposed to interact in a deterministic way.

The signal is completely described by the model coefficients and by the variance σ^2 of the input white noise.

A frequency-domain representation of AR model is generally achieved through the following expression of power spectral density (PSD):

$$PSD = \frac{\sigma^2 \Delta t}{\left|1 + \sum_{i=1}^{N} a_i z^{-i}\right|^2_{z = \exp(j 2 \pi f \Delta t)}} \tag{3}$$

where Δt is the sampling period.

The former expression can be alternatively expressed as a function of the poles of the transfer function:

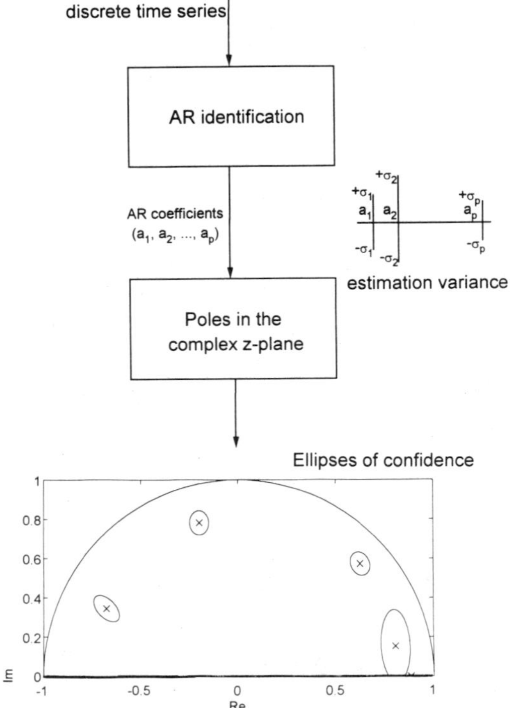

Figure 2. Scheme of the calculation procedure of the confidence ellipses starting from AR parameter identification.

$$PSD = \frac{\sigma^2 \Delta t}{\left| \prod_{l=1}^{N} (z - z_l) \right|^2_{z=\exp(j2\pi f \Delta t)}} \qquad (4)$$

where z_l are the zeroes of the denominator in expression (3). Each estimated pole z_l is a function of the model parameters ($z_l = z_l(a_1, a_2,..., a_p)$), which are estimated with a certain variance. Thus the pole coordinate estimation errors may be approximated by linear combination of the parameter estimation errors, and it is possible to calculate and represent confidence intervals of the poles (ellipses in the complex z-plane), in which the true position of each pole is contained with a given probability (Cerutti, 1989). Fig. 2 summarizes the calculation procedure of the confidence ellipses.

According to the previous definition of confidence ellipses, it is possible to state that, if two consecutive records of the signal lead to different AR models, whose poles are inside the same confidence intervals, they are considered, at a given probability, different realizations of the same process, as no statistical differences are evidenced between them. On the contrary, if at least one pole of the second record is external to the ellipse of the first one, the two records are thought to be the outputs of statistically different generating mechanisms.

The described procedure is employed in order to evaluate the modification induced by sensory stimulation in the EEG signal on a sweep by sweep basis. The analysis can be performed in two different conditions: before or after removing the stimulus related activity through the ARX filter in the post-stimulus record.

Wavelet Approach

Recently, in the study of single-sweep evoked potentials, the application of the wavelet transform was proposed by (Bartnik, 1992).

Wavelet transformation (WT) can be viewed as a signal decomposition into a set of basis functions. The basis functions, called *wavelets*, are obtained from a single prototype wavelet by dilation and contraction (scaling), as well as by time shifts. Therefore in a WT the terms of *scale* is introduced as an alternative to frequency, leading to a so-called time-scale representation.

The signal is thus mapped into a multiresolution time-scale plane: in fact, high frequency resolution is obtained at low frequencies, while high time resolution is obtained at high frequencies. For a complete theoretical description of the WT see (Rioul & Vetterli, 1991).

Fig.3 shows a schematic representation of the discrete WT algorithm: at each step, to achieve finer frequency resolution at lower frequencies, the low-pass filter halves the width of the low-band (increasing its frequency resolution by two), but, due to the undersampling by two, the time resolution is halved. At each iteration, the current high band portion corresponds to the difference between the previous low-band portion and the current one and is called the detail d_j of the signal, at a given resolution j, while A_j is obtained as the projection of the signal on the wavelet at scale j. In summary, a multiresolution analysis is achieved. Fig.4 shows the wavelet decomposition of the average of 100 somatosensory evoked potentials. The contour plot of the details power is represented and evidences the contribution of details 2 and 3 at the time corresponding to the occurrence of the principal waves in the signal (about 200 ms after the stimulus administration), and later contributions at lower frequencies (detail 2) that account for the ending section of the evoked activity.

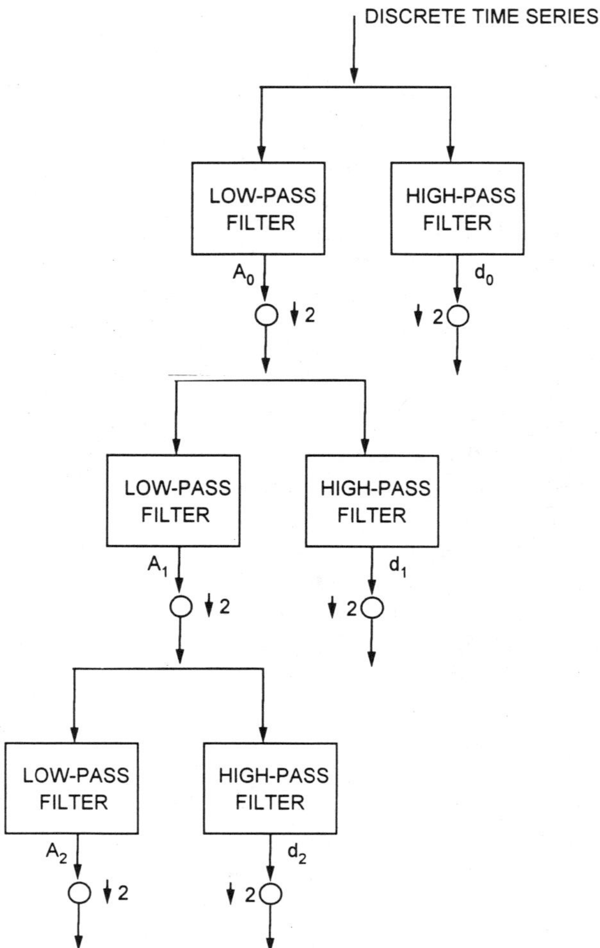

Figure 3. Block diagram of the wavelet transform algorithm.

EXPERIMENTAL RESULTS

A few applications of the previously described methods will be reported in the fields of artifact removal, single-trial event-related responses, photostress recovery, multisensorial evoked potentials, brain mapping.

EOG Artifact Removal

Artifact removal is an important goal in ERP processing (Gratton *et al.*, 1983): especially in cognitive tasks of long duration, eye movements and/or blinking are unavoidable, even if the subject is well trained on how to try to avoid them. The simultaneously recorded electrooculographic (EOG) signal may be used as exogenous input in the ARX model (Cerutti *et al.*, 1988): in fact, it may constitute an estimated template for the EOG noise affecting the ERP. The exogenous branch of the model introduced in Fig.1 thus identifies the B(z)/A(z) transfer function from the EOG to the actual recording. The

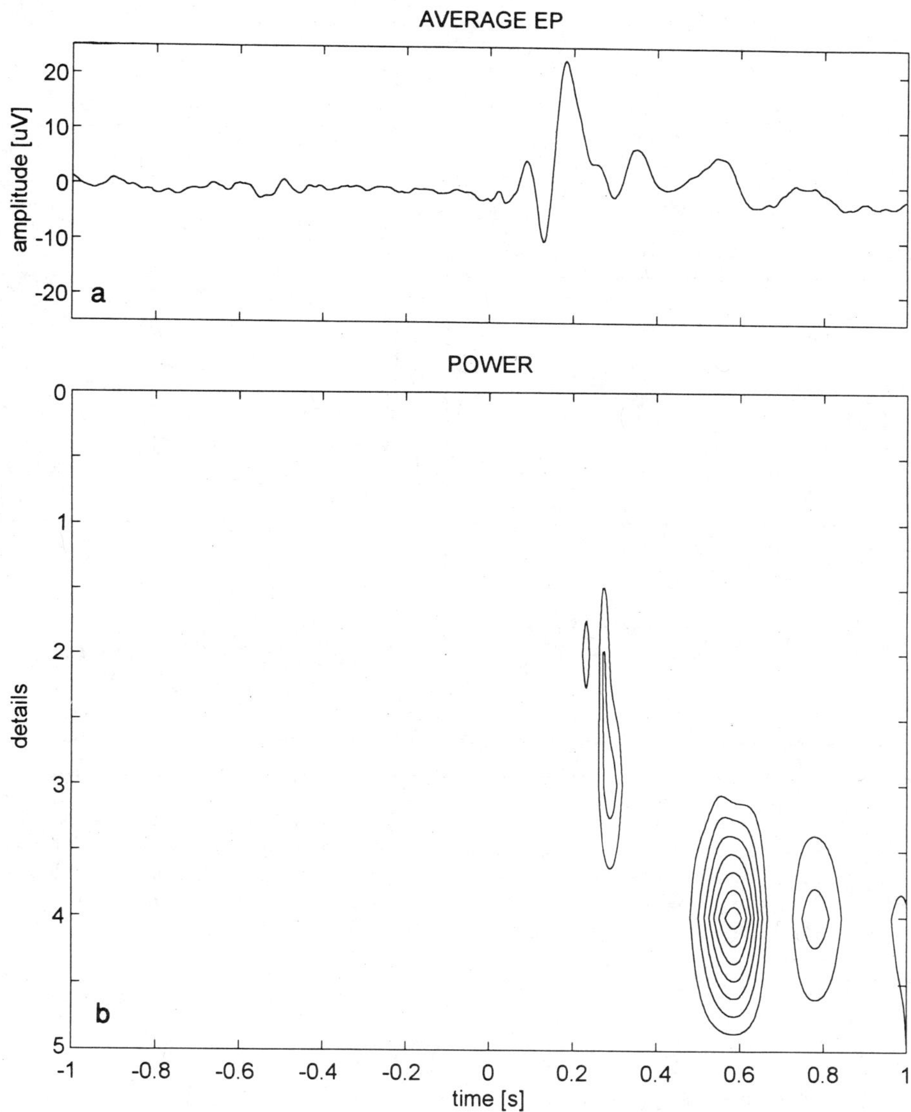

Figure 4. Average of 100 somatosensory evoked potentials. The stimulus time is 0; b) contour plot of the power in the time-scale plane. The time is represented on the horizontal axis, while the detail number is on the vertical one. The corresponding frequency ranges are: 0: 128 - 64 Hz; 1: 64 - 32 Hz; 2: 32 - 16 Hz; 3: 16 - 8 Hz; 4: 8 - 4 Hz; 5: 4 - 2 Hz.

reconstructed contribution of the EOG noise is removed from the recorded potential. Figure 5 illustrates an application in which two exogenous inputs are presented to an ARX^2 filter: one of the two exogenous inputs is the actual recording of the EOG (a), visibly affected by ocular artifacts; while the other one is the template of the searched movement related brain macropotential (MRBM), obtained via conventional averaging and not shown in the figure.

The input signals to the processing procedure are the recorded EOG signal (a) and the unfiltered single-trial MRBM (b) in a normal 6-year old child performing a task consisting in sequentially pushing two buttons (one in each hand), in such a way to start (left

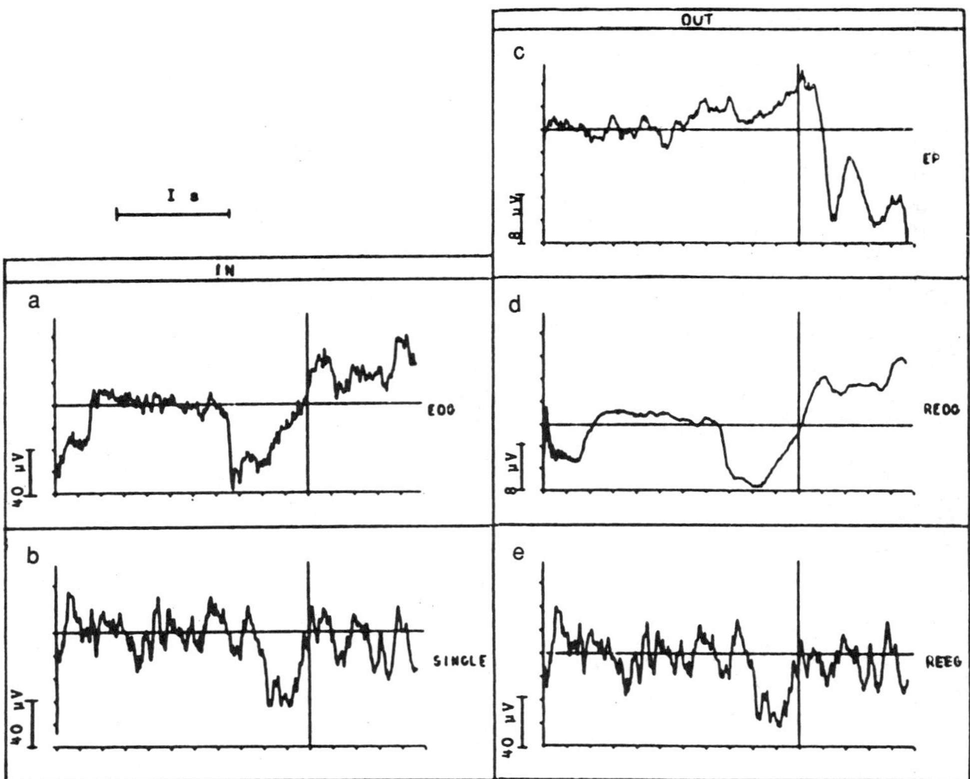

Figure 5. ARX^2 processing of MRBM: a) recorded EOG; b) recorded single sweep potential over the premotor cortex; c) filtered evoked potential extracted from b) via ARX^2 identification; d) residual EOG component extracted via ARX^2 from b); e) residual background EEG component present in b). The vertical bar in each picture marks the time of pressure of the left push button. The signals in the left column are inputs for the algorithm, while the three in the right column are the outputs.

push button) the left-to-right fast movement of a light point over a monitor, and then stop the target (right push button) still within the screen (meaning in less than 100 msec from the first push) (Chiarenza *et al.*, 1987).

Through the ARX^2 parametric identification, it is possible to enhance the single-trial response (c) after removing the background EEG noise (e) and also the EOG signal artifact (d). In this particular recording, the eye movement happens just before the beginning of the task (vertical bar). The recording is accordingly corrupted (a), but the filtered response is able to provide the complex polyphasic brain electrical activity related to the performance, i.e.:

- the negative (upward) Beretschaft potential (BP) before the movement onset, related to the mental preparation of the performance;
- the negative motor cortex potential (MCP) at some 30 msec latency, and the subsequent P200 of the visuo-motor process while executing the task;
- the skilled performance positivity (SPP), as a positive wave at latencies as 300 msec or more, reflecting the cognitive active evaluation of the task.

Trial-to-Trial Variability

Through the previously described algorithm, it is possible to study the dynamics of the attention and the learning capability of the subject through subsequent filtered responses as in Fig. 6:

- the negative (upwards in the figure) BP component before the movement onset (vertical line at time zero), related to the mental preparation to the task, disappears in the second trial from the top, when lack of attention results in a lowered response amplitude;
- the positive SPP potential related to the cognitive active evaluation of the task dynamically varyies along the subsequent trials with the learning and mental strategies of the subject.

A single-sweep analysis may detect possible dynamics in the response from the brain, quantitatively correlated with the scores recorded during the task (Chiarenza *et al.*, 1987; Cerutti *et al.*, 1988; Chiarenza *et al.*, 1994).

Macular Recovery Evaluation

The dynamic behavior of the evoked responses gives important information about the status of a sensory system when studying the evolution of the signal as a consequence of provocative tests. The time taken by the visual system for resetting the basal conditions after a photostress, is strictly related to the capability of restoring the retinal photopigment transformed by the light. In diabetic subjects some metabolic functions are altered; also, the retinal system may present some dysfunctions. An objective measure of the recovery time of the visual function after photostress, is determined by the evaluation of the visual evoked potentials. A quantification of the recovery is obtained by measuring the P_{100} wave amplitude on sweep-by-sweep basis and a recovery time was evaluated as the time taken by the visual system for coming back to the control condition.

In order to evaluate the degree of alteration in the macular recovery, the following protocol was assessed by Magni *et al.* (1994):

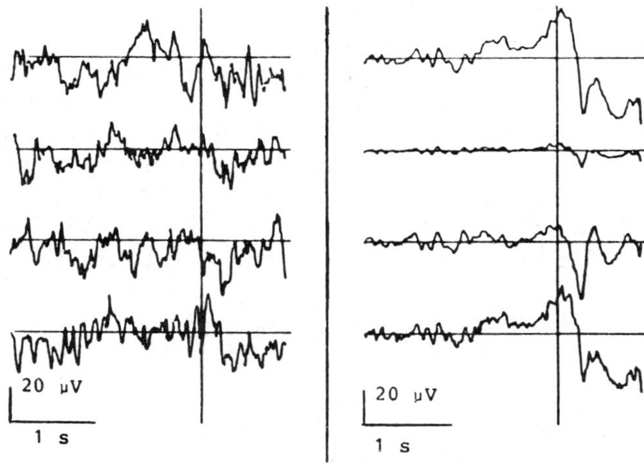

Figure 6. Four recorded (left column) and ARX^2 filtered (right) single consecutive MRBM's.

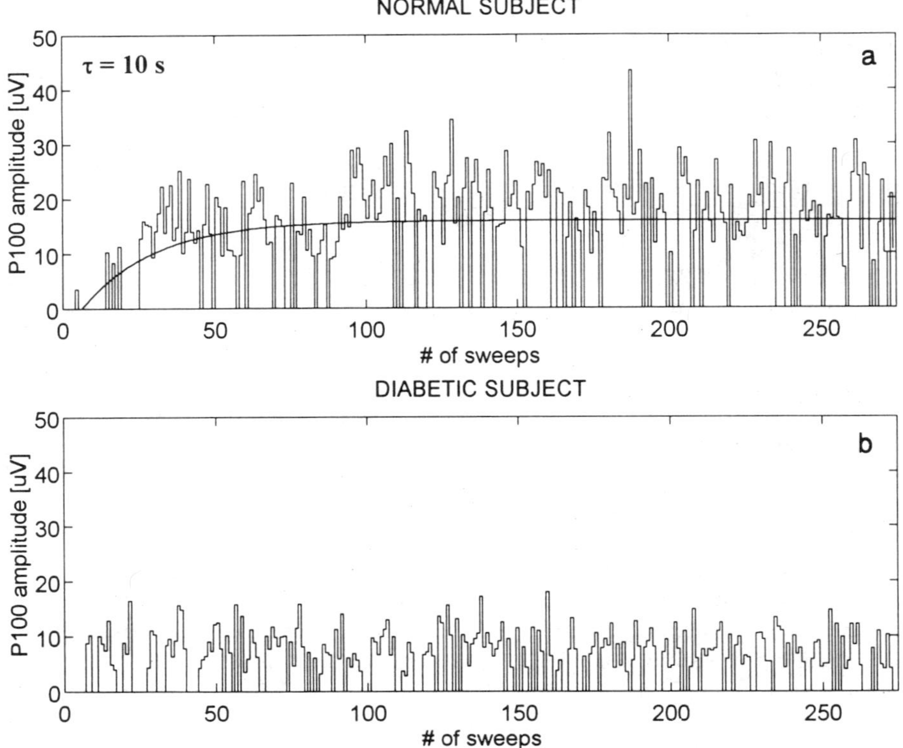

Figure 7. a) Single-sweep P_{100} amplitude sequence in a normal subject after photostress test. An exponential fitting is superimposed and a recovery time constant of 10 s is evaluated. b) Single-sweep P_{100} amplitude sequence in a diabetic subject.

- 1 min of pattern reversal stimulation with a stimulation frequency of 2.32 Hz, in order to assess the basal P_{100} amplitude.
- 2 sec of photostress induced by a luminous flash
- 2 min of pattern reversal stimulation with a stimulation frequency of 2.32 Hz.

The data were then analyzed off line and the sequence of the P_{100} was obtained for normal and diabetic subjects. Figure 7a shows the sequence of the P_{100} amplitudes after the photostress of a normal subject. An exponential curve was fitted to the sequence, and the related time constant t was calculated in order to quantify the recovery time.

In this study a group of 10 normal subjects were analyzed and a value of 9.73 ±2.83 s (mean ±SD) was found for the time constant, leading to a recovery time of about 30 s in accordance to previous data reported in the literature (Polizzi *et al.*, 1985). The P_{100} amplitude sequence for the case of a diabetic subject is plotted in Fig. 7b. In this case, it is not possible to fit a reliable curve to the data and an increased number of missing responses is measured in respect to the normal subjects. An altered recovery mechanism is thus evidenced in diabetic patients.

Multisensory Potential

The study of multisensory potentials, in which two or more sensorial stimuli are simultaneously delivered to the subject, also involves two exogenous inputs as reference signals.

The visual stimulus consists in a pattern reversal checker-board with 60 cd m^{-2} luminance, 100% contrast. The somatosensory electrical stimulus is a rectangular pulse to the right median nerve at the wrist, about 10 dB over sensation threshold. Figure 8b shows the average potential on 100 consecutive parietal responses to the simultaneous delivery of the described visual pattern reversal and somatosensory electrical pulse stimulation. Averaged potentials obtained with only one of the two stimuli are reported in (c) (somato-sensory) and (d) (visual), respectively. Their sum is displayed in (a) (thin line) with the bisensory potential (dark line). The same bisensory potential is shown in (b) for clarity. The difference in the two patterns suggests a non-linear effect when combining the two stimuli, resulting in not a pure superposition of the corresponding potentials. Shipley (1989) already suggested the terms of facilitation or defacilitation to describe such a phenomenon, clearly indicating the no-reductionistic nature of the brain even for such a simple experiment. Interactions between linear and non-linear mechanisms inside the brain are also described in Lopes da Silva *et al.* (1990).

The same ARX^2 model previously described is used to extract from the single-sweep bipotential the contribution related to the visual and somatosensory stimulus, respectively. The two averages shown in Fig. 8c and 8d are the exogenous inputs. Figure 9a shows a single occipital recording of the bisensory response, 9c and 9e are the exogenous inputs; the outputs

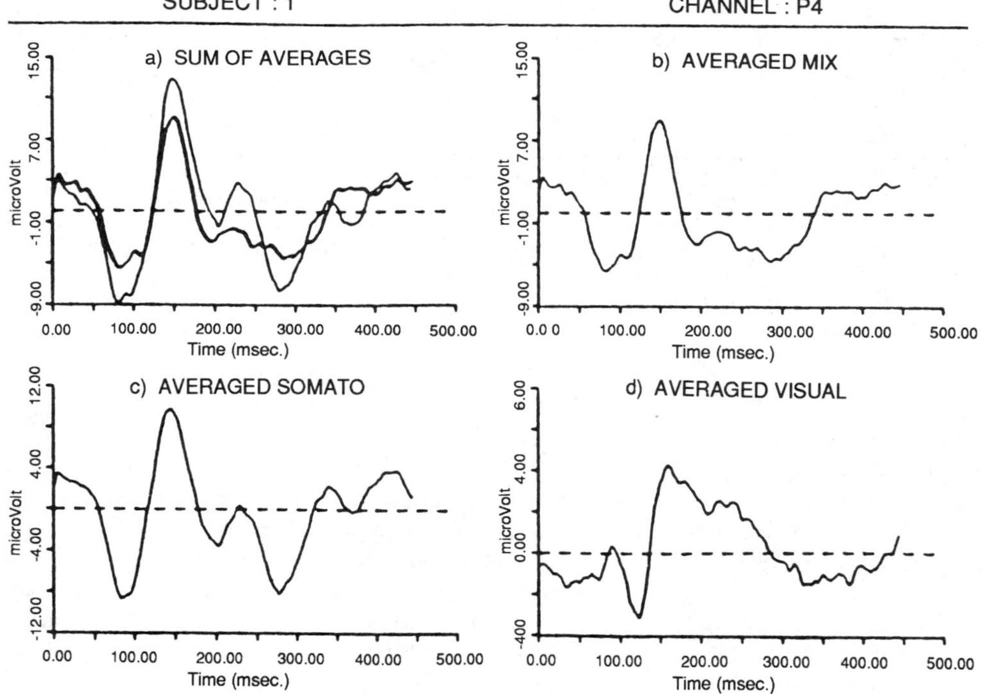

Figure 8. a) Comparison of the averaged evoked bipotential (dark line) with the sum (thin line) of the averaged evoked SEP and the average evoked VEP; b) averaged evoked bipotential; c) averaged evoked SEP; d) averaged evoked VEP.

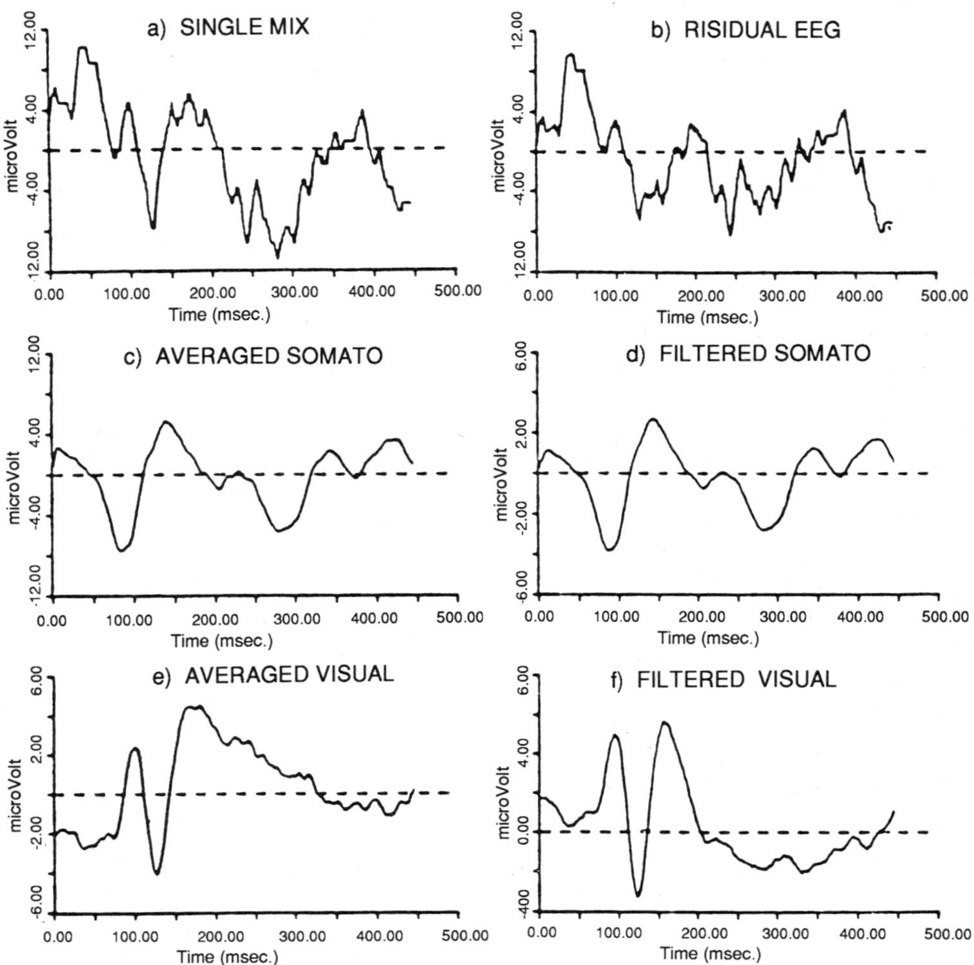

Figure 9. ARX filtering of a single-sweep bipotential a); b) residual background EEG after filtering; c) averaged somatosensory evoked potential (1st exogenous input); d) filtered somatosensory component extracted from a); e) averaged visual evoked potential (2nd exogenous input); f) filtered visual component extracted from a).

of the ARX^2 algorithm are the estimated somato-sensory component (Fig. 9d), the visual component (Fig. 9f) and the residual background EEG (Fig. 9b). As in Fig. 5, Fig. 9 shows input signals in the left column, while the output signals are shown in the right column.

The cross-correlation function indicates a correlation bigger than r = 0.9 between each filtered component and the corresponding average template. The correlation between the two templates is, instead, as low as r = 0.3, like the one between the two filtered components. The procedure thus separate the two contributions. In the visual component of the bipotential over the occipital region (Fig. 9f), the birth of a positive (downwards) wave at latencies around 200 msec, was not present in the purely visual potential (e). Such a relaitonship may be supposed to modulate the visual response when the somatosensory cortex is also active (Liberati *et al.*, 1991b). The spread of the somatosensory potential through the head (d) is instead analogous to the EOG contribution in Fig. 5.

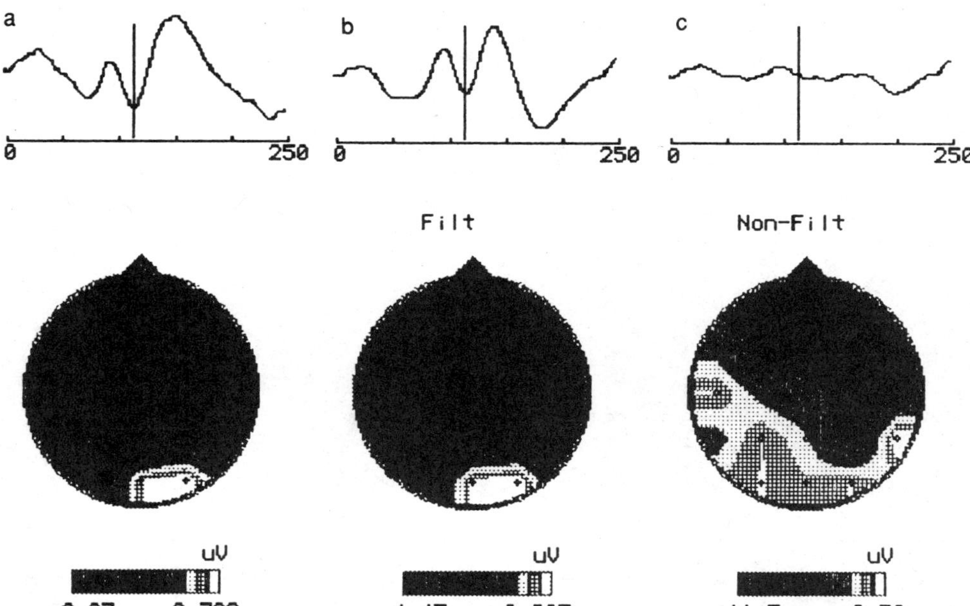

Figure 10. Brain topography mapping (bottom) and corresponding occipital trace for the conventional averaged potential (a), a single-sweep filtered (b) and unfiltered (c) potentials, at the latency of 112 msec for a visual stimulation.

Brain Mapping

The extension of the ARX^2 algorithm to the various channels of a multi-channel recording provides a filtered brain map. In our approach, 16 channels from the international 10-20 system are recorded in the positions represented by the little circles on the maps in Fig. 10. Filtering is performed on every channel via ARX method. Then interpolation depending upon the inverse of the distance from the four closest recordings is performed for each set of 16 samples (sampling frequency = 1 kHz). The same interpolation is used for single-sweep unfiltered data, as well as for conventional averages of 50 sweeps. Figure 10 is taken in correspondence with a latency characterizing the main peak of the visual response. The single-sweep filtered map (b) provides information of the same quality as for a traditional map obtained from the conventional average (a): the P100 wave is focused on the occipital region. The single sweep unfiltered map displayed in (c), is almost meaningless, for the spread potential due to the noise. The possibility to obtain sweep by sweep brain maps (Liberati *et al.*, 1992) opens an important insight for a fast monitoring of brain physiological functions in performing especially designed tasks. In brain mapping studies, the improvement of signal-to-noise ratio is essential in the post-processing of the map, as contour delimitation, determination of regions of interest, computation and positioning of equivalent electrical (or magnetic) dipoles, etc.

EEG Changes before and after Somato-Sensory Stimulation

A group of 10 normal subjects underwent a session of 100 somato sensory stimulations at the median nerve. The EEG was continuously recorded from the Cz lead and 1 s of pre-stimulus and 1 s of post-stimulus signal were analyzed. Figure 11a shows an EEG record

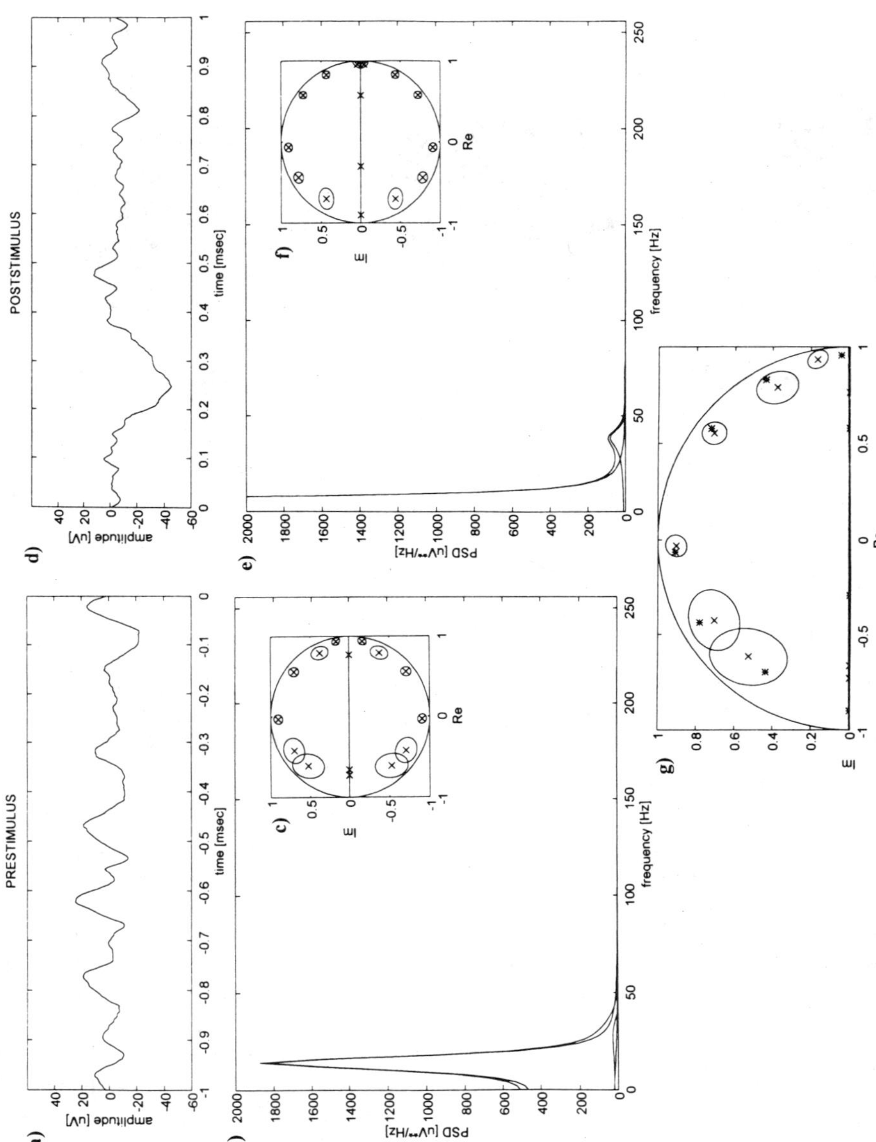

Figure 11. a) pre-stimulus EEG; b) AR PSD of the signal in a); c) AR pole diagram in the complex z-plane of the signal in a); d) post-stimulus EEG; e) AR PSD of the signal in d); f) AR pole diagram in the complex z-plane of the signal in d); g) comparison of the two signals generating models (see text).

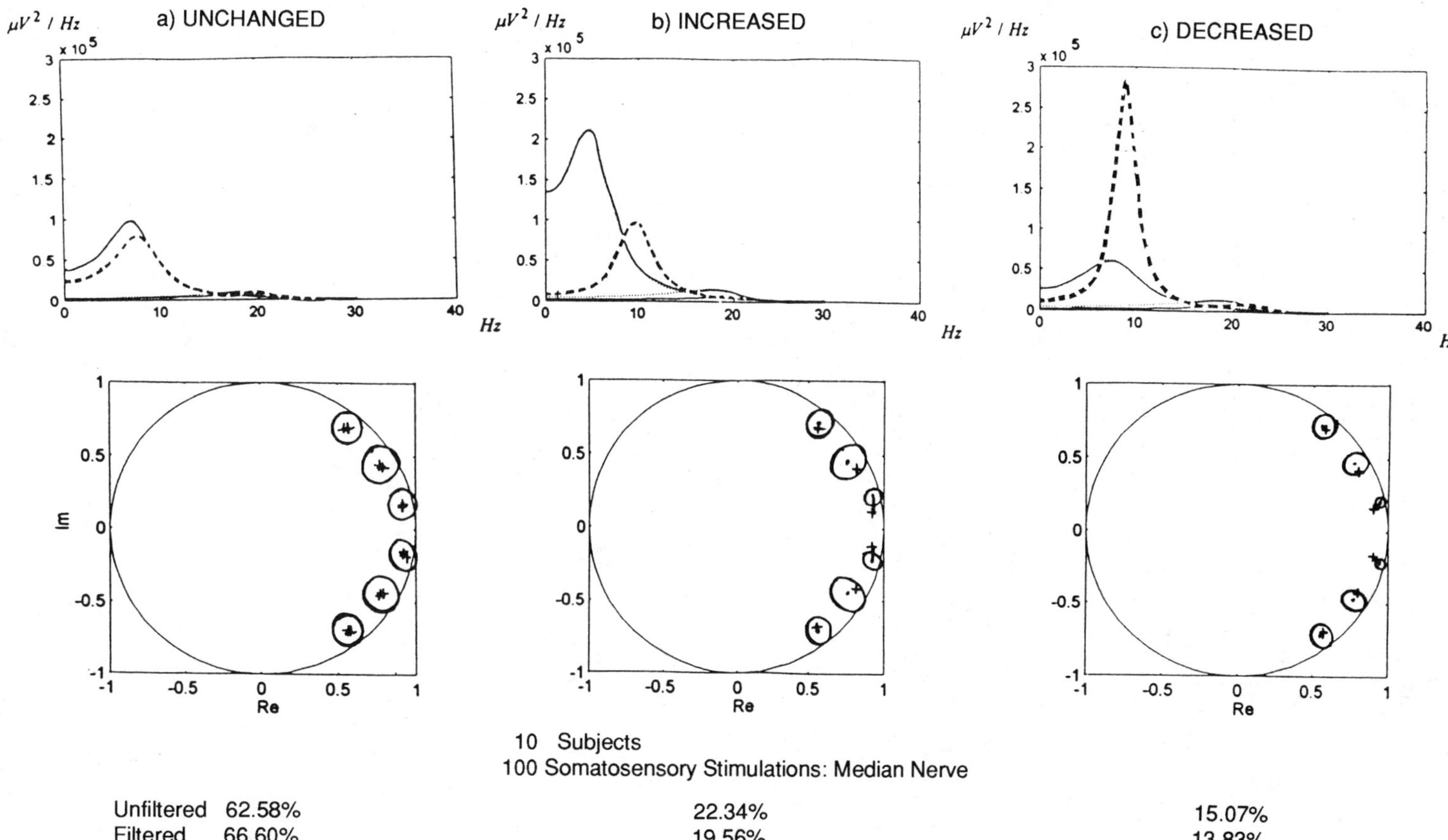

Figure 12. Modifications induced by the somatosensory stimulations: a) no modifications; b) increasing in a power; c) decreasing in a power. Continuous line: post-stimulus; dashed line: pre-stimulus.

of 1 s obtained in basal conditions, i.e. before the stimulation: the corresponding PSD, evaluated by means of AR identification, is in Fig. 11b. The spectrum is decomposed in the sum of spectral components, according to the procedure proposed by (Zetterberg, 1969) and the central frequency and the power associated to each component can be easily calculated through a residual integration algorithm (Baselli *et al.*, 1988). An alternate and equivalent representation of the frequency characteristics of the signal is plotted in Fig. 11c, where the poles, evidenced in eq.4, are plotted inside the unit circle in the z-transform complex plane. Figure 11d shows an EEG record obtained after the somatosensory stimulation, recorded in sequence to the record shown in Fig. 11a. The corresponding PSD and pole diagram are in Fig. 11e and f, respectively.

The comparison between signal generating models of the two records is summarized in Fig. 11g: the poles and the confidence ellipses related to the pre-stimulus record (x) are plotted and the poles related to the post-stimulus record (*) are superimposed. In case of real poles the ellipses degenerate in segments on the real axis. In the example shown, two post-stimulus poles are external to the pre-stimulus ellipses, indicating a significant statistical difference in the signal generating models.

For each subject of the examined group, and for each stimulus, the pre and post stimulus signal have been compared according to the procedure described above. In order to better evaluate the modifications induced in the EEG by the stimulus related activity, the post-stimulus signal was filtered through an ARX filter, and the evoked activity has been removed. In accordance with Fig. 11, the principal amount of variation was in the poles

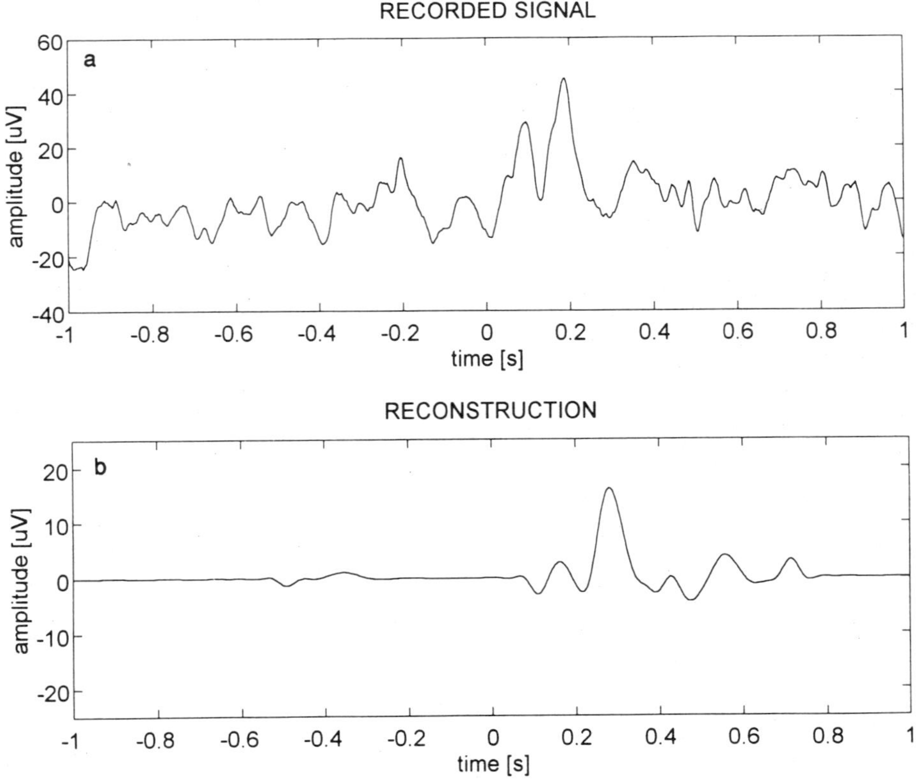

Figure 13. a) Single evoked response to a somatosensory stimulation and b) the same signal after the filtering procedure based on WT (see text).

corresponding to the α-rhythm. However in 62.58% of unfiltered cases and in 66.60% of filtered ones, no differences were evidenced between pre and post stimulus EEG. In the other cases 22.34% (unfiltered) and 19.56% (filtered) evidenced an increase in the power, while in 15.07% (unfiltered) and 13.83% (filtered) there is a decrease.

From a preliminary study of such results, the modifications induced in the signal by the sensory stimulation seem related to the pre-stimulus conditions of the EEG. A summary of these results is in Fig. 12.

Single Sweep Analysis by Means of Wavelet Transform

The single sweep analysis of single evoked potentials by means of the wavelet transform is based on the time-scale power distribution depicted in Fig. 7b. As the average evoked potential seems completely described by details 2 and 3 (corresponding to a frequency range of 32-8 Hz), these details can be employed for the reconstruction of the single evoked potential. In order to well localize in time the evoked response, the details of the single sweeps are set to zero except for the time intervals in which the power of the average overcomes its mean value +3 standards deviations (evaluated in the post-stimulus time). Figure 13a and b shows a single evoked potential before and after the WT filtering. The filtered signal well evidences the wave typical of the somatosensory evoked potential (see Fig. 4a for a comparison with the corresponding average).

Besides the single evoked response, the method allows to obtain a time-frequency (or better, a time scale) representation of the signal, that is impossible to achieve through traditional frequency analysis for the high level of non stationarity of the signal.

CONCLUSION

Single-sweep or single-trial analysis is capable to track the dynamics of the responses inside the protocol of event-related or evoked potentials.

In most of the cases, a linear model of signal-noise interaction is required: the verification of the fitting of the model (as well as its complexity) to the real case has to be verified. Such a "black-box" model of signal processing procedure may be of help in the definition of physiological models (genesis and propagation of the information in the brain, interaction between stimuli of different sources and between stimulus and noises superimposed, studying of the adaptation mechanisms of the brain, etc.).

Various clinical applications are interested in the development of this approach such as for the diagnosis of central and peripheral neurological disorders, the monitoring of anesthesia level as well as of drug infusion effect, the studying of the variability of low amplitude and latency activities, possibly connected to cognitive processes and the extraction of other quantitative parameters which characterize the background EEG activity and the evoked or event-related responses.

REFERENCES

Akaike, H., 1970,Statistical predictor identification, *Ann. Inst. Stat. Math.* 22:203-247.

Auerbach, V. H., Baird, H. W., and, Grover, W., 1977, The clinical EEG - A search for a buried message, *IEEE Trans. BME* 24(4):399-402.

Baselli, G., Cerutti, S., Civardi, S., Malliani, A., Orsi, G., Pagani, M., and Rizzo, G., 1988, Parameter extraction from heart rate and blood pressure variability signals in dogs for the validation of a physiological model, *Comp. Biol. Med.* 18 1:1-16.

Barlow, J.S., 1979, Computerized clinical electroencephalography in perspective, *IEEE Trans. BME* 26:377-391

Bartnik, E. A., Blinowska, K. J., and Durka, P. J., 1992, Single evoked potential reconstruction by means of wavelet transform, *Biol. Cybern.* 67:175-181.

Cerutti, S., Liberati, D., and Mascellani, P., 1985, Parameter extraction in EEG processing during riskful neurosurgical operations, *Signal Processing* 9:25-35.

Cerutti, S., Bersani, V., Carrara, A., and Liberati, D., 1986, Analysis of visual evoked potentials through Wiener filtering applied to a small number of sweeps, *J. Biomed. Eng.* 9:3-12.

Cerutti, S., Baselli, G., and Liberati, D., 1987, Single sweep analysis of visual evoked potentials through a model of parametric identification, *Biol. Cybern.* 56:11-120.

Cerutti, S., Chiarenza, G., Liberati, D., Mascellani, P., Pavesi, G., 1988, A parametric method of identification of single-trial event-related potentials in the brain, *IEEE Trans. Biomed. Eng.* 35(9):701-711.

Cerutti, S., Bianchi, A., Baselli, G., Civardi, S., Guzzetti, S., Malliani, A., Pagani, A., and Pagani, M., 1989, Compressed spectral array for the analysis of 24-h heart rate variability signal: enhancement and data reduction, *Comp. and Biomed. Res.* 22:424-441.

Chiarenza, G., Cerutti, S., Liberati, D., Mascellani, P., Pavesi, G., 1987, Autoregressive-exogenous filters for single-trial analysis of movement-related brain macropotentials in children, *EEG and Clin. Neurophys.* S40:8-12.

Chiarenza, G.A., Cerutti, S., Liberati, D., 1994, Analysis of single trial movement-related brain macropotential, *Int. J. Psychophysiology* 16:163-174.

Gevins, A.S., 1984, Analysis of the electromagnetic signals of the human brain: milestones, obstacles, and goals, *IEEE Trans. Biomed. Eng.* 31:833-850.

Gratton, G., Coles, M. G. H., and Donchin, E. A., 1983, A new method for off-line removal of ocular artifacts, *EEG Clin. Neurophys.* 55:468-484.

Isaksson, A., Wennberg, A, and Zetterberg, L. H., 1981, Computer analysis of EEG signals with parametric models, *Proc. IEEE* 69:451-461.

Jansen, B. H., Bourne, J. R., and Ward J. W., 1981, Autoregressive estimation of short segment spectra for computerized EEG analysis, *IEEE Trans. Biomed. Eng.* 28:630-638.

Kay, S. M., and Marple, S. L., 1981, Spectrum analysis: a modern perspective, *Proc. IEEE* 69(11):1380-1419.

Liberati, D., Cerutti, S., DiPonzio, E., Ventimiglia, V., and Zaninelli L., 1989, The implementation of an autoregressive model with exogenous input in a single sweep visual evoked potential analysis, *J. Biomed. Eng.* 11:285-282.

Liberati, D., Bertolini, L., and Colombo, D., 1991a, Parametric method for the detection of inter- and intra-sweep variability in VEP processing, *Med. & Biol. Eng. & Comput.* 29:159-166.

Liberati, D., Bedarida, L., Brandazza, P., and Cerutti, S., 1991b, A model for the cortico-cortical neural interaction in multisensory evoked potentials, *IEEE Trans. Biomed. Eng.* 38 (9): 879-890.

Liberati, D., DiCorrado, S., and Mandelli S., 1992, Topographic mapping of single-sweep evoked potentials in the brain, *IEEE Trans. Biomed. Eng.* 39(9):943-951.

Lopes da Silva, F. H., Pijn, J. P. M., and Boerijinga P.:, 1990, Interdependence of EEG signals: linear vs. non-linear associations and the significance of time delays and phase shifts. *Brain Topography*, vol. 2, 1/2: 9-18, 1990.

Magni, R., Giunti, S., Bianchi, A., Reni, G., Bandello, F., Durante, A., Cerutti, S., and Brancato, R., 1994, Single sweep analysis using an autoregressive with exogenous input (ARX) model, *Doc. Ophtal* 86:95-104.

McGillem, C. D., and Aunon, J. I., 1977, Measurement of signal components in single visually evoked brain potentials, *IEEE Trans. Biomed. Eng.*, 24:232-241.

Mocks, J., Gasser, T., and Dinh Tuan, P., 1984, Variability of single visual evoked potentials evaluated by two statistical tests, *Electroenphalogr. & Clin. Neurophysiol.* 57:571-580.

Moser, J. M. and Aunon, J. I., 1986, Classification and detection of single evoked brain potentials using time-frequency amplitude features, *IEEE Trans. Biomed. Eng.* 33:1096-1106.

Polizzi, A., Grillo, N., Corallo, G., Rovida, S., Cataldi, L., and Zingirian, 1995, M.: Macular function testing the screening of diabetic retinopathy *Boll. Ocul.* 64:997.

Rioul,O., and Vetterli, M., 1991, Wavelets and signal processing, *IEEE SP Mag.* Oct.:14-38.

Shipley, T., 1989, Intersensory evoked brain potentials and intersensory psychophysics: a critical overview, *IEEE EMBS 11th Int. Conf.*.

von Spreckelsen, M. and Bromm, B., 1988, Estimation of single-evoked cerebral potentials by means of parametric modeling and Kalman filtering, *IEEE Trans. Biomed. Eng.* 35:691-700.

Westerkamp, J. J. and Aunon, J. I., 1987, Optimum multielectrode a posteriori estimates of single-response evoked potentials, *IEEE Trans. Biomed. Eng.* 34:13-22.

Yu, K. B. and McGillem, C. D., 1983, Optimum filters for estimating evoked potential waveforms, *IEEE Trans. Biomed. Eng.* 30:730-737.

Zetterberg, L.H., 1979, Estimation parameters for the linear difference equation with application to EEG analysis, *Math. Bioscience* 5:227-275.

8

SPATIO-TEMPORAL SOURCE ESTIMATION OF EVOKED POTENTIALS BY WAVELET-TYPE DECOMPOSITION

Wavelet-Type Source Estimation of EPs

Amir B. Geva,[1,2] Hillel Pratt,[1] and Yehoshua Y. Zeevi[2]

Technion - Israel Institute of Technology
[1] Evoked Potentials Laboratory, Behavioral Biology, Gutwirth Bldg.
[2] Departments of Electrical and Biomedical Engineering
Haifa 32000, Israel

ABSTRACT

Scalp recording of electrical events allows evaluation of human cerebral function, but contributions of the specific brain structures generating the recorded activity are ambiguous. This problem is ill-posed and cannot be solved without auxiliary physiological knowledge about the spatio-temporal characteristics of the generators' activity. The widely-used model to describe the evoked potentials' sources is a set of current dipoles. It does not include a temporal model of source activity and does not propose a solution of the number of sources that are active simultaneously nor how to differentiate their contributions.

In our multichannel wavelet-type decomposition, scalp recorded signals are decomposed into a combination of physiologically-based wavelets. The coherent activity of a population of neurons may be derived by convolving a single cell's electrical contribution with the population's Gaussian temporal distribution of activity. Thus, we chose the Hermite Functions (derived from the Gaussian function to form mono-, bi- and tri-phasic waveforms) as the mathematical model to describe the temporal pattern of mass neural activity.

For each wavelet we solve the inverse problem for two symmerically positioned and oriented dipoles, one of which attains zero magnitude when a single source is more suitable. We use the wavelet to model the temporal activity pattern of the symmetrical dipoles. By this we reduce the dimension of inverse problem and find a plausible solution. Once the number and the initial parameters of the sources are given, we can apply multiple source estimation to correct the solution for generators with overlapping activity.

Application of the procedure to subcortical and cortical components of short-latency visual evoked potentials (SVEP) in response to high-intensity, strobe flashes, demonstrates its feasibility.

Advances in Processing and Pattern Analysis of Biological Signals, Edited by Isak Gath and Gideon F. Inbar
Plenum Press, New York, 1996

INTRODUCTION

Evoked potentials (EPs) reflect the macroscopic neural mass activity of cell assemblies located in specific regions of the brain. When a number of brain structures are simultaneously active, their potentials superimpose on the scalp (Nunez, 1981). An EP signal, generated by the superimposition of a number of simultaneously active sources, may result in an ambiguous input-output relationship insofar as it's generators' identification is concerned. This ambiguity constitutes the major obstacle to studying brain processes by surface recorded electrical signals (Achim *et al.*, 1991). This so-called 'inverse problem' is ill-posed and cannot be solved without auxiliary physiological knowledge about the spatio-temporal characteristics of the generators' activity.

The scalp recorded signal reflects a linear sum of currents conducted from multiple neural generators to the recording electrodes. The recorded signals can be decomposed into sums of basis functions and associated scalp distributions which suggest possible generators (McGillem & Aunon, 1987; Regan, 1989). The most widely used decompositions employ fixed basis functions such as sine and cosine waveforms (e.g., Fourier transform). Such basis functions have no physiological foundation, are global by their very nature, and are thus inappropriate for inferring contributions from specific localized neural generators. Basis functions that are matched in some sense to the specific signals which are under consideration (e.g., principal component analysis; McGillem & Aunon, 1987; Regan, 1989) are more appropriate for transient phenomena such as EPs, but they are mathematically derived with no physiological context. Whereas both principal component analysis and Fourier transform calculate basis functions which are orthogonal, the contributions of neural generators to surface activity derive from the same physiological processes, and are therefore not orthogonal.

Further, EPs by their very nature are transient and therefore require the application of appropriate techniques. Scherg and Von Cramon (1985) were the first to choose monophasic and biphasic transient waveshapes to model the generators' temporal activity of brain stem EPs. We have generalized and formalized this approach to select the transient waveforms by means of a model-based wavelet-type analysis (Geva *et al.*, 1993). In our wavelet-type analysis, **a prototype function ("mother wavelet")** is modified by dilation (duration) and translation (shift in time) to create a set of temporally well-localized waveforms. This set is then used as a model to represent the analyzed EPs signals (Daubechies, 1988; Mallat, 1989; Rioul & Duhamel, 1992).

In this report we present a model-based wavelet-type decomposition of multichannel scalp recorded electrical activity, based on the spatio-temporal properties of neural mass activity. We use the monophasic and biphasic Hermite families of wavelets as the transient temporal activity pattern of the sources in the inverse problem solution. By this we propose a **general** computational tool that uses both the spatial and the **transient temporal information** for the solution of the source estimation inverse problem.

METHODS

Prototype Functions ("Mother Wavelet") Selection

The shapes of scalp- and intracranially-recorded potentials suggest that the Hermite set of functions may be an appropriate model to describe the temporal template for our EPs wavelet analysis. Hermite functions are derived from the Gaussian function; the first Hermite function is Monophasic, the second is Biphasic, the third is Triphasic and so on (Fig. 1). Similarly, extracellular recording of action potentials may assume a biphasic or triphasic waveform, depending on the relative positions of the recording electrode and the active cell (Plonsey & Fleming, 1969).

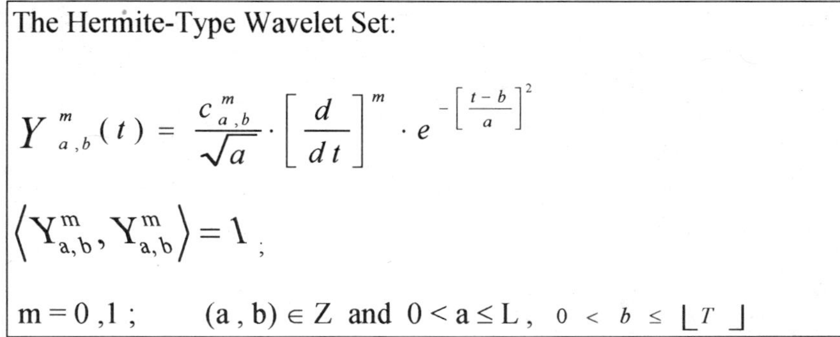

The Hermite-Type Wavelet Set:

$$Y^{m}_{a,b}(t) = \frac{c^{m}_{a,b}}{\sqrt{a}} \cdot \left[\frac{d}{dt}\right]^{m} \cdot e^{-\left[\frac{t-b}{a}\right]^{2}}$$

$$\left\langle Y^{m}_{a,b}, Y^{m}_{a,b}\right\rangle = 1 \; ;$$

$$m = 0, 1 ; \qquad (a, b) \in Z \text{ and } 0 < a \le L, \; 0 < b \le \lfloor T \rfloor$$

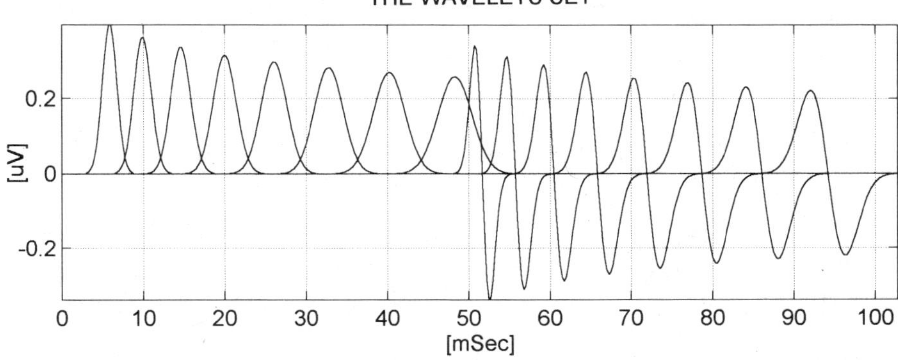

Figure 1. Representative monophasic and biphasic Hermite functions which constitute a wavelet set with representative dilations (durations) and translations (shifts in time).

Surface activity is generated by large populations of neurones which are synchronously active within a limited time period. Statistically, each such population is expected to display a Gaussian temporal distribution of the activity of its constituent neurones. The surface activity is the result of the convolution of each neuron's activity (Fig. 2) with such a Gaussian temporal distribution of neuronal activity within the population. The result of such convolution of a single neuron biphasic activity (compound action potential) and narrow Gaussian temporal distribution (high degree of synchrony within the neural populations activity) can be approximated by a biphasic Hermite function. This biphasic wavelet appears to be particularly appropriate for the short latency whole-nerve potentials (Fig. 3). The result of such convolution of single neuron biphasic activity and wide Gaussian temporal distribution (low degree of synchrony within the neural population activity) can be approximated by a monophasic Hermite function. The monophasic wavelet appears appropriate for the higher levels of sensory pathways and long latency evoked potentials (Fig. 4).

Model-Based Multichannel Wavelet-Type Decomposition

The method is separated into **temporal** and **spatial** stages. In the first stage the signals measured on the scalp are decomposed into a set of wavelets, and in the second stage the sources of each wavelet are estimated. The temporal activity of a given current dipole with fixed orientation and location in a quasi-static volume conductor is reflected by the same wavelet at all locations on the scalp. Only the amplitude of the wavelet will vary across sites on the scalp, depending on the dipole orientation intracranially. Our model-based wavelet-

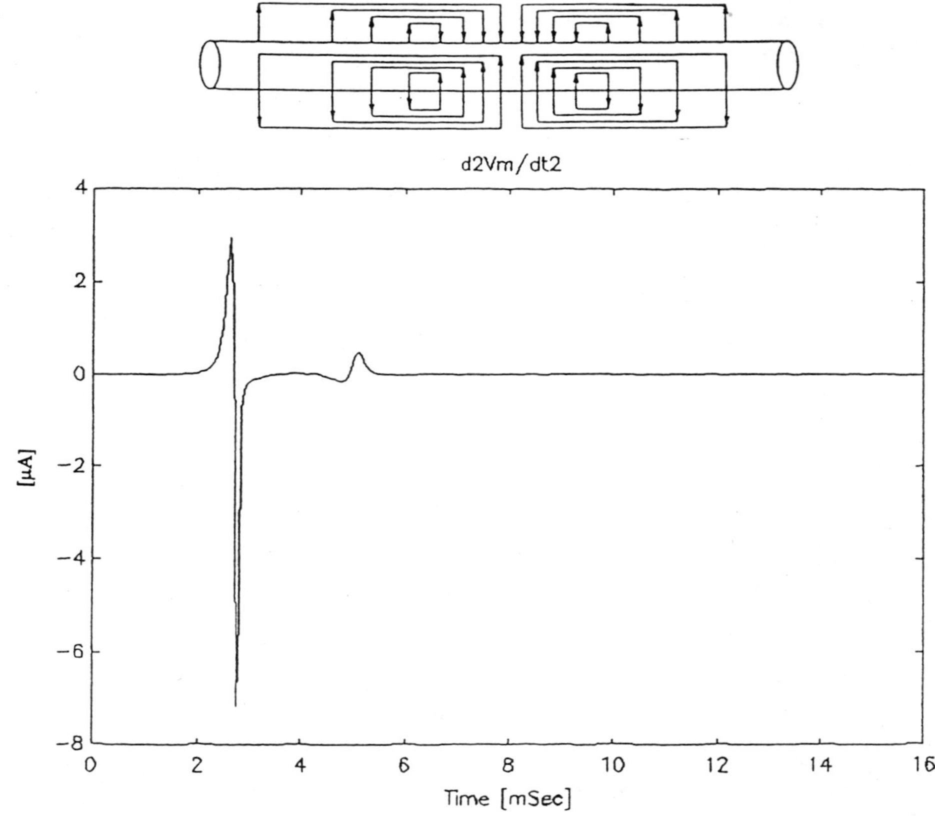

Figure 2. Local circuit membrane current - propagating action potential.

type analysis begins with selection, from the wavelet set, of the wavelet which best estimates the temporal pattern of the generator's activity. This wavelet is selected, from a **multichannel record**, according to the electrode at which the wavelet has the best-match with the signal (Appendix 1). The electrode to which the current dipole is pointing will be the optimal one for recording that dipole's activity (Nunez, 1981; Fender, 1987). Increasing the number of recording electrodes improves the possibility of accurately locating the optimal site for wavelet estimation, and thus better estimation of the temporal activity pattern. By using this optimal **model-based** estimation of the temporal activity pattern, we can **separate overlapping sources with different temporal patterns of activity**. Viewed differently, the wavelet-type decomposition performs a spatio-temporal **filtering** of the EPs signals, such that each wavelet should, ideally, reflect the activity of a generator (Geva *et al.*, 1993).

Spatio-Temporal Multiple Source Estimation by Wavelet-Type Decomposition

We use the physical model of a set of current dipoles with fixed orientations and locations in a quasi-static volume conductor (Mosher *et al.*, 1992). Each fixed dipole can describe the neural mass activity in a small volume of the brain tissue, and the entire brain's electrical activity can be described by a set of such fixed dipoles. Although we use a spherically symmetrical homogeneous volume conductor (Fender, 1987), the procedure can be applied to any head shape and conductivity model, including a finite elements model.

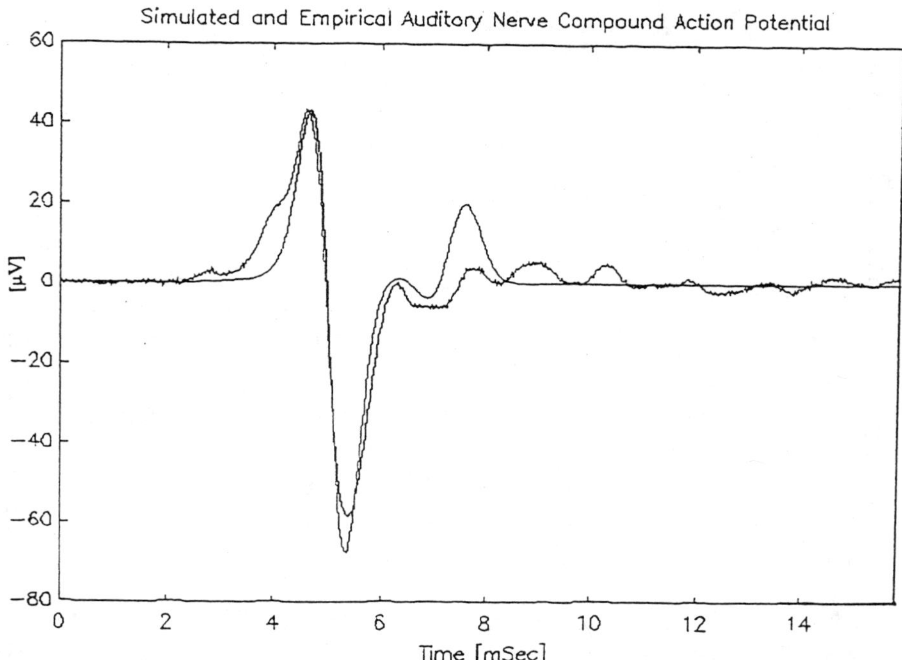

Figure 3. Simulated (by convolution of neuron waveform and Gaussian temporal distribution) and empirical (in human brain surgery) Auditory Nerv Compound Action Potential.

In addition to these conventional models we use the **wavelet-type** set as the physiological **model** of the **temporal activity pattern** of the generators.

Because identical temporal activity in symetrical generators cannot be separated by temporal decomposition, for each wavelet we solve the inverse problem for two dipoles with symetrical positions and orientations about the midsaggital plane. The magnitude of the symetrical dipoles need not be the same, and, when a single source is more appropriate, one of these symetrical **dipoles** will attain zero magnitude (see Appendix 2). We use the wavelet as the temporal activity pattern of the two symmetrical dipoles. By this we use the entire sequence of data points within a wavelet to solve **the inverse problem**. Thus, we increase the number of equations available for the solution of the given number of dipole parameters leading to a physiologically feasible solution. The residual signals that are not reconstructed by the first two symmetrical dipoles are subjected to the same process, selecting a second wavelet. The process is repeated until the energy of the residual signal becomes insignificant (smaller than 5% of the original signal's energy) or until the residual noise cannot be matched by the wavelet set and no longer decreases. Naturally, we try to find the **minimum number of sources** that can represent our EPs according to some predetermined criteria. The analysis is hierarchical, or pyramidal, in nature, whereby the wavelets and their corresponding sources are estimated in the decreasing order of their contribution to signals. By this approach we estimate both the temporal activity pattern and the **number** of generators. Estimating the

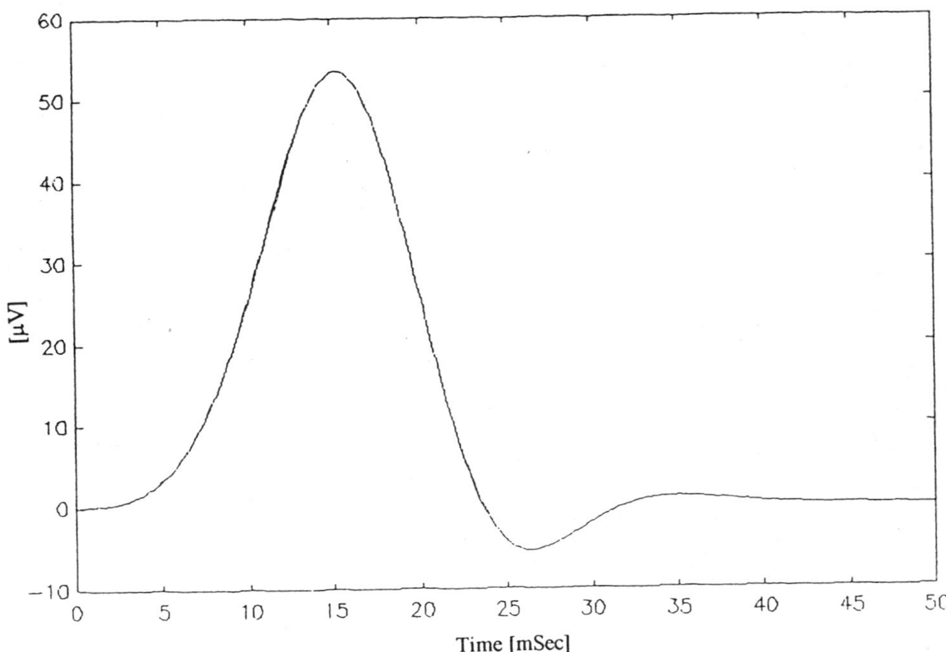

Figure 4. Simulated Compound Action Potential based on summed action potentials with a wide Gaussian distribution.

number of generators is considered one of the major problems in source estimation analysis (Mosher *et al.*, 1992).

After estimating the spatio-temporal parameters of the sources one by one, we can apply multiple source estimation to correct the solution for overlapping activity of generators. Once the number and the initial parameters of the sources is given, multiple source estimation becomes a soluble problem (Mosher *et al.*, 1992).

The next five steps summarize the algorithm for spatio-temporal multiple source estimation by wavelet-type decomposition:

(1) Find the wavelet best-matched, by cross-correlation, to all scalp signals.
(2) Using this wavelet as a model of the temporal pattern of the generator activity, solve the source estimation problem for two symmetrical dipoles by linear and non-linear optimization methods.
(3) Calculate the residual signals not represented by the last two symmetrical dipoles.
(4) If the energy of the residual signal is larger than some criterion, repeat from (1) for the residual signals.
(5) Apply multiple source localization to correct the solution for activity of overlapping generators.

The flow chart in Fig. 5 presents the major steps of the algorithm. A formal description of the procedure is detailed in Appendix 2.

Physiologically-Motivated Time-Frequency Filtering

By excluding certain wavelets from the reconstructed waveform following **decomposition of multichannel EPs into wavelets** (Appendix 1), the scalp recorded signal can be

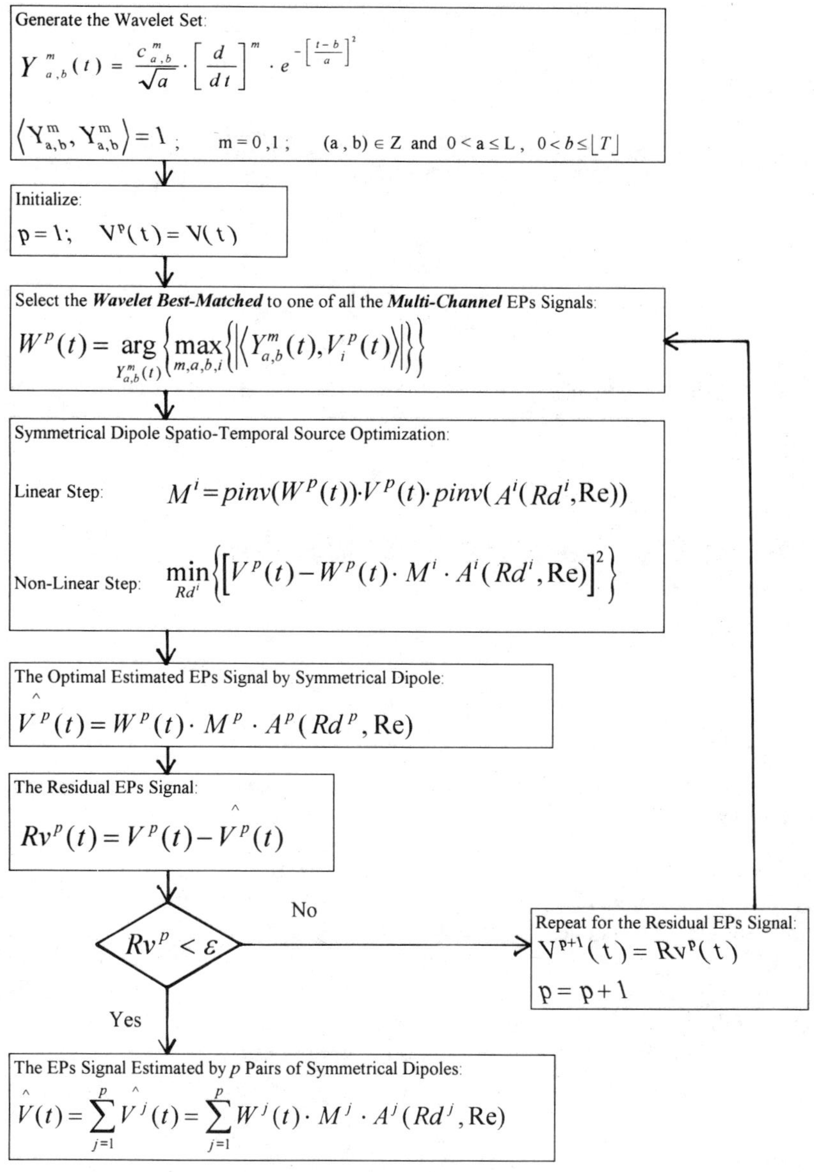

Figure 5. Flow chart of the procedure for spatio-temporal multiple source estimation by wavelet-type analysis.

selectively filtered to enhance EP components of interest (inverse filtering; Tichonov & Arsenin, 1977). This type of physiologically motivated time-frequency filtering eliminates waveform contribution by specific generators and/or background activity. Time-frequency filtering thus contrasts with that accomplished by widely used frequency specific filters (e.g., FFT, FIR, IIR) which affect the signal irrespective of its physiological generators.

Time-frequency filtering may be used to separate EP from the ongoing background EEG on a single trial basis. The prototype function selected for representing the averaged signal can serve as model-based prototypes, and can be applied as such to decompose single trial (non-averaged) signals. Moreover, using physiologically-motivated criteria, prototype

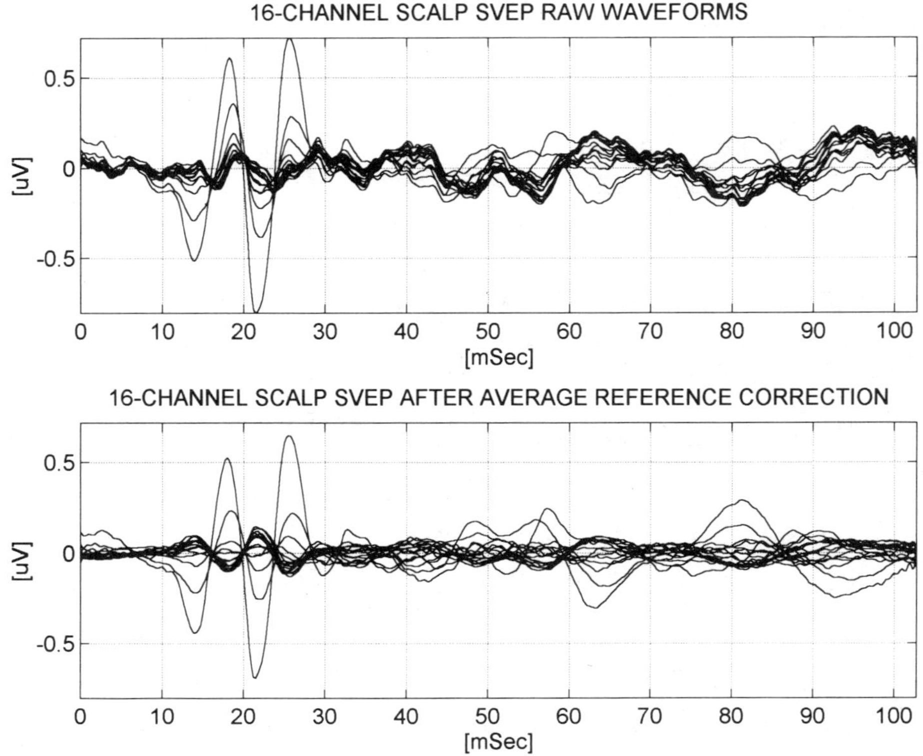

Figure 6. Sixteen channel signals of short latency visual evoked potentials grand averaged from 10 subjects, before (up) and after (bottom) average reference correction.

functions may be selected independent of the averaged signal. Information on the scalp distribution of the component of interest can be used in the selection of the prototype functions for representing the component in the single trial (see Results). In addition, the known timing of such components can narrow the search for their scalp manifestation to limited time frames. These spatial and temporal constraints on Hermite-type function selection are compatible with the criteria for EP component definition (Donchin, 1980).

RESULTS

In this report we demonstrate the application of the procedure to short-latency visual evoked potentials (SVEP) in response to high-intensity flashes (Pratt *et al.*, 1994) and to single-trial event-related potentials.

The 16-channel scalp SVEP signals, grand averaged across 10 subjects, before and after the average reference correction (Lehmann, 1987; Fender, 1987), are presented in Fig. 6. Figure 7 shows the calculated locations of all the estimated sources of activity up to 100 msec after a flash. Note the distribution of sources along the known anatomy of the visual pathway. A representative set of estimated sources for the time period up to 40 msec after the flash is shown in Fig. 8. At 13.75 msec only a single backward pointing dipole at the stimulated eye was sufficient to account for the surface activity. At this time after the stimulus only the retina is known to be active. When activity reaches the visual cortex, at 80 msec,

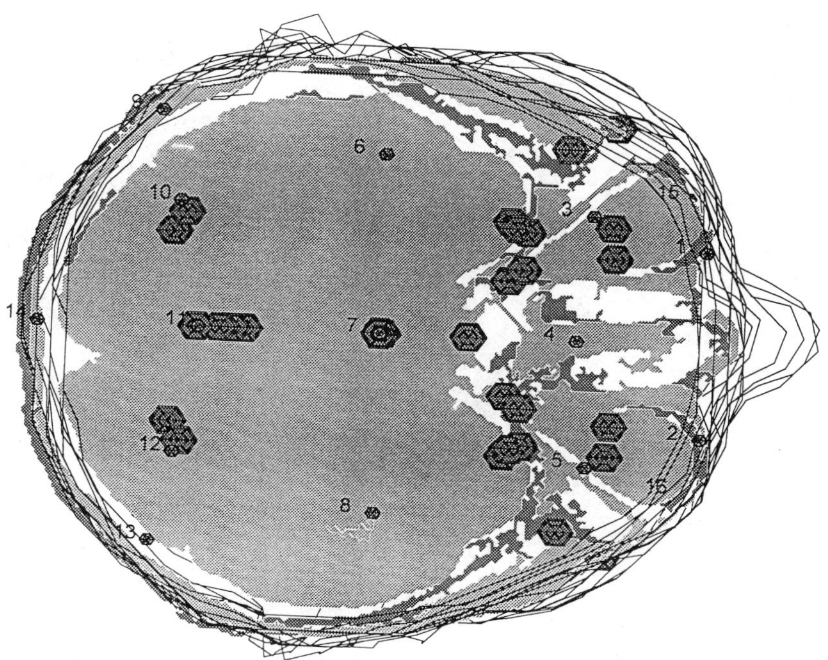

Figure 7. The calculated locations of all the estimated sources of activity up to 100 msec after a flash. The large objects displayed on a Magnetic Resonance Image of the head denote the source locations and the smaller objects are the electrode placements used to record the scalp potentials. Note the distribution of sources along the known anatomy of the visual pathway.

all activity is concentrated at the occipital pole. Indeed, the results of dipole estimation converged on one single dipole deep in the calcarine fissure as can be seen in Fig. 9.

Figure 10 demonstrates application of physiologically-motivated time-frequency filtering to single-trial data of cognitive evoked potentials, to circumvent distortions in the averaged signal due to latency jitters of components across the trials comprising the average. Time-frequency filtering was performed on single-trial auditory cognitive evoked potentials recorded during performance of a memory scanning task (Pratt *et al.*, 1989). Subjects heard a random sequence of spoken digits ('1' to '9'), each of which they had to identify as being, or not being a member of a previously memorized set of 2 digits. Time-frequency filtering was performed by reconstructing only N1, P2 and P3. In (A) note the ommission by this filtering of the slow noise in waveform 1. In (B) note the good correspondence between waveforms and the ommission of the slow noise from single trial 1 (A). No frequency-based filter could have eliminated this type of noise. In (C) note the identical timing and waveform of N1 and P2, which are determined by the acoustic characteristics of the stimulus, and the differences in P3, which is related to processing the speech material. Processing of digits can only begin after their perception, which varies between words presented acoustically (Morton *et al.*, 1976). In (D) Note the multiple peaks of P3, probably because of variability in P3 timing across digits.

DISCUSSION AND CONCLUSIONS

In this report we presented an algorithm for wavelet-type decomposition of evoked potentials. We further demonstrated its utility for source estimation. It is important to

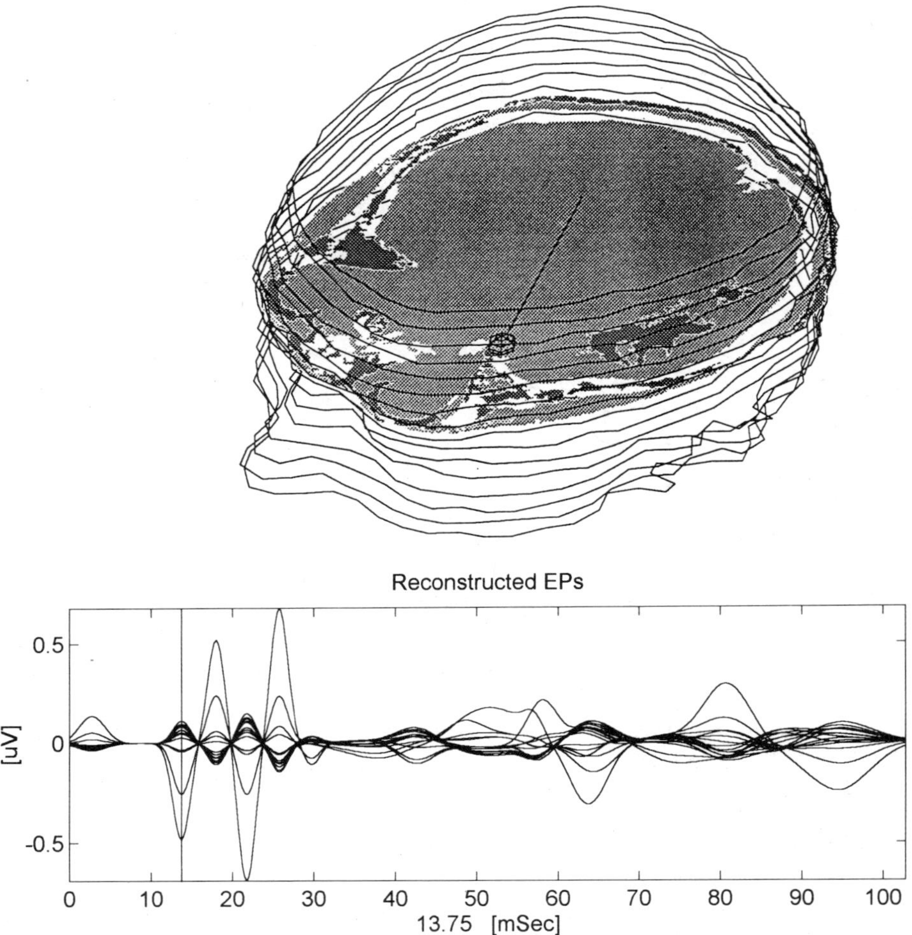

Figure 8. A representative set of estimated sources for the time period up to 40 msec after the flash. At 13.75 msec only a single backward pointing dipole at the stimulated eye was sufficient to account for the surface activity. At this time after the stimulus only the retina is known to be active. The length of each dipole vector is proportional to the dipole's current magnitude, in relative units. The waveforms reconstructed by all dipoles found by the procedure over the entire analysis period are presented at the bottom of the figure. The cursor indicates the latency for which the sources are presented.

emphasize that the source estimation algorithm provides a feasible solution without any additional assumptions nor initial conditions. By using the physiologically-motivated assumption that the temporal pattern of activity from each generator is a wavelet, our algorithm indicates the number of generators that are simultaneously active and separates the total surface activity to the individual contributions of these generators. Once a wavelet is defined in the signal, all the data points during its duration are used for the source estimation, and at each point, only the contribution of that wavelet is used. Thus the algorithm can utilize more data points for each source computation avoiding confounding effects of additional generators that are active at the same time.

In the analysis of SVEPs, without any anatomical constraints, allowing for possible multiple sources at any level, the solutions converged on physiologically and anatomically

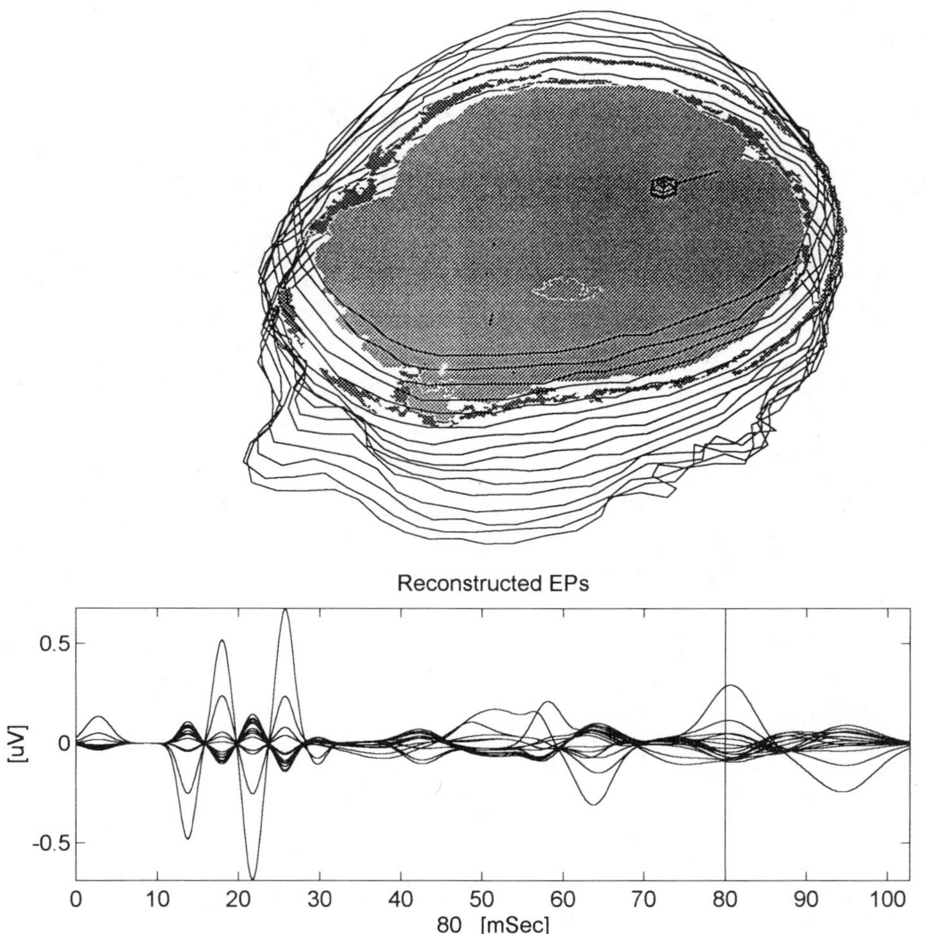

Figure 9. At 80 msec the results of dipole estimation converged on one single dipole deep in the calcarine fissure. At that time activity reaches the visual cortex, and is concentrated at the occipital pole.

reasonable sources that were appropriately distributed along the visual pathway. Similarly, we applied the procedure to Somatosensory Evoked Potentials and arrived at reasonable estimates as well (Geva *et al.*, 1995).

Furthermore, the decomposition procedure may be used for signal enhancement in single-trial analysis of evoked potentials. When we applied it to cognitive event-related potentials, waveform components could be enhanced, demonstrating different sensitivities to stimulus parameters, as established in averaged waveforms. Moreover, variations in the waveform from trial to trial that cannot be accomplished with averaged waveforms, could be followed.

Application of wavelet-type decomposition need not be limited to filtering and source estimation. The wavelet coefficients can be used as features for physiologically-based characterization and classification procedures such as fuzzy clustering (Gath and Geva, 1989; Geva & Pratt, 1994). Whereas principal component analysis coefficients may be used for these purposes, they are not sensitive to variations in specific compo-

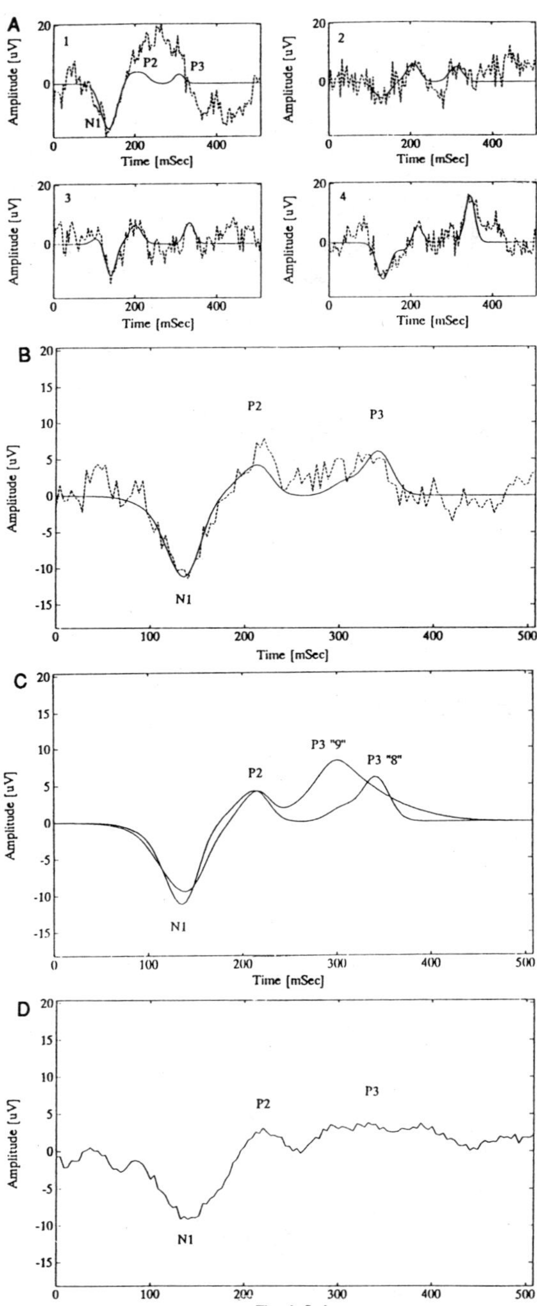

Figure 10. Wavelet-type filtering of single-trial auditory cognitive evoked potentials during a memory scanning task. (A) Unaveraged potentials following 4 auditory presentations of the digit '8'. Unfiltered raw waveforms are presented in broken lines and the wavelet-type filtered (for N1, P2 and P3) waveform in solid lines. (B) Superimposed averages of the raw and wavelet-type filtered single-trials to the digit '8'. (C) Superimposed averaged wavelet-type filtered single trials to digits '8' and '9' (5 trials each). (D) Average of the raw single-trials to digits '1' to '9'.

nents. The very same multichannel wavelet-type decomposition procedure may be used for analysis of other composite waveforms which have model-based constituents (e.g., EEG, EMG, EKG, diastolic heart sounds).

In this report we presented a specific selection of Hermite-type wavelets. However, **wavelet selection** need not be restricted to Hermite-type functions. Gabor functions, which are practically very similar to the Hermite-type wavelets, can also be used (Porat & Zeevi, 1988). Non-symmetrical positive-followed-by-negative pairs of Gaussian functions (which are similar to the first derivative of Gaussians) were also considered by us. Here it is possible to adapt the ratio between the magnitude and dilation of the two Gaussians, which seem to be changing between different recording sites. Poissonian wavelets may be suitable because, similar to neuronal activity, they have an abrupt onset and a gradual decay. Poissonian wavelets, however, also have more parameters and are thus computationally more cumbersome. Again, the advantage of our procedure is that we can use several "mother wavelets" together, but one has to be careful in their selection because the variety can lead to non-physiological results.

Wavelet-type decomposition may be further refined to incorporate the spatial distribution of brain activity (Duffy, 1982; Lehmann, 1987) in wavelet selection. volume conductor theory predicts the surface distribution of known intracranial events. The method need not be limited to scalp events and may be generalized to more invasive methods, bridging the gap between macro- and micro-electrode recordings. Wavelet-type analysis of scalp electrical activity may thus be applied to enhancing the spatial resolution of computed sources of electrical and magnetic signals from the scalp.

The very different spatial and temporal distribution of the sources of the different modalities emphasizes the flexibility and generality of the method. Using methods of structural imaging, like MRI and CT, will contribute to create a more accurate geometrical and physical head model, and will permit using additional assumptions on source location. Combining additional non-invasive measurment methods, like MEG, and functional imaging like functional MRI and PET, will make possible a spatio-temporal imaging of brain activity with accuracy which will make a significant contribution to brain research.

ACKNOWLEDGMENTS

Data acquisition of SVEP by Naomi Bleich, fruitful discussions with Dr. Joseph Segman, MRI data by Ruth Ben-Kish and the graphics visualization programming by Ozi Ben-Ari and Aharon Binur are gratefully acknowledged.

APPENDICES

Definitions

Unlike most of the conventional wavelet techniques, which provide complete orthogonal (or non-orthogonal) bases, each of which is generated from a single mother wavelet, we combine several mother wavelets in order to correlate and better match the waveform used in our analysis with the physiological and physical phenomena underlying the generation of Eps.

Let $Y^m(t)$, m = 1,2,... be a family of "mother" (prototype, model) wavelets. The waveform corresponding to the m−type wavelet with scale (dilation) a and translation b is in the continuous case:

$$Y^m_{a,b}(t) = \frac{1}{\sqrt{a}} \cdot Y^m\left[\frac{t-b}{a}\right]$$

For our physiological realization we use the following family of "Hermite−type mother wavelets" (see the discussion for other options):

$$Y^m(t) = \left[\frac{d}{dt}\right]^m \cdot e^{-t^2}, m = 0,1,2$$

We thus obtain the following NON−orthogonal set of wavelets:

$$S = \left\{Y^m_{a,b}(t) = \frac{c^m_{a,b}}{\sqrt{a}} \cdot \left[\frac{d}{dt}\right]^m \cdot e^{-\left[\frac{t-b}{a}\right]^2} \quad \text{such that}: \right.$$

$$m = 0,1; \quad (a,b) \in Z \quad and \quad 0 < a \leq A, \quad 0 < b < N \},$$

where A is determined by the spectrum (bandwidth) of the EP signals. To make sure that we get the best match with the best fitting signal, we normalize the wavelets by $C^m_{a,b}$ such that:

$$\left\langle Y^m_{a,b}, Y^m_{a,b} \right\rangle = 1$$

where <,> denotes the inner product.

In a multi−channel EP record, each sampled EP signal can be expressed in the form of row vector $V_i(n)$, n = 1,2,...,N; i = 1,2,...,R where N and R are the number of samples in each channel and the number of EP channels respectively. V is the (N x R) EP matrix which consists of the set of $\{V_i\}$ row vectors.

APPENDIX 1

Alogrithm for Multichannel Wavelet-Type Decomposition of Eps

(0) Initialize the system: p = 1, and $V^p = V$.

(1) For each translation b, find the best matching wavelet $W_b \in S$ and an EP signal V^p_j over all channels and waveforms set:

$$T^p(b) \equiv \left| \left\langle W_b, V^p_j \right\rangle \right| = \max_{m,a,i}\left\{\left|\left\langle Y^m_{a,b}, V^p_i \right\rangle\right|\right\}; \quad 0 < b \leq N$$

Thus we obtain a reduced subset of N wavelets best matched to one of the EP signals at translation **b**:

$$S^p_1 = \left\{W_b \in S : \quad T^p(b) = \left|\left\langle W_b, V^p_j \right\rangle\right|; \quad 0 < b \leq N\right\}.$$

(2) Select from the subset S_1^p, all the wavelets that correspond to local maxima of the vector $T^p(b)$:

$$S_2^p \equiv \left\{ W_l \in S_1^p \ : \ T^p(l-1) < T^p(l) > T^p(l+1); \quad 0 < l \le N \right\}.$$

So we get reduced subset S_2^p of L (the number of local maxima in $T^p(b)$) non−orthogonal localized waveforms, which we select for decomposition of our EP signals. We can restrict the number of local minima in $T^p(b)$, and the gap between them. For example, if we choose one local minimun, we get one localized waveform in S_2^p simply by:

$$S_2^p \equiv \left\{ W_l \in S_1^p : \ \left| \langle W_l , V_j^p \rangle \right| = \max_{m,a,b,i} \left\{ \left| \langle Y_{a,b}^m , V_i^p \rangle \right| \right\} \right\}$$

namely, the best maching between all waveforms and all EP signals.

(3) Let B^p denote an (N x L) matrix which consists of the set of all the row vectors $W_l \in S_2^p$. We can use the optimal bi−orthonormal reconstruction theorm (Genossar & Porat, 1992) to reconstruct all the EP signals V^p by the reduced subset of wavelets B^p

$$\hat{V}^p = B^p \cdot pinv\left(B^p\right) V_p$$

where

$$pinv\,(B^p) \equiv \left(\left(B^p\right)^{\mathsf{T}} \cdot B^p \right)^{-1} \cdot \left(B^p\right)^{\mathsf{T}}.$$

We can define a coefficient matrix:

$$C^p = pinv\,(B^p) \cdot V_p$$

so we can write:

$$\hat{V}^p = B^p \cdot C^p.$$

(4) The residual signals are:

$$R_v^p = \hat{V}^p - V^p.$$

If $\left(R_v^p\right)^2 > \varepsilon$, we set $V^{p+1} = R_v^p$, p = p+1,

and start from step (1) again !

(5) The resultant reconstruction (model−based approximation) of the EP signals is given by:

$$\hat{V} = \sum_{j=1}^{p} \hat{V}^j = \sum_{j=1}^{p} B^j \cdot C^j = B \cdot C$$

$$\text{when } B = \begin{bmatrix} B^1, \ldots, B^p \end{bmatrix} \text{ and } C = \begin{bmatrix} C^1 \\ \cdot \\ \cdot \\ \cdot \\ C^p \end{bmatrix}$$

Each row in C represents the scalp distribution of the corresponding waveform in B. Applying any available source localization method (e.g., Fender, 1987) for each row in C can reveal the generators of the corresponding waveforms in B.

APPENDIX 2

Alogrithm for Spatio-Temporal Multiple Source Estimation of Eps by Wavelet-Type Decomposition

(0) Initialize the system: $p = 1$; $V^p(t) = V(t)$.

(1) Of all the wavelets in the set, find the wavelet with the best match to any one of all the EP channels:

$$W^p(t) \in S \text{ such that: } \left| \left\langle W^p, V_j^p \right\rangle \right| = \max_{m,a,b,i} \left\{ \left| \left\langle Y_{a,b}^m, V_i^p \right\rangle \right| \right\}$$

(2) For the wavelet $W^p(t)$ found in the previous step, solve the following source estimation optimization problem (see details in the next paragraph):

$$\min_{\overline{M}^i \cdot \overline{R}_d^i} \left\{ \left(V^p(t) - W^p(t) \cdot \overline{M}^i \cdot A^i \left(\overline{R}_d^i, R_e \right) \right)^2 \right\}$$

where $\overline{M}^i = \begin{bmatrix} M_x^i, M_y^i, M_z^i \end{bmatrix}$ is a vector of the current dipole magnitude,

$\overline{R}_d^i = \begin{bmatrix} R_x^i, R_y^i, R_z^i \end{bmatrix}$ is the current dipole coordinates, R_e represents the electrodes'

coordinate matrix, and $A^i \left(\overline{R}_d^i, R_e \right)$ is the head shape and conductivity model matrix as a

function of the dipole and the electrodes' coordinates (Fender, 1987).

The estimated potential on the scalp generated by the optimal dipole parameters $\overline{M}^p$ and

$\overline{R}_d^p$ that were found in the previous step is:

$$\hat{V}^p(t) = W^p(t) \cdot \overline{M}^p \cdot A^p \left(\overline{R}_d^p, R_e \right)$$

(3) The residual signals are:

$$R_v^p(t) = V^p(t) - \hat{V}^p(t)$$

$$\text{if } \left(R_v^p \right)^2 > \varepsilon \text{ , we set } V^{p+1}(t) = R_v^p(t), \quad p = p + 1,$$

and start from step (1) again !

(4) The resultant reconstruction of the EP signals is given by:

$$\hat{V} = \sum_{j=1}^{p} \hat{V}^j(t) = \sum_{j=1}^{p} W^j(t) \cdot \overline{M}^j \cdot A^j\left(R_d^j, R_e\right)$$

Optimizing Source Estimation by The Temporal Pattern of Activity: Symmetrical Current Dipoles Model for Each Wavelet

Identical temporal activity in two symmerical generators cannot be separated by temporal decomposition to two separate sources. This case is particularly prevalent in sensory pathways and only use of the spatial distribution of potentials and the known anatomical symmetry can help in the separation. The anatomy of sensory pathways is always symmetrical, even if physiologically activity may be lateralized. Therefore we chose to use a symmetrical dipoles model, and in case of asymmetrical activation in the pathway, one dipole is smaller than its counterpart, or even missing. In our symmetrical source estimation analysis we make the following assumptions:

(1) The symmetry is relative to the mid–saggital plane (left/right):

$$R_{d_{2x}} = R_{d_{1x}} \, , \; R_{d_{2y}} = -R_{d_{1y}} \, , \; R_{d_{2z}} = R_{d_{1z}} \, ,$$

(2) The orientation of the two symmetrical dipoles is symmetrical (because of the anatomical symmetry), although the magnitude of the two is not necessarily equal:

$$M_{2x} = b_{12} \cdot M_{1x} \, , \; M_{2y} = -b_{12} \cdot M_{1y} \, , \; M_{2z} = b_{12} \cdot M_{1z} \, ,$$

where b_{12} is a positive scalar, describing the ratio between magnitudes of the two symmetrical dipoles.

The potential generated by such two symmetrical dipoles is:

$$V = M_{1x} \cdot A_{1x}\left(\overline{R}_{d_1}, R_e\right) + M_{1y} \cdot A_{1y}\left(\overline{R}_{d_1}, R_e\right) + M_{1z} \cdot A_{1z}\left(\overline{R}_{d_1}, R_e\right)$$

$$+ b_{12} \cdot M_{2x} \cdot A_{2x}\left(\overline{R}_{d_2}, R_e\right) - b_{12} \cdot M_{2y} \cdot A_{2y}\left(\overline{R}_{d_2}, R_e\right) + b_{12} \cdot M_{2z} \cdot A_{2z}\left(\overline{R}_{d_2}, R_e\right)$$

$$= M_{1x} \cdot \left[A_{1x}\left(\overline{R}_{d_1}, R_e\right) + b_{12} \cdot A_{2x}\left(\overline{R}_{d_2}, R_e\right)\right]$$

$$+ M_{1y} \cdot \left[A_{1y}\left(\overline{R}_{d_1}, R_e\right) - b_{12} \cdot A_{2y}\left(\overline{R}_{d_2}, R_e\right)\right]$$

$$+ M_{1z} \cdot \left[A_{1z}\left(\overline{R}_{d_1}, R_e\right) + b_{12} \cdot A_{2z}\left(\overline{R}_{d_2}, R_e\right)\right]$$

We can define a special "symmetrical dipole head model matrix":

$$A_{12}\left(\overline{R}_{d_1}, b_{12}, R_e\right) = \begin{bmatrix} A_{1x}\left(\overline{R}_{d_1}, R_e\right) + b_{12} \cdot A_{2x}\left(\overline{R}_{d_2}, R_e\right) \\ A_{1y}\left(\overline{R}_{d_1}, R_e\right) - b_{12} \cdot A_{2y}\left(\overline{R}_{d_2}, R_e\right) \\ A_{1z}\left(\overline{R}_{d_1}, R_e\right) + b_{12} \cdot A_{2z}\left(\overline{R}_{d_2}, R_e\right) \end{bmatrix}$$

and can then write a combined equation for both dipole:

$$V = \overline{M}_1 \cdot A_{12}\left(\overline{R}_{d_1}, b_{12}, R_e\right),$$

or

$$V(t) = W(t) \cdot \overline{M}_1 \cdot A_{12}\left(\overline{R}_{d_1}, b_{12}, R_e\right).$$

By using A_{12} instead of A , we can apply the same algorithm that is used for a single dipole – we only have to add the variable b_{12} to the non–linear optimization stage. We take advantage of the linear relation that exists between $V(t)$ and $\overline{M}$ by dividing the optimization algorithm into a linear part and a non–linear part in the following way:

(0) Initialize the system: j = 0; two symmetrical current dipoles at the head center:

$$\overline{R}_{d_1}^{\,j} = [0,0,0] \text{ and } \overline{R}_{d_2}^{\,j} = [0,0,0];$$

starting with one active dipole $b_{12}^{\,j} = 0$.

(1) Calculate the head shape and conductivity model matrix

$$A_1^{\,j}\left(\overline{R}_{d_1}^{\,j}, R_e\right), \ A_2^{\,j}\left(\overline{R}_{d_2}^{\,j}, R_e\right), \text{ and } A_{12}^{\,j}\left(\overline{R}_{d_1}^{\,j}, b_{12}^{\,j}, R_e\right).$$

(2) Calculate the current dipole magnitude M_{1j} by the following linear solution:

$$\overline{M}_1^{\,j} = pinv(W(t)) \cdot V(t) \cdot pinv\left(A_{12}^{\,j}\left(\overline{R}_{d_1}^{\,j}, b_{12}^{\,j}, R_e\right)\right)$$

$$pinv(X) = X^T \cdot \left(X^T \cdot X + I\right)^{-1}.$$

$\overline{M}_2^{\,j}$ is then calculated by $\overline{R}_1^{\,j}$ and $b_{12}^{\,j}$.

The Pseudo Inverse solution is an optimal solution in terms of minimum norm of the current dipoles' magnitude at the given coordinates.

(3) Make the next non–linear optimization step (for example by SIMPLEX, gradient descent or any other optimization method) to minimize:

$$\min_{\overline{R}_{d_1}^{\,j}, b_{12}^{\,j}} \left\{ \begin{array}{l} \left[V(t) - W(t) \cdot \overline{M}_1^{\,j} \cdot A_{12}^{\,j}\left(\overline{R}_{d_1}^{\,j}, b_{12}^{\,j}, R_e\right)\right]^2 \\ + \left[V(t) - W(t) \cdot \overline{M}_1^{\,j} \cdot A_1^{\,j}\left(\overline{R}_{d_1}^{\,j}, R_e\right)\right]^2 \\ + \left[V(t) - W(t) \cdot \overline{M}_2^{\,j} \cdot A_2^{\,j}\left(\overline{R}_{d_2}^{\,j}, R_e\right)\right]^2 \end{array} \right\}$$

The latter two lines prevent creation of a spurious dipole symmetrical to the physiologically feasible dipole.

(4) If

$$\left[V(t)-W(t)\cdot \overline{M}_1^{\,j}\cdot A_{12}^{\,j}\left(\overline{R}_{d_1}^{\,j},b_{12}^{\,j},R_e\right)\right]^2 > \varepsilon \;,$$

set j = j + 1, and go back to stage (1).

Dividing the optimization algorithm into linear and non-linear parts leads to a faster and better conversion into a feasible solution.

Multiple Source Estimation for O Sources

After estimating one by one the spatio-temporal parameters of the sources, we can apply multiple source estimation to correct the solution for overlapping activity of generators. Once the number and the initial parameters of the sources are given, multiple source estimation becomes a soluble problem (Mosher *et al.*, 1992). We use the wavelets as the temporal activity pattern in multiple source estimation by:

$$V(t)=W(t)\cdot M_W\cdot A\left(R_{d_W},R_e\right).$$

This procedure defines the following optimization problem:

$$\min_{M_W,R_{d_W}}\left\{V(t)-W(t)\cdot M_W\cdot A\left(R_{d_W},R_e\right)\right\},$$

where $W(t)$ is a matrix consisting of all the selected wavelets, M_w is a matrix of the corresponding dipoles, and R_{d_w} is a matrix of the dipoles' coordinates. For a more detailed formulation and options for the multiple source estimation analysis after estimating the number and initial spatio-temporal parameters of the sources, see Mosher *et al.* (1992).

REFERENCES

Achim, A., Richer, F., Saint-Hilaire, J.M., 1991, Methodological considerations for the evaluation of Spatio-Temporal Source Models (STAM), Electroenceph. Clin. Neurophysiol., 79:227-240.

Allison, G., Goff. W. R., Williamson, P. D., and Van Gilder, J. C., 1980, On the neural origin of early components of the human somatosensory evoked potential, In Desmedt J.E. (ed.): Clinical Uses of Cerebral, Brainstem and Spinal Somatosensory Evoked Potentials. Progress in Clinical Neurophysiology. Basel, Karger, 7: 51-68.

Daubechies, I., 1988, Orthogonal bases of compactly supported wavelets, Commun. Pure Appl. Math. 41:909-996.

Donchin, E., 1980, Event-related brain potentials: a tool in the study of human information processing. In: Begleiter H. (ed.) Evoked Potentials in Psychiatry. Plenum, New York.

Duffy, F. H., 1982, Topographic display of evoked potentials: clinical applications of brain electrical activity mapping (BEAM) evoked potentials, *Ann. N.Y. Acad. Sci.* 388:183-198.

Fender, D. H.,1987, Source localization of brain electrical activity. In: Gevins, A. S, and Remond, A. (eds.), *Methods of Analysis of Brain Electrical and Magnetic Signals. Handbook of Electroencephalography and Clinical Neurophysiology.* Elsevier, Amsterdam, Vol. 1, 13:355-403.

Gath, I., and Geva, A. B., 1989, Unsupervised optimal fuzzy clustering. *IEEE Trans. Pattern Anal. Machine Intel.* 7:773-781.

Genossar, T., and Porat, M., 1992, Optimal bi-orthonormal approximation of signals. *IEEE Trans. on Systems, Man, and Cybernetics* 22(3):449.

Geva, A. B., Pratt H., and Zeevi Y. Y., 1993, Wavelet Decomposition of multichannel evoked potentials, *Electroenceph. Clin. Neurophysiol.* 87:S25.

Geva, A. B., Pratt, H., 1994, Unsupervised clustering of evoked potentials by waveform, *Med. & Biol. Eng. & Comput.* 32:543-550.

Geva, A. B., Pratt, H., and Zeevi, Y. Y., 1995, Spatio-temporal multiple source localization by wavelet-type decomposition of evoked potentials, *Electroenceph. Clin. Neurophysiol.* In press.

Lehmann, D., 1987, Principles of spatial analysis. In: Gevins,A. S, and Remond, A. (eds.), *Methods of Analysis of Brain Electrical and Magnetic Signals. Handbook of Electroencephalography and Clinical Neurophysiology.* Elsevier: Amsterdam, vol. 1, 12:309-354.

Mallat, S. G., 1989, A theory for multiresolution signal decomposition: the wavelet representation, *IEEE Trans. on Pattern Analysis and Machine Intelligence* 11:674-693.

McGillem, C. D., and Aunon, J. I., 1987, Analysis of event-related potentials. In: Gevins, A. S., and Remond, A. (eds.), *Methods and Analysis of Brain Electrical and Magnetic Signals. Handbook of Electroencephalography and Clinical Neurophysiology.* Elsevier: Amsterdam, vol. 1, 5, 131-169.

Moller, A. R., Jannetta, P. J., and Jho, H. D., 1990, Recordings from human dorsal column nuclei using stimulation of the lower limb, *Neurosurgery* 26: 291-299.

Morioka, T., Shima, F., Kato, M., and Fukui, M., 1991, Direct recording of somatosensory evoked potentials in the vicinity of the dorsal column nuclei in man: their generator mechanisms and contribution to the scalp far-field potentials, *Electroenceph. Clin. Neurophysiol.* 80: 215-220.

Morton, J., Marcus, S., and Frankish, C., 1976, Perceptual centers (P-centers), *Psychol. Rev.* 83:405-408.

Mosher, J. C., Lewis, P. S., and Leahy, R. M., 1992, Multiple dipole modeling and localization from spatio-temporal MEG data, *IEEE Trans. on Biomed. Eng.* 39:541-557.

Nunez, P.L., 1981, *Electric Fields of the Brain. The Neurophysics of EEG.* Oxford: New York.

Papakostopolous, D. A., and Crow, H. J., 1980, Direct recording of the somatosensory evoked potentials from the cerebral cortex of man and the difference between precentral and postcentral potentials. In Desmedt, J.E. (ed.): *Clinical Uses of Cerebral, Brainstem and Spinal Somatosensory Evoked Potentials. Progress in Clinical Neurophysiology.* Karger:Basel, 7: 15-26.

Plonsey, R., and Fleming, D. G., 1969, *Bioelectric Phenomena.* McGraw-Hill: New York, 5:203-275.

Porat, M., and Zeevi, Y. Y., 1988, The generalized Gabor scheme of image representation in biological and machine vision, *IEEE Trans. Pattern Anal. Machine Intel.* 10:452-468.

Pratt, H., Michalewski, H. J., Barrett, G., and Starr, A., 1989, Brain potentials in a memory-scanning task: I. Modality and task effects on potentials to probes, *Electroenceph. Clin. Neurophysiol.* 72:407-421.

Pratt, H., Martin, W. H., Bleich, N., Zaaroor, M., and Schacham, S. E., 1994, A high intensity, goggle-mounted flash stimulator for short latency visual evoked potentials, *Electroenceph. Clin. Neurophysiol.* 92:469-472.

Regan, D., 1989, *Human Brain Electrophysiology. Evoked Potentials and Evoked Magnetic Fields in Science and Medicine.* Elsevier: Amsterdam, pp. 57-66.

Rioul, O., and Duhamel, P., 1992, Fast algorithms for discrete and continuous wavelet transforms, *IEEE Trans. Inform. Theory* IT38:569-586.

Scherg, M., and von Cramon, D., 1985, A new interpretation of the generators of BAEP waves I-V. Results of spatio-temporal dipole model, *IEEE Trans. Inform. Theory* IT62:290-299.

Slimp, J. C., Tamas, L. B., Stolov, W. C., and Wyler, A. R., 1985, Somatosensory evoked potentials after removal of somatosensory cortex, *Electroenceph. Clin. Neurophysiol.* 37:663-669.

Tichonov, A. N., and Arsenin, V. Y., 1977, *Solution of Ill-Posed Problems.* Wiley: New York.

Urasaki, E., Wada, S., Kadoya, C., Yokota, A., Matsuoka, S., and Shima, F.,1990, Origin of scalp far-field N18 of SSEPs in response to median nerve stimulation, *Electroenceph. Clin. Neurophysiol.* 77:39-51.

Urasaki, E., Uematsu, S., and Lesser, R. P., 1993, Short latency somatosensory evoked potentials recorded around the human upper brainstem, *Electroenceph. Clin. Neurophysiol. 88:92-104.*

9

MODELING AND ESTIMATION OF AMPLITUDE AND TIME SHIFTS IN SINGLE EVOKED POTENTIAL COMPONENTS

Daniel H. Lange and Gideon F. Inbar

Department of Electrical Engineering
Technion-IIT
Haifa 32000, Israel

ABSTRACT

It is well established that different components of an evoked potential complex may originate from different functional brain sites, and can be distinguished by their respective latencies and amplitudes. This emphasizes the need and advantage in the ability to estimate single evoked potential components, which would open a window into complex dynamic processes occurring in the central nervous system.

We present a new parametric estimator for single evoked potential components which can detect changes in specific components within the evoked potential complex, extending current methods which only compensate for global latency and magnitude variations with respect to an averaged evoked response. Although the proposed estimator also relies on the average response, it is significantly more versatile, incorporating a novel decomposition method in order to process each evoked potential component separately. The performance of the estimator is analyzed analytically for variable component amplitudes, implying also on variable latencies. The analytical qualities are compared to simulated results confirming the estimator's capacity to track morphological changes of simulated signals. Depending on the extent of allowed variations, the estimator is capable of extracting simulated evoked potentials with signal to noise ratios as low as -15 dB.

The estimator is applied to two evoked potential classes: single movement related brain potentials, recorded during changing experimental conditions and thus expected to yield varying evoked potential components, and cognitive event related potentials recorded during a typical odd-ball type paradigm, with varying performance levels, which are likely to be reflected in the evoked potential morphology. The results of both applications demonstrate the ability to capture features of dynamic processes, which enables tracking of transient properties typical of brain processes throughout an experimental session.

Advances in Processing and Pattern Analysis of Biological Signals, Edited by Isak Gath and Gideon F. Inbar
Plenum Press, New York, 1996

INTRODUCTION

Evoked brain potentials originate from various brain sites, depending on the type of applied stimulus and its associated task. Some evoked potentials such as cognitive and movement related potentials include a complex of partially overlapping components, which represent different processing stages along the neural pathways (1). In order to examine dynamic processes such as learning or adaptation of the central nervous system, using the data recorded from the brain, one needs to be able to distinguish between the different components and track their variations, preferably on a single-trial basis. Such is the case of learning and adaptation in the motor control system (2, 3), where signals at the highest level in this hierarchical interconnected system, have components related to different loops with different time delays and thus are, at least in some cases, separable. The process of learning or adaptation to new loads takes about five repetitions of the same movement, until high and stable performance is achieved (3). The ability to track amplitude and latency variations from one trial to the next throughout the learning and adaptation processes, will facilitate investigation of the role that various brain structures and peripheral proprioceptive loops perform during the process.

Traditionally, evoked potentials are synchronously averaged to enhance the evoked signal and suppress the background brain activity, which contradicts the basic assumption of signal variability (4, 5). Several single-trial estimation techniques were developed during the last decade, usually limited to waveforms similar to the average response, differing only in global latency and scale parameters (6, 7). Others report techniques capable of extracting morphologically variable evoked potentials and give some experimental results to demonstrate the performance (8, 9). We propose a new parametric model for extraction of single evoked potentials from the noisy background activity, by treating the evoked response as a sum of distinctive components rather than a single waveform, and therefore allowing variability to each component of the signal. The suggested procedure estimates the characteristics of the pre-stimulus ongoing EEG, decomposes the evoked potential template, and separates the post-stimulus interval into its two assumed contributions: the spontaneous EEG and the unique components of the single evoked response. The parametric estimator is assessed analytically and through its application on simulated signals. Ultimately it is applied to real evoked potential data, captured during varying experimental conditions and thus expected to also yield morphologically varying evoked potentials.

THE PARAMETRIC ESTIMATOR

Estimator Structure

The suggested new estimator represents the spontaneous EEG as a stochastic ongoing process, realized by an autoregressive (AR) model, on which the evoked potential is superimposed. The AR coefficients are calculated from the pre-stimulus EEG and are assumed to remain constant throughout the post-stimulus epoch (10), which is a common and reasonable assumption noting the relatively short duration of the entire epoch - less than 2 seconds (11).

The single evoked response is obtained by decomposing the averaged response into a set of components, each with a Gaussian amplitude distribution, after which the single evoked response is recomposed allowing the components to have variable latencies and magnitudes. The Gaussian amplitude distribution assumption is based on the large number of neurons that take part in the cluster of neurons responsible for the generation of an evoked

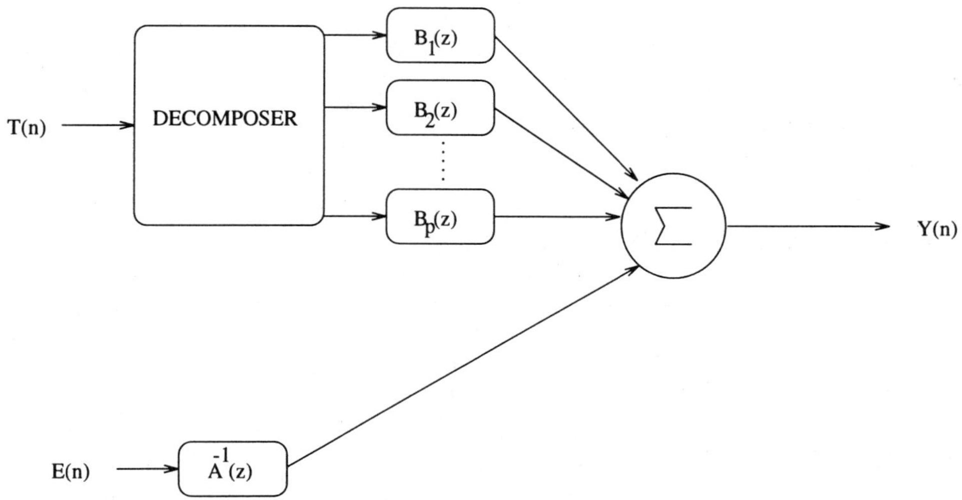

Figure 1. A block diagram of the signal synthesis model. T, E and Y represent the template, a white Gaussian noise series and the measured data respectively.

potential component, however it is not restricted to this distribution and other distributions such as Poisson may be assumed (12). Moreover, different distributions may be assumed for different components of the signal. The set of latency and magnitude correcting filters $B_i(z)$ is determined by optimizing the model parameters in the least-square sense, selecting filter orders according to the maximal latency shift allowed for each component. Finally, the single response is obtained by summing the contributions of all the filtered components. A block diagram of the signal synthesis model is presented in Fig. 1.

Mathematical Formulation

Template Decomposition. It is assumed that the evoked potential waveform is a deterministic signal with stochastic parameters which determine the latency (τ_i) and magnitude (K_i) of each component within the evoked potential complex:

$$EP(t) = \sum_{i=1}^{p} K_i \cdot COMP_i(t - \tau_i) \tag{1}$$

The components are derived from the template using a novel decomposition scheme, which decomposes the evoked potential template into a set of components which sum up exactly to the original waveform. This is an important feature, achieved by dividing each sample point among the various components where the weights are determined by the assumed distribution function. Equation 2 presents the decomposition rule for a general set of distributions $d_i(t)$:

$$COMP_i(t) = TEMP(t) * \frac{d_i(t)}{\sum_{j=1}^{p} d_j(t)} \tag{2}$$

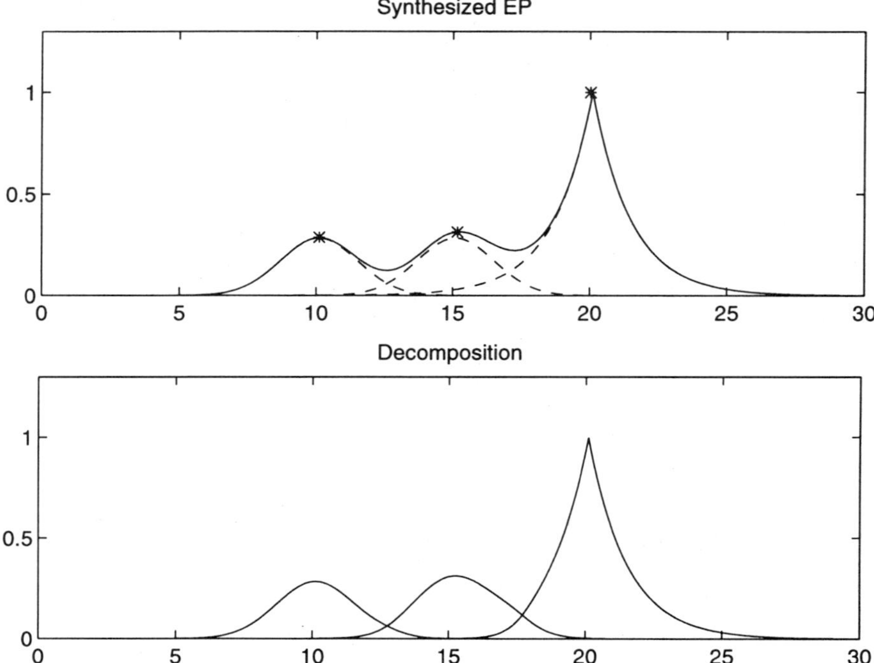

Figure 2. Decomposition of a simulated signal. Top: A synthesized waveform and its underlying components (two Gaussians and one exponent). Bottom: Decomposition results, assuming Gaussian distributions for all components. A slight distortion only appears in the overlapping tail area of the Gaussian and exponential components.

As already mentioned, we used Gaussian distribution laws which were found to be adequate in our applications. Equation 2 gives the decomposition rule for the Gaussian distribution, an example of which can be seen in Fig. 2:

$$COMP_i(t) = TEMP(t) * \frac{\sigma_i^{-1} \exp\left(-\frac{(\tau - \tau_i)^2}{2\sigma_i^2}\right)}{\sum_{j=1}^{p}\left[\sigma_j^{-1}\exp\left(-\frac{(\tau - \tau_j)^2}{2\sigma_j^2}\right)\right]} \tag{3}$$

where τ_i are extracted from the peaks of the signal complex and σ_i are identified according to a minimum least-squares fitting criterion (13). This approach decomposition is mathematically error-free, so that we do not lose any information stored in the evoked response, which also implies robustness of the method, i.e. the decomposition is not sensitive to deviations from the assumed distributions (14). A simulated example is shown in Fig. 2, where a composite waveshape was synthesized from an exponential and two Gaussian waveforms. The decomposition was performed assuming Gaussian components only, yielding an insignificant error in the overlapping tails of the Gaussian and exponent.

We would like to stress here that the assumed distributions used for the decomposition process are not imposed on the resulting decomposition, which would perfectly describe the underlying components if the assumed distributions are correct; however, moderate deviations would not significantly distort peak amplitudes and latencies but may affect the overlapping component tails as demonstrated in the simulation.

The Parametric Estimator. The notations used throughout the formulations are those of Fig. 1. $A(z)$ is determined via autoregressive modeling of a prestimulus interval by minimizing the quadratic residual-error cost function. The template and measured EEG are filtered through the identified $A(z)$, to whiten the EEG channel and thus facilitate closed-form least-square solution of the model. Equation 4 is the autoregressive description of the prestimulus interval, and Eqs. (5) and (6) represent the filtering process of the template and EEG.

$$y(n) = \sum_{j=1}^{q} a_j \cdot y(n-j) + e(n) \tag{4}$$

$$T'(n) = \sum_{j=1}^{q} a_j \cdot T(n-j) \tag{5}$$

$$y'(n) = \sum_{j=1}^{q} a_j \cdot y(n-j) \tag{6}$$

The complete model can thus be expressed by a regression-type formula (T_i represents the i-th component):

$$y'(n) = \sum_{i=1}^{p} \sum_{j=-d}^{d} b_{i,j} \cdot T_i'(n-j) + e(n) \tag{7}$$

To solve the model, we turn to matrix notation. Let $\underline{y}'^{T}$ be the whitened measurements vector, let $\underline{\underline{A}}$ be the input matrix, and let $\underline{b}^{T}$ be the filter coefficients vector, as defined below:

$$\underline{y}'^{T} = \left[y'(d+1), y'(d+2), \ldots, y'(N-d) \right] \tag{8}$$

$$\underline{\underline{A}}^T = \begin{pmatrix} T_1'(2d+1) & T_1'(2d+2) & \cdots & T_1'(N) \\ T_1'(2d) & T_1'(2d+1) & \cdots & T_1'(N-1) \\ \vdots & \vdots & & \vdots \\ T_1'(1) & T_1'(2) & \cdots & T_1'(N-2d) \\ T_2'(2d+1) & T_2'(2d+2) & & T_2'(N) \\ \vdots & \vdots & & \vdots \\ T_2'(1) & T_2'(2) & \cdots & T_2'(N-2d) \\ \vdots & \vdots & & \vdots \\ \vdots & \vdots & & \vdots \\ T_p'(2d+1) & T_p'(2d+2) & \cdots & T_p'(N) \\ \vdots & \vdots & & \vdots \\ T_p'(1) & T_p'(2) & \cdots & T_p'(N-2d) \end{pmatrix} \tag{9}$$

$$\underline{b}^T = [b_{1,-d}, b_{1,-d+1}, \ldots, b_{1,d}, b_{2,-d}, \ldots, b_{p,d}] \tag{10}$$

Using the defined notations, the model can be expressed as follows:

$$\underline{y}' = \underline{\underline{A}} \cdot \underline{b} + \underline{e} \tag{11}$$

which can be solved using the standard least-square solution (13):

$$\hat{\underline{b}} = \left(\underline{\underline{A}}^T \cdot \underline{\underline{A}} \right)^{-1} \cdot \underline{\underline{A}}^T \cdot \underline{y}' \tag{12}$$

Performance Evaluation

We will show that since the ongoing EEG is zero-mean (band passed prior to sampling), the obtained least-square estimator is unbiased, and the estimation variance is proportional to the EEG variance and inversely proportional to the energy of the template. To simplify calculations, we assume $\tau_i = 0$, which limits the following derivation to the varying magnitude - constant latency case; however, a general result for the varying latency case will also be presented.

Assuming constant latency, the estimation problem is stated as follows: the noisy measurement consists of a weak transient signal $s(n)$, on which a strong white noise $e(n)$ is additively superimposed, so that $y(n) = s(n) + e(n)$. Note that the white noise is a whitened version of the ongoing EEG, as described in section 2.2.2. It is assumed that an evoked potential template $T(n)$ is available, differing from $s(n)$ due to a scaling factor K. To estimate K, the following quadratic mean-error cost function is to be minimized:

$$\rho = \sum_n \{K \cdot T(n) - y(n)\}^2 \tag{13}$$

By deriving the error function with respect to K and equating to zero, we get:

$$\frac{d\rho}{dK} = 2\sum_n K \cdot T^2(n) - 2T(n)y(n) = 0 \tag{14}$$

Yielding the estimated K:

$$\hat{K} = \frac{\Sigma_n T(n) \cdot y(n)}{\Sigma_n T^2(n)} \tag{15}$$

Now we calculate the mean and variance of the estimator. The mean is obtained by taking the expectation of $\hat{K}$:

$$E[\hat{K}] = \frac{\Sigma_n\{T(n)\cdot(K\cdot T(n)+E[e(n)])\}}{\Sigma_n T^2(n)} \tag{16}$$

Since the ongoing EEG is zero mean, $E(e(n)) = 0$, we get:

$$E[\hat{K}] = K\cdot\frac{\Sigma_n T^2(n)}{\Sigma_n T^2(n)} = K \tag{17}$$

and therefore the estimate is indeed unbiased.

To calculate the variance, we write the expression for $\hat{K}^2$, and take the expected value:

$$E[\hat{K}^2] = E\left[\left\{\frac{\Sigma_n T(n)y(n)}{\Sigma_n T^2(n)}\right\}^2\right] = E\left[\left\{\frac{\Sigma_n T(n)\cdot[KT(n)+e(n)]}{\Sigma_n T^2(n)}\right\}^2\right] =$$

$$= E\left[\frac{\Sigma_i\Sigma_j T(i)T(j)[KT(i)+e(i)]\cdot[KT(j)+e(j)]}{\{\Sigma_n T^2(n)\}^2}\right] \tag{18}$$

Noting that

$$E[e(i)e(j)] = \begin{cases} 0 & i \neq j \\ \sigma^2 & i = j \end{cases}, \tag{19}$$

we have:

$$E[\hat{K}^2] = \frac{K^2\cdot\{\Sigma_n T^2(n)\}^2+\sigma^2\Sigma_n T^2(n)}{\{\Sigma_n T^2(n)\}^2} = K^2+\sigma^2\cdot\left\{\sum_n T^2(n)\right\}^{-1} \tag{20}$$

Therefore, the estimation variance is given by:

$$E[\hat{K}^2-K^2] = \sigma^2\cdot\left\{\sum_n T^2(n)\right\}^{-1} \tag{21}$$

The strength of this result lies in the obtained relationship between the template's energy, which is affected by the decomposition, and the variance of estimate, quantifying the tradeoff between number of components and accuracy of estimation. Similarly, a general derivation for the varying latency case can be shown to yield (15):

$$E[\hat{b}^2-b^2] = \sigma^2\cdot(A^T A)^{-1} \tag{22}$$

The credibility of the estimate should be assessed by verifying that the obtained estimation variances (Eq. 22) are significantly smaller than the identified parameters (Eq. 12).

SIMULATION RESULTS

First, we study the estimator's performance at varying signal to noise ratios, where the buried response is identical to the template; thus we expect the estimator to yield K≈1, with increasing variances as the SNR decreases. Figure 3 shows the estimation results and

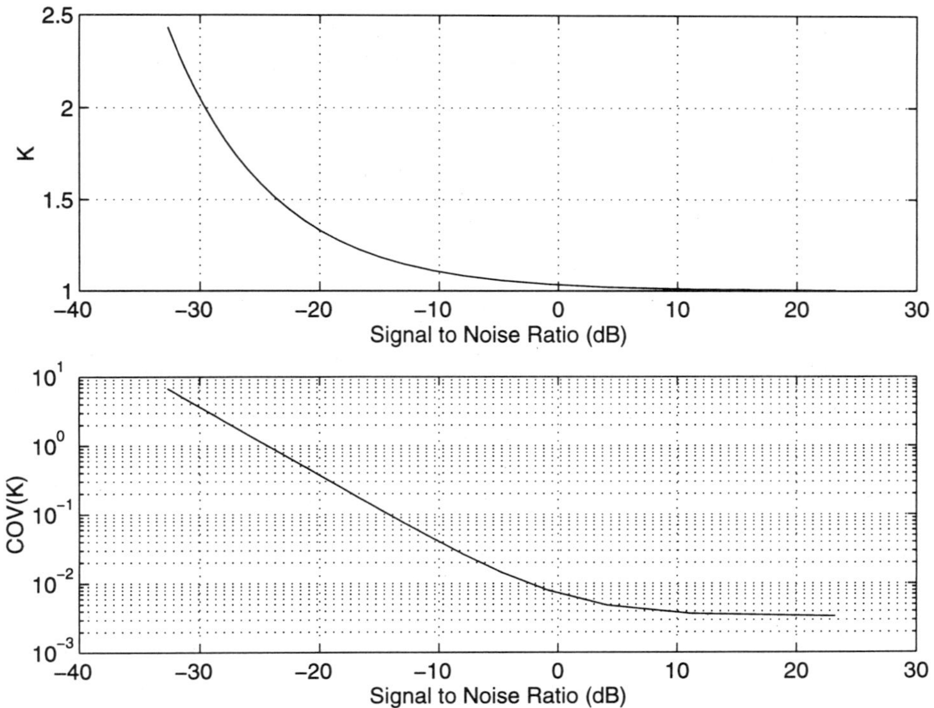

Figure 3. Estimation of K and variance of estimation as a function of the signal to noise ratio. Top: Estimation of K under varying SNR conditions. Bottom: estimation variances.

variances of estimation for a range of SNRs. The estimation variance is acceptable for SNRs higher than -15 dB, at which point the variance reaches 10% and the simulation result is $K = 1.2$.

The effect of decomposition on these results may be evaluated using equation 21, which presents an inversely proportional relation between the signal's energy and variance of estimation. Since the decomposition yields components that inherently are of a lower energy than the total template's energy, a higher SNR is required to obtain equivalent performance. For example, if we divide the template into five components with roughly equal energies, the estimation variances for each component would increase by a factor of 5. Noting that an initial SNR of -15 dB yields variances of *10%* for the full template, an SNR of -10 dB would be required to obtain *10%* variances with five components. Therefore, the next simulation which deals with estimation of five components, is carried out under -10 dB SNR conditions rather than -15 dB.

The following simulation demonstrates the estimator's ability to reconstruct a signal which differs from the template both in latency and in amplitude of a single component. Figure 4 shows a real movement related potential and its component-wise decomposition into five components. This signal is used for synthesizing the test case explored with the following simulation.

Figure 5 presents the outcome of applying the model to a simulated single-trial at a low SNR of -10 dB, where the second peak has been multiplied by 1.5 and advanced by three sample points. In accordance with the results of the previous simulation, the estimator

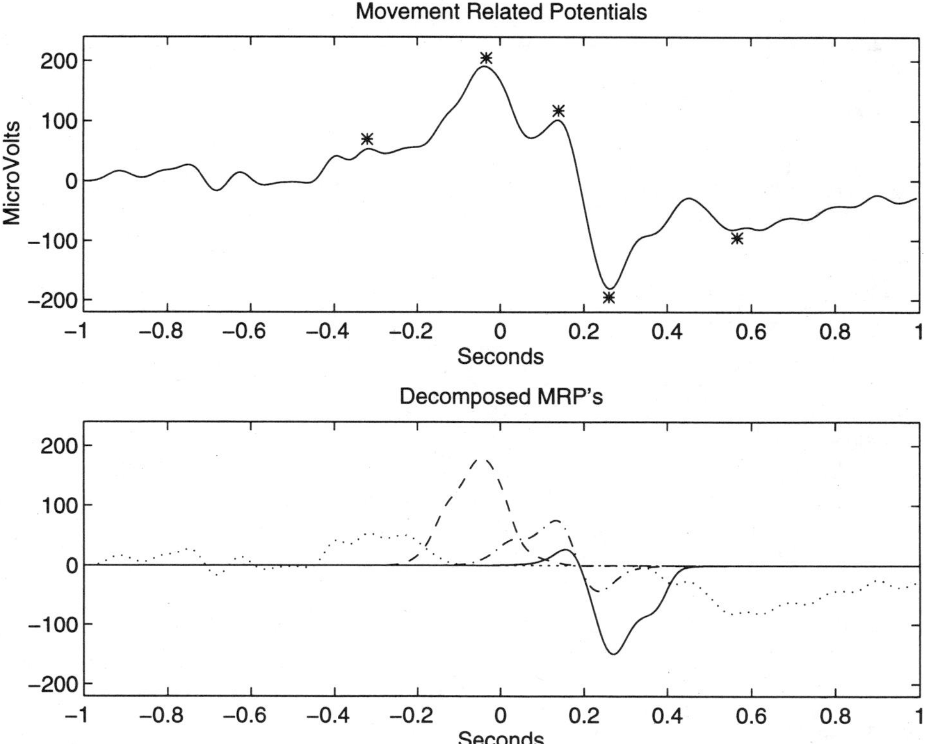

Figure 4. Template decomposition. Top: The asterisks denote "physiological" peaks, according to which the template is decomposed. Bottom: The five decomposed components.

provides a good reconstruction of the morphologically different signal, which nearly coincides with the buried response.

EXPERIMENTAL RESULTS

Movement Related Evoked Potentials (MRPs)

Averaging methods reveal relationships between specific components of MRPs and actual movement parameters (16, 17). We wish to test whether such effects could be described on a single-trial basis, which would obviously improve the existing analysis tools by enabling tracking of variable features. We used evoked potential data recorded during a self-paced finger flexion experiment, using a differential recording between C_3 and C_4 according to the conventional 10-20 electrode placement system. The self-paced movements where occasionally disturbed by a mechanical device without prior knowledge of the subject, who was blindfolded during the experiment. A detailed description of the experiment can be found in (7).

The estimator was applied to the two classes of responses - free and disturbed, which revealed a unique peak appearing at about 150 msec after movement onset with the disturbed movements only (Fig. 6). This unique peak seems to be in response to the reafferent proprioceptive feedback, resulting from the unexpected mechanical disturbance. This result,

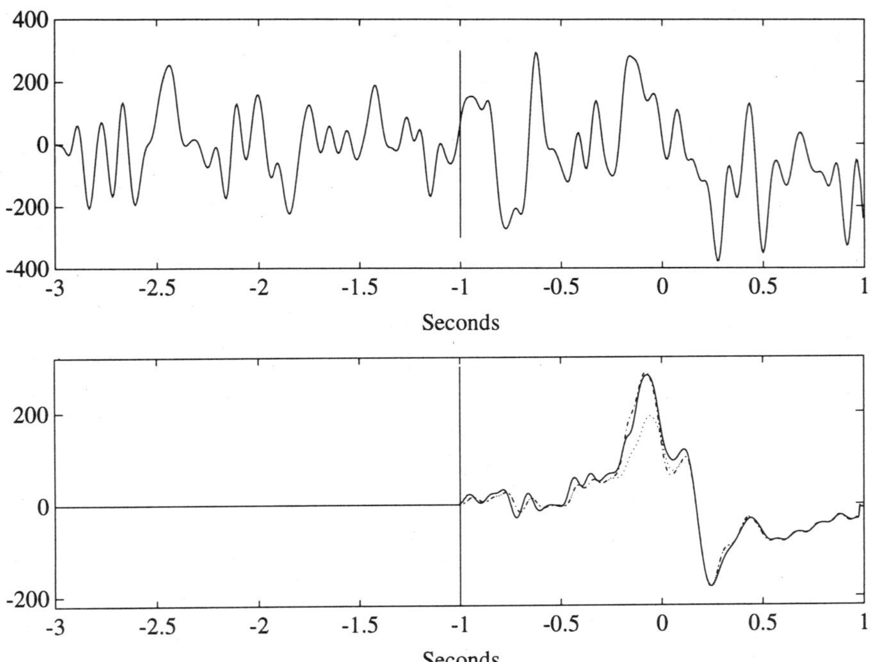

Figure 5. Simulation results. Top: A simulated single-trial. Bottom: Buried signal (solid), estimated signal (dashdotted) and template (dotted). The vertical lines distinguish between pre-stimulus and post-stimulus intervals.

which resembles similar tests carried out with averaging techniques, motivated us to further pursue this line of investigation and try to track signal variations under well-known experimental paradigms. Since cognitive evoked potentials have been studied extensively for a very long time now, and recordings are easily accessible, we chose to use such data in the next test.

Cognitive Event Related Potentials

The main motivation for the development of the proposed estimation scheme was to facilitate tracking capabilities over time-varying evoked potentials. To demonstrate tracking performance we apply the estimator to cognitive evoked potential data, recorded during a typical odd-ball type paradigm (from P_z referenced to the mid-lower jaw). A detailed description of the experiment can be found in (7). In addition to the evoked potential data, reaction times to target stimuli were recorded, and used to demonstrate a relation between the reaction time and the P_{300} complex (18), and to compare the results to previous analysis which we reported in (7). Figure 7, which is reproduced from (7) shows the grand average of 40 single trials. Figure 8 shows 8 sub-ensembles of five trials each enhanced by means of matched filtering (7). The analysis revealed that the latency of the P_{300} component increases with the increase in reaction time, which gives one possible explanation to the appearance of two peaks in the averaged response (see Fig. 7).

However, applying the current estimation procedure, while choosing p_{300a} and p_{300b} to be two distinctive components, yields a different result as can be seen in Fig. 9. In this case, it seems that rather than a change in the latency of P_{300}, we get a change of magnitude

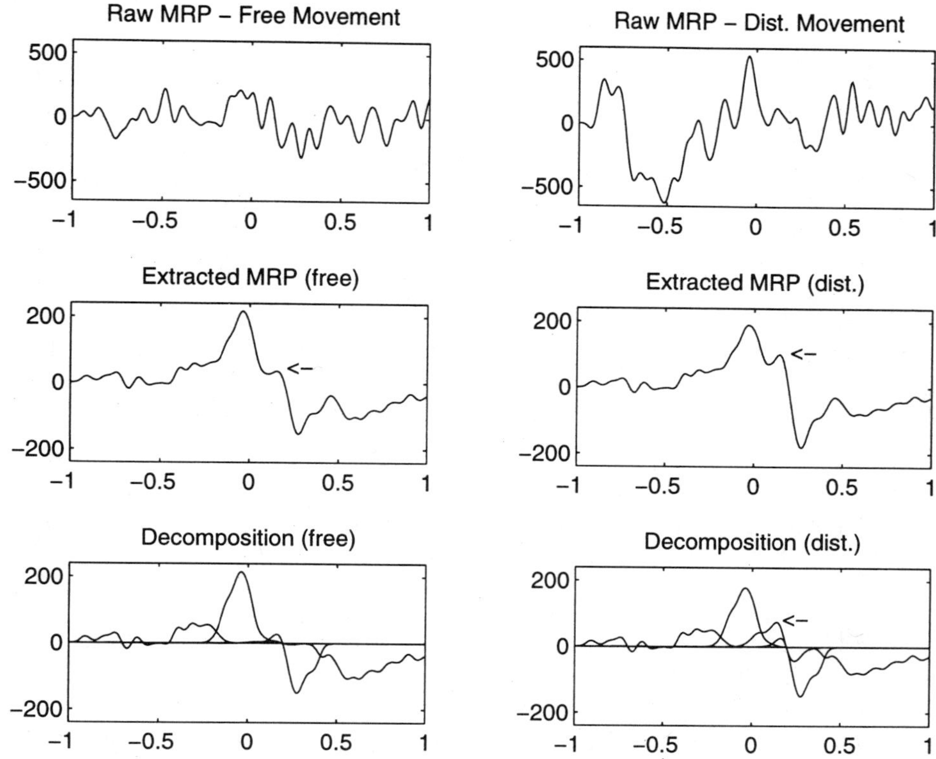

Figure 6. Experimental results. Top row: Raw EEG signals recorded during free and disturbed movements. Middle row: Extracted evoked potentials. Note the unique peak (arrowed) which is significantly larger with the disturbed movement. Bottom row: Evoked potential decomposition; the reafferent peak is dominant with the disturbed movement.

of each sub P_{300} component so that with the increase in reaction time P_{300a} decreases and P_{300b} increases. Although we do not have a physiological interpretation
for this phenomenon, neither do we know whether P_{300} is indeed composed of two distinctive components, the obtained results indicate that there may be a different interpretation to evoked potential variations reflecting dynamic brain processes.

DISCUSSION AND CONCLUSION

A new estimation method for single evoked potential components was presented. The method is based on a novel, error-free decomposition scheme, providing objective latency and magnitude measure to each component which enables estimation of evoked potentials with variable component characteristics. The decomposition pre-processing phase is followed by a parametric single-trial estimation procedure, developed to utilize the outcome of the decomposition phase. The estimation was found to be effective down to an initial SNR of -15dB, depending on the number of required variable components. The tradeoff between template energy and accuracy of estimation was presented analytically, the result of which should be taken into account when considering the number of components allowed to vary from trial to trial.

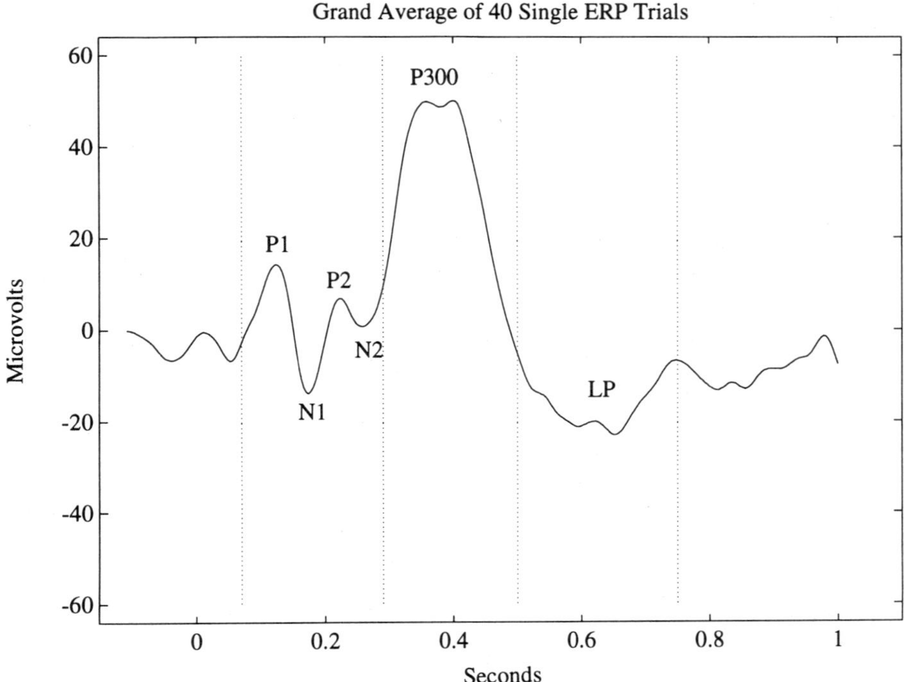

Figure 7. Grand average of 40 ERP target trials recorded during a typical odd-ball type paradigm. The P_{300} component seems to be comprised of two contributions.

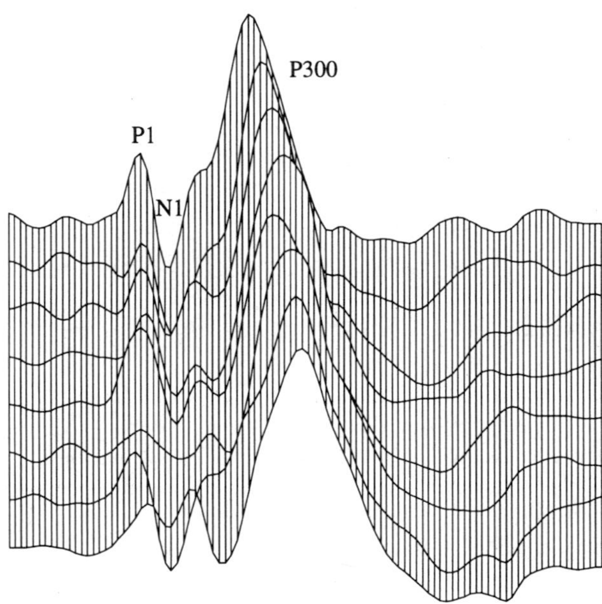

Figure 8. Eight filtered sub-ensembles sorted according to reaction times. The P_{300} latencey increases with the increase in reaction time.

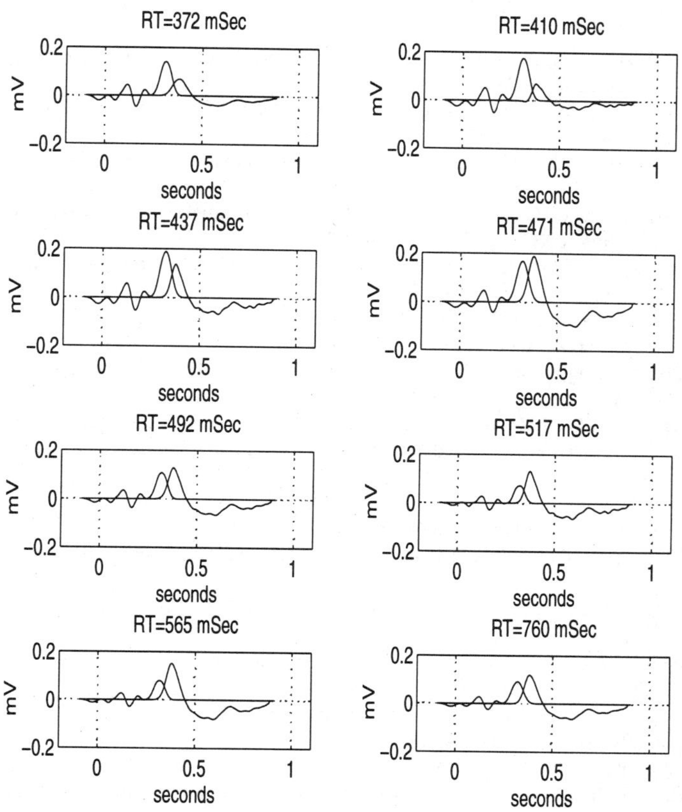

Figure 9. Result of applying the proposed estimator on the ERP data. P_{300a} decreases and P_{300b} increases with the increase in reaction time.

The estimator was demonstrated to compensate for latency and magnitude variations of a single component from an evoked potential complex, displaying a unique difference between two MRPs due to the application of a mechanical disturbance, and a gradual change of relationship between two components of cognitive ERPs depending on physiological parameters. The proposed procedure enables improved evoked potential analysis on a single-trial basis, thus facilitating tracking of amplitude and latency shifts of single evoked potential components; the association of such components with their neuroelectric origins would significantly contribute to the understanding of the genesis of the observed signals, which reflect complicated cognitive processes occurring in the central nervous system.

ACKNOWLEDGMENT

This research was supported in part by the Committee for Research & Prevention in Occupational Safety & Health, Ministry of Labor & Social Affairs, Jerusalem, and by the Ollendorff Center Research Fund.

REFERENCES

1. Gevins, A. S., 1984, Analysis of the electromagnetic signals of the human brain: milestones, obstacles and goals, *IEEE Trans. Biomed. Eng.* 31:833-850.
2. Inbar, G. F., 1972, Muscle spindles in muscle control: Part III: Analysis of adaptive system model, *Kybernetic* 11:130-141.
3. Inbar, G. F., and Yafe, A., 1976, Parameter and signal adaptation in the stretch reflex loop, *Prog. in Brain Res.* 44:317-337.
4. Aunon, J. I., McGillem, C. D., Childers, D. G., 1981, Signal processing in evoked potential research: averaging and modeling, *CRC Crit. Rev. Bioeng.* 5:323-367.
5. Lange, D. H., and Inbar, G. F., 1994, Estimation of morphologically varying evoked brain potentials," In: *Proc. 16th IEEE EMBS Conf.*, Baltimore.
6. Spreckelsen, M. V., and Bromm, B., 1988, Estimation of single-evoked cerebral potentials by means of parametric modeling and kalman filtering, *IEEE Trans. Biomed. Eng.* 35:691-700.
7. Lange, D. H., Pratt, H., and Inbar, G. F., 1995, Segmented matched filtering of single event related evoked potentials, *IEEE Trans. on Biomed. Eng.* 42:317-321.
8. Cerutti, S., Chiarenza, G., Liberati, D., Mascellani, P., and Pavesi, G., 1988, A parametric method of identification of single trial event related potentials in the brain, *IEEE Trans. on Biomed. Eng.* 35(9):701-711.
9. Bartnik, E. A., Blinowska, K. J., and Durka, P. J., 1992, Single evoked potential reconstruction by means of wavelet transform, *Biol. Cybern.* 67:175-181.
10. Gersch, W., 1970, Spectral analysis of EEG's by autoregressive decomposition of time series, *Math. Biosc.* 7:205-222.
11. Madhaven, P. G., 1992, Minimal repetition evoked potential by modified adaptive line enhancement, *IEEE Trans. on Biomed. Eng.* 39:760-764.
12. Pratt, H., and Geva, A., 1995, Spatio-temporal source estimation of evoked potenatials by wavelet-type decomposition, *Proc. Bat-Sheva Seminar*, in press.
13. Haykin, S., 1986, *Adaptive Filter Theory*. Prentice-Hall: NJ.
14. Lange, D. H., and Inbar, G. F., 1995, A robust and error-free decomposition method for transient evoked brain potentials, submitted.
15. Lange, D. H., and Inbar, G. F., 1995, Modeling and estimation of single evoked brain potential components, submitted.
16. Kristeva, R., Cheyne, D., Lang, W., Lindinger, G., and Deecke, L., 1979, Movement related potentials accompanying unilateral and bilateral finger movements with different inertial loads, *Electroencephalogr. Clin. Neurophysiol.* 75:410-418.
17. Deecke, L. Grozinger, B., and Kornhuber, H. H., 1976, Voluntary finger movement in man: cerebral potentials and theory, *Biol. Cybernetics* 23:99-119.
18. McCarthy, G., and Donchin, E., 1981, A metric for thought: A comparison of P_{300} latency and reaction time, Science 211:77-80.

10

TESTING FOR SYNCHRONIZATION IN EVOKED POTENTIALS USING HIGHER ORDER SPECTRA TECHNIQUE

Miram Furst, Irit Sha'aya-Segal, and Hagit Messer

Department of Electrical Engineering-Systems
Faculty of Engineering, Tel Aviv University
Tel Aviv 69978, Israel

ABSTRACT

Ensemble averaging is commonly employed to estimate brain activity evoked by known stimulus. Similar distorted average responses will be obtained whether there is a partial blocking of the nerve, or synchronization disorders. Two algorithms for detecting the responses and the amount of synchronization, which are insensitive to asynchronization problems, are introduced. The first is based on estimating the spectrum of the recorded evoked response, and the other on estimating its bispectrum. In the spectral method, the background noise spectrum should be known or estimated. In the bispectral method, no additional estimation is required, if the signal has a zero dc-component, and the background noise is zero-mean and white, of any symmetric distribution.

INTRODUCTION

Ensemble averaging is commonly employed to estimate brain activity evoked by known stimulus. Generally, when evoked potentials (EP) are considered for clinical applications, the amplitude and the latency of the distinctive waves are measured and compared to mean standard values (Chiappa, 1990). For example, auditory brainstem evoked potentials (BAEPs) in normal hearing people are well characterized by five distinguished peaks that are labeled as waves I-V. The relative amplitude and latency of those waves are routinely used to diagnose dysfunction of the lower auditory pathway. The standard values have been obtained by averaging the brain activity following a given stimulus in hundreds of healthy people. For BAEP, for example, the signal to noise ratio is about -30 dB, which requires at least a thousand sweeps, before the averaged response is sufficient to recognize the desired signal.

Averaging technique is justified when the evoked response is fully synchronized to the stimulus and there is no time shifts between successive EPs. However, it is well known that a time shift jitter exists (e.g., Woody, 1967; McGillem *et al.*, 1985; von Spreckelsen and Bromm, 1988; Nakamura *et al.*, 1991), but because it is relatively small in healthy people, it is usually ignored. When measuring EP of patients, the assumption that the time shift jitter is negligible, is not always justified. For example, in a patient with multiple sclerosis (MS), it is conceivable that some of the expected waves are missing due to the averaging procedure. Since MS is a disease that mainly affects the axons' myelin sheets, which ultimately slows the nerve transmission. One of the outcome of such a disease can be an increase in the time shift jitter. In other diseases such as acoustic neuroma, that involves a tumor on the auditory nerve, the nerve is blocked and BAEP can not be generated. Averaging of successive BAEPs can not distinguish between those two types of lesions.

In the following paper we introduce two methods to detect EP, also when the time averaging fails to detect the response. When the algorithm indicates that the response exists, the jitter of the time shift is estimated.

PROBLEM FORMULATION

A typical evoked potential measurement which is recorded for a period of time T can be described as a random process x(t),

$$x(t) = s(t + \theta) + n(t) \qquad 0 \le t \le T \tag{1}$$

where n(t) represents the ongoing EEG and is approximated as a stationary random process, and θ is a random variable. It is possible to represent x(t) in the frequency domain as,

$$X(\omega) = S(\omega)e^{-j\omega\theta} + N(\omega). \tag{2}$$

The mean of the ongoing EEG was measured and found to be zero, thus, $E\{N(\omega)\} = 0$, and the expected value of $X(\omega)$ is obtained by

$$E\{X(\omega)\} = S(\omega) \cdot \Phi_\theta(\omega) \tag{3}$$

where $\Phi_\theta(\omega)$ is the characteristic function of θ. It is obvious from Eq. (3) that when the first moment of θ is zero, and its second moment is relatively high, then $\Phi_\theta(\omega) \approx 0$ and therefore also $E\{X(\omega)\} \rightarrow 0$.

When time averaging is applied on successive records of evoked potentials, the time average response is given by

$$\overline{x}(t) = \frac{1}{M} \sum_{i=1}^{M} s_i(t) + n_i(t) \tag{4}$$

where

$$\begin{cases} s_i(t) = s(t + \theta_i) & (i-1)T \le t < iT \\ n_i(t) = n(t) & i = 1, M \end{cases}$$

θ_i are I.I.D random variables with characteristic function $\Phi_\theta(\omega)$. Representing the time average response in the frequency domain yields an unbiased and consistent estimator

$$\overline{X}(\omega) = \frac{1}{M}\sum_{i=1}^{M} X_i(\omega).$$ In particularly,

$$E\{\overline{X}(\omega)\} = S(\omega)\cdot \Phi_\theta(\omega),\tag{5}$$

which indicates the time averaged signal is not an adequate estimator of s(t) when a large jitter in the time shift exists. Therefore other estimators are required.

Estimating S(ω) from the Spectrum of X(t)

The spectrum of x(t) can be represented in the frequency domain as follows:

$$S_{xx}(\omega) = E\{X(\omega)\cdot X^*(\omega)\} = S_{ss}(\omega) + S_{nn}(\omega),\tag{6}$$

when $X(\omega)$ is as defined in Eq. (2), whereas $S_{ss}(\omega)$ and $S_{nn}(\omega)$ are the spectrum of s(t) and n(t), respectively. Equation (6) is obtained only when θ and n(t) are statistically independent, which is a reasonable assumption for EP measurements. Since s(t) is a deterministic process,

$$S_{ss}(\omega) = S(\omega)\cdot S^*(\omega) = |S(\omega)|^2.\tag{7}$$

It is, therefore, possible to obtain $|S(\omega)|$ from Eqs. (6) and (7) if both $S_{nn}(\omega)$ and $S_{xx}(\omega)$ are known.

In a session of evoked response recording, in addition to the time averaging of the individual responses $x_i(t)$, it is possible to estimate $S_{xx}(\omega)$; following each T seconds of recording, the Fourier transform can be applied on $x_i(t)$ to yield $X_i(\omega)$. A consistent and unbiased estimator of $S_{xx}(\omega)$ will be

$$\hat{S}_{xx}(\omega) = \frac{1}{M}\sum_{i=1}^{M} X_i(\omega)\cdot X_i^*(\omega)\tag{8}$$

In the same manner one can estimate $S_{nn}(\omega)$ either from the pre-stimulus time, or preferably from EEG recording, when no stimulus is introduced to the subject. In either way, the hidden assumption is that the ongoing EEG added to the synchronized potential is equal to the EEG when no stimulus is given. The variance of the estimated $|S(\omega)|$ equals to the sum of the variances of each of the two estimators, since it was obtained as a difference of two independent estimators.

Estimating S(ω) from the Bispectrum of X(t)

In order to avoid the need of estimating the noise properties independently from the evoked potential measurements, an estimator that is based on the bispectrum of x(t) is proposed.

The bispectrum of x(t), in the frequency domain is given by

$$B_x(\omega_1,\omega_2) = E\{X(\omega_1)\cdot X(\omega_2)\cdot X^*(\omega_1+\omega_2)\}.\tag{9}$$

Substituting $X(\omega)$ from Eq. (2), in Eq. (9) yields

$$B_x(\omega_1,\omega_2) = S(\omega_1) \cdot S(\omega_2) \cdot S^*(\omega_1+\omega_2) +$$

$$+E\{N(\omega_1) \cdot N(\omega_2) \cdot N^*(\omega_1+\omega_2)\} +$$

$$+S(\omega_1) \cdot \Phi_\theta(\omega_1)E\{N(\omega_2) \cdot N^*(\omega_1+\omega_2)\} +$$

$$+S(\omega_2) \cdot \Phi_\theta(\omega_2)E\{N(\omega_1) \cdot N^*(\omega_1+\omega_2)\} +$$

$$+S^*(\omega_1+\omega_2) \cdot \Phi_\theta^*(\omega_1+\omega_2)E\{N(\omega_1) \cdot N(\omega_2)\}. \qquad (10)$$

The bispectrum of a random process with a symmetric distribution equals to zero (e.g., Nikias and Petopulu, 1993). Testing the distribution of the ongoing EEG reveals that its distribution is symmetric. Therefore, the second term in Eq. (10) is equal to zero. The last three terms in Eq. (10) will be equal to zero as well, if the following two conditions are satisfied: (i) the background noise $n(t)$ is a white random process; (ii) the signal $s(t)$ has a zero dc component. When the first condition is satisfied it is easy to show that

$$E\{N(\omega_1) \cdot N^*(\omega_1+\omega_2)\} = 2\pi\sigma_n^2\delta(\omega_1)$$

$$E\{N(\omega_1) \cdot N(\omega_2)\} = 2\pi\sigma_n^2\delta(\omega_1+\omega_2), \qquad (11)$$

where $S_{nn}(\omega) = \sigma_n^2$. Thus Eq. (10) becomes

$$B_x(\omega_1,\omega_2) = B_s(\omega_1,\omega_2) +$$

$$+S(0) \cdot \Phi_\theta(0) \cdot 2\pi\sigma_n^2\delta(\omega_1) +$$

$$+S(0) \cdot \Phi_\theta(0) \cdot 2\pi\sigma_n^2\delta(\omega_2) +$$

$$+S(0) \cdot \Phi_\theta(0) \cdot 2\pi\sigma_n^2\delta(\omega_1+\omega_2) . \qquad (12)$$

When $s(t)$ has a zero dc component (condition ii), $S(0) = 0$, then

$$B_x(\omega_1,\omega_2) = B_s(\omega_1,\omega_2). \qquad (13)$$

Nakamura (1993) has proposed a recursive algorithm to estimate $|S(\omega)|$ from $B_s(\omega_1, \omega_2)$ based on algorithm suggested by Sundaramoorthy *et al.* (1990). From Eq. (13) it follows, that it is sufficient to estimate $B_x(\omega_1, \omega_2)$ instead of $B_s(\omega_1, \omega_2)$ in order to estimate $|S(\omega)|$.

If $s(t)$ does not have a zero dc component, $|S(\omega)|$ can be still estimated from $B_x(\omega_1, \omega_2)$, when the lines in which $B_x(\omega_1, \omega_2) \neq B_s(\omega_1, \omega_2)$ are ignored in the reconstruction algorithm.

A consistent and unbiased estimator of the bispectrum of $x(t)$ is given by

$$\hat{B}_x(\omega_1,\omega_2) = \frac{1}{M}\sum_{i=1}^{M}X_i(\omega_1)X_i(\omega_2)X_i^*(\omega_1+\omega_2) \qquad (14)$$

It can be shown that $E\{\hat{B}_x(\omega_1,\omega_2)\} \xrightarrow[M\to\infty]{} B_x(\omega_1,\omega_2) = B_s(\omega_1,\omega_2)$, and

$$\mathrm{VAR}\{\hat{B}_x(\omega_1,\omega_2)\} \propto O\!\left(\frac{1}{M}\right).$$

SIMULATION RESULTS

In order to test the two estimators presented above, we have chosen to model BAEP as a sum of five impulses about 1 msec apart one from another with various amplitudes, that are passing through a zero-phase band pass filter $H(\omega)$,

$$s(t) = \sum_{i=1}^{5} A_i \delta(t - t_i + \theta) * h(t). \tag{15}$$

The random variable, θ, was chosen to be uniformly distributed in the interval $[-\Delta. \Delta]$, and the background noise n(t) was a white Gaussian process with zero mean. The simulated signal s(t) is presented in Fig. 1a, for $\theta = 0$. Its representation in the frequency domain, $|S(\omega)|$, is presented in Fig. 1b, whereas the bispectrum of s(t), $B_s(\omega_1, \omega_2)$ is presented in Fig. 1c.

Figure 2 represents estimation results for signal to noise ratios: SNR = -4.8 and -8.5 dB, with $\Delta = 1$ msec. For both SNR values, there is no indication for the signal in the time averaging (panels a and d), and both estimators are successfully detect the existence of s(t). Although, the error in the reconstruction of $|S(\omega)|$ seems to be bigger according to the spectrum estimation than according to the bispectrum estimation. As can be expected, the reconstruction improves with increasing SNR.

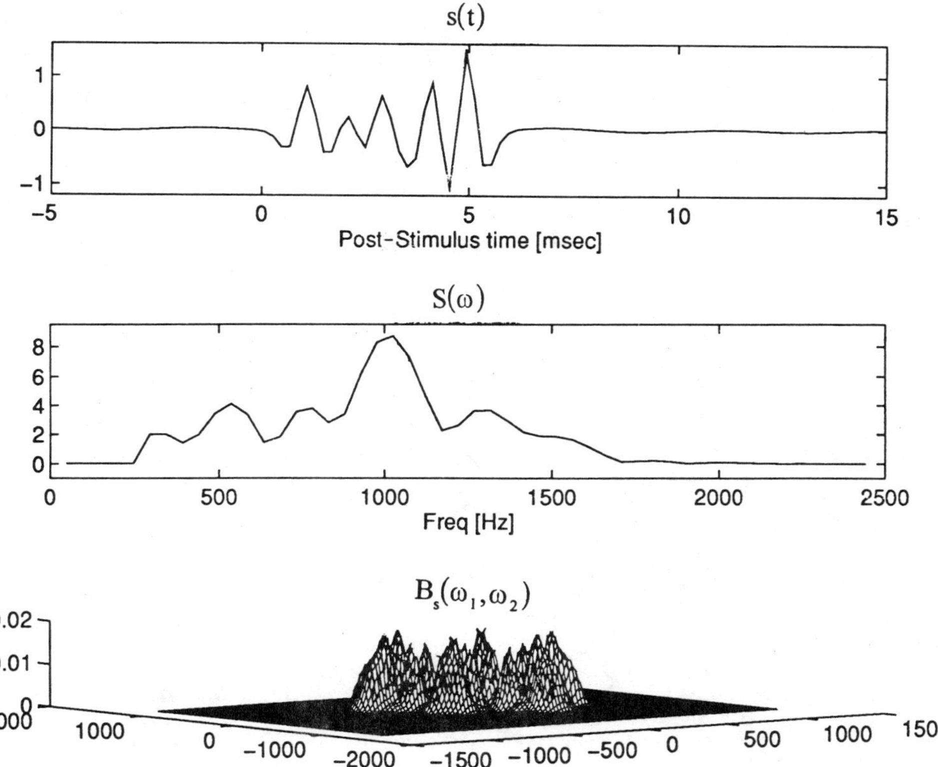

Figure 1. (a) The signal s(t) as a function of the post-stimulus time; (b) its representation in the frequency domain, $|S(\omega)|$; and (c) the bispectrum of s(t), $B_s(\omega_1, \omega_2)$.

Estimating the Time Shift Jitter

Since, in the simulated signal $\theta \sim U[-\Delta, \Delta]$, it follows that $\Phi_\theta = \mathrm{sinc}(\omega\Delta)$ and thus

$$E\{\overline{X}(\omega)\} = S(\omega) \cdot \mathrm{sinc}(\omega\Delta) \tag{16}$$

Based on of the estimator of $|S(\omega)|$ described in the previous section, let's define:

$$\Gamma_\Delta(\omega) = \left(|S(\omega)| \cdot |\mathrm{sinc}(\omega\Delta)| - |\overline{X}(\omega)|\right)^2 \tag{17}$$

Substituting different values for Δ, reveals a family of functions $\Gamma_\Delta(\omega)$. Since $X(\omega)$ was obtained while applying FFT, which provides L different samples in the frequency domain. Let's define the mean-square-error as

$$MSE(\Delta) = \frac{1}{L}\sum_{i=1}^{L}\Gamma_\Delta(\omega_i), \tag{18}$$

Δ that provides the minimal value for MSE is determined as the actual time jitter. Δ was obtained by using a gradient-decent method.

Figure 3 represents the estimation of Δ, for different values of time shifts and two SNR values, as a function of number of sweeps, when $|S(\omega)|$ was either estimated from the spectrum of x(t) (Eq. 8), or from the bispectrum of x(t) (Eq. 14). For relatively high SNR (SNR = -4.8 dB) and relatively small time shift jitter (Δ = 0.4 msec), both estimators yield

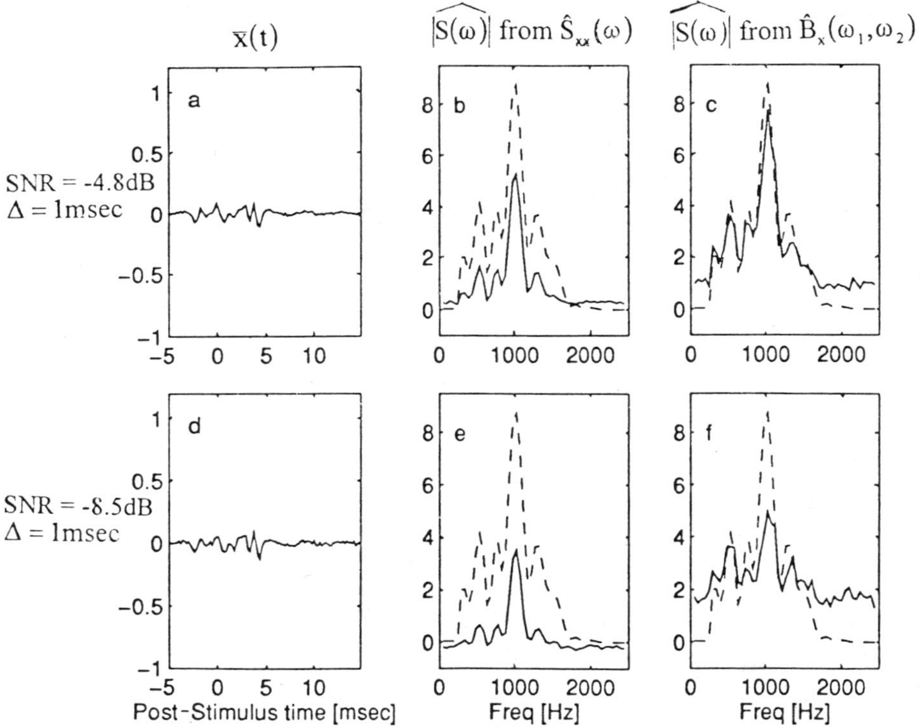

Figure 2. Time averaged signal (panels a and d), and estimation of the signal according to the spectrum (panels b and e) and bispectrum estimation (panels c and f). The continuous lines represent the estimated signals, and the broken lines represent the original signal.

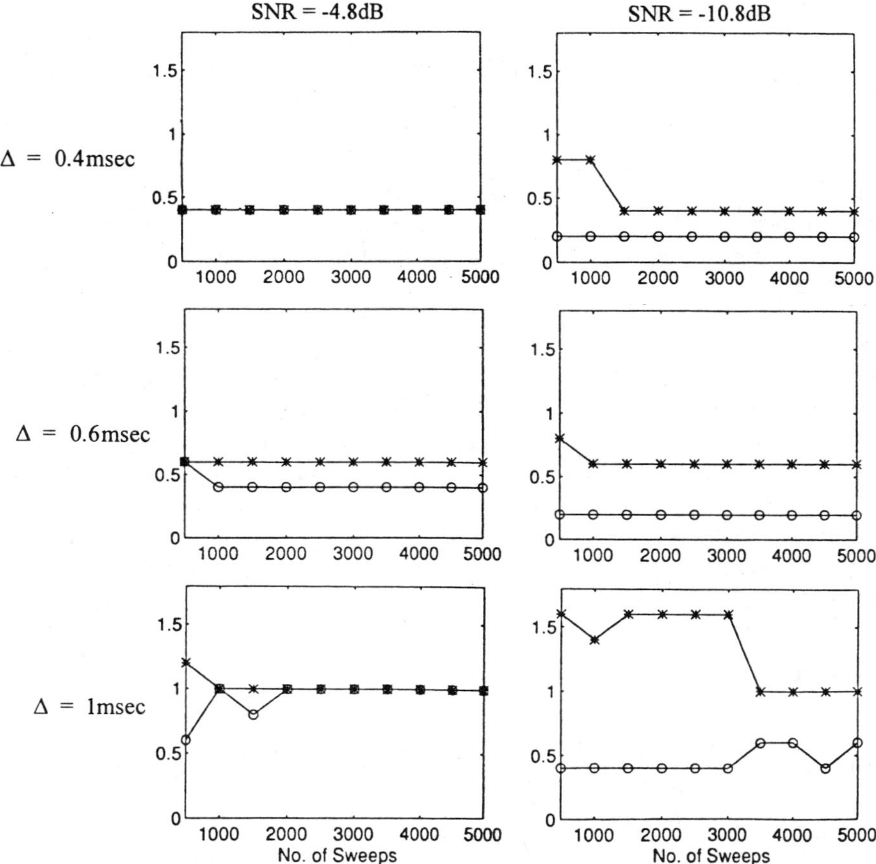

Figure 3. Estimation of the time shift jitter for Δ = 0.4, 0.6, and 1 msec, calculated for SNR = -4.8, and -10.8 dB as a function of number of sweeps. The open circles represent the spectrum estimation method, and the crosses the bispectrum estimation.

a correct estimation for Δ even for as low as 500 sweeps. However for the lower SNR value (SNR = -10.8 dB), a correct Δ was yielded only when the estimation was based on the bispectrum and the number of sweeps was high enough. The minimal number of sweeps required for a correct estimation was increased with Δ. In the simulation presented in Fig. 3, the maximal number of sweeps was 5000, which was not adequate for estimating Δ on the basis of the spectrum of x(t).

DISCUSSION

In the present study we have introduced an algorithm to evaluate the time shift jitter while recording evoked response on basis of estimating $|S(\omega)|$. Two methods of estimating $|S(\omega)|$ were presented; the first was based on estimating the spectrum of the recorded signal and the second was based on the estimation of the recorded signal bispectrum. We have shown that the bispectral method performance is superior to the spectral method, especially for relatively low SNR (Fig. 3). This result is expected, since the spectrum method requires knowledge of the noise spectrum. It is possible to estimate the noise spectrum from a separate recording, or from the pre-stimulus time, but the need for additional estimation eventually effects the precision of the method. The bispectrum method, however, can be applied only if the background noise is white with a symmetric distribution. In EP recording, the background noise is the EEG signal, whose spectrum is not white, but its distribution is symmetric. It is thus recommend to use in an EP recording, a filter that will whiten the ongoing EEG. In most EP recordings, such a filter is reasonable, since the spectrum of the synchronous response does not include low frequencies that compose the EEG spectrum.

In most clinical applications, where there is a need for estimating the time shift jitter, we can not expect that jitter will be the same for all waves. In BAEP, for example, in many cases waves I-III are in the normal range, and only waves IV and V have unrecognized shapes. It is, therefore, not conceivable to apply the algorithm suggested in this paper for the whole signal, but to divide the response to its normal and abnormal parts, and to apply the algorithm only on the abnormal part. The division to normal and abnormal parts can be performed according to the time averaging, which prevents the possibility to perform the algorithm in real time and requires storing the individual responses for an off-line analysis. A better procedure will be dividing the recording session into two subsections. First only a small number of averages will be recorded in order to determine the normal and abnormal parts. At the second step the recording will include more averages, and the algorithm will be applied in real time separately on each of the pre-determined parts.

REFERENCES

Chiappa, K. H., 1990, *Evoked Potentials in Clinical Medicine*, Raven Press, NY.

McGilem, C. D., Aunon, J. I., and Yu, K., 1985, Signals and noise in evoked brain potentials, *IEEE Trans. Biomed. Eng.* 32, 1012-1016.

Nakamura, M., 1993, Waveform estimation from noisy signals with variable signal delay using bispectrum averaging, *IEEE Trans. Biomed. Eng.* 40, 118-127.

Nakamura, N., Nishida, S., and Shibasaki, H., 1991, deterioration of averaged evoked potential waveform due to asynchronous averaging and its compensation, *IEEE Trans. Biomed. Eng.* 38, 309-312.

Nikias, C. L. and Petopulu, A.P., 1993, *Higher-Order Spectra Analysis: A Nonlinear Signal Processing Framework*, Prentice Hall, NJ.

von Sperckelsen, M., and Bromm, B., 1988, Estimation of single-evoked cerebral potentials by means of parametric modeling and Kalman filtering, *IEEE Trans. Biomed. Eng.* 35, 691-700.

Sundaramoothy, G., Raghuveer, M. R., and Dianat, S. A., 1990, Bispectral reconstruction of signals in noise: Amplitude reconstruction issues, *IEEE Trans. Acoust., Speech, Signal Process.* 38, 1297-1306.

Woody, C. D., 1967, Characterization of an adaptive filter for the analysis of variable latency neuroelectric signals, *Med. & Biol. Eng. & Comput.* 5, 539-555.

11

ANALYSES OF TRANSIENT AND TIME-VARYING EVOKED POTENTIALS FOR DETECTION OF BRAIN INJURY

Nitish V. Thakor[1] and Xuan Kong[2]

[1] Biomedical Engineering Department
Johns Hopkins School of Medicine
Baltimore, Maryland 21205
[2] Electrical Engineering Department
Northern Illinois University
DeKalb, Illinois

ABSTRACT

In clinical situations, such as high-risk surgical or neurological critical care, evoked potential (EP) monitoring may help identify incidence of brain injury. Experimental models of brain injury, such as cerebral hypoxia or ischemia, have been created to help understand the cerebral physiology and to develop algorithms to detect such events. The problem here is to identify transient or time-varying changes in EP signals during continuous monitoring of brain in noisy environments. Several signal processing techniques have been developed to identify injury-related changes in EP signals: adaptive filtering, adaptive Fourier series modeling, adaptive delay estimation and adaptive coherence estimation. The adaptive filtering algorithm analyzes time-varying changes in EP signals while improving the signal-to-noise ratio. Application of this algorithm helped capture, in controlled experimental studies and intraoperatively, transient events such as clipping of cerebral artery and response to anesthetics, respectively. The adaptive Fourier series modeling algorithm constructs a model of normal EP signal from which injury-related changes can be inferred. This algorithm showed that neurological injury, such as caused by cerebral hypoxia, results in frequency dispersion; this observation may be used as an early indicator of brain injury in critical care settings. The adaptive coherence estimation algorithm is employed to determine if the injury-related response causes nonlinear transformation of the measured signals. A linearity index, derived from the coherence estimates, helps identify at an early stage the irreversible course of brain injury. In conclusion, several advanced signal processing algorithms have been developed to demonstrate that transient and time-varying changes in neurological signals can be used to monitor the brain's response to injury in critical care situations.

Advances in Processing and Pattern Analysis of Biological Signals, Edited by Isak Gath and Gideon F. Inbar
Plenum Press, New York, 1996

INTRODUCTION

Considerable effort and technology has been brought to bear on the problem of monitoring the heart in critical care situations, but a comparable understanding of the problem and corresponding technology is not available for monitoring the brain. For example, extensive technology is available for monitoring cardiac rhythms in cardiac intensive care units. Comparable neurological critical care units are only now emerging (1, 2). The problem of monitoring the brain in such critical care settings is poorly understood, and technological solutions are limited. Similar problems exist for monitoring the brain during high-risk surgical procedures, particularly the ones that may pose direct invasive risk to the brain or indirect risk through compromise to circulation (3, 4) or brain oxygenation (5, 6) (see Fig. 1). Therefore, novel techniques are needed for monitoring the structure and function of the brain during high risk patient monitoring. Similarly, intraoperative monitoring poses high risk situations for the patients (7-11). Here, classically, the cardiovascular system received primary attention. In surgeries that pose risk to the nervous system, including spinal cord (12), subcortical (13, 14) or cortical structures (7), specialized neurological monitoring techniques are desired. Examples and anecdotes of the value of intraoperative neurological monitoring abound. Case studies and applications have been reported: such as detection of cerebral hypoxia (15), brainstem injury (13), injury during aneurysm surgery (16, 17), and cerebral artery occlusion (18). While various imaging techniques, such as x-ray computed tomography, magnetic resonance imaging and positron emission tomography, provide information about the structure of the brain and, to a limited extent - function, these imaging modalities are not useful for continuous diagnostic monitoring of the patient. For obtaining structural as well as functional information, some other techniques, such as intracranial pressure, Doppler ultrasound, and near infrared spectroscopy, are being investigated and, to a limited extent, used in clinical situations. These instruments sense the normal and pathological levels of the brain perfusion and pressure, fluids and blood flow, and oxygenation, which when abnormal, pose risk to the integrity and function of the brain. Despite all these techniques now available to neuroscientists, it is the measurement of electrical function that provides the most direct indication of normal or pathological function of the brain and eventual outcome for the patient when being treated for injury or disease (19).

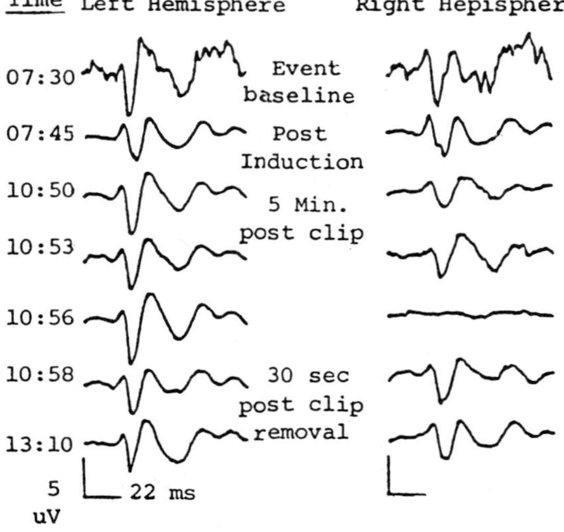

Figure 1. An application of intraoperative EP signal monitoring. During a surgical procedure, right middle cerebral artery is temporarily compressed by a clip. This results in a rapid loss of amplitude on one side. Removal of the clip results in restoration of the EP signal. Continuous intraoperative monitoring of EP signals, thus, is potentially useful as a diagnostic tool which helps to avert serious brain injury (adapted from (9)).

Two dominant methods for measuring the electrical function of the brain prevail: electroencephalogram (EEG) and evoked potentials (EP) (19, 20). These electrical signals are generated by spontaneous or triggered activity of the brain's electrically active functional mass (EEG and EP, respectively) (21). EEG signals represent composite, ongoing or spontaneous activity of the brain structure underlying the measuring electrodes. EP signals are elicited in response to externally applied stimuli, such as auditory clicks (brain stem auditory EP, or BAER), light flashes or patterns (visual EP, or VEP), electrical stimulation of the periphery (somatosensory EP, or SEP) (13, 22). The electrical activity is recorded most commonly from the scalp, although in some limited situations direct recording of electrical activity from the brain surface, known as electrocorticogram, may be possible. Surface electrodes, made of gold plated contacts, are securely attached to the scalp with the aid of adhesives and conductive gels. Spontaneous EEG may be recorded from numerous locations on the scalp, preferably following a standardized arrangement, called the 10-20 grid system. For many high-risk clinical situations or for animal studies, even a single sensing electrode or an electrode pair (preferably situated on either side of the head) often suffices. EP signals are analogously acquired, although electrodes are cited in the somatosensory area being simulated. The external stimulation elicits a functional electrical response which is then acquired, along with the extraneous ongoing activity of the rest of the brain.

Both EEG and EP signals require extensive signal analysis (23). EEG signals have a spontaneous and noncoherent pattern without significant characteristic features that are apparent when viewed directly (an exception being certain patterns such as spikes that arise from epileptic seizures). EP signals have well defined morphologies, but these signals are small and invariably corrupted by very high levels of background noise (24). Even when the problem is to detect the normal from abnormal status of the brain, or an acute incidence of injury, analysis of the brain's electrical activity requires extensive signal processing. EEG signals can be analyzed in time or frequency domains, where various parametric or nonparametric methods are employed to identify signal trends indicative of the brain's response to injury. EEG signal analysis will not be covered in this article but is referenced elsewhere. EP signal analysis poses formidable signal analysis problems (25, 26), of which two are the most significant: 1) low signal-to-noise-ratio (SNR) of the recordings implies that some form of coherent averaging or filtering is needed, and 2) injury to the brain may occur in an unpredictable manner so that signal analysis must capture transient or time-varying events that are specific to the injury. Solution of these and related problems have necessitated development of powerful new signal processing approaches (reviewed in (26, 27)).

This article briefly reviews the subject of EP signal analysis and then proceeds to several methods recently developed to analyze transient or time-varying signals. These methods include adaptive filtering (28, 18), adaptive Fourier series modeling (29, 30), adaptive delay estimation (31), and adaptive coherence estimation (32). This article presents the development of algorithms, certain mathematical results, simulations that demonstrate performance of these algorithms, and then numerous experimental studies.

EP Signals

A classical problem in EP signal analysis can be formulated as follows. In response to the external stimulation, a coherent signal of the responsive pathway in the brain is obtained. For example, SEP signals are elicited in response to repetitive stimulation of the peripheral limb by an electrical current. The signal of interest, or the desired signal, is d while the added noise is n. In most experimental situations, it would be reasonable to assume that the signal d and noise n are uncorrelated (signal responds to the stimulus while noise, representing the background activity of the rest of the brain, is presumed not to). Thus, an assumption is made here that the signal and noise are additive and uncorrelated. The

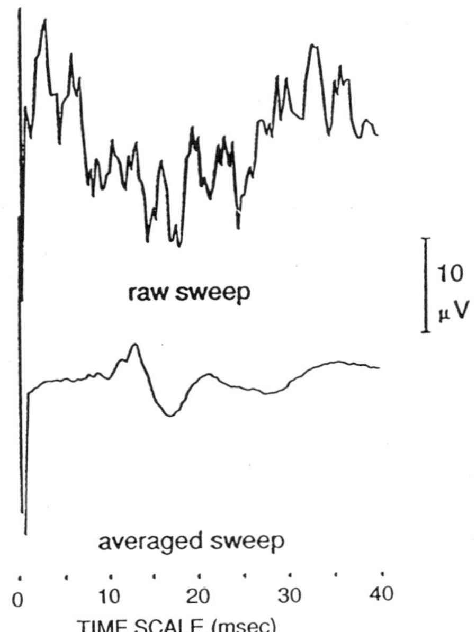

Figure 2. An unaveraged and an ensemble averaged EP waveform. Averaging improves the SNR and brings out the characteristic features (amplitude, latency, morphology) of the waveform.

composite signal is expressed as $s = d + n$. Most of the applications and data analysis are carried out on a computer, for analyses in which the data are digitized, so the discrete signal is represented by $s_k = d_k + n_k$. Let us consider two trials: so that $s1_k = d_k + n1_k$ and $s2_k = d_k + n2_k$.

Ensemble Averaging. In general, then, say i trials are carried out, obtaining $s(i)_k$. If the signal remains constant throughout, a convenient way to extract the desired signal from noise would be to obtain an ensemble average. Noise throughout the acquisition would be expected to generate different realization of the same underlying random process. Therefore, ensemble averaging would result in a reduction of the variance of this noise. If noise is assumed to be zero mean, then ensemble averaging would effectively reduce noise as more responses are averaged. Sufficient number of averages can reduce the noise to levels below that of the signal, or, in other words, increase the signal-to-noise ratio (SNR) to a satisfactory level (Fig. 2). It is usual to need ensemble averages of hundreds of responses to achieve a satisfactory SNR.

Exponentially Weighted Averaging

For the problem of monitoring the brain's response to injury, one cannot assume that the signal is constant or even stationary. That is, the desired signal may be expected to vary unexpectedly in response to changes in the brain's physiological response, as and when injury might occur, or when recovery might take place. Therefore, it would be desirable to forget the old data as new data are acquired. This is accomplished by using a moving or weighted averaging scheme (33). In the moving average, each time a new response is obtained, the oldest response is discarded, keeping in the data buffer a finite number of responses from which the average is obtained. This approach, however requires an extensive amount of memory, and also weighs the oldest data and the newest data equally and the data outside of the buffer are forgotten completely. Therefore, an alternate scheme that also works well is to use exponential weighing, i.e., a forgetting factor is used to proportionally weight the new response and the older weighted average. It can be shown that this process results in an exponentially weighted averaging (34). One problem with this technique is that it is not dynamically responsive to changes in the signal or noise statistics. For example, abrupt changes in signal cannot be accommodated without affecting the SNR. Further, none of the averaging schemes result in a direct interpretation of injury-related response. For this reason, several of techniques based on the principle of adaptive filtering, have been developed (28, 34).

SIGNAL PROCESSING METHODS

Adaptive Filtering

Recall our assumption that the desired signal and noise are additive and uncorrelated. Indeed, for each successive response, the noise is also assumed to be uncorrelated. Therefore, if we minimize the mean-squared error (MSE) between two successive responses, the uncorrelated noise can be eliminated (28). Consider the analysis of two successive sweeps: $p = s + n_1$ *and* $r = s + n_2$, where, for the purpose of notation, we will call p the primary signal and r the reference signal, i.e., two successive responses consist of underlying signal s and two different realizations of noise n_0 and n_1. We assume that the signal is identical between the two sweeps while noise is additive but uncorrelated (with signal as well as between successive realizations). For the filter depicted in Fig. 3, the filter error is $e = (s + n_1) - y$, where y is the output of the filter which is intended to be the best least squares estimate of the signal. The filter error is $e^2 = (s + n_1)^2 - 2y(s + n_1) + y^2 = (s - y)^2 + n_1^2 + 2n_1(s - y)$. Then, the mean squared error (MSE) between the filter estimate of the signal and the measured signal is

$$E[\varepsilon^2] = E[(s-y)^2] + E[n_1^2] + 2E[n_1(s-y)] \tag{1}$$

Since n1 and *(s - y)* are uncorrelated, $E[n_1(s - y)] = 0$. Then, minimizing the MSE, results in

$$E[\varepsilon^2] = Min\{E[(s-y)^2]\} + E[n_1]\} \tag{2}$$

Therefore, when the MSE is minimized, the filter output y is the least squares estimate of signal s. Further derivations of this concept are presented in (28). One requirement for achieving a low SNR is to have a low noise reference signal (*i.e.,* n_2 be small). This can be accomplished by, for example, using a running average as the reference. While this may be

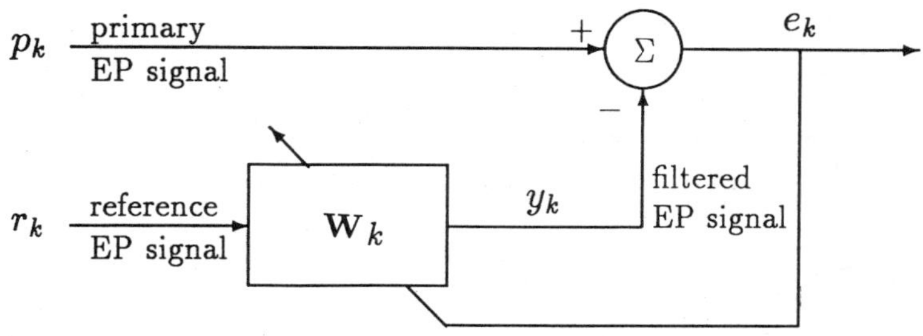

Figure 3. A basic adaptive filter scheme with primary and reference inputs (p and r). Filter weights **W** are adapted to minimize the error e.

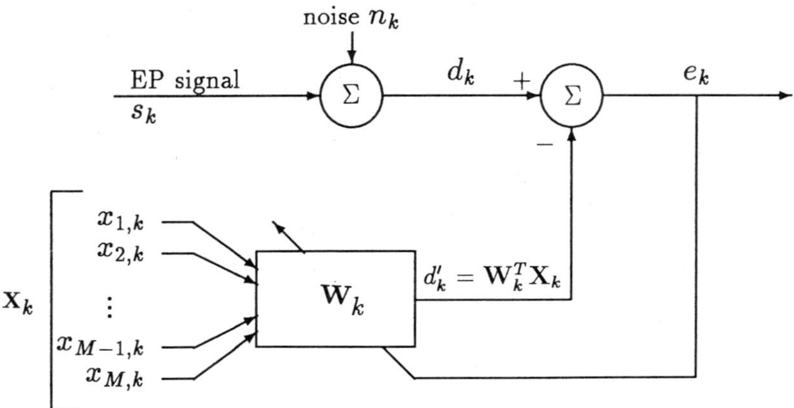

Figure 4. A model referenced adaptive filter. The reference vector $\mathbf{X}$ serves as the signal model (such as Fourier or Walsh), and when weight $\mathbf{W}$ is adapted, the filter output d' serves as the best (in the minimum MSE sense) estimate of the signal s.

convenient in certain low noise situations, such as event-related visual EP signals, in high noise environments, an alternate strategy works well when signal or noise can be parameterized (35). The basic idea is to construct a signal "model" which can serve as the reference (Fig. 4). A significant advantage of this method is that the signal model can be utilized to interpret pathological disturbances, such as might occur as a result of injury (36).

The signal in this application is thought to change transiently or be time-varying. Therefore, the signal analysis algorithm needs to be data adaptive. As shown above, minimizing the MSE results in a reduction in noise, but at the same time, the signal model is allowed to adapt. The signal template generated by a transversal filter, for example, can now be adapted by an algorithm that searches for the error minimum at each sample, and therefore effectively at each response. The adaptive filtering scheme works by minimizing the MSE by a gradient search technique (37). In the gradient search algorithm, the gradient of the expected value of the error is calculated, and the error search is allowed to proceed along the path of steepest descent (Fig. 5). A numerically simple algorithm results if the expected value of the error gradient is approximated by the instantaneous value (38). This is the least mean-squared (LMS) algorithm.

Therefore, we need to calculate the filter weights that minimize the MSE, $\xi = E\,[e_k^2]$. We can do this by calculating the gradient, i.e., by the method of steepest descent. Weight vector $\mathbf{W}_k$ is iteratively recalculated with the intention of minimizing ξ.

$$\mathbf{W}_{k+1} = \mathbf{W}_k + \mu\left(-\frac{\partial \xi}{\partial \mathbf{W}}\right)_k \tag{3}$$

Here, μ determines the gradient, or the rate at which error is minimized (this parameter is empirically determined). A numerically simple algorithm emerges if, instead of taking the gradient of the statistical expectation of the error, we accept the instantaneous estimate of the error, i.e., $\hat{\xi} \cong e_k^2$.

Substituting this approximation into (3), and noting that

$$\varepsilon_k = d_k - \mathbf{X}_k^T\mathbf{W}_k = d_k - \mathbf{W}_k^T\mathbf{X}_k \tag{4}$$

the LMS or the least-mean squared estimation algorithm is obtained

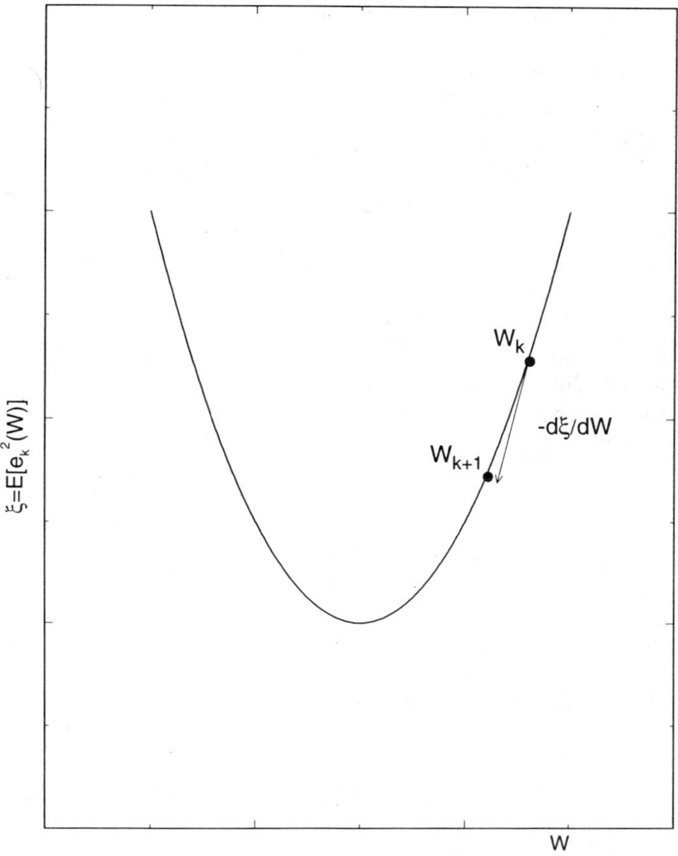

Figure 5. The steepest descent algorithm. MSE is minimized (by adapting weights $\mathbf{W}_k$) along the path of the maximum negative gradient of error.

$$\mathbf{W}_{k+1} = \mathbf{W}_k + 2\mu\varepsilon_k\mathbf{W}_k \tag{5}$$

The advantage of the LMS algorithm, besides its mathematical simplicity, is that it takes advantage of the statistical independence of the signal and noise, while it allows for a dynamic adjustment of the rate at which the adaptation is allowed to proceed. A high value of the adaptation (or convergence) parameter results in a faster adaptation to dynamic changes in the signal but at the expense of a higher steady-state MSE. A small value of the adaptation parameter achieves the opposite result. Thus, an effective improvement in SNR is achieved while transient or time-varying changes in the signal are tracked by a judicious selection of the rate of adaptation. The adaptive LMS algorithm, in steady-state, converges to the Wiener filter

$$\mathbf{W}^* = \mathbf{R}^{-1}\mathbf{P} = E[X_k X_k^T]^{-1}\, E[p_k X_k] \tag{6}$$

ensuring that a minimum MSE is reached. Adaptation rate is also a trade-off between the minimum MSE reached (the smaller the μ, the lower the steady-state MSE) and the ability to follow transients (higher the value of μ, the faster the adaptation). This trade-off needs to be made judiciously, with the knowledge of the desired SNR and the time constant of the transients (18) (Fig. 6).

Adaptive Impulse Correlated Filter. A simple approach to adaptive filtering of EP signals is to provide a unit impulse as a reference (39). It can be shown that this filter acts in a manner equivalent to an exponentially weighted averager, with a simple one parameter weighting of past responses to stimuli (34). Thus, the primary input to the adaptive filter is the newly acquired response, while the reference is a unit impulse coincident with the stimulus. The adaptive filter is a transversal filter, or a tapped delay line with the number of taps or weights equal to the number of samples spanning the EP signal. The MSE between the incoming signal and the output of the filter is minimized adaptively. At each sample, only one weight is adjusted, since the reference input is 1 only once, with all the other reference inputs to the other weights being 0. At each iteration, depending on the adaptation parameter selected, the weights incrementally adapt to a steady state.

For this filter,

$$\mathbf{R} = (1/L)\mathbf{I} \text{ and } \mathbf{P} = (1/L)[s_1\ s_2\ ...\ s_L]^T \tag{7}$$

where L are the number of taps/weights. Therefore, upon convergence when the optimum filter weights are the signal sample values:

$$\mathbf{W}^* = \mathbf{R}^{-1}\mathbf{P} = [s_1\ s_2\ ...\ s_L]^T \tag{8}$$

Further criteria for selecting the adaptation rate, steady-state misadjustment error, and so on are derived in Laguna *et al.* (34). The signal, in the steady-state is seen to be

$$s_k' = \left\{ 1 - \left(1 - 2\frac{\mu}{L} \right)^k \right\} s_k \tag{9}$$

As $k \to \infty$, and for small μ, s'_k becomes s_k. The improvement in SNR can be shown to be

$$\Delta SNR = \frac{1}{\mu} \frac{\left(1 - (1 - 2\mu)^N \right)^2}{\left(1 - (1 - 2\mu)^{2N} \right)} \tag{10}$$

where N is the number of responses analyzed. When N is small enough to satisfy $\mu N \ll 1$, then $\Delta SNR \cong N$, which would be the result obtained for exponentially weighted averaging.

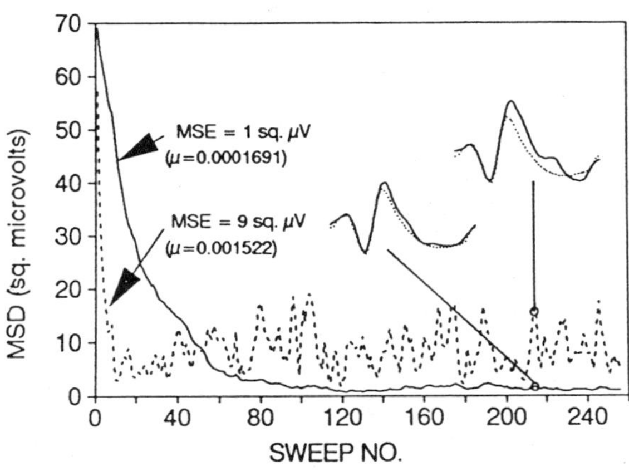

Figure 6. Adaptive filtering reduces the MSE; the larger the adaptation step μ, the faster the adaptation, but the higher the steady-state MSE, and vice versa.

Thus, under limiting conditions, the adaptive impulse correlated filter acts as an exponentially weighted averager. The adaptation parameter μ gives the ability to select the adaptation rate as well as the weighting needed for SNR enhancement.

Modeling

As mentioned earlier, it may be desirable to construct a model of the reference signal. The signal model would be used to obtain changes in the entire signal morphology or appropriate quantitative characteristics to determine the brain's response to anesthetics (40), drugs, or injury. The signal model can conveniently be constructed from any orthonormal expansion, such as Fourier or Walsh functions (36, 41). Thus, a certain number of harmonics may be used to reconstruct the signal (Fig. 7a), and following the scheme described above, minimizing the MSE between the signal model and the newly acquired EP signal results in a noise-free output (29). Thus, a digitized signal (sampling indicated by index k and time period T) can be described by

$$s_k = \sum_{r=1}^{T/2-1} a_r \sin\left(2\pi\frac{k}{T}\right) + b_r\cos\left(2\pi\frac{k}{T}\right) \tag{11}$$

and the signal model, defined here as the reference vector $\mathbf{X}_k$ can be described as

$$\mathbf{X}_k^T = \begin{bmatrix} \sin\left(2\pi\frac{k}{T}\right) \sin\left(2\pi\frac{2k}{T}\right) ...\sin\left(2\pi\frac{ik}{T}\right)...\sin\left(2\pi\frac{(T/2-1)k}{T}\right) \\ ...\ [\cos\left(2\pi\frac{k}{T}\right) \cos\left(2\pi\frac{2k}{T}\right) ...\cos\left(2\pi\frac{ik}{T}\right)...\cos\left(2\pi\frac{(T/2-1)k}{T}\right) \end{bmatrix} \tag{12}$$

Here, $\mathbf{X}_k$ essentially describes a truncated Fourier series. A significant advantage of this algorithm is that the Fourier coefficients, that is the weights $\{a_r, b_r\}$ associated with the harmonics, become indicators of the time-varying changes in the EP signal (30), such as might result from injury.

Note that other signal models may be convenient depending on the application. For example, Walsh function model (Sal and Cal are functions analogous to Sin and Cos functions of the Fourier series) (Fig. 7b)

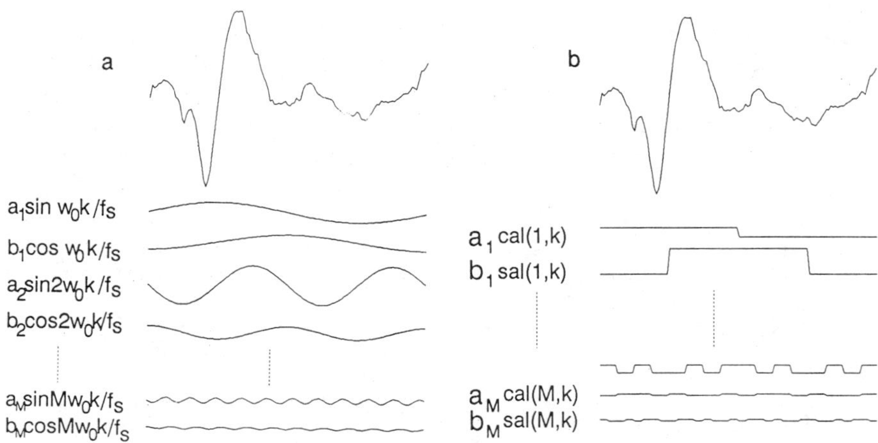

Figure 7. (a) Fourier model of EP signal. (b) Walsh model of EP signal.

$$s_k = \sum_{m=1}^{M} a_m Sal(m, k) + b_m Cal(m, k) \tag{13}$$

would result in computational benefits (36).

Estimation

Delay Estimation. Further applications of adaptive filtering of EP signals are possible, such as when signals are time varying, or when we are interested in the time-varying features of the EP signals, such as the latency (42) (that is, the period between the stimulus and a signal peak) are desirable. For such applications, we can develop an adaptive delay estimation algorithm (31). Here, we define delay as the difference in the latency of a signal peak from its initial or control value and a new value resulting during the experimental procedure. Thus, we obtain two signal realizations:

$$x1_k = s_k + n1_k \quad \text{and} \quad x2_k = \lambda S_{k-\theta_k} + n2_k \tag{14}$$

where s_k is the noise-free signal and $s_k - \theta_\kappa$ is the delayed version of the signal s_k, with θ_k as the delay. If the delay is time-varying, a similar adaptive algorithm can be employed to obtain its time-varying estimates.

$$\theta_{k+1} = \theta_k - \mu \partial e_k^2(\theta_k, \lambda_k) / \partial \theta_k$$
$$\lambda_{k+1} = \lambda_k - \mu \partial e_k^2(\theta_k, \lambda_k) / \partial \lambda_k \tag{15}$$

An iterative algorithm can be derived as before. Adaptive delay estimation is helpful in following time-varying features of the EP signal, such as amplitude and latency. This may be helpful in determining, for example, the brain's response to anesthetics or injury.

Coherence Estimation

Signal modeling is particularly helpful in determining if there are changes from *a priori* or unknown earlier status. Signal modeling may be helpful, for example, when monitoring response to an anesthetic. However, in clinical applications, we may wish to identify changes that are specific to a disease or injury. In such situations, we may wish to determine the degree of association between normal, or control signal, and the altered or disease-related signal. This degree of association can be determined using the coherence function (Fig. 8). This approach works quite conveniently when using the Fourier series model of the EP signal (36). The basic idea is to use the coherence estimate to determine whether the association between control and diseased states is linear, which can be implied by determining the coherence function (32). Coherence function values close to one imply a linear association, while values close to zero imply a nonlinear association (43).

As applicable to our earlier signal model, let us consider two measurements: the primary signal p (obtained, for example, after the experiment) and the reference signal r

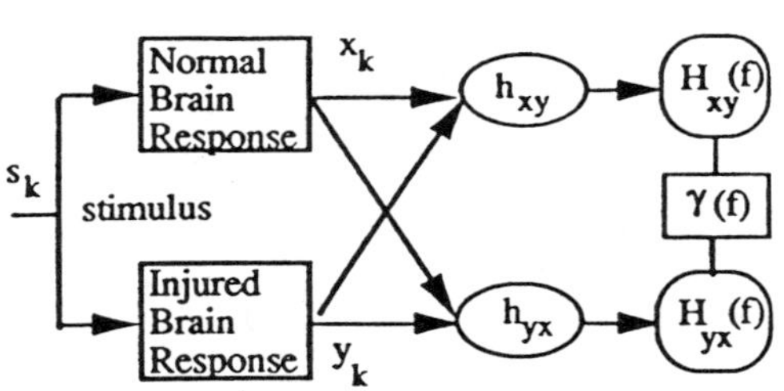

Figure 8. A schematic of the EP measurement system. An input of s results in a measured output x from a normal brain, and an output y from an injured brain. By building the underlying impulse responses h and transfer functions H, coherence function indicative of a linear association between the two measurements can be developed. A low coherence implies that the measurements from an injured brain does not show a linear association between the two signal generating systems.

(obtained, for example, as a control before the experiment). From a systems theory point of view, we can present the primary measurement as an adaptation or an alteration of the reference signal as a result of the experiment (the experiment here, in a general sense, is the perturbation of the brain by any experimental or clinical procedure). The two signals can be related to each other by

$$p(k) = h_{rp} \otimes r(k) + n(k) \tag{16}$$

where $h_{pr}(k)$ is the finite length impulse response of the linear system with input $r(k)$ and output $p(k)$, with $n(k)$ as the noise representing that part of the newly measured signal $p(k)$ that cannot be described by this linear transformation. In vector form,

$$p(k) = H_{rp}^T R(k) + n(k) \tag{17}$$

where $R(k) = [r(k-M)...r(k)...r(k+m))]^T$. Let $\hat{H}_{rp}$ be the estimate of H_{rp} (in the following, $\wedge$ denotes an estimate of the parameter). We use an adaptive algorithm, as before (44).

$$\hat{H}_{rp}^T(k+1) = \hat{H}_{rp}^T(k) - \frac{\mu}{2} \frac{\partial e^2(\hat{H}_{rp}(k))}{\partial \hat{H}_{rp}(k)} \tag{18}$$

which can be shown to result in

$$\hat{H}_{rp}^T(k+1) = \hat{H}_{rp}^T(k) + \mu(p(k) - \hat{p}(k))R(k) \tag{19}$$

with $\hat{p}(k) = \hat{H}_{rp}^T R(k)$.

For a linear invertible system, analogous transformation from $p(k)$ to $r(k)$ would be possible, resulting in a systems description

$$\hat{r}(k) = \hat{H}_{pr}^T P(k) \tag{20}$$

The impulse response $h_{pr}(k)$ can be estimated by

$$\hat{H}_{pr}^T(k+1) = \hat{H}_{pr}^T(k) + \mu_p(r(k) - \hat{r}(k))P(k) \tag{21}$$

The magnitude squared coherence between these two signals is defined as

$$\Gamma(f,k+1) = \hat{H}_{rp}(f,k+1)\hat{H}_{pr}(f,k+1) \tag{22}$$

where Γ is the magnitude squared coherence (MSC) function, and $\hat{H}_{rp}(f, k+1)$ and $H_{pr}(f, k+1)$ are the Fourier transforms of the respective impulse responses of the two systems transforming the primary to the reference and the reference to the primary signal.

Once again, this is an adaptive algorithm for calculating the MSC function for primary (experimental) and reference (control) signals. The MSC function takes a value of 1 when the measurements arise from a linear, invertible system (43), such as might occur when EP signals are recorded in relatively noise-free conditions in an experimental study that does not alter the brain's response (note that the LMS algorithm allows adaptation to a new signal even in the presence of some additive, uncorrelated noise). In some experiments, the brain's response may be linearly altered, such as a rise in amplitude caused by an anesthetic. In this case, the adaptive coherence estimation would yield a relatively constant MSC value of about 1 (small fluctuations may arise due to estimation error or measurement noise). However, experiments that alter the brain's response, say, as a result of injury, would yield MSC values approaching zero. A low value of the MSC function indicates that the measurements have departed from their linear state, presumably due to injury. Thus, the MSC function becomes a convenient single indicator of linearity of the measurements in an EP signal acquisition system (32).

EXPERIMENTAL STUDIES

EP signals are acquired by delivering external stimuli to the sensory system; these can be auditory, visual, or electrical. For clinical monitoring in critical care settings, the stimuli are delivered repeatedly and successively. EP signals reported in this study are mainly somatosensory EP signals obtained in anesthetized animals or intraoperatively in anesthetized patients. To continuously obtain the signals, repetitive stimuli need to be delivered to the subject's limbs (45). Experimental animal studies were performed on anesthetized cats (18, 36). The recording site was prepared by first retracting scalp, skin, and muscles from the skull. Somatosensory EP signals were acquired by placing electrodes in burr holes drilled into the cranium near the somatosensory cortex (1 cm lateral to mid-line and just posterior to the coronal suture on each side of the cranium). Silver ball electrodes were placed in each hole. The reference and ground electrodes were placed on the nose and tongue, respectively. Stimulation was done on the forelimb using a Nicolet stimulator operating under computer (IBM compatible portable PC) control. The stimulator delivered electrical pulses at a current twice the motor threshold (~6 mA) and duration of 200 ms at a rate of 5.9/s (higher rates are thought to fatigue the nervous system, and in awake subjects would be discomforting). The amplifiers for EP acquisition were Nicolet or Grass low noise ac preamplifiers set at bandwidths of 30 - 1500 Hz (in certain experiments bandwidth was expanded to 5 - 1500 Hz). Prestimulus signal of 40 ms and poststimulus signal of 40 ms were acquired, with prestimulus signal used to estimate sweep-by-sweep the noise levels. Data were digitized using a 12-bit, 4 channel data-acquisition system interfaced to an IBM compatible portable PC. A sampling rate of 3200 Hz was used, giving 256 samples of pre and post-stimulus data each sweep. Up to 20,000 sweeps were acquired in many experiments. Several experiments performed on anesthetized animals or in human subjects, are reported here.

Cerebral Artery Occlusion

One very important application of intraoperative monitoring is to rapidly identify any injury-related change (46). Such changes in EP signals might occur as a result of a transient occlusion of a cerebral artery or possibly inadvertent occlusion or clipping. This kind of an event demands rapid interpretation of the data and corrective action. We simulated this situation in experimental animals. In anesthetized cats, middle cerebral artery or common carotid arteries were temporarily occluded (18). Somatosensory EP signals were acquired continuously throughout this experiment. Occlusion was later removed and time course of the neurological response was observed. The signals were analyzed using the adaptive filtering and Fourier series modeling algorithms described above. Amplitude and latency measures were derived from the filtered and modeled signals. Figure 9 shows the trend of amplitude and latency. The MCA occlusion causes a very rapid drop in the amplitude and a modest change in the latency. A surprising observation here is how short the time constant of this response is; in more than a dozen animals studied, the time constant of decline in amplitude following MCA occlusion was of the order of 10 seconds. This observation has important ramifications in intraoperative monitoring. Injury-related changes can occur unexpectedly and can cause rapid changes in the electrical response with a potential for significant neurological damage. Adaptive filtering of time-varying EP signals serves as a helpful tool for detecting transient events in intraoperative monitoring.

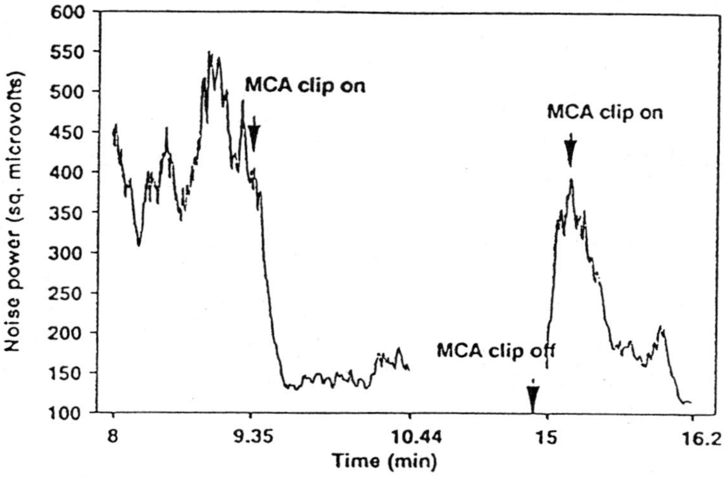

Figure 9. An experimental study of response of somatosensory EP signal to temporary clipping of middle cerebral artery (MCA).

Intraoperative Monitoring

Intraoperative monitoring has many uses. Besides the more serious situation of capturing critical injury-related change, the anesthesiologists and surgeons are interested in responses to anesthetics, drugs or circulatory compromise. Therefore, monitoring of transient events can have many valuable applications. One example of such a study was the response of the drug etomidate. Etomidate is a short acting anesthetic drug which is known to cause a rise in amplitude of the EP signals (40). We monitored several patients in whom a bolus of this drug was given. Drug causes a transient, step change in the amplitude (Fig. 10). Amplitudes typically increased about three-fold while latency also showed a slight increase (47). In this successful illustration of intraoperative monitoring, we observed the

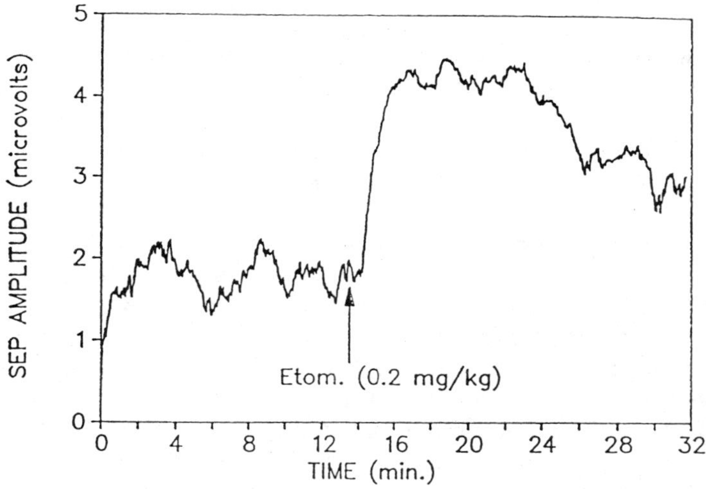

Figure 10. Intraoperative monitoring of response to administration of etomidate in a human subject.

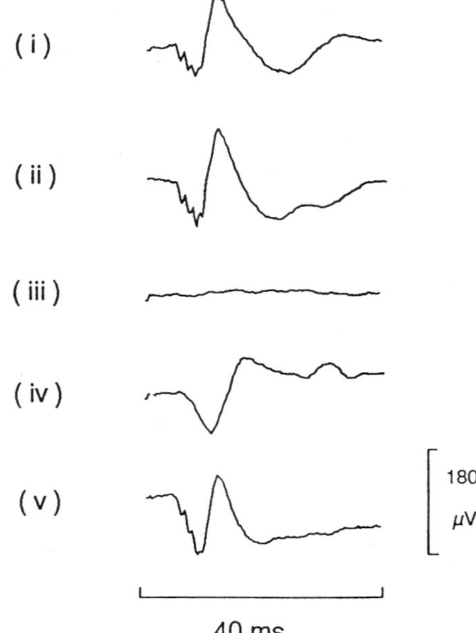

Figure 11. Changes in EP signal morphology with cerebral hypoxia at various experimental stages: i) normal or control, ii) early hypoxia, iii) late hypoxia, iv) early recovery, v) late recovery.

time constant of the step response to the drug administration to be of the order of 150 seconds (47). An interesting application here is the use of this drug etomidate to "enhance" the EP signal amplitude when certain other anesthetics or operative procedures might significantly reduce the measured amplitude.

Cerebral Hypoxia

Brain activity in response to cerebral hypoxia would analogously expect to be compromised and an altered activity should manifest itself as a change in signal amplitude, latency, or shape (6, 15). In our preliminary studies we observed significant morphological changes that are not easily and completely characterized by amplitude and latency measurements alone (Fig. 11). Here, the signal modeling approach was found to be particularly useful. EP signals were modeled by the Fourier series (36). Power spectral analysis revealed that significant energy was restricted to the first 10 harmonics, with higher harmonics or fine details not carrying any interpretable level of information. Experimental animals inhaled 7 to 8% oxygen gas mixture, causing a gradual decline in the measured signal amplitude (Fig. 12a). When amplitude was reduced to less than 10% of control, the inspired gases were returned to 100% O_2 to facilitate recovery. Throughout, EP signals were monitored as before, and amplitude, latencies and magnitudes of the harmonics were calculated. Our aim was to observe if any features of the waveform exhibited enhanced sensitivity to the early stage of injury. This was found to be the case; the lower harmonics showed a transient initial elevation, while the higher harmonics showed a rapid decline in magnitude. The reverse appeared to be true during recovery (Fig. 12b) (36). The lower harmonics were quick to recover, while the higher harmonics never fully recovered in our experiments. Note that the signal amplitude does not show a similar rich pattern. The recovery process showed a wide disparity in the pattern for different harmonics. It would appear that changes in the pattern of decline and recovery of harmonics may serve as indicators of injury. In fact, a dispersion in the recovery pattern, that is, separate time course of recovery of different harmonics, is a strong indicator of injury.

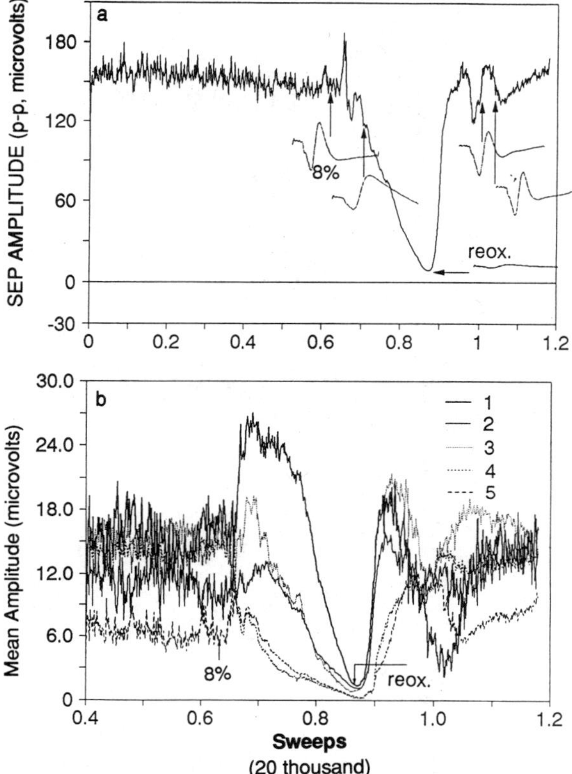

Figure 12. Trends in somatosensory EP a) signal amplitude and b) mean amplitudes of each of the harmonics in response to cerebral hypoxia (inhalation of 8% O_2 air) and subsequent recovery (reoxygenation with 100% O_2). Note the time-varying changes in amplitude and a complex pattern of decline and recovery of various harmonics (1 through 5).

Irreversible Injury

Previous studies clearly show that the adaptive techniques are helpful in capturing transient or time-varying changes in the EP signals, such as might occur in response to anesthetics. Further, signal modeling - such as by Fourier series - helps identify incidences of injury, such as resulting from cerebral ischemia. However, it is desirable to discriminate benign changes, such as responses to anesthetics from the pathological changes such as responses to ischemia. Signal modeling gives such an indication, in that response of different harmonics to injury manifests in the form of spectral dispersion (Fig.12b). Adaptive coherence estimation serves as a single index of such spectral changes (32). The coherence function is estimated from primary signals observed during injury-states and reference signals derived from the control measurements. Note that such a spectral dispersion is not seen during the early stages when anesthetics are administered. Initial spectral dispersion results in a rapid loss of coherence of the higher frequencies. Similarly, coherence of the higher frequency never fully recovers (Fig. 13). A loss of coherence is an indicator of departure from linearity in such measurements. In other words, an injury-causing event produces nonlinear changes in the EP signal measurements. A single indicator is derived to quantitate this observation. The linearity index, Λ_k, is derived as follows:

$$\Lambda_k = \frac{\hat{\Gamma}(k)}{\hat{\Gamma}(k) + \sum_f \left\{ \Gamma(f,k) - \hat{\Gamma}(k) \right\}}$$

(23)

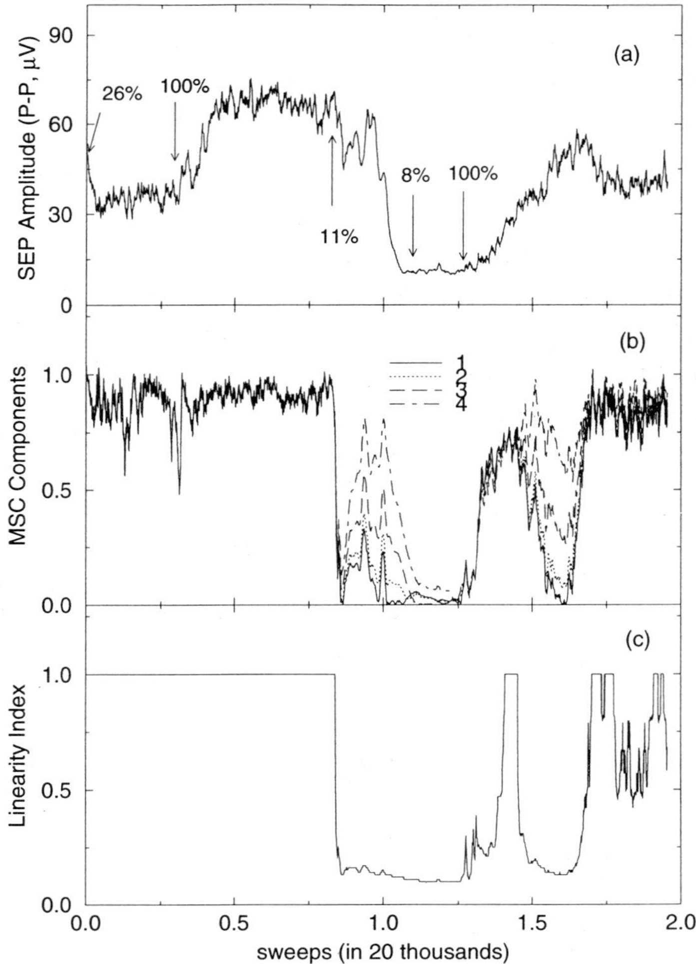

Figure 13. Somatosensory EP amplitude, coherence at various harmonics, and the linearity index in response to inhalation of 100% O_2 and anesthetic, subsequent hypoxia (11% and 8% O_2), and attempts at recovery by reoxygenation (100% O_2). A sharp drop in the linearity index to below 0.5 signals potentially irreversible brain injury.

where $\hat{\Gamma}(k)$ is the average of the coherence $\Gamma_{(f,k)}$ over all f. Thus, the linearity index Λ_k at each time instant k includes the summary representation of the contribution coherence at all frequencies f. A linearity index value close to 1 indicates that the measured response has a linear association with the control measurement. An index value below 0.5 results very soon after the injury-causing event (Fig. 13). This drop in the linearity index is quite abrupt and specific. During recovery, once again the linearity index serves as a good indicator of the process. While the amplitude was often seen to recover fully in a short time, the linearity index never fully recovered in these experiments. A sharp drop in the linearity index certainly suggests an injury-causing event. A partial recovery of the linearity index might suggest residual injury and incomplete recovery of the electrical function. The linearity index, thus, serves as a single but specific indicator of the brain's response to injury.

DISCUSSION

EP signals are useful in monitoring specialized pathways, such as somatosensory, auditory or visual, of the brain (21, 22). These signal indicate electrical functions of the spinal, subcortical as well cortical structures. EP signals, through coherent averaging or estimation, result in a unique waveform or signal "model." This model waveform is useful in diagnosing response to anesthetics, drugs, or injury-causing events (29, 36). Typically, EP signals are interpreted through an analysis of wave peaks; that is, amplitudes and latencies of the characteristic wave peaks are analyzed. In many clinical situations, however, complex changes in the waveform morphologies occur as a result of interventions. In these situations, the data analysis demands mathematically more comprehensive approaches. Such rigorous analysis of waveforms can be accomplished, for example, by using orthonormal basis functions, such as Fourier, Walsh, Hermite, or Principal Components (41). Orthonormal basis functions produce compact representations and confer some useful mathematical properties such as improved convergence to the model and independence of the model parameters. The most useful aspect of model building is that normal or control signals can be first quantitated and then the results of any intervention can be interpreted as changes in the model parameters. Thus, changes in the power at specific harmonics (Fourier basis) can be interpreted as a response to injury (29, 36). EP signal modeling, in this manner, serves as a very useful diagnostic tool.

EP signals are invariably corrupted by noise - mostly background brain activity. Therefore, it is imperative that some filtering or noise enhancement is done. Spectrum of the most dominant component of noise, the background brain activity, typically overlaps that of the signal being measured. Therefore, conventional filtering alone does not suffice. The most classical approach is to carry out ensemble averaging. Ensemble averaging enhances the SNR with increasing number of EP sweeps being averaged (27). Some improvements are possible when the experimental conditions differ, such as when the underlying signals are thought to be time-varying (48). In other situations we may have some a priori knowledge of the underlying signal. Based on such a priori knowledge of the signal, optimum filtering approaches have been devised. *A posteriori* Wiener filtering is one such example (49). Other least squares estimation approaches have been developed since these take advantage of certain correlational properties of the signal and noise (50). For example, the background noise can be considered to be independent or uncorrelated with the signal being measured. Further, the background noise can also be assumed to be additive. If these assumptions hold true (as commonly believed, because of the localized nature of the EP signal source and broad undifferentiated nature of the background activity), then the least square estimation techniques offer valuable tools for extracting signals from noise (51, 52). In the clinical situations discussed earlier, further complications arise due to transient or time-varying nature of the signal. In such applications, the signal model needs to be time-varying, or in other words the time-varying signal may be compared with the signal model to arrive at a diagnosis. In such situations, adaptive filtering algorithms are useful. Adaptive algorithms presented here help improve SNR by the least-squares estimation technique while allowing for adaptation to the changes in the underlying signal.

Both experimental research and clinical studies can benefit from the methods described here. In experimental research one is interested in studying the fundamental mechanism or pathways responsible for processing the sensory stimuli or cognition of the stimulus. In pathological or clinically driven studies one is interested in the response to drugs or anesthetics, or measuring the response of the nervous system (brain or spinal cord) to injury, etc.. In these diverse applications the common question is: how is the sensory-cortical response of the nervous system altered by injury? The classical problem has been formulated in terms of "single trial" analysis (25, 27, 51), i.e., can one decipher the response at each individual stimulation of the nervous system? A study of such event-related, single-trial signals would be helpful in understanding, for example, the cognitive or higher order functional response. In clinical or pathological studies, more often than not, one is interested in how the response changes upon injury, intervention, or therapy. Such problems demand techniques that interpret transient or time-varying changes. In particular, these methods must detect changes that signal critical injury or recovery from pathological conditions. For such applications, adaptive filtering algorithms reported here (28, 29) along with estimation of signal models (such as Fourier and Walsh) (36, 47) or model-derived functions (such as delay and coherence) (31, 32) are particularly useful.

Future work in this area needs to concentrate on improved algorithm designs and specific applications. There are important and powerful signal processing methods that may be applicable to the problems outlined here. For example, block adaptive filtering or direct frequency domain filtering may help deal with large data sets (53). Exact least squares estimation algorithms can improve the convergence rates, especially when eigenvalue disparities might exist. *A priori* knowledge of the signal is helpful in devising methods such as employed in single trial analysis (48, 49). *A priori* knowledge of noise in a similar manner can be helpful in setting the adaptation rates or devising optimal filters. Prediction of noise process, for example, derived from the prestimulus signal can result in improved prediction of the poststimulus signal (51, 52). Noise in neurological signal analysis is typically assumed to have Gaussian statistics. However, newly developed methods for calculating higher order moments and spectra can provide a more accurate description of the noise processes (54). Temporal features of the waveform, such as amplitude and delay, may specifically be estimated. Improved algorithms may improve estimation based on a priori knowledge of such parameters. Signals measured in the present studies have clearly defined morphologies. One can take advantage of the newly emerging time-frequency (55) and time-scale analysis algorithms (56). Time-frequency analysis methods, such as short-term Fourier transform or reduced interference distributions, can identify signal changes in both temporal and spectral (or scale) domains. Thus, localized frequency changes or signal details may identify the injury-specific changes (57, 58). Localized information can also become available through detection of fine or coarse features of the waveforms. This can be done by time-scale or wavelet analysis (59). In such a study, fine features or details were found to be sensitive to oxygen deprivation injury. Specialized algorithms that quantitate signal transformation, such as coherence analysis, are particularly helpful (43, 44). Coherence and the linearity index developed in this article are helpful in defining injury-related nonlinear transformations. However, this approach is based on linear systems' view point. A direct nonlinear systems analysis may be more revealing. For example, the signal of interest may be decomposed into linear and nonlinear components, each responsive to different aspects of the experimental measurement. Here, Wiener or Volterra kernels may be applicable (60). Neurological signals have been shown to involve phase coupling and other nonlinearities (54). Therefore, higher order spectral analysis techniques may be applicable (61), and preliminary studies support this idea (58).

Experimental models of response to anesthetics or hypoxic-ischemic brain injury are explored in this paper. A vast array of experimental and clinical studies can benefit from the approaches to EP signal analysis now being developed. Future work should include generating other experimental models of drug response, or brain injury. These might include studies of spinal, subcortical as well as cortical responses. Experiments need to define how specific or localized can the identification of injury site be. Can EP signals quantitate the extent of drug's action or injury? It remains to be shown whether changes in coherence or linearity directly correspond to the extent of brain injury. Equally useful would be the demonstration that these parameters predict whether the brain injury is irreversible or what is the extent of such irreversible damage. If injury is irreversible, then methods are needed to demonstrate what is the extent of permanent damage.

Finally, armed with this technical information as well as experimental demonstrations, it should be feasible now to construct a new generation of instruments that can be employed in sensitive clinical studies or patient monitoring. Such instruments should be capable of on-line analysis and interpretation of the signals and assisting with the decision to alter the patient therapy. Such instruments should particularly be useful in patient monitoring in intraoperative or neurocritical care settings.

SUMMARY

Several algorithms presented here successively demonstrate the technique of evaluating transient or time-varying brain's response to externally applied stimuli. Adaptive algorithms are needed to estimate a signal present in additive (but uncorrelated) noise. Minimizing the MSE is shown to increase the SNR in a simple filter, such as the impulse correlated filter. Minimizing the MSE between pairs of successive EP signals, or alternately, a noisy measured signal and a relatively noise-free average also helps extract the underlying signal from noise. Adaptive delay estimation algorithm is useful in tracking time-varying parameters, such as amplitude and latency of the EP waveform. The adaptive coherence estimation algorithm is helpful in situations where the brain's status is altered so that successive measurements cannot be considered to arise from the same system. In which case, we can utilize the magnitude squared coherence function to quantitate departure from linearity of the measured system. A low coherence estimate would indicate that a drug or injury has caused alteration in the EP measurement system. Applications of these techniques to monitoring neurological response to anesthetics in patients and to simulated brain injury in anesthetized animals are presented. Time constants of brain injury are found to be surprisingly low and strongly support the need for continuous monitoring in high-risk clinical procedures. A single indicator, such as the linearity index, should be helpful in diagnosing episodes of brain injury. Experimental studies demonstrate the usefulness of such techniques in obtaining responses to drugs and anesthetics as well as in identifying incidence of brain injury. Overall, this work demonstrates that transient or time-varying responses of the brain can be obtained in critical care settings.

ACKNOWLEDGMENTS

Research reported in this paper was supported by grants over several years from the National Institutes of Health and the National Science Foundation. The authors also acknowledge efforts of Christopher Vaz, Peter Poon, Jeff Braun, Guo Xin-rong, Yi-Chun Sun, Pablo Laguna and others for their contributions to algorithm development and data analysis. The authors thank Daniel Hanley, Robert McPherson and Raymond Kohler for their cooperation with experimental and clinical studies.

REFERENCES

1. Bricolo, A., Faccioli, J. C. , Grosslercher, M. L. , Pasut, M. L., Pinna, G. P., and Turazzi, S., 1987, Electrophysiological monitoring in the intensive care unit, *Electroenceph. Clin. Neurophys.* Suppl. 39: 255-263.

2. Kawahara, N., Tanaka, H. Sakamoto, T. Sakasi, M., Aruga, T., Mii, K., and Takakura, K., 1988, Clinical application of the multidimensional neurological monitoring system and the cerebrosystemic hemodynamic profile for critically ill neurosurgical patients, Neurol. Med. Chir. (Tokyo) 28: 27-33.

3. Crawford, E. S., Mizrahi, E. M., Hess, K. R., Kossli, J. S., Safi, H. J., and Patel, V. M., 1988, The impact of distal aortic perfusion and somatosensory evoked potential monitoring on prevention of paraplegia after aortic aneurysm operation, *J. Thoracic Cardiovsc. Surg.* 95: 357-367.

4. Matsumiya, N. , Koehler, R. C., and Traystman, R. J., 1990, Consistency of cerebral blood flow and evoked potential alterations with reversible focal ischemia in cats, *Stroke* 21: 908-916.

5. Colin, F., Bourgain, R., and Manil, J., 1978, Progressive alteration of somatosensory evoked potential waveforms with lowering of cerebral tissue pO2 in rabbit, *Arch. Int. Physiol. Biochim.* 86: 677-679.

6. McPherson, R. W., Zeger, S., and Traystman, R. J., 1986, Relationship of somatosensory evoked potentials and cerebral oxygen consumption during hypoxic hypoxia in dogs, *Stroke* 17: 30-36.

7. Grundy, B. L., 1983, Intraoperative monitoring of sensory evoked potentials, *Anesthesiol.* 58: 72-87.

8. Dinner, B. L., 1986, Intraoperative spinal somatosensory evoked potential monitoring, *J. Neurosurg.* 65: 807-814.

9. Mcpherson, R. W. , 1987, Intraoperative monitoring of evoked potentials, *Prog. Neurol. Surg.* 12:146-163.

10. Moller, A. R., 1988, *Evoked Potentials in Intraoperative Monitoring*, Williams and Wilkins:Baltimore.

11. Nuwer, M. R. , 1986, *Evoked Potential Monitoring in the Operating Room*, Raven Press: New York, N.Y.

12. McCallum, J. E. and Bennett, M. H., 1975, Electrophysiological monitoring of spinal cord function during intraspinal surgery, *Surg. Forum* 26: 469-471.

13. Hagerdine, J. R., and Snyder, E., 1982, Brain stem and somatosensory evoked potentials: application in the operating room and intensive care unit, *Bull. Los Angeles Neurol. Soc.* 47: 62-75.

14. McPherson, R. W., Mahla, M., Szymanski, J., and Rogers, M. C., 1984, Somatosensory evoked potential changes in position-related brain stem ischemia, *Anesthesiol.* 61: 88-90.

15. Grundy, B. L., Heros, R. C., Tung, A. S., and Doyle, E., 1981, Intraoperative hypoxia detected by evoked potential monitoring, *Anasth. Analg.* 60: 437-439.

16. McPherson, R. W., et al., 1983, Correlation of transient neurological deficit and somatosensory evoked potentials after intracranial aneurysm surgery, *J. Neurosurg.* 59: 146-149.

17. Momma, F., Wang, A. D., and Symon, L., 1987, Effects of temporary arterial occlusion on somatosensory evoked responses in aneurysm surgery, *Surg. Neurol.* 27: 343-352.

18. Poon, P., Koehler, R. C., and Thakor, N. V., 1995, Somatosensory evoked potential response to cerebral artery occlusion, *Med. & Biol. Eng. & Comput.*33:396-402.

19. Vaz, C. A., and Thakor, N. V., 1986, Monitoring brain electrical and magnetic activity, *IEEE Eng. Med. Biol. Mag.* 5: 11-16.

20. Gevins, A., 1984, Analysis of electromagnetic signals of the human brain: milestones, obstacles, and goals, *IEEE Trans. Biomed. Eng.* 31: 833-850.

21. Epstein, C. M., and Andriola, M. R.. 1983, *Introduction to EEG and Evoked Potentials*, Lippincott Co., Philadelphia.

22. Spehlmann, R. S., 1985, *Evoked Potential Primer - Visual, Auditory and Somatosensory Evoked Potentials in Clinical Diagnosis*, Butterworth Publ., Boston.

23. Glaser, E., and Ruchkin, D., 1976, *Principles of Neurobiological Signal Analysis*, Acad. Press: New York, N.Y.

24. Mocks, J. , Gasser, T., Kohler, W., and de Weerd, J. P. C., 1986, Does filtering and smoothing average evoked potentials really pay? A statistical comparison, *Electroenceph. Clin. Neurophys.* 64: 469-480.

25. McGillem, C. D., and Aunon, J. I., 1977, Measurements of signal components in single visually evoked brain potentials, *IEEE Trans. Biomed. Eng.* 24: 232-241.

26. McGillem, C. D., Aunon, J. I., and Childers, D. G., 1981, *Signal Processing in Evoked Potential Research: Applications of Filtering and Pattern Recognition*, CRC Press: Cleveland, OH.

27. J. I. Aunon, C. D. McGillem, and D. G. Childers, 1981, *Signal Processing in Evoked Potential Research: Averaging, Principal Components, Modeling*, CRC Press: Cleveland, OH.

28. Thakor, N. V., 1987, Adaptive filtering of evoked potentials, *IEEE Trans. Biomed. Eng.* 34: 6-12.

29. Vaz, C. A. and Thakor, N. V. , 1989, Adaptive Fourier estimation of time-varying evoked potentials, *IEEE Trans. Biomed. Eng.* 36: 448-455.

30. Vaz, C. A., Kong, X., and Thakor, N. V., 1994, Adaptive estimation of periodic signals using a Fourier linear combiner, *IEEE Trans. Sig. Proc.* 42: 1-10.

31. Kong, X., and Thakor, N. V., 1996, Adaptive delay estimation: theory and application to evoked potentials, *IEEE Trans. Biomed. Eng.*, in press.

32. Thakor,N. V., Kong, X., and Hanley, D. F., 1995, Nonlinear evoked response of brain to injury as measured by adaptive coherence estimation, *IEEE Trans. Biomed. Eng.* 42: 42-51.

33. Hoke, M., Ross, B., Wickesberg, R.,and Luetkenhoener, B., 1984, Weighted averaging - theory and application to electrical response audiometry, *Electroenceph. Clin. Neurophysiol.* 57: 484-489.

34. Laguna, P., Jane, R., Meste, O., Poon, P., Caminal, P., Rix, H., and Thakor, N. V., 1992, Adaptive filter for event-related bioelectric signals using an impulse correlated reference input: comparison with signal averaging techniques, *IEEE Trans. Biomed. Eng.* 39: 1032-1044.

35. Cerutti, S., Chiarenza, G., Liberati, D., Mascellani, P., and Pavesi, G., 1988, A parametric method of identification of single-trial event-related potentials in the brain, *IEEE Trans. Biomed. Eng.* 35: 701-711.

36. Thakor, N. V., Xin-rong, G., Vaz, C. A., Laguna, P., Jane, R., Caminal, P., Rix, H., and Hanley, D. F., 1993, Orthonormal (Walsh and Fourier) models of evoked potentials in neurological injury, *IEEE Trans. Biomed. Eng.* 40: 213-221.

37. Haykin, S., 1984, *Introduction to Adaptive Filters*, Macmillan, New York, N.Y.

38. Widrow, B., Glover, J. R., McCool, J. M., Kaunitz, J., Williams, C. S., Hearn, R. H., Zeidleer, J. R., Dong, E., and Goodlin, R. C., 1975, Adaptive noise canceling: principles and applications, *Proc. IEEE* 63: 1692-1716.

39. Thakor, N. V., 1994, Adaptive filters for analysis of intra-cardiac signals, *Med. & Biol. Eng. & Comput.* 32: S19-S24.

40. McPherson, R. W., Sell, B., and Traystman, R. J., 1986, Effects of thiopental, fentanyl, and etomidate on upper extremity somatosensory evoked potentials in humans, *Anesthesiol.* 65: 584-589.

41. Ahmed, N., 1975, *Orthogonal Transforms for Digital Signal Processing*, Springer Verlag, New York, N.Y.

42. Coyer, P. E., Simeone, F. A., and Michele, J. J., 1987, Latency of the cortical component of the somatosensory evoked potential in relation to cerebral blood flow measured in the white matter of the cat brain during focal ischemia, *Neurosurg.* 497-502.

43. Cadzow, J., and Solomon, O., 1987, Linear modeling and the coherence function, *IEEE Trans. Accoust. Speech. Signal Proc.* 31: 137-142.

44. Youn, D., Ahmed, N., and Carter, G., 1983, Magnitude-squared coherence function estimation: an adaptive approach, *IEEE Trans. Acoust. Speech, Signal Proc.* 31: 137-142.

45. Sherbune, E., 1985, Continuous somatosensory evoked potential monitoring in the ICU, *J. Neurosurg. Nurs.* 17: 247-252.

46. Steinberg, G. K., Gelb, A. W., Lam, A. M., Manninen, P. H., Peerless, S. J., Rassi., N. A., and Floyd, P., 1986, Correlation between somatosensory evoked potentials and neuronal ischemic damage following middle cerebral artery occlusion, *Stroke* 17: 1193-1197.

47. Thakor, N. V., Vaz, C.A., McPherson, R.W., and Hanley, D.F., 1991, Adaptive Fourier series modeling of time-varying evoked potentials: study of human somatosensory evoked response to etomidate anesthetic, *Electroenceph. Clin. Neurophys.* 80: 108-118.

48. de Weerd, J. P. C., 1981, A posteriori time-varying filtering of averaged evoked potentials. II: mathematical and conceptual aspects, *Biol. Cybernetics* 41: 223-234.

49. de Weerd, J. P. C., and Kap, J. I., 1978, Theory and practice of a posteriori 'Wiener' filtering of average evoked potentials, *Biol. Cybernet.* 30: 81-94.

50. Yu, K., and McGillem, C. D., 1983, Optimum filters for estimating evoked potential waveforms, *IEEE Trans. Biomed. Eng.* 30: 730-737.

51. Davila, C. E., Abaye, A., and Khotanzad, A., 1994, Estimation of single sweep steady-state visual evoked potentials by adaptive line enhancement, *IEEE Trans. Biomed.* 41: 197-200.

52. Madhavan, P. G., 1992, Minimal repetition evoked potentials by modified adaptive line enhancement, *IEEE Trans. Biomed. Eng.* 37: 650-652.

53. Mansour, D., and Gray, Jr., A. H., 1982, Unconstrained frequency domain adaptive filter, *IEEE Trans Acoust., Speech, Signal Proc.* 30: 726-734.

54. Sherman, D. L., 1993, Novel techniques for the detection and estimation of three-wave coupling with application to human brain waves, *Ph.D. Dissertation*, Purdue University, Lafayette, IN.

55. Cohen, L., 1989, Time-frequency distributions - a review, *Proc. IEEE* 77: 941-981.

56. Mallat, S. G., 1989, A theory of multiresolution signal decomposition: the wavelet presentation, *IEEE Trans. Pattern Analysis Machine Intell.* 11: 674-693.

57. Moser, J. M., and Aunon, J. I., 1986 Classification and detection of single evoked potentials using time-frequency amplitude features, *IEEE Trans. Biomed. Eng.* 33: 1096-1106.

58. Braun, J. C., 1994, Neurological monitoring using time-frequency and bispectral analysis of evoked potential and electroencephalographic signals, *M.S.E. Dissertation*, Johns Hopkins Univ., Baltimore, MD.

59. Thakor, N. V., Xin-rong, G., Yi-chun, S., and Hanley, D. F., 1993, Multiresolution wavelet analysis of evoked potentials, *IEEE Trans. Biomed. Eng.* 40: 1085-1094.

60. Bard, Y., 1974, *Nonlinear Parameter Estimation*, Acad. Press., New York.

61. Nikias, C. L., and Raghuveer, M. R., 1987, Bispectrum estimation: a digital signal processing framework, *Proc. IEEE* pp. 869-891.

Processing and Pattern Analysis of Neuronal Cell Activity

12

DETECTION AND QUANTIFICATION OF CORRELATIONS IN NEURAL POPULATIONS BY COHERENCE ANALYSIS

Constantinos N. Christakos

Department of Basic Sciences, Medical School
University of Crete
Heraklion, Greece
Center for Neurobiology and Behavior
College of Physicians and Surgeons
Columbia University, New York

ABSTRACT

The behavior of the unit-to-aggregate and the aggregate-to-aggregate coherence function as tools for detection and quantification of synchrony in neural populations showing partial correlations is examined by mathematical analysis and computer simulations. The results indicate that the former function provides a suitable tool for both purposes, whereas the latter function is best suited for detection of population synchrony.

INTRODUCTION

The detection and quantification of correlations in neural populations are issues that often arise in the study of neural systems. Such synchrony may, for example, reflect interactions between elements of a neuronal network and thus be characteristic of the structure and function of a system. Alternatively, it may constitute the basis of a certain phenomenon or behavior. Therefore, determining the presence and "magnitude" of population synchrony, and studying the changes in synchrony as conditions change, may be of critical importance for understanding a neural mechanism.

This, however, is a difficult task, particularly if a population comprises many units (more than 100). A measure of such synchrony has to be expressed in terms of the following parameters, which constitute the three characteristics of population synchrony:

A) The extent of synchrony within a population (i.e., the fraction of correlated units);

B) the strength of synchrony (i.e., the strengths of the unitary correlations);

C) the phase relations of the correlated units.

Advances in Processing and Pattern Analysis of Biological Signals, Edited by Isak Gath and Gideon F. Inbar
Plenum Press, New York, 1996

For a population that is large in numerical size, the number of combinations of units, taken two at a time, is a few orders of magnitude larger than the number of the units. Therefore, traditional correlation analysis (Perkel *et al.*, 1967) applied to pairs of simultaneously and independently recorded units is not efficient for detecting synchrony and for estimating its extent and strength. For the sample to be representative of the situation within the entire population, one has to record and analyze a very large number of different pairs of unitary activities. But the parallel and separate recording of many units is not an easy task. Also, the result of such analyses will be in the form of a very large number of cross-correlograms, which complicates the representation of unitary correlations for any quantification of synchrony. Other, more general problems of correlation analysis in the case of neural activities have been discussed by Houk *et al.* (1987) and Rosenberg *et al.* (1989).

Coherence analysis applied to pairs of population activities, is used in different areas of Neurophysiology for assessing population synchrony (e.g., the EEG area). However, the interpretation of coherence observations is performed on the basis of general arguments and guidelines which stem from early studies (e.g., Elul, 1972a; Lopes da Silva, 1974). Specific analyses of how the above defined characteristics of population synchrony are reflected in the estimated coherence between population activities are lacking. Therefore, essential information for any quantification of such synchrony is not available.

On the basis of theoretical considerations and computer simulations, a recent study (Christakos, 1994a) indicated that the coherence function between unitary and population-aggregate activity, i.e., the unit-to-aggregate (UTA) coherence function, can be used as a tool for analysis of population synchrony (see also Appendix in Christakos *et al.*, 1991). Specifically, that study revealed that the UTA coherence function is almost zero for uncorrelated units in a large population, but has a non-zero value at the frequency (or frequency band) of synchrony for units that are correlated with other units. This dual property provides a basis for discriminating between correlated and uncorrelated units in a population.

That study further revealed that the value of the UTA coherence for the correlated subset reflects the three characteristics of synchrony (the strength and extent of synchrony and the similarity of the phases of the correlated units) and also a fourth parameter, not related to synchrony, namely, the numerical size of the population. As indicated by computer simulations of a uniform population, this value remains substantial in a very wide range of conditions (e.g., the extent of synchrony within the population can be as low as 5%).

Thus, UTA coherence computations on a sample of recorded unit/population activities provides a means of identifying correlated units in a population and obtaining information on the extent and strength of synchrony.

Current mathematical analysis using a more general model of population synchrony has provided detailed predictions on the behavior of the UTA coherence. This analysis has also dealt with the behavior of the coherence function between two (sub)populations, i.e., the behavior of the aggregate-to-aggregate (ATA) coherence function. The results of the computer simulations and the mathematical analysis are summarized below, and an evaluation of the UTA and the ATA coherence as tools for detection and measurement of population synchrony is also described. These results have previously been presented in the form of an abstract (Christakos, 1994b).

ANALYSIS

The coherence function is obtained from the cross-spectrum and the auto-spectra of two signals and provides a measure of linear correlation between the signals as a function of frequency. Its value at each frequency reflects the strength of the correlation between the signals, and has a maximum of 1 in the case of a perfect correlation. Uncorrelated components present in the signals, such as noise, reduce this value, and at the extreme, for signals showing no correlation at all, the coherence is minimal and equal to zero. Thus, the coherence function is normative (values lying between 0 and 1).

In what follows, the behavior of the coherence function is examined for the pairs unit/unit (UTU), unit/aggregate (UTA) and aggregate/aggregate (ATA), in the case of uniform populations showing partial unitary correlations. In the simulations, unitary correlations are assumed to exist around each distinct frequency, and the above three types of coherence function are studied. In the mathematical analysis, expressions are obtained for the UTA and ATA coherence as functions of the UTU coherence at every frequency.

A Model of Synchrony in a Uniform Population Comprising a Correlated and an Uncorrelated Subset of Units

In the simulations, a subset of spike trains (event sequences) $x_i(t)$ that are correlated around a frequency F are created from a rhythmic reference train $x_r(t)$ with central frequency F in the following way: Spikes in x_i occur with random delays $v_i = d_i + \varepsilon_i$ following spikes in x_r, where d_i is a constant delay and ε_i is zero-mean Gaussian jitter. The so produced trains therefore have the same frequency and statistical characteristics, they are linearly correlated around F with a correlation strength that is determined by the variance of the jitter variable, and their phases are determined by the d_i's. It is important to note that these apply to the more general event sequence, where the spikes are replaced by (convolved with) identical pulses having any shape.

Consequently, the cross-correlation function $R_{ik}(\tau)$ of any two trains (or event sequences) x_i and x_k contains an oscillation at F, with an amplitude that reflects the correlation strength of x_i and x_k and a phase that equals the phase difference of x_i and x_k. Equivalently, the UTU coherence function shows a clear peak at F with an amplitude reflecting this correlation strength, and it is zero at other frequencies where there is no synchrony (see Fig. 1b).

For the study of the UTA coherence, a uniform population of spike trains is created. The population comprises a subset of correlated trains generated as above, and a subset of uncorrelated trains (generated as in Christakos, 1982) with the same central frequency F and autospectra similar to those of the correlated trains. In order to thoroughly test the effects of the phases of the correlated units on the UTA coherence, the d_i's are generated as random variables uniformly distributed over fractions D of the mean period of the trains (e.g., D = $T/4 = 1/4F$). The Gaussian distribution is also used for the d_i's, as a more realistic distribution for many neural populations. To enable the comparison between the results for the two distributions, the variance of the Gaussian delays is the same as that of the uniformly distributed delays.

For the study of the ATA coherence, two uniform populations are created as above, each comprising a subset of correlated units that are also correlated to those of the other population. The additional assumption that the two (sub)populations have the same numerical size and composition simplifies the analysis. At the same time, it results in a realistic model of contralateral activities in EEG's, neurograms, EMG's, etc.

UTA Coherence for Members of the Correlated Subset

As indicated by simple theoretical considerations (Christakos, 1994a), the key rules that determine the behavior of the UTA coherence for the correlated subset are:

(a) Each unit of this subset, being linearly correlated with the other units of the subset, is generally correlated with the aggregate activity of the subset. The strength of this correlation reflects the strengths of the correlations of the given unit to the other units of the subset and the similarity of the phases of the units in this subset.

 (b) In a very wide range of conditions, the aggregate activity of the correlated subset dominates, or is at least a significant part of, the total aggregate activity of the population at the frequency (band) of synchrony. The relative strength of the activity of the correlated subset reflects the strengths of the unitary correlations, the degree of phase similarity for the correlated units, and the numerical size of this subset, both absolute and relative to that of the uncorrelated subset (Christakos, 1990).

Therefore, unless synchrony is very limited, any unit of the correlated subset is also correlated to a significant degree with the total aggregate activity of the population. Equivalently, the UTA coherence for members of the correlated subset is nonzero at the frequency (band) of synchrony, with a value that reflects the three characteristics of synchrony (strength, phase similarity for the correlated units, and extent within the population) and the numerical size of the population. (Note that the activity of the uncorrelated subset acts like pure noise that tends to reduce the UTA coherence.)

By a similar reasoning one sees that any member of the uncorrelated subset is only correlated with its own contribution to the total aggregate, with all other units acting like strong noise. Therefore, the UTA coherence for this subset is almost zero at all frequencies.

The computer simulations confirm the above predictions. Accordingly, the UTA coherence for the correlated subset shows, in a very wide range of conditions, a peak at the frequency of synchrony. The value of this coherence peak indeed reflects the above identified four parameters, and is substantial, even if the extent and strength of synchrony are limited. In contrast, the UTA coherence for the

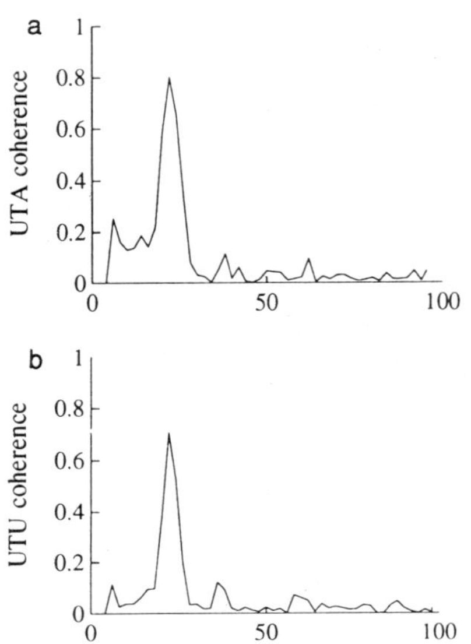

Figure 1. Top: UTA coherence function for a correlated subset of 200 rhythmic spike trains at 20 Hz, in a simulated uniform population that comprises another 200 uncorrelated rhythmic trains at the same frequency. Bottom: UTU coherence function for the correlated subset. Note the clear peaks in both plots at the frequency of synchrony (20 Hz) and the low coherence at other frequencies. Also note that the peak UTA coherence is higher than the peak UTU coherence.

uncorrelatedsubsetisalmostzeroatallfrequencies(seeChristakos(1994a)formoredetails).

In the example of Fig. 1a, the simulated uniform population comprises 200 correlated and 200 uncorrelated rhythmic trains at 20 Hz. The delays for the correlated subset are uniformly distributed within one half of the mean period of the trains; and the unitary correlations are quite strong, the peak UTU coherence at 20 Hz being 0.7 (see Fig. 1b). The UTA coherence also shows a clear peak at the frequency of synchrony. The value of this peak (0.8) is larger than that of the peak UTU coherence, even though only a fraction of the units are correlated. This is indicative of the sensitivity of the UTA coherence in reflecting synchrony.

The mathematical analysis that follows provides a better picture of the effects of the above identified parameters on the UTA coherence, and of the performance of this coherence as a tool for analysis of population synchrony.

Consider a set $\{x\}$ of n unitary processes, in which a subset of the x_i's are linearly correlated (subscripts $i = 1, 2... n_c$) and the remaining x_i's are uncorrelated (subscripts $i=n_c+1,$ ...n).

Assume that the correlated subset is uniform in the sense that: (a) all correlated x_i's have the same auto-spectrum $S_a(\omega)$; and (b) the cross-spectra of all correlated x_i's have the same modulus $S_c(\omega)$, i.e., $S_{ik}(\omega) = S_c(\omega)\exp(-j\omega d_{ik})$, where $d_{ik} = d_i-d_k$ is the difference of the units' delays from a reference point on the time axis. Assumption (a) implies that all x_i's have the same "amplitude" and second order statistical properties. Assumption (b) implies that the various cross-correlation functions $R_{ik}(\tau)$ are the same even function shifted by different intervals d_{ik} along the time axis. Consequently, the only difference between the various R_{ik}'s, or S_{ik}'s, is due to the differing delays for the units of the subset. The event sequences of the simulations (see Sect. 2.1) have these properties.

Let $x(t) = \Sigma x_i(t)$ denote the total aggregate activity of set $\{x\}$. $x(t)$ may, for example, be the electrical activity recorded from a nerve or an area in the cortex. It may also be the force of a contracting muscle, where the x_i's are the force contributions of the individual motor units.

The UTA coherence is then defined as:

$$Q_{ix}^2(\omega) = |S_{ix}(\omega)|^2 \, / \, [S_{ii}(\omega)S_{xx}(\omega)] \quad ; \quad i = 1, 2, ...n$$

By setting $i = 1$, for convenience, the cross-spectrum of any of the correlated units and of the total aggregate activity is given as (Papoulis, 1965):

$$S_{1x}(\omega) = S_a(\omega) + \sum_{k=2}^{n_c} S_c(\omega) \exp(-j\omega d_{1k}) \quad ; \quad \omega \neq 0 \tag{1}$$

Therefore:

$$|S_{1x}(\omega)|^2 = S_a^2(\omega) + 2S_a(\omega)S_c(\omega)\left(\sum_{k=2}^{n_c} \cos\omega d_{1k}\right) +$$

$$+ S_c^2(\omega)\left[\left(\sum_{k=2}^{n_c} \cos\omega d_{1k}\right)^2 + \left(\sum_{k=2}^{n_c} \sin\omega d_{1k}\right)^2\right] \quad ; \quad \omega \neq 0 \tag{2}$$

But the quantity in square brackets is equal to:

$$(n_c - 1) + 2 \sum_{k=2}^{n_c-1} \sum_{m=k+1}^{n_c} (\cos\omega d_{1k}\cos\omega d_{1m} + \sin\omega d_{1k}\sin\omega d_{1m}) =$$

$$= (n_c - 1) + 2 \sum_{k=2}^{n_c-1} \sum_{m=k+1}^{n_c} \cos\omega d_{km} \quad ; \quad d_{km} = d_{1m} - d_{1k} = d_k - d_m \quad ,$$

where d_k and d_m are unit delays, and can be written as: $(n_c-1) + 2C_1$, where $2C_1$ is the sum of the cosines of the phase differences between all pairs of x_i's, excluding x_1.

Denoting $B_1 = \Sigma \cos\omega d_{1k}$ the sum of the cosines of the phase differences of x_1 to all other correlated x_i's, Eq. (2) becomes:

$$|S_{1x}(\omega)|^2 = S_a^2(\omega) + 2B_1 S_a(\omega)S_c(\omega) + (n_c - 1 + 2C_1)S_c^2(\omega) \quad ; \quad \omega \neq 0 \tag{3}$$

The auto-spectrum of the total aggregate activity $x(t)$ equals the sum of all auto- and cross-spectra of the x_i's (Papoulis, 1965). On the assumption that the auto-spectra of the uncorrelated units are also $S_a(\omega)$, this is written (Christakos, 1990):

$$S_{xx}(\omega) = nS_a(\omega) + 2S_c(\omega) \sum_{i=1}^{n_c-1} \sum_{k=i+1}^{n_c} \cos\omega d_{ik} \quad ; \quad \omega \neq 0 \tag{4}$$

since the imaginary parts of the cross-spectra cancel out. Denoting the sum of the cosines of the phase differences between all pairs of correlated x_i's as $2A$, Eq. (4) takes the form:

$$S_{xx}(\omega) = nS_a(\omega) + 2AS_c(\omega) \quad ; \quad \omega \neq 0 \tag{5}$$

Note that $A = B_1 + C_1$.

Therefore, the UTA coherence is given from Eqs. (3) and (5) as:

$$Q_{1x}^2(\omega) = \frac{S_a^2(\omega) + 2B_1 S_a(\omega)S_c(\omega) + (n_c - 1 + 2C_1)S_c^2(\omega)}{S_a(\omega)[nS_a(\omega) + 2AS_c(\omega)]} \quad ; \quad \omega \neq 0$$

or, in terms of the UTU coherence, $Q^2(\omega)$:

$$Q_{1x}^2(\omega) = \frac{1 + 2B_1 Q(\omega) + (n_c - 1 + 2C_1)Q^2(\omega)}{n + 2AQ(\omega)} \quad ; \quad \omega \neq 0 \tag{6}$$

From Eq. (6) it is evident that the value of the UTA coherence at any frequency depends on the respective value of the UTU coherence, and on the distribution of the unit delays (or phases), which determines parameters A, B_1, C_1. It is also influenced by n_c and n, or equivalently, by the extent of synchrony and by the numerical size of the entire population. These dependencies have also been deduced by the theoretical considerations at the beginning of this Section. It is interesting to note that as the UTU coherence tends to 0, the UTA coherence tends to $1/n$, which is almost 0 for a large population. In other words, at frequencies where there is no synchrony the UTA coherence is almost zero.

The influence of the various parameters on the value of the UTA coherence at the frequency of synchrony can be studied using the analytic expression of Eq. (6). The unit delays in the model are assumed to be random variables, as they have to be seen in any real situation. Therefore, quantities A, B_1 and C_1 in Eq. (6) are also random variables. In order to simplify the study of the UTA coherence, sets of pseudorandom delays are generated on the computer for different "seeds", and the distribution and statistical moments (mean and standard deviation) of the UTA coherence are estimated under different conditions.

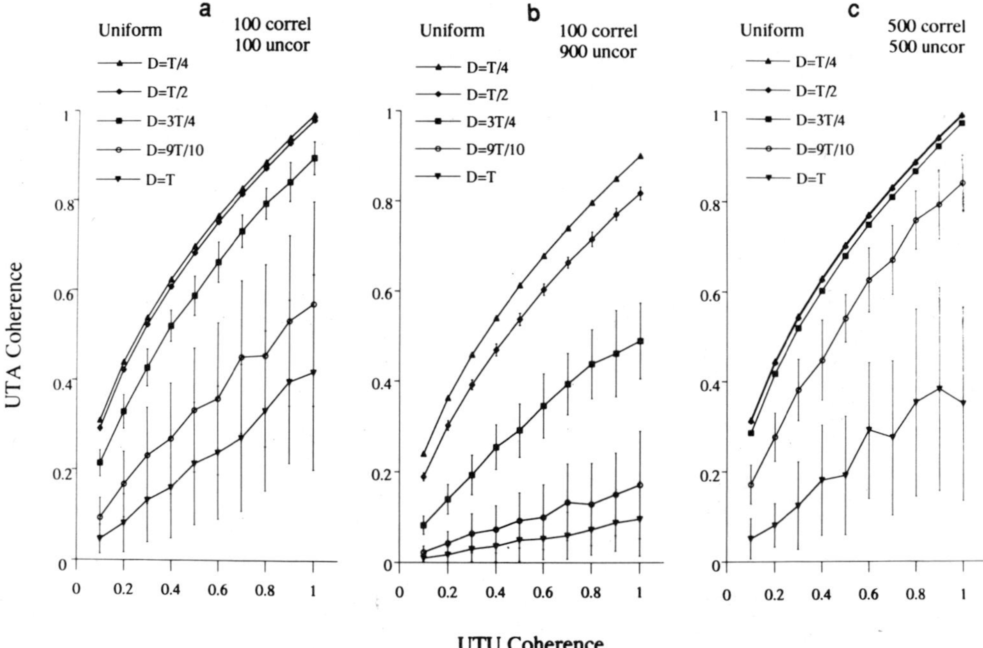

Figure 2. UTA coherence (mean ± one standard deviation) at the frequency of synchrony for the correlated subset, as a function of the UTU coherence at the same frequency. The different curves in each plot correspond to different delay ranges (i.e., ranges of unit phases), as indicated at top. Note that for delay ranges that are not too wide (<3T/4) the UTA coherence is larger than, or at least a large fraction of, the UTU coherence, even if the correlated subset is a smallfraction of the population (10%). Also note that the UTA coherence becomes higher, as the numerical size of the population increases.

Figure 2 shows an example of such computations for uniformly distributed delays over intervals that are fractions of the mean period of synchrony, T = 1/F (e.g., D = T/2). The following interesting features are seen in this Figure:

a) The value of the UTA coherence is an increasing function of the UTU coherence, in a fairly linear way, and of the extent of synchrony (compare Figs. 2b and 2c). In addition, it is a decreasing function of the width of the range of the uniformly distributed phases of the units (compare the different curves in any of the three plots in this figure), and an increasing function of the numerical size of the population (compare Figs. 2a and 2c).

b) Of the three characteristics of synchrony, the strength (UTU coherence) and the uniformly distributed phases have strong effects on the UTA coherence, whereas the effect of the extent is limited, except when the range of the phases is very broad (near a cycle at F). (Note that for broadly and uniformly distributed phases, the representation of the correlated subset in the total aggregate is limited due to cancellations of the unitary rhythms at each frequency of synchrony.)

c) Except for phases that are uniformly distributed over a broad range of values, the UTA coherence is substantial and hence detectable, even if the strength and extent of synchrony are limited. In particular, for small UTU values (<0.4) and phase ranges not exceeding one half of a cycle, the UTA coherence is larger than, or

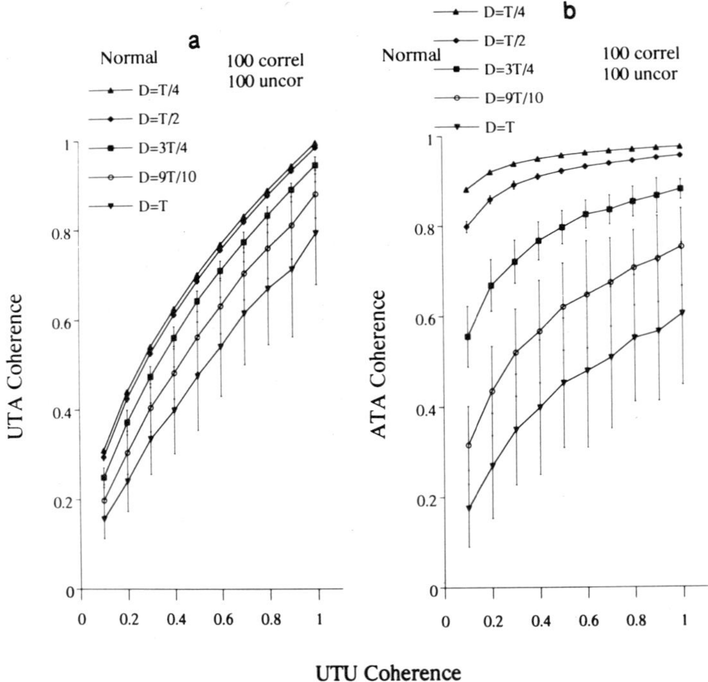

Figure 3. UTA and ATA coherences, as functions of the UTU coherence, for Gaussian unit delays. The plots are as those in Figs. 2 and 4. Note that the various curves in these plots have been shifted upwards, compared to the respective curves for the uniformly distributed delays.

comparable to, the UTU coherence for an extent of synchrony of the order of only 10% (see e.g. Fig. 2b).

These are typical features, and are in agreement with the results of the simulations. Therefore:

1. The UTA coherence, being substantial in a wide range of conditions, can efficiently be used for detection of synchrony. Even a single observation of a significant UTA coherence implies that the given unit is linearly correlated with at least a fraction of the units of the population, which in turn are correlated among themselves. Equivalently, such an observation reveals the presence of a correlated subset of units, to which this unit belongs.
2. UTA coherence computations applied to a sample of recorded unit/population activities:

   - Enable the estimation of the extent of synchrony within the population, as being the fraction of recorded units that show a non-zero coherence to the aggregate;
   - Provide a measure of the strengths of the unitary correlations, in the form of a histogram of UTA coherence values for the different units in the sample.

Therefore, the UTA coherence behaves well, in a wide range of conditions, as a tool for detection and quantification of population synchrony. A combination of weak unitary correlations, broadly and uniformly distributed phases and very limited extent has to occur in order for the UTA coherence to fail to detect the synchrony. Unless the phases and the extent combine in this way, this coherence may even be larger than the UTU coherence, thus enabling the identification of correlations for units whose coherence with other units is too low to be detected as a significant UTU coherence value.

It should be stressed that the negative effects on the UTA coherence of phases that are uniformly distributed over broad ranges of values are not severe if the distribution shows some concentration around a certain value. For Gaussian phases, similar computations show that the curves for the different delay ranges are very close, and the UTA coherence is substantial even for broadly distributed phases. In the example of Fig. 3a, the UTA coherence is larger than, or at least close to, the UTU coherence for all phase ranges.

ATA Coherence for Two Uniform Populations of Partially Correlated Units

As indicated by simple theoretical considerations, the key rules that determine the behavior of the ATA coherence are similar to those for the UTA coherence. Specifically:

a) The aggregate activities of the linearly correlated subsets of the two (sub)populations are also correlated.
b) These activities dominate the total aggregate activities of the two (sub)populations, in a wide range of conditions.

Consider two sets of n unitary processes, $\{x\}$ and $\{y\}$, each comprising a subset of n_c correlated processes, as before. The members of the correlated subsets in the two populations are linearly correlated, the sets are again uniform, and $S_a(\omega)$ and $S_c(\omega)$ are as before. Let $x(t)$ and $y(t)$ denote the total aggregate activities of the two sets. The ATA coherence is defined as:

$$Q_{xy}^2(\omega) = |S_{xy}(\omega)|^2 / [S_{xx}(\omega)S_{yy}(\omega)]$$

The cross-correlation function of $x(t)$ and $y(t)$ is given as (Papoulis, 1965):

$$R_{xy}(\tau) = \sum_{i=1}^{n} \sum_{k=1}^{n} R_{ik}(\tau)$$

where subscript i is for set $\{x\}$ and subscript k is for set $\{y\}$.

Therefore, the cross-spectrum of $x(t)$ and $y(t)$ is:

$$S_{xy}(\omega) = S_c(\omega) \sum_{i=1}^{n_c} \sum_{k=1}^{n_c} \exp[-j\omega(\gamma_i - \delta_k)] \quad ; \quad \omega \neq 0$$

where $\{\gamma_i\}$ and $\{\delta_k\}$ represent the delays for the units in $\{x\}$ and in $\{y\}$, respectively. Its squared modulus is then:

$$|S_{xy}(\omega)|^2 = S_c^2(\omega)\left[\left(\sum_{i=1}^{n_c}\sum_{k=1}^{n_c}\cos\omega(\gamma_i - \delta_k)\right)^2 + \left(\sum_{i=1}^{n_c}\sum_{k=1}^{n_c}\sin\omega(\gamma_i - \delta_k)\right)^2\right] \quad ; \quad \omega \neq 0 \tag{7}$$

After some calculations Eq. (7) takes the form:

$$|S_{xy}(\omega)|^2 = S_c^2(\omega)\{n_c^2 + 2[n_c(A_x + A_y) + 2A_xA_y]\} \quad ; \quad \omega \neq 0 \tag{8}$$

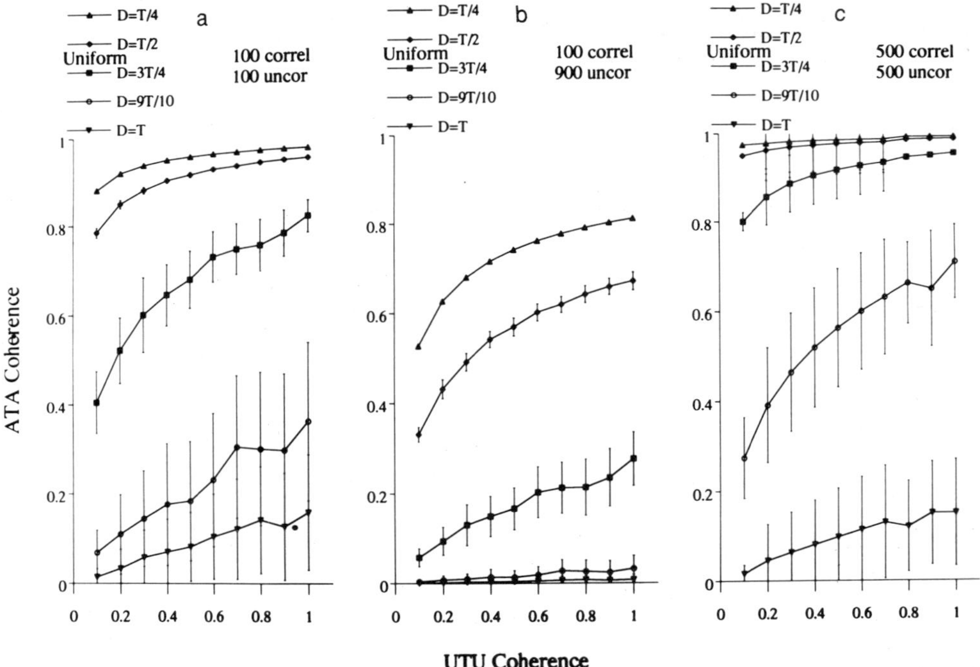

Figure 4. ATA coherence (mean ± one standard deviation) at the frequency of synchrony for two partially correlated populations, as a function of the UTU coherence at the same frequency. The different curves in each plot correspond to different delay ranges, as indicated at the top. Note that for delay ranges that are of the order of T/2 or less, the ATA coherence can be larger than the UTU coherence, even if the correlated subsets are small fractions of the respective populations (e.g., 10%). Also note that for delay ranges that are broader than 3T/4, the ATA coherence is very low if the correlated subsets are small fractions of the populations. Finally, note that the ATA coherence becomes higher as the numerical size of the populations increases.

where, as before, $2A_x$ and $2A_y$ represent, respectively, the sums of the cosines of the phase differences between all pairs of x_i's and between all pairs of y_k's .

Taking into account Eq. (5) for the auto-spectra of $x(t)$ and $y(t)$, the ATA coherence is written:

$$Q_{xy}^2(\omega) = \frac{S_c^2(\omega) \{n_c^2 + 2[n_c(A_x + A_y) + 2A_xA_y]\}}{[nS_a(\omega) + 2A_xS_c(\omega)] \; [nS_a(\omega) + 2A_yS_c(\omega)]} \quad ; \quad \omega \neq 0$$

or, in terms of the UTU coherence:

$$Q_{xy}^2(\omega) = \frac{Q^2(\omega) \, (n_c + 2A_x) \, (n_c + 2A_y)}{[n + 2A_xQ(\omega)] \; [n + 2A_yQ(\omega)]} \quad ; \quad \omega \neq 0 \tag{9}$$

Equation (9) indicates that the value of the ATA coherence is also influenced by the three characteristics of synchrony and by the numerical size of the populations. It also reveals that as the UTU coherence tends to zero, the ATA coherence also tends to zero.

The study of the effects of these parameters on the ATA coherence is again easily done by generating on the computer sets of pseudorandom delays γ_i and δ_k for different "seeds". Figure 4 shows an example of such computations for delays that are uniformly distributed over fractions of the period T of synchrony. The following interesting features are seen in this Figure:

a) As in the case of the UTA coherence, the ATA coherence is an increasing function of the UTU coherence and of the extent of synchrony (compare Figs. 4b and 4c). In addition, it is a decreasing function of the width of the range of the uniformly distributed phases of the units (compare the different curves in any of the three plots). Finally, it is an increasing function of the numerical size of the populations, but saturates for concentrated phases and for sizes of the order of 1000 or more (compare Figs. 4a and 4c).

b) Unless the range of the uniformly distributed phases is broad (3/4 of a cycle or more) the ATA coherence is substantial, even if the extent and strength of synchrony are limited. For small UTU coherences, it may even be larger than both the UTU and the UTA coherence. For very broad phase ranges, on the other hand, the value of the ATA coherence falls, and it is smaller than both the UTU and the UTA coherence (compare Figs. 2 and 4).

Finally, the negative effects of broadly distributed phases on the ATA coherence are less severe if the distribution shows a concentration around some value. In Fig. 3b, for Gaussian phases, the different curves are much closer than for the uniformly distributed phases of Fig. 4a.

These are typical features of the ATA coherence. Therefore, the ATA coherence, being substantial in a wide range of conditions, and often larger than the UTU and the UTA coherence, can be efficiently used for detection of synchrony.

With respect to the measurement of synchrony, however, this approach has serious weaknesses. Specifically:

A) The result of ATA coherence analysis is one value per frequency, in which the effects of the four parameters that influence this coherence are lumped together. Clearly, from this single value the characteristics of synchrony cannot be separately determined. Rather, the value of the ATA coherence provides a general index of synchrony, on the condition that the numerical size of the two (sub)populations remains practically constant.

B) For narrow phase ranges (half a cycle or less), the ATA coherence is very sensitive in reflecting synchrony and saturates easily. In Fig. 4c, for example, the value of this coherence is almost the same for all UTU coherences. Consequently, the ATA coherence may be very high in cases of limited synchrony. In Figs. 3b, 4a and 4c, for narrow phase ranges, the value of the ATA coherence is of the order of 0.9, even for a UTU coherence that is as low as 0.2 and for an extent of synchrony of the order of 50%. These "nonlinearities" make the ATA coherence a poor index of synchrony.

APPLICATIONS

The principles of the approach of Sect. 2.2, which uses the UTA coherence as a tool for detection and measurement of population synchrony, have so far been applied to: (a) the study of fast rhythms in the discharges of inspiratory (I) motoneurons and nerves (Christakos, 1990; Christakos *et al.*, 1991; Christakos *et al.*, 1994) (b) the analysis of correlations of motor unit activities during time-varying muscle contractions (Iyer *et al.*, 1994), and (c) the study of fast sympathetic rhythms (Cohen *et al.*, 1992). In these same studies ATA coherence

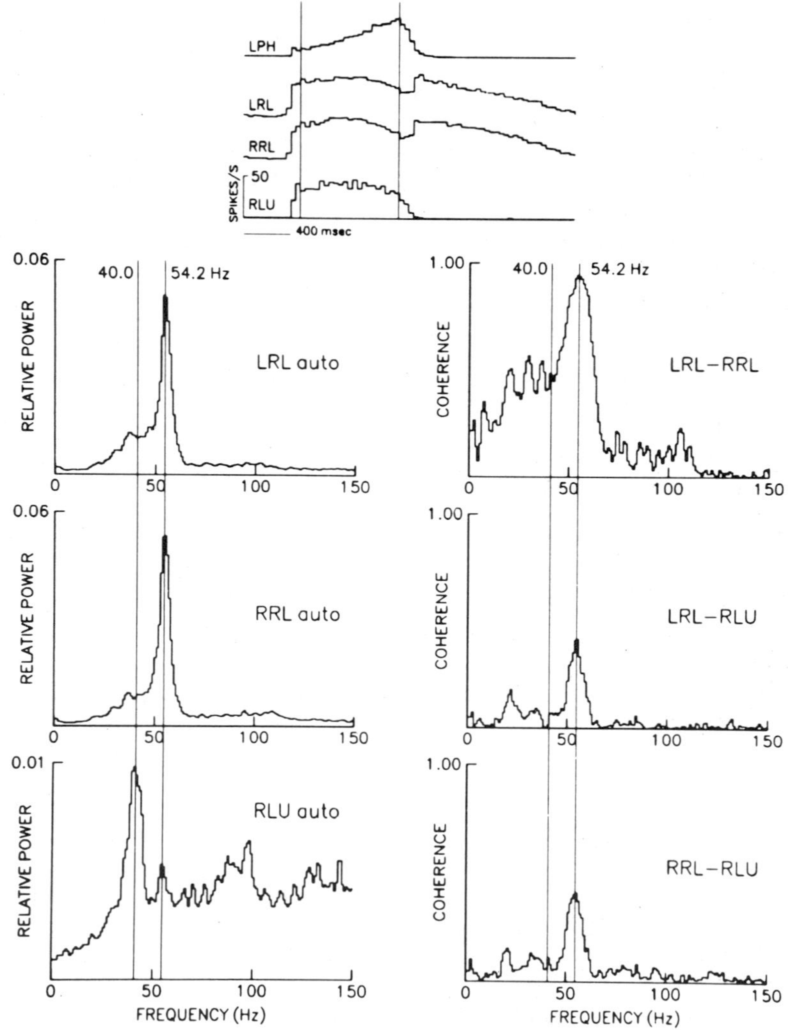

Figure 5. Top: Cycle-triggered histograms of activities of PHR nerve, right RL and left RL nerves, and a medullary RL unit. Note that the unit's plateau-type discharge is at 40 Hz. Bottom, left column: Auto-spectra of RL nerve activities and of the unit's discharge. Note the small and broad MFO deflection (band 30 to 45 Hz) in the auto-spectra of both nerves, and the clear MFO peak in the unit's auto-spectrum at 40 Hz (left vertical line). Also note the HFO peaks in all three auto-spectra at 54 Hz (right vertical line). Bottom, right column: Coherences between the different activities. Note the broad-band MFO coherence between the nerves and the nearly zero unit/nerve MFO coherence at 40 Hz. Also note the very high HFO coherence (0.94) between the nerves, and the moderately high unit/nerve HFO coherence (0.41) (from Christakos *et al.*, 1994).

(neurogram-to-neurogram and EMG-to-EMG) computations were also performed in order to detect synchrony (see Sect. 2.3). An example from study (a), dealing with fast rhythms in recurrent laryngeal (RL) activities (see Christakos *et al.*, 1994) , is here briefly presented.

Two fast rhythms exist in I discharges of RL nerves. They are termed medium-frequency oscillations (MFO, range 20-50 Hz) and high-frequency oscillations (HFO, range 50-100 Hz). Similar rhythms exist in the activities of other I nerves, such as the phrenic (PHR), and HFO are correlated between different nerves. Also, correlated HFO are widespread in the activities of medullary I neurons which drive the RL and PHR motoneurons (Christakos *et al.*, 1988; Mitchel & Herbert, 1974). These facts led Cohen (1973) to suggest that HFO is a characteristic of I activity, reflecting neuronal interactions in the medulla, and could furnish a possible means of studying network interactions within the central I-pattern generator. A similar hypothesis was forwarded for RL and PHR nerve MFOs by Richardson and Mitchell (1982).

For either aggregate rhythm in RL nerve activities, a central premotor origin at the level of the I-pattern generator implies widespread correlations between the unitary activities composing it. The presence of correlated HFOs and MFOs in RL motoneurons was tested by performing spectral, as well as unit-nerve (i.e., UTA) and nerve-nerve (i.e., ATA) coherence, analyses on a sample of 35 RL units (27 single RL fibers and 8 identified RL motoneurons in the medulla) and the simultaneously recorded activities of one or both RL nerves.

HFO: These were present in one half of the sample, and had the same frequency as the nerve HFO. [In Fig. 5, the HFO is at 54 Hz.] The unit-nerve (i.e., UTA) coherences were significant (range 0.22 to 0.65) for all motoneurons showing HFO, indicating the presence of widespread correlations between unitary rhythms of this type. (In Fig. 5, the UTA HFO coherence has a value of 0.41.) The very broad range of observed UTA coherences reveals that the strengths of HFO correlations vary a lot within the motoneuron population. It is interesting to note that the nerve-nerve (i.e., ATA) HFO coherences usually were very high, even though only a fraction of the RL motoneurons seem to have this type of rhythm in their discharges and the UTA HFO coherences for RL units in the sample did not reach such large values. [In Fig. 5, the ATA HFO coherence has a value of 0.94.] This is in agreement with the results of Sect. 2.3, regarding the high sensitivity of the ATA coherence in reflecting synchrony.

MFO: These were present in all 35 units, reflecting the rhythmic discharges of the cells, and their frequencies were in the band of the broad MFO spectral deflection for the RL nerve. [In Fig. 5, the unit MFO spectral peak is at 40 Hz, which is the nearly constant discharge rate of the motoneuron (see histogram at top), whereas the MFO spectral deflection for the nerves is in the band 30-45 Hz.] However, the respective UTA MFO coherences were near zero (<0.10) for 60% of the units and very low (range 0.10 to 0.15) for the remaining units, indicating that correlations between unitary rhythms of this type are uncommon and weak. [In Fig. 5, the UTA MFO coherence has a value of 0.09.] At the same time, various types of RL nerve-nerve (i.e., ATA) MFO coherence were observed, including (a) very low values over the entire MFO range, (b) a distinct peak at an HFO submultiple, and (c) a broad, moderately high deflection (as in Fig.5). Of the two types of significant MFO range ATA coherence, type (b) can be explained as reflecting correlations of the discharges of RL motoneurons skipping cycles of the HFO in their membrane potential. Type (c), on the other hand, seems to reflect temporary synchrony of a few non-plateau-type motoneurons that fire at similar rates over parts of the I phase. As indicated by the results of Sect. 2.3, such limited correlations are magnified at the level of the ATA coherence.

Therefore, in this case, UTA coherence analysis revealed an important difference between the two rhythmic components of RL activities, which reflects the different origins of these components. These are: (a) For the MFO, the refractoriness of the motoneurons, which causes a rhythmic component in the discharges of the cells. Note that the frequency of this rhythm for a non-plateau-type discharge changes in the course of inspiration, and the cell's MFO spectral deflection is then broad with a peak at the highest discharge rate of the motoneuron. (b) For the HFO, the correlated rhythmic medullary inputs (system HFO) to the cells, which, upon superposition, cause an HFO in the cells' membrane potential.

ATA coherence analysis, on the other hand, correctly identified the existence of HFO synchrony in RL motoneuron populations, but often gave the false impression of a substantial MFO synchrony (as in Fig. 5).

DISCUSSION

The principal merits of the UTA coherence approach for detection and measurement of population synchrony are its simplicity and efficiency: (a) It does not require separate recordings of multiple unitary activities, but rather uses two signals that are readily recorded simultaneously. (b) It enables the easy identification of units that are correlated to a number of other units, and thus provides information on the existence of synchrony and on how widespread synchrony is. (c) It provides a simple representation of estimated strengths of unitary correlations within the population, in the form of a distribution of UTA coherences for the units of the sample.

Consider, for example the case of a correlated subset comprising 30% of the units of a population. The probability of simultaneously recording two units from this subset, and thus establishing by UTU correlation (or coherence) analysis the existence of synchrony between just two units in the population, is only 9%. In contrast, the probability of recording one such unit, and establishing by UTA coherence analysis the presence of a whole correlated subset to which this unit belongs, is much higher, 30%.

As follows from the arguments at the beginning of Sect. 2.2, the value of the UTA coherence at the frequency of synchrony is larger for units that show stronger correlations to other units. Moreover, as argued in (Christakos, 1994a), the effects of the extent, the phases, and the numerical size of the population, on the UTA coherence are the same for all members of the correlated subset. Therefore, the distribution of UTA coherences for a sample of recorded units is indicative of the differing strengths of the unitary correlations within the population.

In summary, UTA coherence analysis is a positive approach, well suited for detection and quantification of population synchrony, taking into account the complexity of the problem. Moreover, this analysis can be speedily performed via the FFT or other such algorithms, and provides results in a compact form (which is another advantage over the traditional UTU correlation analysis). It also facilitates the study of changes in synchrony as conditions change. Even if the numerical size or composition of the population changes, the extent of synchrony can still be estimated in the same way, and other changes may then be deduced. However, for the study of changes in strength, it is important to record, if possible, each unit under the different conditions. Finally, it is noteworthy that this method allows comparisons with respect to synchrony between physiological and pathological situations.

Turning to the ATA coherence approach for analysis of population synchrony, an obvious advantage over UTU correlation and UTA coherence analysis is that it uses two signals that are both very easy to record. Another merit with regard to the detection of such synchrony, is that this coherence reflects the total amount of correlations that exist between members of the two populations at any time. Therefore, it can even detect synchrony that is present over short periods of recording time or in small shifting subsets of units.

As an example, in the study of Elul (1972b) coherence computations were performed between unitary wave activity and the EEG alpha as a quantitative test of the relationship between the two types of activity. The recorded units exhibited an apparent correlation to the EEG, over short time intervals (1-2 sec). This implies the existence of unitary correlations within the population, which are generally manifested as a significant EEG/EEG (i.e., ATA) coherence. However, Elul's UTA coherence computations failed to detect the synchrony, presumably because of its short time duration for any particular unit, relative to the recording time that is required for spectral analysis (more than 10 sec).

It should be noted, however, that in the same manner, temporary and limited unitary correlations may be identified as substantial synchrony by ATA coherence analysis. This seems to be the case with the MFO range ATA coherence of Fig. 5.

With respect to the use of the value of the ATA coherence as a general index of synchrony (see Bullock and McLune, 1989), the results of Sect. 2.3 clearly indicate the possibility of serious problems in specific situations. For concentrated unit phases, very large ATA coherence values can occur even if the extent and strength of synchrony are limited, thus giving the misleading impression of widespread and strong synchrony. Moreover, the influence on the ATA coherence of the populations' size, which is not a characteristic of synchrony, is significant in most cases. The index could thus be of use for comparisons of synchrony only if the populations' size remains practically fixed as conditions change.

The case of uniform populations (i.e., equally sized units and equally strong unitary correlations) was considered in the present study as a simple way of examining the behavior of the UTA and ATA coherence in an average sense. For non-uniform populations, both the model and the analysis would be more complicated, and the results would have to be specific to particular systems.

In most real situations, size differences between units do exist and may affect the behavior of both the UTA and the ATA coherence (see Discussion in Christakos, 1994a). A preliminary study that considered the extreme case of two types of unit in a population, namely, small and large, revealed that problems with the UTA coherence may arise if: (a) the correlated subset consists of the small units; (b) the numerical size of the population is relatively small (<500) and the correlated units form the minority; and (c) the unit phases are broadly distributed. In such situations, the UTA coherence for the correlated subset will be low at the frequency of synchrony, and possibly not detectable as a significant coherence, particularly when the UTU coherence is low.

Otherwise, and as long as the activity of the uncorrelated subset does not dominate the total aggregate activity, the UTA coherence for both large and small units of the correlated subset is substantial at the frequency (band) of synchrony and can be used for detection and measurement of synchrony, as described above.

Finally, it should be emphasized that the case of uniformly distributed unit phases has been considered in the present analysis as a stringent test of the limitations of the UTA and ATA coherence approach. Other distributions showing a concentration around some value, such as the Gaussian, are more realistic for most neural populations.

ACKNOWLEDGMENTS

I thank Sophia Erimaki for excellent technical assistance and valuable discussions. This research was supported in part by The Special Account For Research (Univ. of Crete).

REFERENCES

Bullock, T. H., and McClune, M. C., 1989, Lateral coherence of the electrocorticogram: a new measure of brain synchrony, *Electroenceph. clin. Neurophysiol.* 73:479-498.
Christakos, C. N., 1982, A linear stochastic model of the single motor unit, *Biol. Cybern.* 44:79-89.
Christakos, C. N., 1990, Modeling aggregate rhythms in neural populations: The stochastic approach, *Trends Biol. Cybern.* 1:271-280.
Christakos, C. N., 1994a, Analysis of synchrony (correlations) in neural populations by means of unit-to-aggregate coherence computations, *Neurosci.* 58:43-57.

Christakos, C. N., 1994b, Quantification of synchrony in neural populations by coherence analysis, *Soc. Neurosci. Abstr.* 20:1202.

Christakos, C. N., Cohen, M. I., Barnhardt, R. and Shaw, C.-F., 1991, Fast rhythms in the discharges of phrenic motoneurons and nerves, *J. Neurophysiol.* 66:674-687.

Christakos, C. N., Cohen, M. I., Sica, A. L., Huang, W.-X., See, W. R. and Barnhardt, R.,1994, Analysis of recurrent laryngeal inspiratory discharges in relation to fast rhythms, *J. Neurophysiol.* 72:1304-1316.

Cohen, M. I., 1973, Synchronization of discharge, spontaneous and evoked, between inspiratory neurons, *Acta Neurobiol. Exp.* 33:189-218.

Cohen, M. I., Christakos, C. N., Barnhardt, R., Huang, W.-X. and See, W. R., 1992, State-dependent synchronized fast rhythms in neural networks (inspiratory and sympathetic), *Soc. Neurosci. Abstr.* 18:317.

Elul, R., 1972a, The genesis of the EEG, *Int. Rev. Neurobiol.* 15:227-272.

Elul, R., 1972b, Randomness and synchrony in the generation of the electroencephalogram. In: *Synchronization of EEG Activity in Epilepsies*, Petsche, M. and Brazier, M. A. B. (eds.), Springer, New York, pp. 59-77.

Houk, J. C., Dessem, D. A., Miller, L. E. and Sybirska, E. H., 1987, Correlation and spectral analysis of relations between single unit discharge and muscle activities, *J. Neurosci. Meth.* 21:201-224.

Iyer, M. B., Christakos, C. N. and Ghez, C., 1994, Coherent modulations of human motor unit discharges during quasi-sinusoidal isometric muscle contractions, *Neurosci. Lettr.* 170:94-98.

Lopes da Silva, F. H., Hoeks, A., Smits, H. and Zetterberg, L. H., 1974, Model of brain rhythmic activity. The alpha rhythm of the thalamus, *Kybernetik* 15:27-37.

Mitchell, R. A. and Herbert, D. A., 1974, Synchronized high frequency synaptic potentials in medullary respiratory neurons, *Brain Res.* 75:350-355.

Papoulis, A., 1965, *Probability, Random Variables and Stochastic Processes*, McGraw-Hill, New York.

Perkel, D. H., Gerstein, G. L. and Moore, G. P., 1967, Neuronal spike trains and point processes. II. Simultaneous spike trains, *Biophys. J.* 7:419-440.

Richardson, C. A. and Mitchell, R. A., 1982, Power spectral analysis of inspiratory nerve activity in the decerebrate cat, *Brain Res.* 233:317-336.

Rosenberg, J. R., Amjad, A. M., Breeze, P., Brillinger, D. R. and Halliday, D. M., 1989, The Fourier approach to the identification of functional coupling between neuronal spike trains, *Prog. Biophys. Molec. Biol.* 53:1-31.

13

SYSTEM IDENTIFICATION OF SPIKING SENSORY NEURONS USING REALISTICALLY CONSTRAINED NONLINEAR TIME SERIES MODELS

Michael G. Paulin

Department of Zoology and Centre for Neuroscience
University of Otago, Box 56
Dunedin, New Zealand

ABSTRACT

Gaussian local rate coding (GLR) transforms spike train data into time series data, making it possible to use time series models for neural system identification. A simple computational model is used to represent the dynamics of peripheral electrosensory system of an elasmobranch, and a maximum entropy criterion is used to simultaneously optimize the coding bandwidth and the structure and parameters of the computational model. The computational model may be suitable for other neural systems. The coding model is general and provides a natural definition of neural firing rate.

INTRODUCTION

System identification is the art of building simple models that can predict the behavior of dynamical systems. System identification methods have been developed by engineers and applied mathematicians for specific practical goals involving prediction and control. Other things being equal, more realistic models can be expected to give more accurate predictions, but in neurobiology other things are not equal. The reality is extremely complex and realistic models carry computational complexity and inefficiency as baggage. Modelling is hampered by the absence of theory to guide choices about what details of neurons need to be modelled in order to replicate the computational properties of neural networks.

From recent work on invertebrate neural networks there is little doubt that biophysical details of individual neurons are sometimes crucial for network function (Selverston, 1993). However, in certain cases it may be possible to design simple, efficient models that are computationally equivalent to but otherwise unlike real neurons. It should be feasible to model neural networks on a large scale using a hybrid approach in which different neural

Advances in Processing and Pattern Analysis of Biological Signals, Edited by Isak Gath and Gideon F. Inbar
Plenum Press, New York, 1996

elements are modelled in different ways. In particular, models may be obtained by using system identification methods to mimic the behavior of real neurons without detailed attention to biophysics. My aim in this paper is to show how information theory and time series analysis techniques can be combined to provide an effective system identification technique for modelling spiking sensory neurons. The method is illustrated by applying it to electrosensory neurons recorded from the New Zealand carpet shark, *Cephaloscyllium isabella*.

SPIKE CODING

From a reverse-engineering point of view the question in spike coding is not how neurons code or represent messages using spike trains, but how we should code or represent spike trains in order to analyze neural computation. I have approached this problem using information theory. Consider an idealized representation of a spike train as a sequence of precise spike times. Real spike trains must carry information at a finite rate, but arbitrary timing precision would allow for infinite information transmission rates. Therefore an idealized timing representation is unrealistic. Realism aside, parsimony considerations suggest that to analyze the role of spikes in neural computation we should use a representation that hides all information not relevant to that role (See Jaynes, 1979). Hidden or unavailable information in a state description is called entropy (Jaynes, 1979; Usher, 1984). For analyzing neural computation, the most parsimonious realistic model of a spike train has maximum entropy consistent with the spike train's role in neural computation.

The standard way to add entropy to a statistical model is to replace deterministic terms with probability density functions (Usher, 1984). For the timing model this means representing each spike using a pdf whose mean is the spike time. Another approach would be to represent spike times using fuzzy sets (Paulin, 1993; Kosko, 1992). The fuzzy approach has the advantage that it can be interpreted in terms of real timing uncertainty in neural computation, while the probabalistic model suffers from the problem that neurons actually transmit spikes not spike probabilities. I will not pursue this further here because within the scope of the present paper the analysis is formally the same whether the functions representing spikes are interpreted as pdfs or fuzzy membership functions.

From a well known result in information theory the Gaussian pdf has maximum entropy for a given variance (Usher, 1984). This means that the least squares best approximation to a spike with a specified entropy is a Gaussian curve. The smaller the variance, the lower the entropy and more information is transmitted by the spike. I will assume that all spikes in a train carry the same amount of information. This is a dubious assumption in general but it may be reasonable for the electrosensory neurons that I deal with here. Relaxing this assumption makes the representation much more difficult to deal with and, fortunately for my little brain, it would be unwise to deal with the general case without first exploring the simpler one. It follows in this case that a spike train should be represented as a sum of Gaussians centered on the spike times. In this maximum entropy model spikes are interpreted as discrete events distributed over a localized temporal region. The precise timing code is an unrealistic limiting case corresponding to an infinite amount of information per spike.

Now consider another popular model of spike coding, average rate. Rate is inherently defined over an interval or window in time and only well defined in the limit as the window gets large. In that limit the rate is constant and the information transmission rate is zero. To transmit information at a finite rate the window width should match the time scale of computationally relevant changes in interspike interval. From the result used above, Gaussian smoothing defines the rate of a spike train in a way that most closely approximates a point process but transmits information at a finite rate. Therefore firing rate can be defined

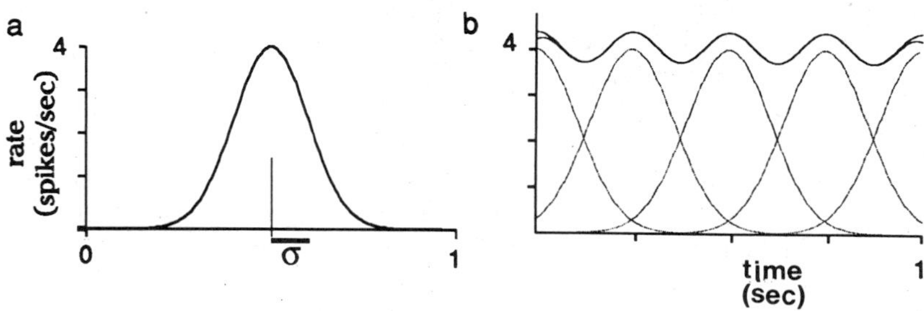

Figure 1. (a) A Gaussian filter with standard deviation σ seconds limits a single spike to rate $1/\sigma\sqrt{2\pi}$ spikes per second and distributes its power over time. (b) A GLR filtered spike train is not limited to the firing rate of the individual spikes; it can reach arbitrary levels if spikes are closely spaced. When a sequence of spikes occurring at regular intervals of 1/r seconds is rate limited to r spikes per second, the GLR representation is a pure sine wave. This figure shows the sum of five Gaussians positioned every 0.25 seconds between t = 0 and t = 1 with $\sigma = 4/\sqrt{2}\pi$, superimposed with the signal c = 4 + αcos(2π4t). Calculation and plot using MATLAB.

in terms of a Gaussian smoothing window with a parameter s that specifies the information in each spike. In this maximum entropy model spike rate is interpreted as a continuous variable that is a property of the spike train over a localized temporal region. The average rate code is an unrealistic limiting case corresponding to zero information per spike.

In practice, to encode a spike train using the Gaussian local rate code (GLR) you need to know the spike train's information transmission rate. The information capacity of a spike train is potentially much higher than that of a binary transmission line that can send bits as fast as the neuron can fire. The information content of a spike depends on the target's ability to determine and make use of small variations in the time at which the spike arrives, and this may be somewhat larger than one bit per spike. If you overestimate the information production rate of a spiking neuron you will need an unnecessarily complex computational model to predict the firing pattern, while if you underestimate the information production rate you will miss important details of the firing pattern and will not be able to analyze the computational role of the neuron in terms of firing rate. It is easy to see that this is true by considering the limiting cases.

If the neurons under consideration transmit information in the temporal pattern of spikes there is no loss of generality when GLR is applied because in the low entropy limit timing is preserved to arbitrary precision. The question of whether neurons might represent particular objects or states using particular patterns of spikes is not a question about coding as defined here. It is a question about neural dynamics or computation, and GLR excludes nothing. On the other hand, it may be true that in some cases a certain amount of spike timing variability is computationally meaningless. In such cases GLR may significantly reduce the complexity of computational models by hiding this variability in the coding model.

A spike train represented using GLR can be converted to a sequence of instantaneous pulses using integral pulse frequency modulation (IPFM) or an integrate-and-fire mechanism (Paulin, 1992). The precision with which spike times can be recovered depends on how large the Gaussian parameter σ is relative to the interspike intervals in the train. The general properties of IPFM can be understood by considering limiting cases. When σ is small the spike times can be recovered to arbitrary precision. When σ is large IPFM will give a spike train with constant interspike intervals. In general, GLR followed by IPFM lowers the interspike interval variability of a spike train without altering the mean spike rate. If coding and decoding are phase locked by setting t = 0 at the time of the first spike in the train then the expected difference between the time of a spike in the original train and the time of the corresponding spike in the recovered train is zero. That is, recovered spikes may occur sooner or later than in the original spike train, but on average they occur at the same time and the net effect is that recovered spikes tend to be more evenly spread.

GLR allows spike trains to be treated as continuous signals rather than as point sequences. For modelling low entropy-high bandwidth neural computations, such as coincidence in spatial hearing, GLR may not be very helpful although it has the virtue of not making the problem any worse. On the other hand it may considerably simplify the modelling of systems with high entropy by converting a sequence of pulses into a relatively slowly-varying continuous signal. Outputs of computational models that use a GLR representation can be expressed as pulse sequences using IPFM, if required.

How much smoothing is required in a particular case? Some guidance in this matter can be provided by working from the limiting cases. Suppose a spike train is represented as a sequence of precise firing times but interpreted in terms of firing rate. Then firing rate is infinite at the spike times and zero elsewhere. GLR maintains the power of each individual spike but distributes it over time to give finite firing rates. By setting $\sigma = 1/r \sqrt{2}\pi$ each spike is rate-limited to r spikes per second (Fig. 1a). This is the bandwidth of the GLR filter. When spikes in a train with constant interspike interval $1/r$ are limited to rate r sec^{-1} using a GLR filter, the result is a signal with dc level r sec^{-1} and an oscillation at rHz. Surprisingly, the oscillation is almost exactly a pure sine wave (Fig. 1b).

The loss of interspike interval variability when GLR is applied corresponds to the entropy of the model. The question is, how much variability in a given spike train is computationally relevant? How much entropy can be added, how much detail can be thrown away as noise, without losing anything that is necessary for modelling neural computation? As far as I can see this question can not be answered without also considering the question of how neural computation should be modelled. There seems little likelihood that a general theoretical framework for modelling neural computation will be developed in the near future but models of specific systems seem feasible. I now turn to the problem of modelling computation in the elasmobranch electrosensory system.

SYSTEM IDENTIFICATION BY TIME SERIES MODELLING

Assuming that the entropy, or equivalently the information transmission rate or bandwidth, of a spike train can be determined, GLR can be applied to transform the spike train into a continuous signal which can be sampled to give a time series. There is an important practical reason for doing this, namely that there is a large literature on system identification using time series models, with many and varied successful applications including nonlinear and time-varying systems (Ljung, 1987; Diggle, 1990). Software for time series modelling is readily available, and extensible high level numerical languages such as MATLAB can be used to rapidly develop custom software for specific applications.

For system identification it is necessary to specify a class from which the system model is to be selected, and so I proceed by specifying a particular class of dynamical models for neurons. However, the procedure that I use to identify maximum entropy models of neural computation and coding is general, and can be applied whenever there are algorithms for fitting models of the specified class to data.

Firing rates can not be less than zero and so the class of time series models appropriate for modelling spiking neurons can be realistically constrained to include only those that have non-negative outputs. The easiest way to do this is to use a cascade model consisting of a general dynamical element followed by a half wave rectifier. Models of this kind can be fitted by fitting the dynamical model with no penalty on the residual at points where the dynamic response is hidden by the rectifier. The fitted model is biased because the data do

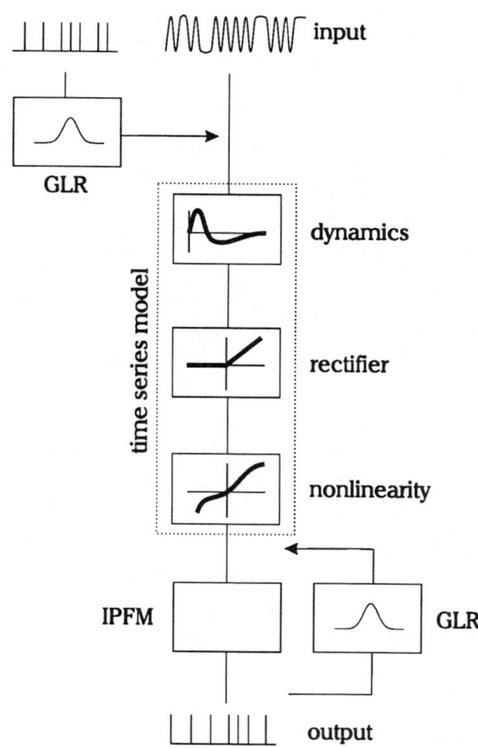

Figure 2. General model for coding and computation in the elasmobranch peripheral electrosensory system. The system receives continuous inputs - electrical potentials across the receptor epithelium - and produces spike trains as output. To analyze the system the output is demodulated to a continuous signal by GLR and a nonlinear time series model is fitted to sampled data. To recover or predict spike trains the output of the time series model may be fed through an IPFM encoder. The method can be applied to systems with spike inputs by using GLR on the input spike train (upper left).

not provide the initial conditions or time-history of the response near rectified regions. However, the model gives biased estimates of this history. Successive iterations in which the current model is used to estimate the unknown dynamic response hidden by the rectifier provide progressively more accurate fits.

Firing rates tend not to change linearly with changes in stimulus strength, in particular they tend to saturate at high levels. This can be represented by placing a polynomial static nonlinearity at the output. An iterative procedure similar to that described above, involving a generalized inverse of the current estimate, can be applied to fit the nonlinearity (Paulin, 1993a). There is a major advantage in separating the saturation nonlinearity from the zero firing rate nonlinearity. Saturation tends to involve a gentle roll-off in the stimulus-response curve at high stimulus amplitudes, while rectification involves a sudden transition to a flat response at zero firing rate. The saturation can be modelled accurately by a low order polynomial, or some other family of curves with a small number of parameters, while the rectification would require a rather higher order polynomial. Since it is necessarily true that firing rates are not less than zero, a rectifier can be included in a model of a spiking neuron without losing degrees of freedom. Doing this has the potential to greatly reduce the number of parameters required to describe the output nonlinearity of the neuron.

I have used linear autoregressive - moving average (ARMA) time series models for the dynamical element. This choice was guided by the historical success of such models in a variety of applications and is ultimately justified by their success in the particular application described below. ARMA models can be fitted quickly and easily using matrix methods, and can be implemented in compact, computationally efficient ways. An important advantage in using linear ARMA models is that the dynamics of such models can be characterized and described using well-established mathematical methods that are widely used in engineering. In particular, they can be represented graphically using zero-pole plots, Bode plots and unit impulse response plots.

The complete neural model that I use is a cascade made up of an ARMA filter followed by a rectifier followed by a saturation nonlinearity followed by an IPFM encoder (Fig. 2). The rectifier and the saturation nonlinearity are interchangeable because these transformations commute. It is possible fit a model of this kind because GLR coding makes it possible to invert the IPFM encoder in an information-preserving fashion. The internal state variables of the model, i.e. the signals passing between the elements, can be estimated and therefore each element can be fitted.

Once it is possible to fit models of a given class to data, system identification becomes a matter of selecting a particular model within the class. The aim is not to fit the data that you have, but to use it to predict data that you do not have. In general a model's fit to data can be improved by increasing its complexity or order, but whenever there is noise in the data you reach a point where better fitting models give worse predictions. Goodness of fit is not a useful criterion for system identification. Information statistics are widely used to solve this problem (Sakamoto *et al.*, 1986). An information statistic measures the tradeoff between improving the goodness of fit and improving the complexity of a model, and a model chosen by minimizing such a statistic is called a maximum entropy model. Maximum entropy models have the property that if the available data are representative of the system's behavior then the model is expected to be the best fit to all data generated by the system, including data not yet seen.

Models can be evaluated by crossvalidation, which simply means testing them on novel data sets. The best model is the one that best predicts novel data. Crossvalidation may be difficult in some applications because it can be difficult to replicate circumstances under which the data were collected. However, it is relatively easy to collect extra data sets in neurophysiological experiments. A model selected by crossvalidation is a maximum entropy model because crossvalidation optimizes the same tradeoff, between complexity and accuracy in fitting, as does an information statistic.

MAXIMUM ENTROPY IN CODING AND COMPUTATION

From above we have a class of simple computational models of peripheral sensory function and an algorithm for fitting these models to spike train data. However, there is an unspecified parameter. This parameter is the variance of the GLR filter, corresponding to the information transmission rate of the spike train. It is necessary to know this parameter so the spike train can be appropriately filtered before a time series model is fitted. However, the information content of a spike train depends on the receiver's ability to measure and make use of the precise timing of spikes and therefore it would seem necessary to model central electrosensory processing before this parameter can be estimated. This is unfortunate, because it would seem necessary to describe what the shark's electroreceptors tell the shark's brain before analyzing electrosensory computation in the shark's brain.

To get around this problem I assume that the peripheral electrosensory system of an elasmobranch only provides the brain with information about electric fields that is actually useful for subsequent computations. This is an optimality assumption; I am assuming that the peripheral electrosensory system is functionally adapted to its role in electroreception. Afferents must provide at least this much information, so if the assumption is wrong then the models may be more complex than necessary for understanding central electrosensory computation. Of course, in this case the behavior of the real system is also more complex than it needs to be, so I am erring towards reality.

From the optimality assumption it follows that the best model maximizes the rate at which information about the stimulus is transmitted to the brain. The information transmission rate of a continuous signal can be computed from Shannon's formula, $I = \log_2(1+S/N)f_{max}$ (Usher, 1984), where S/N is the signal to noise ratio (SNR) and f_{max} is the signal bandwidth. In the present case, S is the model output, N is the residual and the bandwith is (see Section 2)

SNR can be made arbitrarily large by lowering the bandwidth. In the limit the signal is constant and all of the variability in the spike train is accounted for as noise. Despite the fact that the signal can be predicted precisely, no information is transmitted because the bandwidth is zero. Assuming that sensory spike trains do transmit information to the brain, and this seems to me to be a fair assumption, the information transmission rate of the model will increase as the bandwidth increases. Beyond a certain point, however, SNR will rise more rapidly than bandwidth and the information transmission rate will fall. Intuitively it can be seen that this is the point at which pulses generated by the model fail to significantly overlap corresponding pulses in the data. In the limit as the pulses become narrow the probability of coincident events in the model and the data goes to zero, the signal to noise ratio is zero and so is the information transmission rate. It follows that the information transmission rate is a maximum at some finite bandwidth.

The optimum point can be determined by including the GLR parameter σ in the fitting procedure used for system identification. That is, as seemed intuitively obvious from the start, you need to optimize the coding and computational models simultaneously. I have not developed an efficient way of doing this. Instead, I search over a range of bandwidths and identify the maximum entropy - best predictor model at each bandwidth by crossvalidation. Then I select the model with the highest information transmission rate using Shannon's formula.

APPLICATION TO ELECTROSENSORY AFFERENTS

The New Zealand carpet shark, *Cephaloscyllium isabella*, is a small, sedentary elasmobranch. Studies of this and related species indicate that it feeds on benthic invertebrates that it is able to detect using its acute electric sense, even if the prey are buried in sand. A closely related species, *C. ventriosum*, predates small teleost fishes by lying motionless on the substrate at night and using its electric sense to detect when a prey fish has drifted close enough to be drawn into the mouth (Tricas, 1982). It is not known whether *C. isabella* shares this behavior. Elasmobranchs may also use their electric sense for navigation by measuring electric fields induced by motion in the earth's magnetic field (Murray, 1960; Kalmijn, 1982; Paulin,1985). Elasmobranch electroreceptors do not have efferent nerves, and so modelling peripheral sensory processing is not complicated by neural interactions with the central nervous system, as are other octavolateralis sensory systems. *C. isabella* is particularly suitable for experiments of this kind because the cranial vault is somewhat larger than the brain and sections of the sensory cranial nerves including ganglia can be exposed by simply removing the soft cartilage over the brain.

Single unit spike trains were recorded from the ganglion of the anterior lateral line (supraophthalmic) nerve using glass 10-20 MΩ microelectrodes filled with 4 M NaCl. An electric field stimulus was presented using a dipole electrode. When a sensitive unit was identified the dipole was moved to locate the pore of the receptor canal from which that unit projected. The dipole was then positioned over the pore with its axis aligned towards the

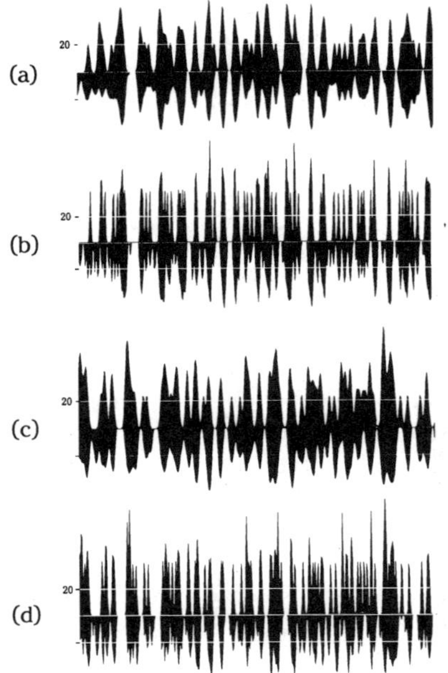

Figure 3. Each trace in this plot shows a GLR-filtered spike train recorded from an electrosensory primary afferent neuron, and a time series model prediction of the same spike train reflected about the time axis. This display method makes it easy to judge the quality of model predictions visually, despite the complexity of the signal waveforms. If the model is a perfect predictor then the corresponding plot will be symmetric about the time axis. In each case the upper waveform is experimental data while the lower waveform is model-generated. The horizontal lines indicate firing rates of 20 spikes per second. There are two plots showing fitting data and model responses fitted to them (a,b), and two plots showing how well the fitted model predicts the neuron's responses to novel stimuli (c,d). Each waveform is five seconds in duration. (a) Section of fitting data with GLR rate-limiting to 20/sec. (b) Same fitting data at 40/sec. (c) Section of validation data at 20/sec (d) Same validation data at 40/sec.

pore, at a distance to give a field gradient at the pore of approximately 1 μV·cm^{-1}. This is well above threshold but below saturation for the afferent neurons (Montgomery, 1984). Then a pseudorandom binary sequence stimulus was applied by switching the dipole polarity (pseudo-)randomly under computer control at intervals of 20 msec for 20 seconds. Afferent spike times were recorded to a precision of 2 msec.

The identification procedure was applied to neural responses by splitting the 20 second stimulus-response data into two, 10-second segments. The first segment was used for fitting and the second for crossvalidation. I report results from one neuron here. With GLR filtered series rate-limited to 20 sec^{-1} and 40 sec^{-1} respectively, I scanned ARMA model structures up to order (64,64). Crossvalidation identified the optimal model as MA(25) in the first case and MA(22) in the second. SNR falls from 2.16 to 1.69 when the rate limit rises from 20 Hz to 40 Hz, meaning that the best predictor at 40 Hz has a larger relative error than the best predictor at 20 Hz. However, it is clear from Fig. 3 that the 40 Hz model is able to predict detail in the 40 Hz response that is not present in the 20 Hz response, indicating that there is information in the 20-40 Hz band. The information transmission rates are 33.2 bits

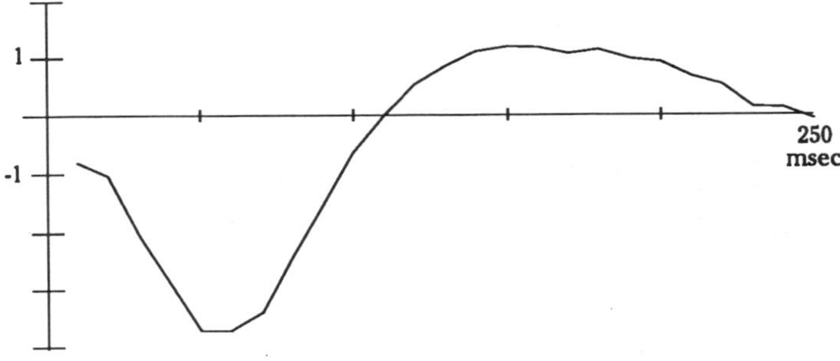

Figure 4. Unit impulse response of the optimal 20 Hz predictor. These are the MA parameters of the filter.

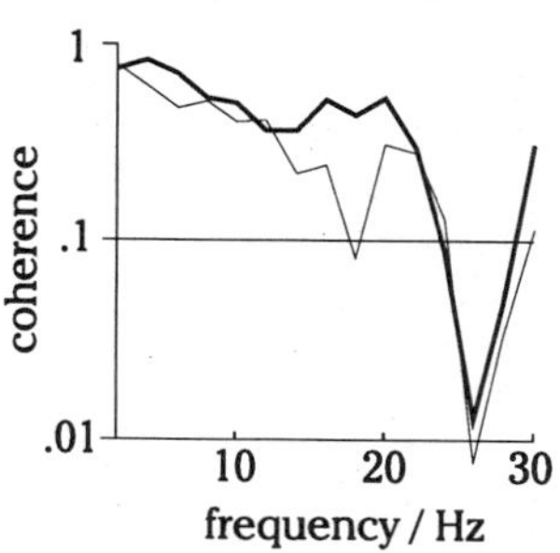

Figure 5. Linear coherence between stimulus and response (light) and between optimal 40 Hz model predictions and response (heavy). The model is deterministic - its predictions depend 100% on the stimulus - therefore the discrepancy between the coherence plots indicates non-linearity correctly characterized by the model.

per second for the 20 Hz model and 57.2 bits per second for the 40 Hz model. Given that the mean firing rate of this neuron is 16.17 sec^{-1}, the information transmission rates are 1.99 bits per spike and 3.54 bits per spike respectively. There is a lot of information above 20 Hz.

Both models have quite high order, more than 20 parameters in each case. However, both are pure MA so the parameter sequence is the unit impulse response (UIR) of the dynamic element of the neural model (Fig. 4). The UIRs superficially resemble those of a second order linear system, but this is not the case because the identification procedure would have quickly discovered this. I have confirmed this by showing that the UIR waveforms can not be fitted by sums of exponentially decaying sinusoids, as would be the case for a low order linear autoregressive process. A low-order model, with perhaps three to five parameters, should exist because the parameters of these models are highly correlated. The form of the UIRs suggests that an adapting second order linear autoregressive model (i.e., with time-varying natural frequency and damping ratio) may fit well.

Figure 5 is a plot of linear coherence between stimulus and response, and between model predictions and response, for the 40Hz model. Coherence is a frequency domain analog of correlation indicating the proportion of variability at a given frequency that can be accounted for by a linear dynamical relationship between two signals (Bendat and Piersol, 1986). The plot shows that the model prediction has consistently greater coherence with the data than the stimulus in the 0-20 Hz bandwidth. Above 10 Hz the stimulus-response coherence falls rapidly. A coherence roll-off is expected in a linear system at the system band limit, because in the presence of white noise the SNR of a linear system rolls off with the system frequency response. Linear systems analysis has previously indicated a bandwidth near 10 Hz for the elasmobranch peripheral electrosensory system (Montgomery, 1984), whereas my analysis has indicated a bandwidth greater than 20 Hz. The coherence analysis indicates that this discrepancy may be largely due to nonlinearities that are usefully taken into account by my cascade model. Zero firing rate rectification can be expected to reduce amplitudes of signal components within the system bandwidth and introduce harmonic distortions, thus biasing the linear analysis and spuriously reducing bandwidth.

DISCUSSION

A previous investigation of rate-estimating filters (Paulin, 1992) was based on the observation that filtering a point process sequence into a time series seems like a good idea because it allows powerful, readily available time series modelling techniques to be applied in neural modelling. I investigated Gaussian filters because they have the desirable property that they preserve locality in time and frequency in a least-squares optimal fashion. Filtering distributes information across time and frequency in such a way that precision in the time domain leads to uncertainty in the frequency domain, and vice versa. The uncertainty principle, developed initially in relation to quantum theory, places a fundamental limit on the precision with which a single representation can simultaneously specify times and rates of events. The filtering procedure that I have called GLR in the present paper attains the limiting precision (Hamming, 1983; Paulin, 1992). Numerical analysis using simulated data shows that the performance of GLR is not likely to be significantly better in practice than the performance of similar rate estimating filters that are easier to implement and faster to apply (Paulin, 1992). I concluded that the optimality property of GLR is not worth the effort required to implement and compute using Gaussian filters.

In this paper the Gaussian was introduced because of a different optimality property, that it is in a least squares sense the most localized representation containing a specified finite amount of information about the location of an event. This is a desirable property because it produces a spike train representation that formally has minimum complexity consistent with its intended use (see Jaynes, 1979). The representation leads to the identification of an elegant predictive model of electroreceptor primary afferent response properties. While simulations indicate that this much could have been obtained using simpler, non-Gaussian filters (Paulin, 1992), the importance of the Gaussian -maximum entropy approach is that it has provided an information-theoretic basis for selecting the parameter(s) of the filter. Previously I have had to estimate signal bandwidth prior to time series analysis, based on published reports and guesswork (Paulin, 1993b). The maximum entropy argument allows bandwidth to be computed from data.

The result that a sequence of Gaussian pulses closely approximates a sine wave has two interesting corollaries. Firstly, it follows from this result and Fourier's theorem that Gaussian pulse trains approximate a basis set for all signals. Therefore, via GLR, spike trains can represent the operands and results of any dynamic computation. Secondly, the result indicates that GLR forms a natural link between pulse rate coding and the more extensively analyzed FM and AM coding schemes used in communication systems (Connor, 1982). The implication is that GLR is not just a convenient way to represent spike trains for data analysis, but may provide a fundamental framework for analyzing neural computation.

I suggest that the distinction made in this paper between coding and computation is important for making progress in analyzing neural computation. Coding is about how information is received and transmitted by agents, while computation is about how it is transformed by those agents. Claims that neurons use more sophisticated coding schemes than firing rate either fail to make this distinction, or use inadequate and inconsistent definitions of rate (see Miller, 1994). I suggest that GLR in combination with a maximum entropy criterion for identifying associated computations provides a natural definition of rate that will eliminate confusion surrounding this issue.

The identified model accurately predicts responses of real neurons and shows that previous descriptions are biased and fail to characterize much of the meaningful information transmitted to the brain. The bandwidth of these receptors has previously been reported to be as low as 10 Hz, but the present analysis indicates that band-limiting the spike trains to 20 Hz may eliminate more than 75% of the information that they contain. The information estimates given are not precise because they do not take into account variations in SNR across the system frequency band. Nonetheless, it is clear from Fig. 3 that there is predictable variation in the spike train filtered at 40Hz that is not present when the train is filtered at 20 Hz, and therefore the conclusion is qualitatively correct. It has not been possible to use currently available data to check for higher information transmission rates at higher bandwidths, because the stimulus had a very steep roll-off in power above 40Hz.

Published estimates of information transmission rates in sensory neurons indicates a rate of about 3 bits per spike, which is remarkably uniform across vertebrates and invertebrates (Bialek *et al.*, 1993). The estimated information transmission rate in elasmobranch electrosensory neurons is about 15% higher than that found by Bialek *et al.* in fly visual movement detecting neurons. However, they used a linear estimation procedure and did not seek conditions under which the information transmission rate would be optimized.

I have presented a general approach to modelling neural computation and coding by system identification, a top-down approach that seeks elegant ways to mimic the behavior of neurons rather than a bottom-up approach that seeks to explain neural function in terms of neural structure. The 'neurons' developed here do not closely resemble real neurons structurally, but it should be noted that they are a data-based generalization of the 'neurons' used in connectionist theory (Rumelhart & McClelland, 1986), and may provide connectionist modelers with an opportunity to take a step towards biological reality. The approach is not limited to the particular dynamic linear-static nonlinear cascade model that I have used here. Any computational model of a neuron could be substituted. While there is no doubt a strong relationship between the structure of neurons and their function, at this point it is not clear to what extent neural biophysics is necessary for neural network function and to what extent it reflects historical contingencies and biological constraints. Given that computational neurobiology has neither the same opportunities nor the same constraints as the historical process of evolution, it seems sensible to pursue a hybrid modelling approach that allows for detailed biophysics to be included where it appears necessary, but otherwise avoids it in favour of simplicity and computational efficiency.

ACKNOWLEDGMENTS

John Montgomery and David Bodznick assisted with data collection. Supported by NZ MoRST:RST-08-2-1-c:SP under the US-NZ cooperative science programme.

REFERENCES

Bialek, W., Rieke, F., De Ruyter van Steveninck, R.R., Warland D., 1991, Reading a neural code, *Science* 252:1854-1856.

Bendat, J.S. and Piersol, A.G., 1986, *Random Data: Analysis and Measurement Techniques*, Wiley -Interscience: New York.

Connor, F.R., 1982, *Modulation,* Edward Arnold Ltd: London.

Hamming, R.W., 1983, *Digital Filters*, Prentice-Hall: Englewood Cliffs, NJ.

Jaynes, E.T., 1979, Where do we stand on maximum entropy? In: Levine, R.D. and Tribus, M., eds., *The Maximum Entropy Formalism*, MIT Press: MA, pp. 15-118.

Kalmijn, A.J., 1982, Electric and magnetic field detection in elasmobranch fishes. *Science* 218:916-918.

Kosko, B., 1992, *Neural Networks and Fuzzy Systems*, Prentice-Hall: NJ.

Miller, J.P., 1994, Neurons Cleverer than we Thought? *Current Biology* 4:818-820.

Montgomery, J.C., 1984, Frequency response characteristics of primary and secondary neurons in the electrosensory system of the thornback ray. *Comp. Biochem. Physiol.* A79:189-195.

Murray, R.W., 1960, Electrical sensitivity of the ampullae of Lorenzini. *Nature*, 187:957.

Paulin, M.G., 1992, Digital filters for neural firing rate estimation. *Biol. Cybernetics* 66, 525-531.

Paulin, M.G., 1993a, A method for constructing data-based models of spiking neurons using a dynamic linear - static nonlinear cascade. *Biological Cybernetics* 69:67-76.

Paulin, M.G., 1993b, Neural system identification applied to modelling dogfish electrosensory neurons. In: Eekman, F., ed., *Computation in Neurons and Neural Systems*, Kluwer Academic Press: Boston, pp. 191-196.

Paulin, M.G., 1985, Electroreception and the compass sense of sharks. *J. Theor. Biol.* 174:325-339.

Rumelhart, D.E. and McClelland, J.L., 1986, *Parallel Distributed Processing: Explorations in the Microstructure of Cognition*, MIT Press: MA.

Sakamoto Y., Ishiguro M. and Kitagawa G, 1986. *Akaike Information Criterion Statistics*, KTK Scientific Press: Tokyo.

Selverston, A.I., 1993, Modelling of neural circuits: What have we learned? *Ann. Rev. Neurosci.* 16:531-46.

Tricas, T.C., 1982, Bioelectric mediated predation by swell sharks, *Cephaloscyllium ventriosum. Copeia*, 4:948-952.
Usher, M.J., 1984, Information theory for information technologists. Macmillan Inc.: London.

14

TEMPORAL ENCODING OF VISUAL FEATURES BY CORTICAL NEURONS

Lance M. Optican and John W. McClurkin

Laboratory of Sensorimotor Research
National Eye Institute, NIH
Bethesda, Maryland

ABSTRACT

Neurophysiological studies have shown that visual information is carried not only by which subset of neurons is active, but also by the temporal messages they carry. Recent theories of visual processing look at both temporal and spatial properties of cortical neurons to elucidate some of the higher order mechanisms that underlie perception. The spatio-temporal approach to neuronal encoding should lead beyond reductionist approaches, that look only at neuronal selectivity, to integrative approaches, that model visual information processing to understand visual function.

INTRODUCTION

Visual stimuli are a rich composite of many features, such as form, color, texture, luminance, contrast, depth, etc. The primary way of studying how the brain processes visual information has been to focus on the selectivity of individual neurons for each of these features. This has led to the classical view of the visual system as a hierarchy of cortical areas, some for form, some for color, etc. Feature representation within these areas is thought of as the activation of a subpopulation of the neurons. Changing the feature, say the color of an object from red to green, would change the activation to a different subpopulation. The problem with this population-encoding approach is that it does not explain why visual neurons show temporally modulated activity that changes when visual features change (Eckhorn & Pöpel, 1974; Eckhorn & Pöpel, 1975; Maffei & Fiorentini, 1973; Optican & Richmond, 1987; Richmond *et al.*, 1987; Richmond & Optican, 1987, 1990; Richmond *et al.*, 1990). Nor does it deal with the problem of feature binding, e.g., how the many features of one object are all linked together (Richmond & Optican, 1992). Finally, the classical approach is strictly reductionist, and does not make any predictions about the type of operations or computations that the brain performs to achieve an integrated visual perception.

Advances in Processing and Pattern Analysis of Biological Signals, Edited by Isak Gath and Gideon F. Inbar
Plenum Press, New York, 1996

Recent work on visual processing has looked at temporal and spatial properties of cortical neurons to elucidate the higher order mechanisms that underlie perception (Golomb *et al.*, 1994; Middlebrooks *et al.*, 1994; Reid & Shapley, 1992). The studies reported here focus on the possibility that the brain uses temporal modulation within neurons, in addition to population coding across neurons, to encode properties of visual stimuli. This work has shown that there are specific temporal waveforms that can act as codes for specific visual features. Temporal codes have been found that are the same for visual and remembered stimuli in IT cortex (Eskandar *et al.*, 1992a, 1992b). Temporal codes have also been found for form and color that are the same across different cortical visual areas, and are similar in different monkeys (McClurkin & Optican, 1996; McClurkin *et al.*, 1994, 1996). A recent study of stimuli with different contrasts and textures has also shown evidence for temporal encoding (Purpura *et al.*, 1993).

These studies of temporal modulation go beyond the search for harmonic oscillations in visual responses (Engel *et al.*, 1992). In awake monkeys performing a visual task with stationary stimuli, oscillations at fixed frequencies are not seen (Gawne *et al.*, 1991). Rather, the distribution of power in the frequency spectrum shifts with the stimulus. Thus, it is better to regard the response of neurons as being temporally encoded, and to look for temporal messages, rather than synchronized oscillations. Interestingly, the mechanism used to look for synchronized oscillations, temporal cross-correlation, is the same mechanism used when looking for similar temporal messages (Eskandar *et al.* 1992a; Richmond & Optican, 1992). Another advantage of using a more general model of temporal encoding is that it is possible to ask whether or not the temporal responses of neurons are in any way predictable or decodable. In fact, a simple nonlinear model that multiplies separate temporal codes to make neuronal responses accurately predicts novel responses in both vision/memory tasks and form/color tasks. It is also possible to design simple detectors that decode stimulus features from temporally modulated neuronal responses with high correlations.

INFORMATION THEORETICAL ANALYSIS

The study of visual information encoding requires the ability to measure the amount of stimulus-dependent information present in putative neuronal codes. Methods that make it practical to compute the amount of information transmitted by neurons needed to be developed. First, it was necessary to find a statistic for transmitted information that was insensitive to bias and sample-size and could be applied to multivariate data. A sufficient boot-strap statistic for unbiased estimation of information in neuronal data is (Chee-Orts & Optican, 1993; Optican *et al.*, 1991):

$$T_u = \left[1 - \left[\frac{T_b}{T} \right]^{1.5(1+d)} \right] T$$

$$\tag{1}$$

where T_u is the unbiased information, T is the biased measure of information, T_b is the bootstrap of T, and d is the dimension of the response code. To calculate the amount of information transmitted by the temporal codes, it was necessary to apply this formula in dimensions as high as 64 (to get joint information at each time sample in a response). As the dimension increases, however, computation of this statistic by standard methods becomes impossible. Thus, a new statistical method for calculating T_u from very high dimensional data, based on response clustering, was developed (Chee-Orts & Optican, 1993). The cluster method showed that neurons carry much more information about stimulus features in the temporal waveform of their responses than in the strength of those responses.

ENCODING OF VISUAL AND REMEMBERED INFORMATION

Information analysis was applied to neuronal responses of IT neurons during a visual recency memory task (match-to-sample) (Eskandar *et al.* 1992b). The result was that neurons in posterior parts of IT (area TE) carried information about both the current and the remembered targets. This suggested that the function of area TE in recency memory is to compare the internal representation of current and remembered images. A mechanism for this comparison, based on temporal modulation, was hypothesized (Eskandar *et al.* 1992a). The model represents both the visual stimulus and the remembered stimulus by temporal codes. Comparison is then achieved through a temporal correlation measure. The temporal codes for current and remembered stimuli in the model turned out to be the same, suggesting that temporal waveforms could contribute to the encoding of visual and remembered information.

This interpretation of IT cortical function is somewhat at odds with the "novelty filter" idea of Desimone and co-workers (Li *et al.*, 1993; Miller & Desimone, 1994; Miller *et al.*, 1991, 1993), who find that activity in IT neurons decays during memory tasks, except when a novel stimulus appears. The differences in the two studies may lie in the much shorter time course of our experiments (less than 1 sec), in the nature of our stimuli (very familiar), or in the region of IT cortex from which recordings were made (posterior rather than anterior). Nonetheless, in these experiments, the responses to matching stimuli could be larger or smaller than responses to non-matching stimuli, so a simple novelty filter would not be sufficient to explain these results (Eskandar *et al.* 1992a). Interestingly, although these two interpretations of IT function differ markedly, the structure of the models used to represent them is quite similar (cf. Fig. 1A, although Desimone's group fails to make this point; Miller *et al.*, 1993, Fig. 19; Eskandar *et al.*, 1992b, Fig. 22; Eskandar *et al.*, 1992a, Fig. 2).

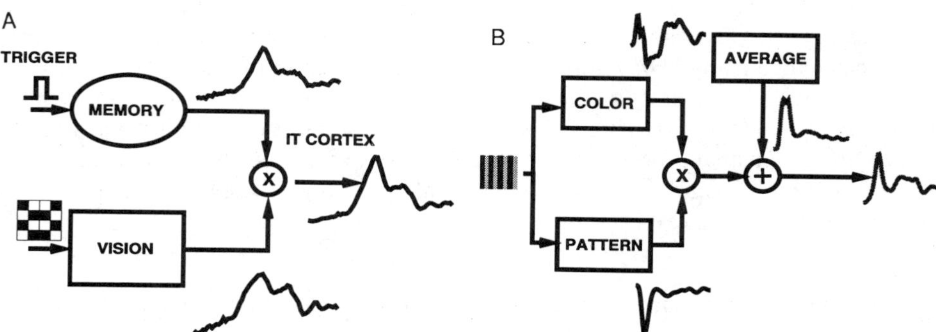

Figure 1. Multiplex temporal code model. A. A visual stimulus gives rise to temporally modulated activity. A synchronous temporal signal is elicited (TRIGGER) from memory. The product of these two signals gives rise to a temporally modulated signal on IT neurons that is related to the similarity of the current and remembered stimulus (after Eskandar *et al.*, 1992a). B. A colored pattern gives rise to a temporal waveform on cortical visual neurons that is the product of a separate code for color and a separate code for pattern. The final response includes an average waveform, that is determined for each neuron, and plays no further role in signal transmission (from McClurkin *et al.*, 1994).

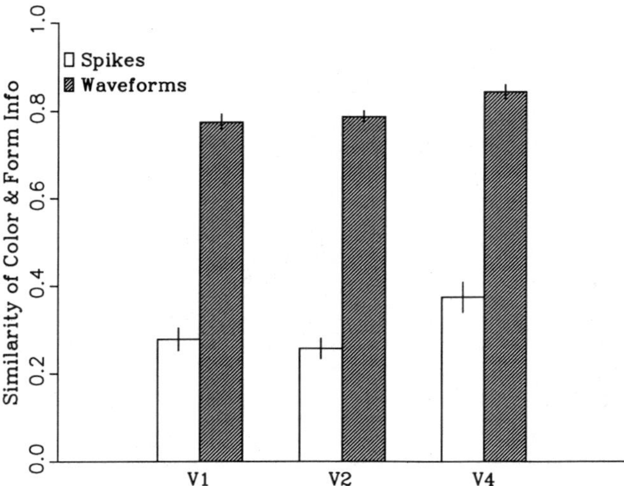

Figure 2. Relative amounts of information on color and pattern in each cell, averaged across all cells within an area (V1, V2, or V4). Error bars show mean ± one standard error of the mean. For each neuron, the amount of information encoded about color and about pattern was determined for both spike and waveform codes. The similarity value is the ratio of the smaller to the larger of the color and pattern information. Thus, similarity varies from zero to one. When the spike count code was used, the similarity values were between 0.3 and 0.4, indicating that, on average, the neurons carried two to three times as much information about one feature as about the other. In contrast, when the waveform code was used, the ratio was about 0.8 in all three areas. Thus, the neurons actually carry almost the same amount of information about both features.

Encoding of Color and Form Information

A visual discrimination task was used to study the encoding of color and form information in cortical neurons (McClurkin & Optican, 1996). It has been proposed that color and form information are divided into separate channels (i.e., discrete subpopulations of neurons) in the cortex. When the amount of information transmitted about either color or pattern alone is calculated for each neuron using a spike count code, the results from neurons in areas V1-V4 are consistent with the idea that neurons encode information about only one feature (Fig. 2, open bars). However, when the amount of information about either color or pattern is calculated using a waveform code, the results from those same V1-V4 neurons gives a much different result. If a temporal code is used, the amount of information about both features is about the same (Fig. 2, filled bars). Such a result is not consistent with the idea that information about form and color is grouped into separate channels in cortex, but rather suggests that all neurons participate in visual processing, irrespective of the type of visual parameter involved (McClurkin *et al.* 1996).

MULTIPLICATIVE MODEL OF NEURONAL INFORMATION ENCODING

It is possible that every combination of features modulates the responses of visual neurons in some unique way, i.e., the effects of color and pattern on the response waveform are inseparable. As many different features can modulate neuronal responses, this would lead to a large combinatoric set of possible responses, and would make decoding their meaning very difficult. An alternative is to use separable codes, one for each feature. Then, the total

Linear Independence   **Uniqueness with Minimum Energy**

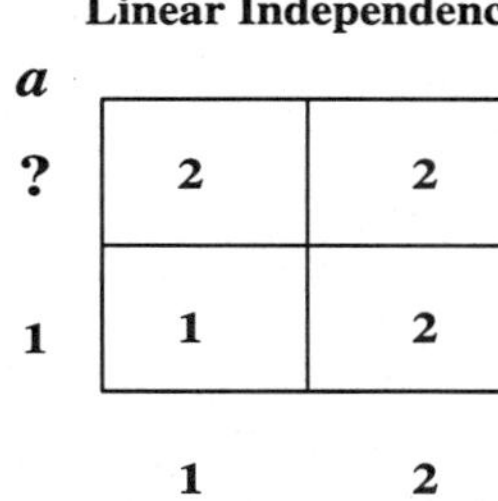

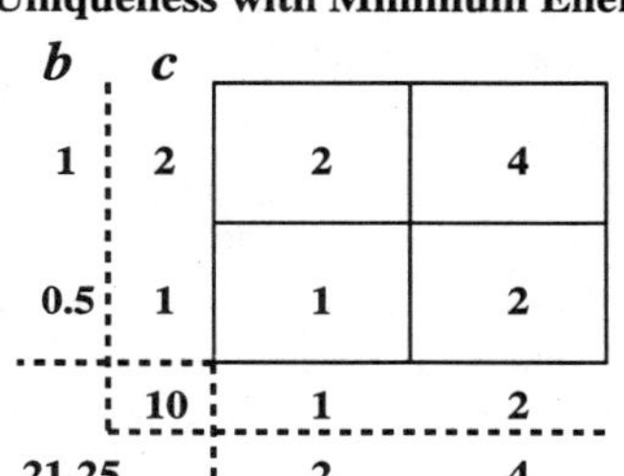

Figure 3. Separable codes require linear dependence of responses. If responses were independent, then no set of parameters for color (column) or pattern (row) could specify a response (e.g., value of the first time sample of the response, shown inside the squares). As shown in column a, no single value can be multiplied by the row values (1, 2) to produce the upper set of response samples (2, 2). If the responses are linearly dependent, then many solutions exist. When the parameters for color and pattern have the values in column b, the sum of the energies is 21.25. When they have the values in column c, the sum of the energies is 10. By constraining the code waveforms to have minimum total energy, it is possible to obtain a unique set of codes for generating neuronal responses.

number of codes needed would be small (the sum of the number of codes for each feature, rather than their product). Using separable codes would also make decoding messages easier. The simplest linear separable codes would be additive, and the simplest nonlinear separable codes would be multiplicative. Surprisingly, the data from both the vision/memory and the color/form experiments could be fit with the same multiplicative model, and that model accurately predicted the responses to novel stimulus combinations better than did a linear, additive model.In these models, the waveform of a neuronal response was assumed to be the product of two temporal codes, one for current and one for remembered targets, or one for stimulus color and one for stimulus form, plus an invariant waveform determined for that neuron:

$$R(t) = X(t) \cdot Y(t) + \bar{R}(t) \tag{2}$$

This model predicts that there are sets of code waveforms ($X(t)$ and $Y(t)$) that represent stimulus parameters. The application of this model is similar for recency memory tasks in IT cortex (Fig. 1A) and for color/pattern discrimination tasks in striate and pre-striate cortex (Fig. 1B). In a detailed study of the stimulus codes in awake primates performing a visual discrimination task, matching colors or patterns of colored Walsh stimuli (McClurkin *et al.* 1996), the codes for color were found to be invariant, i.e., the same across different areas. Both the color and pattern codes were found to be universal, i.e., the same across monkeys. This suggests that visual information processing would benefit from the increased amount of information available in these temporal codes.

For the codes to be separable, the responses of visual neurons must not be independent. As shown in Fig. 3, linear independence means no set of codes, one for color and one for pattern, exists that could determine the responses. Fortunately, the responses of visual neurons were found to be linearly dependent (Eskandar *et al.* 1992a; McClurkin *et al.* 1996). However, linear dependence alone can not constrain the choices of separable codes. The problem is that if one code set is doubled in amplitude, and the other is halved, their product would be the same. A unique solution to the problem of determining the separable codes requires imposing a constraint, that of minimizing the total energy of all the codes (Fig. 3). With the minimum energy constraint, a unique solution could be found by gradient descent using a network structure for the model (Fig. 4).

The average spike-density functions of the responses of a V4 neuron to a set of 36 stimuli (formed from six colors and six patterns) is shown in Fig. 5. Note that both features modulate the response of the neuron. For example, going down the rows in the first column shows how the response to the vertical bar pattern changes as the color of the bars change. Also, going across the columns of the bottom row shows how the response to a pink stimulus changes as its pattern changes (although the luminance remains constant).

If the responses of this neuron form a linearly dependent set, then a set of separable feature codes should exist whose products fit the responses (Eq. (2)). The first step in testing this hypothesis is to fit the parameters of the model shown in Fig. 4 to the data. As shown in Fig. 6, the fit of the model is very good. As the model has a very large number of degrees of freedom, this is not surprising.

The important test of the separable code hypothesis is to use the model in Fig. 4, with the parameters fixed, to predict the responses to a data set that was not used in the training. To accomplish this, the model parameters were fitted using only 30 of the 36 color/pattern combinations. The six unused combinations were distributed in such a way that each color and pattern were used in training five times. The model, with parameters fixed, was then used to predict the responses of the neuron to the six novel color/pattern combinations. As

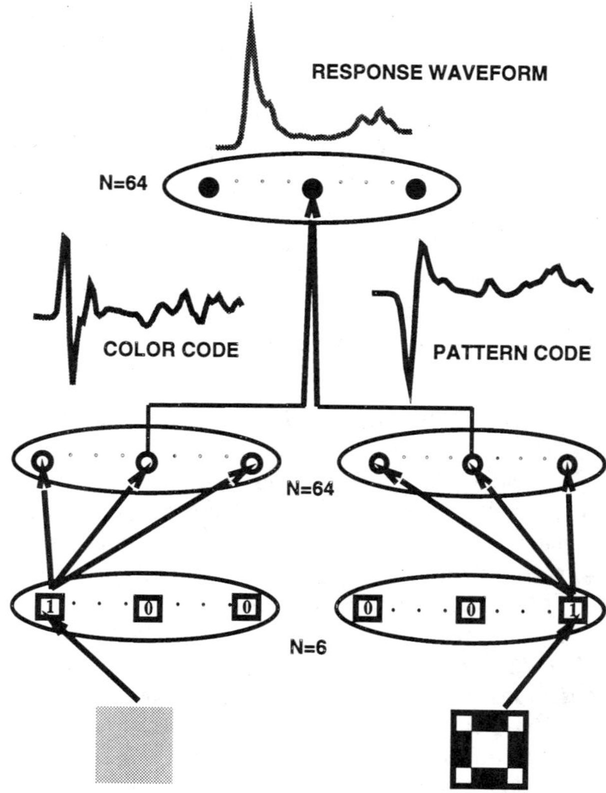

Figure 4. Two channel neural network for implementing models in Fig. 1. The left branch is a two-layer neural network for converting color (or remembered) stimuli into a temporal waveform. The right branch is a two-layer neural network for converting pattern (or current) stimuli into a waveform. These two internal waveforms are combined by multiplying them point-by-point on the output layer, and adding a bias term. A modification of back-propagation, using an error term that includes both goodness of fit and minimum energy, gives a gradient descent mechanism for finding the separable codes. (From McClurkin *et al.*, 1994.)

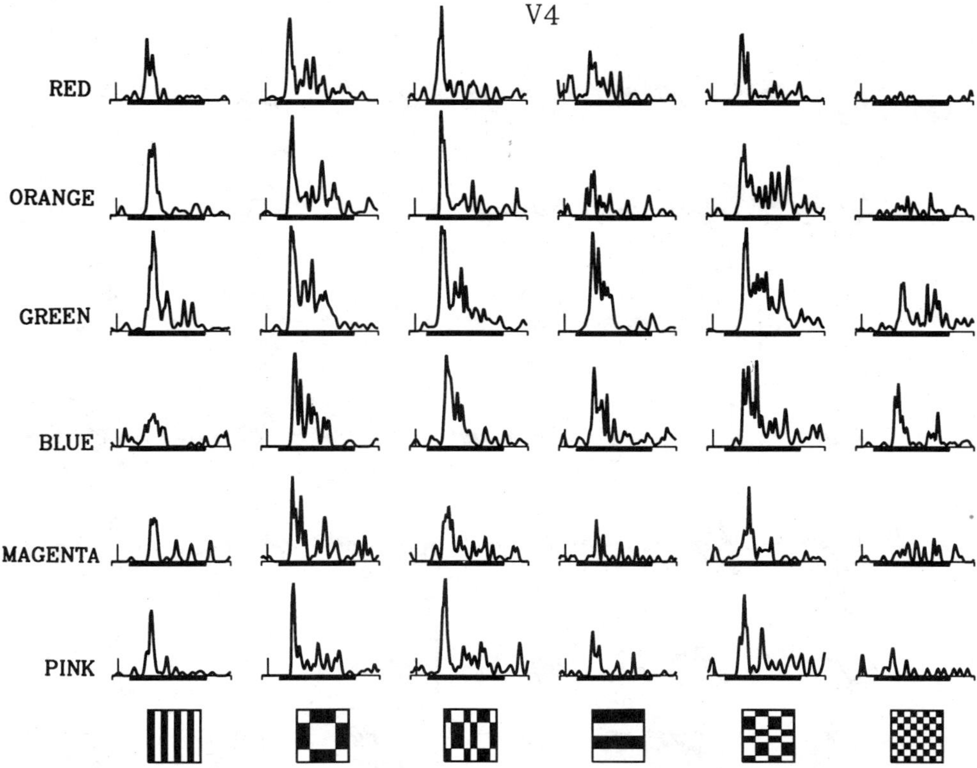

Figure 5. Average spike-density functions from a neuron recorded in area V4 of an awake monkey performing a visual discrimination task. Each trace is the response to a colored Walsh pattern (row label gives color, and column icon gives pattern). Vertical tick mark indicates a probability of an action potential of 0.10. Thick horizontal line indicates the stimulus presentation time of 128 msec (from McClurkin & Optican, 1996).

shown in Fig. 7, the model did very well at predicting those responses. It is possible to infer from results such as those shown in Fig. 7 that the responses of visual neurons can be represented by a composition, or multiplexing of separable, temporal codes.

INTRACORTICAL FEEDBACK MODEL OF TEMPORAL CODES

The presence of universal codes for visual features gives rise to the question of how they are generated. One study looked at the encoding of different texture types by neurons in V1 (Purpura *et al.* 1993). Several different response types were characterized. A nonlinear model of retina and primary visual cortex was constructed using simplified membrane biophysics. Surprisingly, this simple model spontaneously (i.e., without training) reproduced the same types of response waveforms seen experimentally.

Of course, without training the waveforms did not encode any stimulus-dependent information, but the point here is that the waveforms themselves arose naturally. The key to the generation of these waveforms was the presence of intracortical inhibitory feedback. Random variations in the network connections, and the type of inhibitory feedback connections, provided the different waveforms. This study suggests that the temporal codes observed in visual system neurons arise naturally through simple biophysical and network

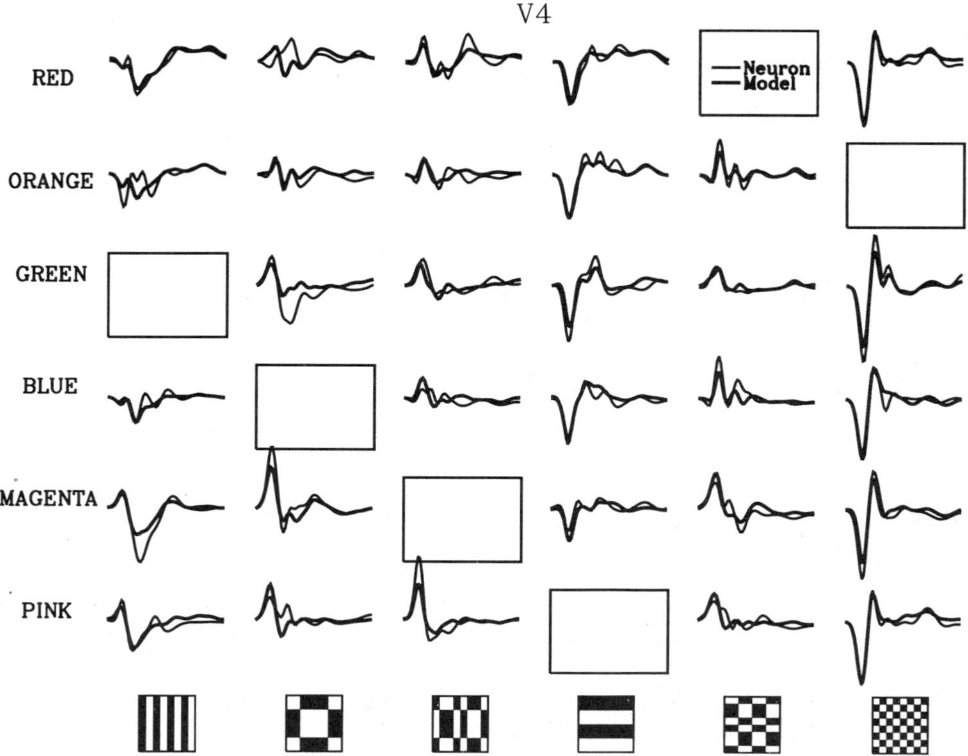

Figure 6. Fit of the model shown in Fig. 4 to data from a V4 neuron. Layout is the same as in Fig. 5, except that only the first 64 msec epoch of the response was used. Traces are the stimulus-dependent part of the response (i.e., the average response has been subtracted off). Thin lines are the responses of the neuron. Thick lines are the fitted responses of the model. The model was fitted to 30 of the 36 responses. Empty boxes represent the six color/pattern combinations not used to train the model (from McClurkin *et al.*, 1996).

Figure 7. Predicted responses of a V4 neuron based on the model in Fig. 4 as trained in Fig. 6. The six color/pattern combinations shown here were not seen by the model during training. High correlations between the neuronal and predicted waveforms implies that the model of separable feature codes (Eq. (2)) is a good representation of the encoding scheme used by the neuron (from McClurkin *et al.*, 1996).

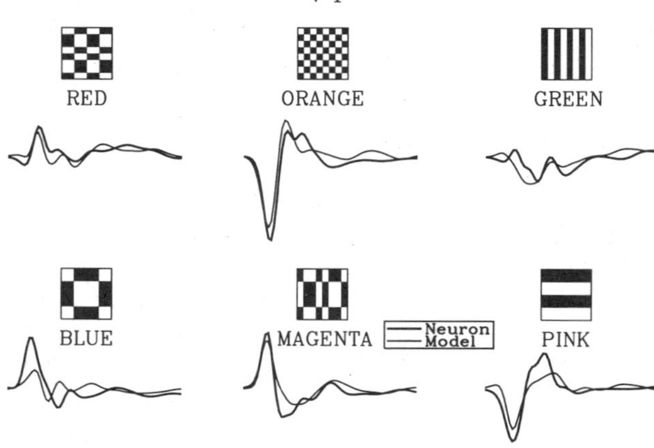

mechanisms. A functional information processing system would merely need to learn to encode specific visual features using those waveforms.

CONCLUSIONS

The multiplex-filter hypothesis suggests that there are different sets of code waveforms for each stimulus feature. The brain represents stimuli by multiplexing these feature codes together in the temporally modulated response of neurons. Such an encoding scheme avoids the geometric growth of subpopulations required by reductionist models to account for the many different possible stimulus combinations. Rather than a hierarchical set of cortical areas, each encoding one feature, the temporal encoding hypothesis states that many features can be represented in each area. This suggests that many visual computations, such as recognition and feature binding, could be carried out by temporal operations such as cross-correlation.

Visual information is carried not only by which subset of neurons is active, but also by the temporal messages they carry. Furthermore, many cortical areas are connected by feedback, as well as feedforward connections. Thus, temporal encoding allows the different visual areas in cortex to be viewed as concurrent, rather than hierarchical, processors (McClurkin *et al.*, 1991). Such a rich encoding scheme seems more compatible with the needs of the visual system, which must deal with complex and diverse scenes. The spatiotemporal approach to neuronal encoding should lead beyond reductionist to integrative approaches to modeling visual system function.

REFERENCES

Chee-Orts, M.-N., and Optican, L. M., 1993, Cluster method for analysis of transmitted information in multivariate neuronal data, *Biol. Cybern.* 69:29-35.

Eckhorn, R., and Pöpel, B., 1974, Rigorous and extended application of information theory to the afferent visual system of the cat. I. basic concepts, *Kybernetik* 16:191-200.

Eckhorn, R., and Pöpel, B., 1975, Rigorous and extended application of information theory to the afferent visual system of the cat. II. experimental results, *Kybernetik* 17:7-17.

Engel, A. K., König, P., Kreiter, A. K., Schillen, T. B., and Singer, W., 1992, Temporal coding in the visual cortex: new vistas on integration in the nervous system, *Trends in Neurosci.* 15:218-226.

Eskandar, E. N., Optican, L. M., and Richmond, B. J., 1992a, Role of inferior temporal neurons in visual memory. II. Multiplying temporal waveforms related to vision and memory, *J. Neurophysiol.* 68:1296-1306.

Eskandar, E. N., Richmond, B. J., and Optican, L. M., 1992b, Role of inferior temporal neurons in visual memory. I. Temporal encoding of information about visual images, recalled images, and behavioral context, *J. Neurophysiol.* 68:1277-1295.

Gawne, T. J., Eskandar, E. N., Richmond, B. J., and Optican, L. M., 1991, Oscillations in the responses of neurons in inferior temporal cortex are not driven by stationary visual stimuli, *Soc. Neurosci. Abstr.* 17:443.

Golomb, D., Kleinfeld, D., Reid, R. C., Shapley, R. M., and Shraiman, B. I., 1994, On temporal codes and the spatiotemporal response of neurons in the lateral geniculate nucleus, *J. Neurophysiol.* 72:2990-3003.

Li, L., Miller, E. K., and Desimone, R., 1993, The representation of stimulus familiarity in anterior inferior temporal cortex, *J. Neurophysiol.* 69:1918-1929.

Maffei, L., and Fiorentini, A., 1973, The visual cortex as a spatial frequency analyzer, *Vision Res.* 13:1255-1267.

McClurkin, J. W., and Optican, L. M., 1996, Primate striate and prestriate cortical neurons during discrimination: I. simultaneous temporal encoding of information about color and pattern, *J. Neurophysiol.* 75:481-495.

McClurkin, J. W., Optican, L. M., Richmond, B. J., and Gawne, T. J., 1991, Concurrent processing and complexity of temporally encoded neuronal messages in visual perception, *Science* 253:675-677.

McClurkin, J. W., Zarbock, J. A., and Optican, L. M., 1994, Temporal codes for colors, patterns, and memories. In: Peters, A., and Rockland, K. S., eds., *Cerebral Cortex, Vol. 10, Primary Visual Cortex in Primates,* Plenum Publ. Corp.: New York, pp. 443-467.

McClurkin, J. W., Zarbock, J. A., and Optican, L. M., 1996, Primate striate and prestriate cortical neurons during discrimination: II. separable temporal codes for color and pattern, *J. Neurophysiol.* 75:496-5007.

Middlebrooks, J. C., Clock, A. E., Xu, L., and Green, D. M., 1994, A panoramic code for sound location by cortical neurons, *Science* 264:842-844.

Miller, E. K., and Desimone, R., 1994, Parallel neuronal mechanisms for short-term memory, *Science* 263:520-522.

Miller, E. K., Li, L., and Desimone, R., 1991, A neural mechanism for working and recognition memory in inferior temporal cortex, *Science* 254:1377-1379.

Miller, E. K., Li, L., and Desimone, R., 1993, Activity of neurons in anterior inferior temporal cortex during a short-term memory task, *J. Neurosci.* 13:1460-1478.

Optican, L. M., Gawne, T. J., Richmond, B. J., and Joseph, P. J., 1991, Unbiased measures of transmitted information and channel capacity from multivariate neuronal data, *Biol. Cybern.* 65:305-310.

Optican, L. M., and Richmond, B. J., 1987, Temporal encoding of two-dimensional patterns by single units in primate inferior temporal cortex. III. Information theoretic analysis, *J. Neurophysiol.* 57:162-178.

Purpura, K., Chee-Orts, M. N., and Optican, L. M., 1993, Temporal encoding of texture properties in visual cortex of awake monkey, *Soc. Neurosci. Abstr.* 19:771.

Reid, R. C., and Shapley, R. M., 1992, Spatial structure of cone inputs to receptive fields in primate lateral geniculate nucleus, *Nature Lond.* 356:716-717.

Richmond, B. J., and Optican, L. M., 1987, Temporal encoding of two-dimensional patterns by single units in primate inferior temporal cortex: II. Quantification of response waveform, *J. Neurophysiol.* 57:147-161.

Richmond, B. J., and Optican, L. M., 1990, Temporal encoding of two-dimensional patterns by single units in primate primary visual cortex. II. Information transmission, *J. Neurophysiol.* 64:370-380.

Richmond, B. J., and Optican, L. M., 1992, The structure and interpretation of neuronal codes in the visual system. In: Wechsler, H., ed., *Neural Networks for Perception Vol. 1: Human and Machine Perception,* Academic Press: Boston, pp. 105-131.

Richmond, B. J., Optican, L. M., Podell, M., and Spitzer, H., 1987, Temporal encoding of two-dimensional patterns by single units in primate inferior temporal cortex: I. Response characteristics, *J. Neurophysiol.* 57:132-146.

Richmond, B. J., Optican, L. M., and Spitzer, H., 1990, Temporal encoding of two-dimensional patterns by single units in primate primary visual cortex. I. Stimulus-response relations, *J. Neurophysiol.* 64:351-369.

15

COHERENT DYNAMICS IN THE FRONTAL CORTEX OF THE BEHAVING MONKEY

Experimental Observations and Model Interpretation

Ad Aertsen,[1] Michael Erb,[2] Iris Haalman,[3] and Eilon Vaadia[3]

[1] Department of Neurobiology, The Weizmann Institute of Science
Rehovot 76100, Israel
[2] Department of Biophysics, Philipps-Universität
Renthof 7, D-35037 Marburg, Germany
[3] Department of Physiology, Hadassah School of Medicine
and The Center for Neural Computation, The Hebrew University
Jerusalem 91010, Israel

ABSTRACT

We have explored the hypothesis that cortical function is mediated by dynamic modulation of coherent firing in groups of neurons. Multiple single neuron activity was recorded in the frontal cortex of behaving monkeys, and analyzed for signs of correlation, using recently developed tools of dynamic correlation analysis. Correlation between neurons changed frequently within a fraction of a second, and in relation to stimuli and behavioral events. Moreover, the dynamic patterns of correlation depended on the distance between neurons. These findings support the notion that, in order to perform a computational task, neurons can associate rapidly into a functional group, while dissociating from concurrently activated competing groups. Possible mechanisms underlying such dynamic organization were investigated in spiking neural network models. We found that similar dynamic organization can be accomplished in these models, even without associated modifications of the synaptic connections. We discuss the consequences of these findings for the spatio-temporal organization of cortical cell assemblies.

INTRODUCTION

Measurement of the firing rate of single neurons is the method most widely used to understand the role of neuronal activity in brain function. This approach assumes that the relevant neuronal messages are carried by modulation of firing rates (e.g., Barlow, 1972, 1992; Newsome *et al.*, 1986). The vast connectivity among neurons in the central nervous

Advances in Processing and Pattern Analysis of Biological Signals, Edited by Isak Gath and Gideon F. Inbar
Plenum Press, New York, 1996

system, however, suggests that neuronal interactions are equally involved in computational processing. In particular, it has been proposed that coherent activity in groups of cortical cells ('cell assemblies') plays a major role in the organization of perception and behavior (Hebb, 1949; see also James, 1890). Hebb's hypothesis concerning the functional role of cell assemblies in central processing was adopted widely among theoretical researchers, and lay the conceptual foundation for most of the current work on 'neural network theories' of brain function. Almost all these models assume that 'learning' is based on selective modifications of the connectivity among neurons, and that 'computations' are carried out by selective activation of groups of interconnected neurons.

The experimental evaluation of the cell assembly hypothesis poses a formidable task. According to this hypothesis, the neural code resides in the correlated activity within and between neuronal groups. Hence, neither the study of the overall activity in large populations of neurons, nor the recording of single-neuron activity, one at a time, allow for a critical test of the assembly concept. Rather, the experimental methodology must involve simultaneous recording of many neurons in parallel, preferably in awake, behaving animals. This enables one to observe simultaneously and separably the activities of many neurons, while they are participating in meaningful processing and computational tasks (Gerstein *et al.*, 1989). The task is then to analyze these activities for possible signs of (dynamic) interactions. Results of such analyses may be used to draw inference regarding the processes taking place within and between hypothetical cell assemblies. Indeed, recent years have shown a growing interest in the simultaneous recording and analysis of spike trains from groups of up to some 10-30 neurons (Abeles, 1982, 1991; Gerstein *et al.*, 1983; Krüger, 1983; Eggermont, 1990). Associated with this development, new computational tools were designed to analyze and interpret the large flow of information coming from such 'multi-neuron' experiments.

Using this approach, we have studied the dynamics of firing correlations between pairs of neurons in the frontal cortex of monkeys that were engaged in a sensori-motor association task. Complementary to these experimental studies, we have also investigated the correlation dynamics in networks of spiking model neurons. The aim of these model studies was to provide guidance in the interpretation of the experimental findings, to study the nature of the underlying mechanisms, and to explore their functional implications and computational potential. In this paper we review results of these various studies, and discuss their functional interpretation by drawing a comparison between the correlation dynamics observed in the cortical activity and in the model networks.

FIRING CORRELATIONS BETWEEN CORTICAL NEURONS

Neuronal cooperativity is commonly investigated by crosscorrelation analysis of simultaneously recorded pairs of spike trains, especially when direct measurement of the synaptic potentials is impossible. Hereby, it was usually assumed that crosscorrelograms reflect the underlying synaptic connectivity (e.g., Toyama *et al.*, 1981; Ts'o *et al.*, 1986; Fetz *et al.*, 1991). However, the anatomy of the cortex indicates that neurons are interconnected in very large networks (Braitenberg & Schüz, 1991). This implies that the crosscorrelogram between any two neurons reflects the net effect of direct and indirect synaptic connections between them, as well as contributions from their connections with the remainder of the network. Therefore, recent studies emphasize the interpretation of crosscorrelation results in terms of functional (or effective) coupling between neurons, characterizing the instantaneous organization of the cortical network (Aertsen *et al.*, 1989; Aertsen & Gerstein, 1991; Ahissar *et al.*, 1992a).

Interestingly, firing correlations among neurons do not present a static phenomenon. Instead, they can change rapidly and in systematic relation to sensory stimuli or motor

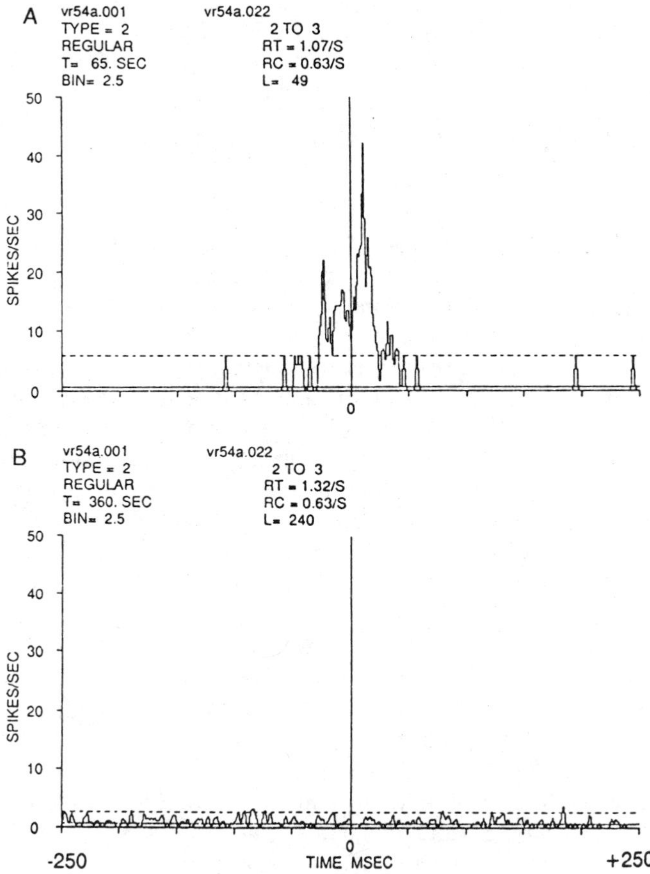

Figure 1. Modification of interactions in relation to behavioral state. Two crosscorrelograms of one pair of neurons recorded in the monkey frontal cortex during performance of a delayed localization task. The cross-correlogram in A was computed from sections of activity, recorded during intervals of 1 sec, during and immediately following the presentation of a visual stimulus. The cross-correlogram in B was computed from sections of inter-trial intervals (ITI; 3 sec each). Observe the marked difference in the interactions during the two behavioral conditions (Reprinted from Vaadia & Aertsen, 1992, with permission).

behavior (Vaadia & Aertsen, 1992; Ahissar *et al.*, 1992a,b; Murthy and Fetz, 1992; Eggermont *et al.*, 1993; Sanes and Donoghue, 1993). Such modifications would, in fact, be expected, if these correlations were to play the functional role that is assigned to them according to the cell assembly hypothesis. We present here one illustrative example of the modification of interactions by behavioral events, obtained from recordings in the frontal cortex of the behaving monkey (Vaadia & Aertsen, 1992).

Figure 1 shows the correlation between the spike trains of two simultaneously recorded neurons in the frontal cortex, collected from two different time sections while the monkey was engaged in a delayed-localization task. The spike trains in Fig. 1A were recorded during intervals of 1 sec, during and immediately following the presentation of a visual stimulus. The data in Fig. 1B were collected from sections of inter-trial intervals (ITI; 3 sec each). Comparison of these two correlograms shows that there is a marked difference in the interactions during the two behavioral conditions. An irregular, but very distinct, two-sided narrow peak can be observed when the monkey was presented with the visual stimulus, while the spike trains recorded during the inter-trial intervals show no correlation at all. We note that in the actual trials, these two conditions were separated in time by at most a few seconds.

MEASURING THE DYNAMICS OF FIRING CORRELATION: JOINT-PSTH

It is quite conceivable that changes of neuronal interaction as shown in Fig. 1 are due to rapid modulations of discharge synchronization among the neurons, induced by (or at least temporally linked to) sensory and behavioral events. Unfortunately, conventional crosscorrelation analysis is not an adequate tool to study the temporal evolution of coherent firing, since it measures the time-averaged rate of near-coincident spikes over a long time interval, typically on the order of several seconds or minutes. For this reason we used dynamic crosscorrelation analysis (Joint Peri-Stimulus Time Histogram JPSTH; Aertsen *et al.*, 1989). This tool highlights the detailed time structure of firing correlation between two neurons and its time-locking to a third event, such as the onset of a stimulus or the initiation of a movement. Moreover, appropriate normalization of the Joint-PSTH enables us to distinguish between contributions due to stimulus- or behavior-induced modulations of the single-neuron firing rates, and those from interneuronal correlation, the latter reflecting the 'effective' or 'functional coupling' between the neurons involved.

The starting point for our considerations is the Joint Peri Stimulus Time Scatter Diagram (Gerstein & Perkel, 1969, 1972). This presents a two-dimensional display, in which the firing of one neuron relative to a third event (e.g., a repeating stimulus presentation of a behavioral event) is measured along the x-axis, while that of the other neuron is measured along the y-axis. Points are entered on this plane at positions corresponding to all the (delayed) coincidences of the firings of each of the two neurons for each successive stimulus or behavioral trial. The resulting scatter diagram may show increased point densities parallel to the axes, representing stimulus/behavior related firing modulations of one or both neurons. There may also be increased point densities along and near the diagonal of the plane, representing correlated firing of the two neurons. With appropriate binning of the scatter diagram one obtains a matrix histogram which measures the point density as a function of the location in the scatter diagram: the Joint-PSTH. (For technical details and mathematical formalisms we refer to Aertsen *et al.*, 1989; Palm *et al.*, 1988). Various summations and marginal distributions recover the two ordinary single neuron PST histograms (along the x- and y-axes), the PST coincidence histogram (along the diagonal), and finally the conventional crosscorrelogram (perpendicular to the diagonal). The PST-coincidence histogram is of particular interest here, and represents the stimulus time-locked average of near-coincident firing of the two neurons in the same sense that the ordinary PST-histogram represents the stimulus time-locked average of firing by each individual neuron (Aertsen & Gerstein, 1991).

These are all 'raw' measurements, i.e., they do not take into account the effects of possible stimulus- or behavior-locked modulation of the two individual neuron firing rates. Time-locked changes in these rates through the repeating cycle would necessarily lead to corresponding changes in the near-coincidence rates. For the present purpose this is an uninteresting 'background' effect that we would like to eliminate from the correlation, in order to study those components which indicate temporal changes in the underlying neuronal interaction. We have devised the appropriate normalizing calculations as well as a confidence test (Aertsen *et al.*, 1989; Palm *et al.*, 1988). This normalization computes the excess of spike correlation above the expected correlation due to the rate co-variations and expresses it as correlation-coefficients. The normalization involves subtracting from each bin in the JPSTH plane the product of the two corresponding bins in the individual PST histograms, and dividing the result by the product of the standard deviations of these same PST bins. Subsequent summations over appropriate diagonal and para-diagonal bins in the plane give us the corrected, 'neuronal interactions only', coincidence-time histogram and crosscorrelogram.

Examples of normalized JPSTHs are shown in Figs. 3 and 4. Each JPSTH in the figure is composed of four displays:

- Peri-stimulus time histograms (PSTs; horizontal and vertical histograms), depicting the individual firing rate of each of the two neurons, relative to the time of occurrence of a stimulus or behavioral event (the trigger event);
- A grey-coded matrix, with each bin depicting the correlation coefficient, expressing the amount of correlation of the two neurons at the corresponding pair of time delays relative to the trigger event; The dark grey range represents co-firing at a rate lower than expected for uncorrelated firing, whereas white represents the highest co-firing rate above the expected rate;
- A coincidence-time histogram (diagonal histogram), showing the temporal evolution of the correlation-coefficient relative to the trigger event;
- The normalized version of the conventional, time-averaged crosscorrelogram (top-right histogram, perpendicular to the diagonal).

To facilitate the interpretation, we also express the correlation strength by the relative deviation of the observed coincidence rate from the expected rate for independently firing neurons. The maximal absolute value of this ratio, measured along the coincidence-time histogram, and expressed as a percentage of the expected rate, is denoted as the *'maximal modulation depth'*.

BEHAVIOR-RELATED MODULATIONS OF FIRING CORRELATION

Using the computational tool of the Joint-PSTH, we have examined neural activity in the frontal cortex of behaving monkeys for the presence of stimulus- or behavior-related modulations in the firing correlations among pairs of simultaneously recorded neurons (Vaadia *et al.*, 1995).

Neuronal activity of 6-16 neurons in the frontal cortex of a Rhesus monkey was recorded simultaneously, during performance of a spatial delayed response task (Fig. 2). Two

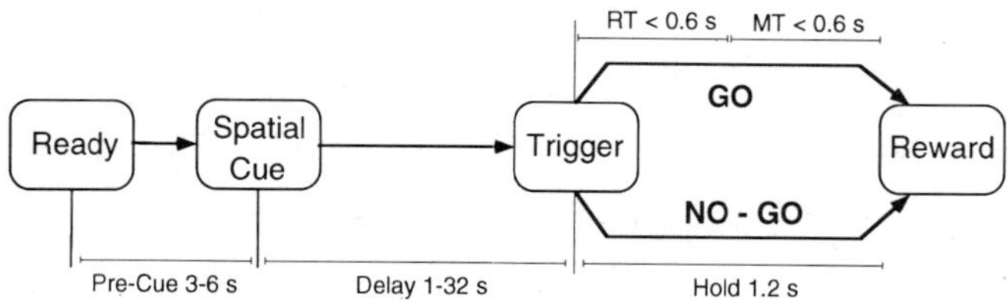

Figure 2. The spatial delayed-response task. Trials were initiated when the monkey touched a central key and a central red light was turned on ("ready"). After a variable delay (3-6 s), one of two target keys was illuminated for 200 ms ("cue"). After a delay of 1-32 s, the central red light was dimmed ("trigger") instructing the monkey to select the appropriate behavioral response. In one paradigm ("GO") the monkey was rewarded for releasing the center key within 0.6 s (RT = reaction time) after the trigger signal, and touching the correct target key within the next 0.6 s (MT = movement time). In the second paradigm ("NO-GO") the monkey was rewarded for continuing to touch the center key for at least 1.2 s after the trigger. Special peripheral lights instructed the monkey to switch paradigms every four correct trials (Reprinted from Vaadia *et al.*, 1995, with permission).

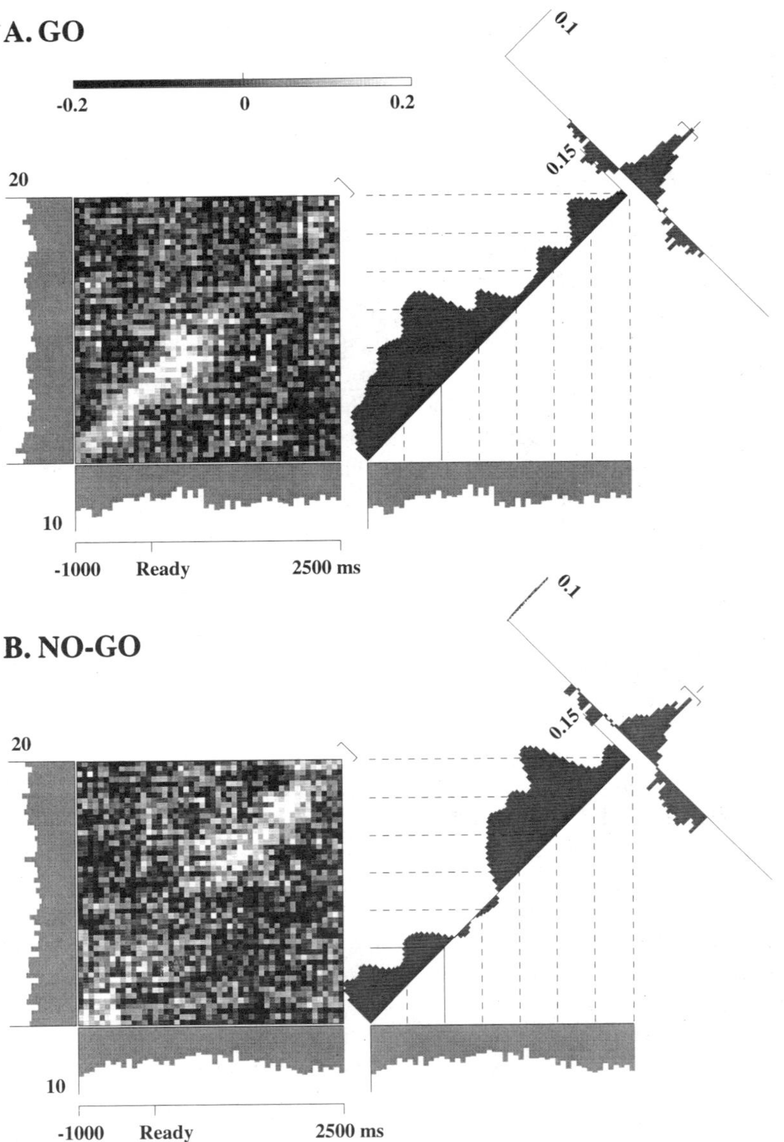

Figure 3. Dynamic modification of correlated firing of two frontal cortex neurons in relation to a visual stimulus ("ready" signal) as revealed by the normalized JPSTH. The *left half* of each plot (A and B) shows the correlation matrix (grey scale shown in A), and two ordinary PST-histograms along the X-axis (unit 1, full scale: 10 spikes/s) and the Y-axis (unit 2, full scale: 20 spikes/s). Binwidth: 70 ms (PST histograms) and 70ms x 70ms (matrix). The *right half* of each JPSTH shows the coincidence-time histogram (full scale: 0.15), and the crosscorrelogram (full scale: 0.1). The coincidence-time histogram measures the excess correlation within the time-window marked by the] sign (at the upper-right corner of the matrix) as a function of time around the trigger event. The coincidence-time histogram was smoothed by convolution with a gaussian (σ = 140 ms, two bins). The PST of unit 1 is duplicated along the X-axis of the right side for clarity. The distance between the grid lines is 500 ms. The onset of the ready signal is marked by solid lines in the grid. **A:** JPSTH constructed from 221 GO trials, with 4414 spikes of unit 1 and 10121 spikes of unit 2. Observe a gradual build-up of excess correlation, which starts before the onset of the ready signal, reaches a peak 400 ms after the signal and then decays. **B:** JPSTH constructed from 194 NO-GO trials for the same neurons (unit 1: 3764 spikes; unit 2: 8715 spikes). Note the different evolution of the correlation in comparison to A. For more details see text (modified from Vaadia *et al.*, 1995, with permission).

monkeys (Maccaca Mullata, female, 3-4 kg) were trained to perform the task. Recording sessions began after the monkey was over-trained. In each session we recorded simultaneously the activity of 6-16 neurons during performance of 400-800 correct trials with 100-200 alternations between the two behavioral paradigms. Six microelectrodes (glass-coated tungsten, 0.2-1.5 MΩ impedance) were inserted into the frontal cortex in a circular array, with 500-1000 μm inter-electrode distance. Spike-sorters were utilized to isolate the activity of 1-3 single units from each electrode. Typically, we recorded in each session 6-10 well-isolated units and 6-10 clusters with a mixture of 2-3 spike-shapes in each cluster. Eye movements were monitored by Ag-AgCl cup-electrodes. The locations of electrodes tracks in the cortex were reconstructed by standard histological techniques. Neuronal activity was examined by raster displays, auto- and crosscorrelograms, time-interval histograms and JPSTHs.

As we demonstrate below, modulation of the firing correlation between neurons in the frontal cortex is not a rare phenomenon: JPSTH analysis revealed that about 60% of the neuronal pairs we analyzed showed clear modulation of firing correlation in relation to stimuli or behavioral events (Vaadia *et al.*, 1995). Examples of such modulations are shown in Figs. 3 and 4. Figure 3 shows JPSTHs for two neurons in the premotor cortex. In this case, the JPSTH reveals behavioral dependent modulation of co-firing that could neither be predicted from the individual firing rates (responses) of the two neurons, nor from the time-averaged crosscorrelograms. The two JPSTHs in A and B show the joint activity of the two neurons in the two behavioral paradigms: GO (A) and NO-GO (B). The time scale spans 3500 ms, beginning 1000 ms before the ready signal. During this entire period, the monkey's hand was touching the central key. The firing rate of one unit does not change throughout the interval (histogram along the Y axis of plots A and B). The firing rate of the second unit shows a weak modulation, with a small decrease following the onset of the ready signal (histogram along the X axis of plots A and B). Moreover, the remarkably similar PST histograms in A and B show no significant difference between the firing rates of both neurons during performance in GO paradigm trials and performance in NO-GO trials. Thus, PST analysis alone would indicate that these two neurons taken individually neither respond to the ready signal, nor contribute to the discrimination between the two behavioral paradigms. Evaluation of the time-averaged crosscorrelation (histograms on the upper-right of each plot) does not change this inference. Both crosscorrelograms are characterized by a relatively broad peak around the origin. This peak indicates that the probability that each of the two neurons will emit a spike is highest when the other neuron fires a spike as well, and decays with a time constant of about 300 ms (as reflected by the peak width). Comparison of the crosscorrelograms in plots A and B, however, does not reveal any qualitative difference. Thus, the time-averaged correlation does not discriminate between the two behavioral paradigms either.

The situation is quite different, though, when we examine the dynamics of correlation. The time course of the amount of co-firing contributing to the crosscorrelogram peak is shown along the main diagonal (from bottom-left to upper-right) of the grey-coded matrix. This time course of excess near-coincident firing is emphasized in the (diagonal) coincidence time histograms. The most outstanding feature in the top panel (A) is the gradual change of co-firing along the principal diagonal. Co-firing is evident and increasing from the beginning of the selected interval, reaches a peak after the onset of the ready signal (maximal modulation depth = 50%), and then decays to the expected (uncorrelated) level. By contrast, the correlation in the NO-GO condition (B) shows a very different time course. Here, the co-firing decays from the beginning, and reaches the zero-level just after the onset of the ready signal, after which the firing of the two neurons remains uncorrelated for almost a second, and then increases to a high level of correlation, lasting for about 1.5 sec.

This detailed dynamic analysis indicates that the correlation between these two neurons is strongly modulated and temporally linked to the onset of the ready signal. Moreover, the figure illustrates that the evolution of co-firing is practically opposite in the two behavioral conditions. During performance of the GO paradigm, the co-firing is maximal about 500 ms after the onset of the ready signal, whereas during the NO-GO trials the co-firing at that time is minimal. We stress that this difference in the evolution of co-firing could not be predicted from the responses of the two neurons alone, nor was it reflected in the time-averaged crosscorrelograms. However, the dynamics of the correlation between these two neurons reveal that they participate in the coding of the sensory event and of the behavioral context.

We examined the temporal patterns of correlated activity of pairs of neurons in relation to the onset of visual stimuli (ready, cue, trigger) and in relation to the motor behavior of the monkey (initiation of arm movements and of saccadic eye movements). Correlated activity was found in 499 neuronal pairs. The crosscorrelograms of these pairs had a peak ('positive' correlation), a trough ('negative' correlation) or a mixed pattern including both peaks and troughs ('compound' correlation). Peaks and troughs were always located near time 0 of the averaged crosscorrelogram, and decayed within tens to a few hundred milliseconds (cf. Figs. 3 and 4). All 499 neuronal pairs were analyzed in detail by the JPSTH method. We found that the dynamic pattern or the strength of the correlation was modified significantly in relation to behavior in 308/499 cases (61%).

CORRELATION DYNAMICS DEPEND ON THE DISTANCE BETWEEN NEURONS

Several previous studies demonstrated that correlation of firing exists between cortical neurons, even when these neurons are located several millimeters apart (e.g., Engel et al., 1991; Gray et al., 1992). Moreover, studies of visual cortex neurons indicated that the strength and pattern of the correlation may depend on the locations of the two neurons in the cortex (Ts'o et al., 1986; Eckhorn et al., 1988; Gray et al., 1989; Krüger, 1990).

We have investigated the dependence of dynamic correlation in the frontal cortex on the distance between neurons using the JPSTH analysis (Vaadia et al., 1995). Figure 4 shows examples of correlation dynamics in two pairs of neurons recorded simultaneously by two microelectrodes in the frontal eye field, near the genu of the arcuate sulcus. The first two neurons, 0 and 1 [0,1], were recorded by a single microelectrode and, hence, we refer to them as 'neighboring' neurons. The second pair [0,6] consists of one neuron from the first pair (neuron 0) and another neuron (neuron 6) which was recorded by a different electrode, located about 400 μm deeper and 500 μm lateral to the first electrode. Hence, we refer to neurons 0 and 6 as 'distant' neurons.

Figures 4A,B show JPSTHs for the neighboring neurons [0,1]. The top plot (A) shows the JPSTH around the initiation of saccades to the right. The bottom plot (B) shows the JPSTH for the same pair around saccades to the left. The peri-event time histograms in the plots along the X- and Y-axes show that in both cases the firing rates of the neurons were strongly modulated before and during the saccades. The peaks of the crosscorrelograms in A and B indicate that, on the average, the amount of correlation between the two neurons is similar in the two conditions. The JPSTH analysis, however, clearly reveals that these time-averaged crosscorrelations are misleading: while the average correlation is indeed constant, dramatic modifications of co-firing occur near the initiation of the saccades; just as the rightward saccade

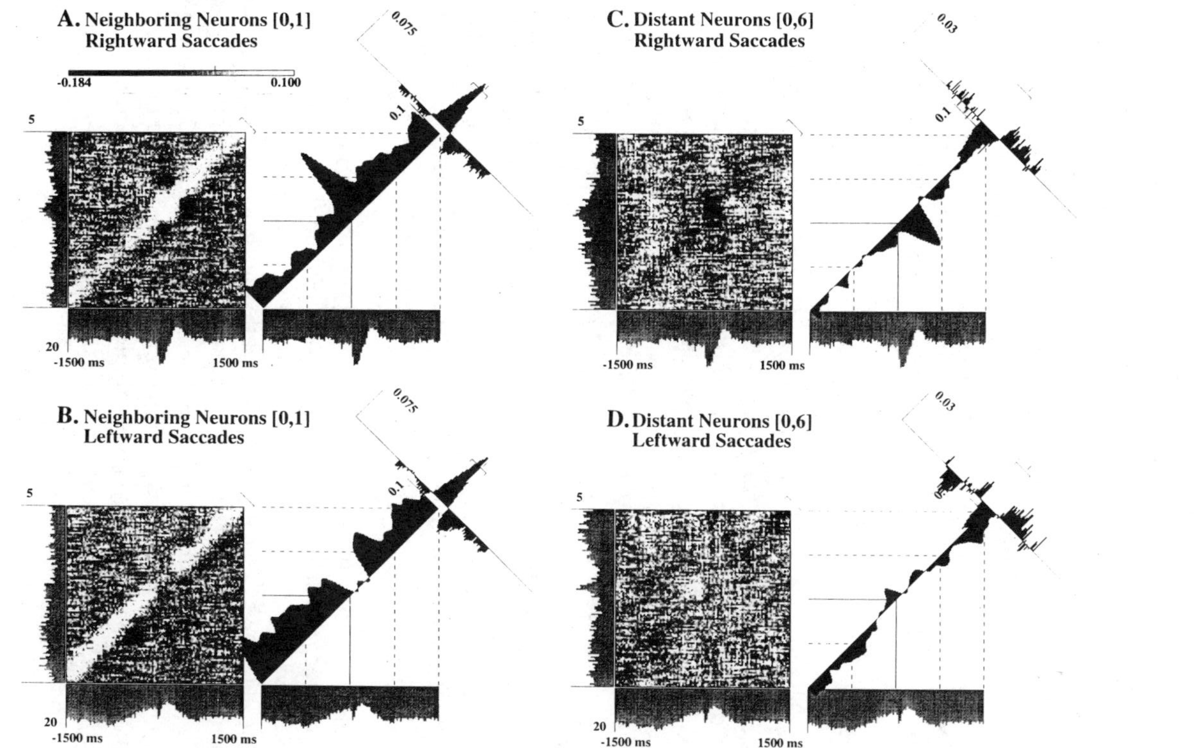

Figure 4. Dynamic modification of correlated firing between frontal cortex neurons in relation to the onset of saccadic eye movements. **A,B:** JPSTHs for units 0 and 1 that were recorded by the same microelectrode (neighboring neurons [0,1]). **C,D:** JPSTHs for two distant neurons [0,6]. Unit 0 is the same unit shown in A and B. Unit 6 was recorded by another electrode. The normalization and format of the JPSTs are the same as in Fig. 3. Bin size in all plots is 30 ms. The JPSTs around onsets of rightward saccades (A and C) were constructed from 776 saccades, with 33882 spikes of unit 0 (A and C), 4299 spikes of unit 1 (A) and 6927 spikes of unit 6 (C). The JPSTs around onsets of leftward saccades (B and D) were constructed from 734 saccades, with 32621 spikes of unit 0 (B and D), 4167 spikes of unit 1 (B) and 5992 spikes of unit 6 (D). Note that: (1) The averaged crosscorrelograms for a given pair are similar. However, as shown by the matrix and the coincidence histograms, the dynamics are temporally linked to the saccades and depend strongly on their direction (A vs. B and C vs. D); (2) The correlations between neighboring neurons (A and B) are positive, whereas the correlations between distant neurons are negative (C and D) (modified from Vaadia *et al.*, 1995, with permission).

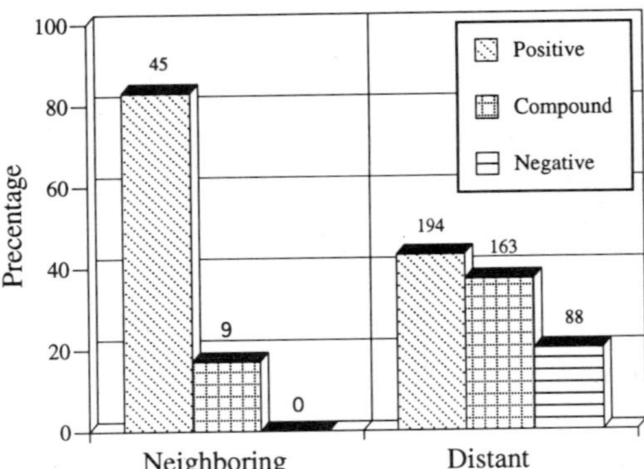

Figure 5. Distribution of crosscorrelogram patterns in neighboring vs. distant neurons. Cross correlograms between pairs of neighboring neurons are, in most cases, characterized by positive peaks. Note that negative crosscorrelograms are completely missing in the population of neighboring neurons. The number of neuronal pairs in each class is given above the bars. The difference between the two population is statistically significant at a level of 0.001 (χ^2 = 318.77) (modified from Vaadia *et al.*, 1995, with permission).

starts, co-firing increases to its highest level within a few tens of milliseconds (white cluster in the center of the matrix, *maximal modulation depth* = 110%).

The opposite change takes place for saccades in the leftward direction (B), where the correlation decreases to zero (no-correlation) as soon as the eye movement starts. The JPSTHs in C and D show the same analysis for the distant pair [0,6]. In this case, the average correlation is negative, i.e., the probability that either one of the neurons will fire a spike is lower around the times the other neuron fires. This type of correlation can result from reciprocal effects of correlated inputs to the two cells (while one neuron is excited by the net effect of the correlated inputs, the other is inhibited). Alternatively, it may reflect mutual inhibition between the two neurons. Again, the JPSTH shows that the average correlation is misleading. In fact, the negative correlation is only evident just after the initiation of saccades to the right (C; *maximal modulation depth* of 70%), and is completely missing around the onset of saccades to the left (D). Identical patterns of correlation dynamics were found between neurons 1 and 6 (not shown).

Although the three neurons 0,1,6 do show modulation of their firing rates in relation to saccade initiation, the changes of correlations could not be predicted from the firing rates of the neurons. For example, the correlations either increased (4A), or decreased (4C) near the time of saccade initiation, while the firing rates of all three neurons increased around the onset of saccades, regardless of its direction.

The plots in Fig. 4A,B depict the correlation dynamics between two neighboring neurons, that show a modulation of positive correlation. By contrast, the plots in Fig. 4C,D describe modulations of negative correlation, found between two distant neurons. The distance between the cell bodies of neurons that can be monitored simultaneously by a single electrode, is estimated at less than 100 μm (Abeles, 1982), while the horizontal distance between two adjacent electrodes in this study was about 500-1000 μm.

The distribution of correlation patterns was examined by comparing JPSTHs of neuronal pairs that were recorded by the same electrode, to JPSTHs of neuronal pairs that were recorded by different electrodes. Figure 5 summarizes the results of this comparison. The Figure illustrates that positive correlation was observed between distant neurons as well as between neighboring neurons, but more frequently between neighboring ones. By contrast, negative correlation was only found between distant neurons, and never between neighboring ones.

MECHANISMS OF DYNAMIC FIRING CORRELATION: MULTIPLE SINGLE-NEURON RECORDINGS FROM A SPIKING NEURAL NETWORK MODEL

These findings indicate that the usual concept of cortical neurons with static inter-connections of fixed or only slowly changing efficacy (during learning, for example) is no longer appropriate. Instead, one should distinguish between *structural connectivity* on the one hand and *functional coupling* on the other (Aertsen & Gerstein, 1991; Aertsen & Preissl, 1991). Whereas the former can presumably be described as (quasi) stationary, the latter may be highly dynamic and context-sensitive. These findings raise a two-fold question: what is the nature of the underlying mechanisms, and what are the functional implications?

In a number of theoretical studies we have sought to elucidate these issues. Several different mechanisms may be invoked to mediate the transition from static, anatomic connectivity to dynamic, functional coupling. On the one hand, the underlying mechanism may be local, as in von der Malsburg's proposal of rapid modulation of synaptic efficacy (von der Malsburg, 1986, 1991). On the other hand, the observed modulations of correlation may reflect more global effects, residing in the cortical network dynamics. In order to obtain more insight into these different alternatives, we investigated the correlation dynamics in various physiology-oriented networks of spiking model neurons (Aertsen & Preissl, 1991; Erb & Aertsen, 1992). We review here the results obtained in one of these studies (Aertsen *et al.*, 1994), in which we simulated a feedback network of 100 spiking model neurons with fixed synaptic connections. These simulations were designed to test the hypothesis that the modulations of firing correlation in cortical recordings are the result of the global activity dynamics in the network, without the need to invoke rapid synaptic modifications.

A detailed description of the network model and the associated equations is given elsewhere (Erb & Aertsen, 1992; Aertsen *et al.*, 1994). It was designed to capture in a greatly simplified structure the principal features of real cortical networks, both in terms of anatomy and of physiology. Briefly, the spiking model neurons were connected by excitatory synapses, inspired by neuroanatomical findings that about 90% of the cortical synapses are of this type (Braitenberg & Schüz, 1991). The synapses were modeled as lowpass filters with delayed response, transforming the incoming spike activity into EPSP's. These were summated linearly to yield the instantaneous value of the membrane potential at the cell body. The probability of spike generation was described by a sigmoid function of this membrane potential; this probability, in turn, was modulated by a refractory mechanism, driven by the recent spike history of the neuron. The final spike output was obtained by using the firing probability as the instantaneous rate of a stochastic event generator. In addition to this basic network of spiking neurons (the 'pyramidal cells'), the model comprised a global inhibitory mechanism for activity regulation, consisting of two parallel, linear branches: a fast one, with a time constant similar to that of the excitatory synapses (5 ms), and a slow one with a considerably larger time constant (200 ms). The fast branch mimicked the inhibitory action of stellate cells, the slow one served to regulate the overall activity in the network towards a preset value (threshold control) (Braitenberg, 1978; Palm, 1982). Parameter values (e.g., synaptic strengths, time constants, refractoriness) were chosen such that they conform to values from the physiological literature, at the same time ensuring that the network was operating in a regime of stable, sparse firing, with an overall activity in the order of 10-20 spikes per neuron per second (for a discussion on the problems of maintaining stability in such networks, see Erb & Aertsen (1992)).

To investigate the influence of the network dynamics on the functional coupling between neurons embedded in the network, we used a fixed connectivity matrix, in which were stored the previously learned memory traces of a set of 10 sparsely occupied, randomly

3-1	3-2	3-3
2-1	2-2	2-3
1-1	1-2	1-3

Figure 6. Connectivity matrix for a subset of 16 out of 100 neurons in the model network. The left diagram shows the strength of the 'synaptic' connections, coded in grey (white: zero, black: maximum). Observe that several groups of strongly interconnected model neurons ("cell assemblies") pop out, with weaker connections between the groups. This is summarized in the simplified diagram (right) of the three dominant cell assemblies (1, 2, 3), present in this group of 16 neurons. Note, however, that the actual connectivity matrix is more complex, and that more than three, partly overlapping, assemblies can be discerned. Particularly, group 2 consists of two strongly overlapping cell assemblies: group 2a (neurons 8-13) and group 2b (neurons 10-14). We investigated the dynamics of correlated firing between pairs of these 16 neurons by Joint-PST analysis of the simultaneously recorded spike trains in response to a set of continuously varying stimuli (modified from Aertsen et al., 1994, with permission).

distributed input patterns. In order to obtain insight into the functional associations hidden in such a distributed connectivity matrix the neurons can be re-sorted, such that the input patterns become more or less optimally segregated (Aertsen et al., 1994). The resulting 'permuted' connectivity matrix was characterized by a clear block-like structure, straddling the diagonal. Each of these compact blocks (containing on average some 7 neurons) denoted a heavily interconnected 'cell assembly', preferentially associated with one of the input patterns. The segregation was, however, not complete, since the input patterns were random, not orthogonal. Hence, associations did overlap, resulting in connections between the assemblies; these were, however, usually weaker than the intra-assembly connections (cf. Fig. 6). A fraction of almost 25% of the neurons was not dedicated to any one of the input patterns in particular. Instead, they provided a diffuse background connectivity to each of the assemblies, while being driven by several of them. On average, each assembly involved some 10 neurons, having an overlap of 2 to 3 neurons with one or more of the other assemblies.

We investigated the response of this model network to a set of continuously varying stimuli, by recording simultaneously the spike activity of 16 of the 100 model neurons, in much the same way as a physiological multi-neuron recording would pick up a fraction of the entire population of neurons. By adopting this approach, we could apply the same analytical tools (here: the JPSTH) as used for the cortical recordings and, thus, directly confront the outcome of the theoretical study with that of the physiological experiments. We refer to such experiments on theoretical constructs as 'in virtu' recordings, to distinguish them from their in vivo and in vitro physiological counterparts (Aertsen et al., 1994). The 'synaptic' connectivity matrix for the selected subset of 16 model neurons (after re-sorting as described above) is shown in Fig. 6. Notice the block-like connectivity structure, with strong connections within and weaker interconnections between the various 'cell assemblies'. This is outlined in the simplified membership diagram to the right, depicting the three dominant assemblies (numbered 1, 2, 3) within the group of 16. Note, however, that the actual connectivity matrix is more complex, and that more than three, partly overlapping, assemblies can be discerned.

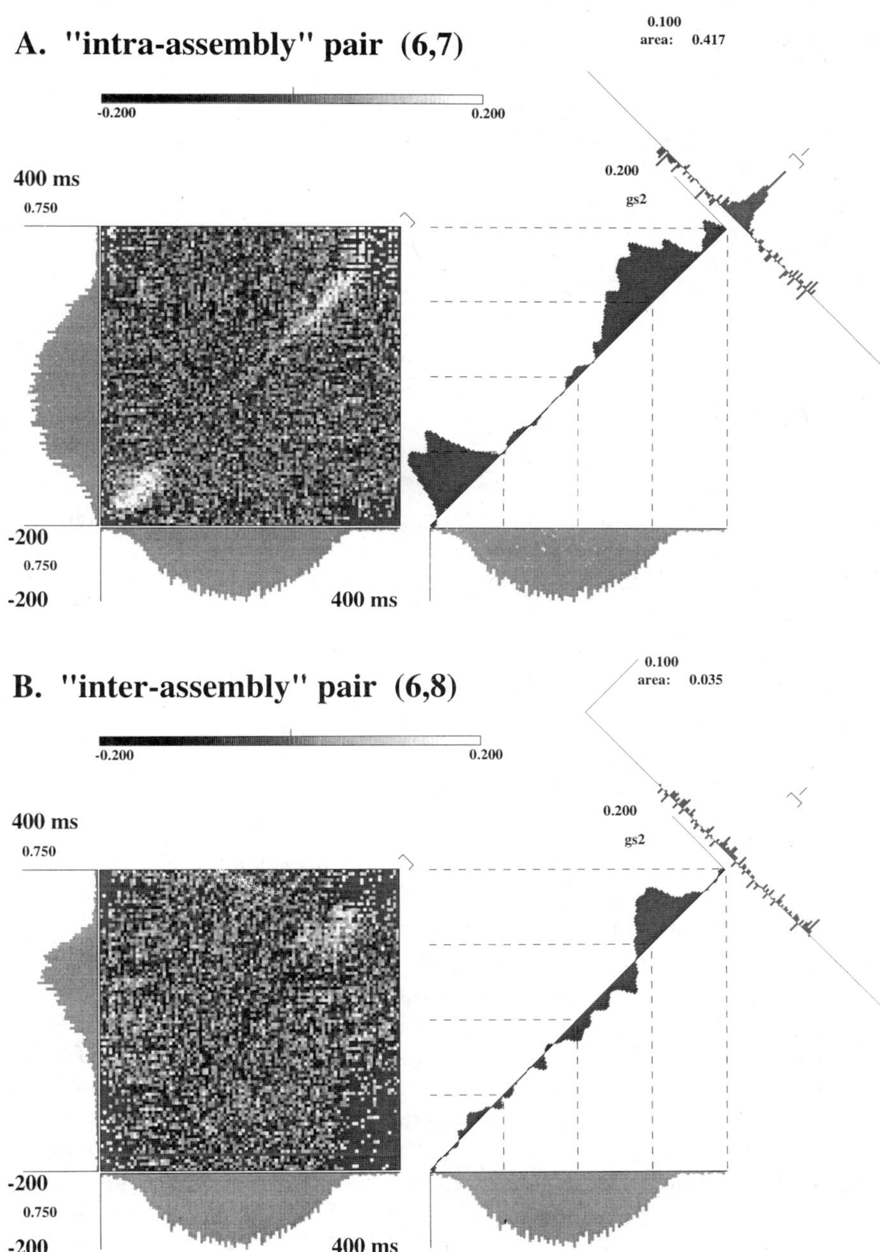

Figure 7. Stimulus-locked dynamic correlation of firing between selected pairs of neurons from the model network during repeated presentation of a stimulus. Normalized Joint-PST histograms were determined for an 'intra-assembly' pair (A) and for an 'inter-assembly' pair (B). The 'intra-assembly' pair consists of the neurons 6 and 7, both members of a single cell assembly (group 1 in Fig. 6). The 'inter-assembly' pair consists of one neuron from the first pair (neuron 6) and another neuron (neuron 8) from a different assembly (group 2 in Fig. 6). Observe the strong modulation of correlated firing in both pairs, however, with quite different coupling dynamics within the assembly (A) as compared to across assemblies (B). The format of these Figures is the same as in Figs 3 and 4. Numbers of spikes: 9244 (neuron 6), 8426 (neuron 7), 4136 (neuron 8), recorded during 250 stimulus trials of 600 ms each (Reprint from Aertsen *et al.*, 1994, with permission).

DYNAMICS OF FIRING CORRELATION IN A MODEL NEURAL NETWORK

Spike trains were recorded simultaneously from pairs of model neurons, chosen such that each of them belonged to one or more of the cell assemblies embedded in the network. This allowed us to study the dynamic interactions both within and between cell assemblies. In order to obtain reliable statistics, stimuli were presented repeatedly over 250 consecutive trials. Spike trains were then inspected for the time course of correlated firing, using the Joint-PSTH analysis.

Figure 7 displays the normalized JPSTHs of two selected pairs of model neurons in the identical format as used for the cortical recordings (Figs. 3 and 4). The normalized JPSTH for a pair of neurons from a single model assembly is shown in Fig. 7A. By contrast, Fig. 7B shows the normalized JPSTH for a pair of neurons that belong to different assemblies. Both Joint-PSTHs are displayed using the same scales to facilitate quantitative comparison.

Inspection of the normalized Joint-PSTH for the "intra-assembly" pair (Fig. 7A) reveals clear stimulus time-locked responses of the two neurons (the PST histograms along the x- and y-axes), with a positive and symmetric time-averaged spike correlation (the peak in the crosscorrelogram on top-right). Interestingly, however, the PST-coincidence histogram along the diagonal shows that the rate of correlated firing between the two model neurons is not constant throughout the trial. Instead, there is a strong modulation of correlated firing, which deviates considerably from the single neuron firing rates. Correlation is positive in the rising and falling phases of the single neuron responses, while it breaks down to zero in between, and stays down for most of the duration of the peak response rates. Comparison with further pairs taken from a single assembly confirmed that the behavior exhibited by these two neurons is representative for the interactions taking place within a model assembly. As neurons are recruited by the incoming activation, there is a brief phase of enhanced spike correlation during the ignition of the cell assembly. This correlation ceases to exist as the firing rates increase further, and returns shortly before the activation dies out.

The situation is quite different, though, for the interactions between model neurons that belong to different assemblies. This is illustrated for the "inter-assembly" pair (6,8) in Fig. 7B. While both neurons again exhibit clear time-locked variations in their response rates, the time-averaged crosscorrelation (top right) now is essentially zero. This absence of time-averaged firing correlation, however, does not imply that there is no correlation between the spiking of the two neurons throughout the stimulus trial. Rather, the PST-coincidence histogram along the diagonal exhibits a distinct alternating trajectory. It passes through a prolonged, shallow negativity during most of the elevated single neuron responses, before rising to a brief burst of positive spike correlation as the responses have almost decayed down to background level. This result once more demonstrates the dangers of taking the conventional, time-averaged crosscorrelogram as evidence for the presence or absence of neural interactions. Contributions of opposite sign may cancel in the time integral and, hence, leave no trace in the crosscorrelogram (for an example of this phenomenon in visual cortex recordings see Aertsen & Gerstein (1991)). More interestingly, though, are the functional implications of this type of interaction. Apparently, the two model assemblies compete with each other during most of the time they are both strongly activated by the stimulus. As a result, there is a tendency for anti-synchronous firing during this stage. This competition is presumably mediated by the global inhibition in our model network. It only ceases to exist, or, rather, it is overridden by the positive cross-coupling between the assemblies, after the firing rates have decayed enough, that the simultaneous co-existence of two active assemblies no longer poses a challenge to the activity regulation mechanism.

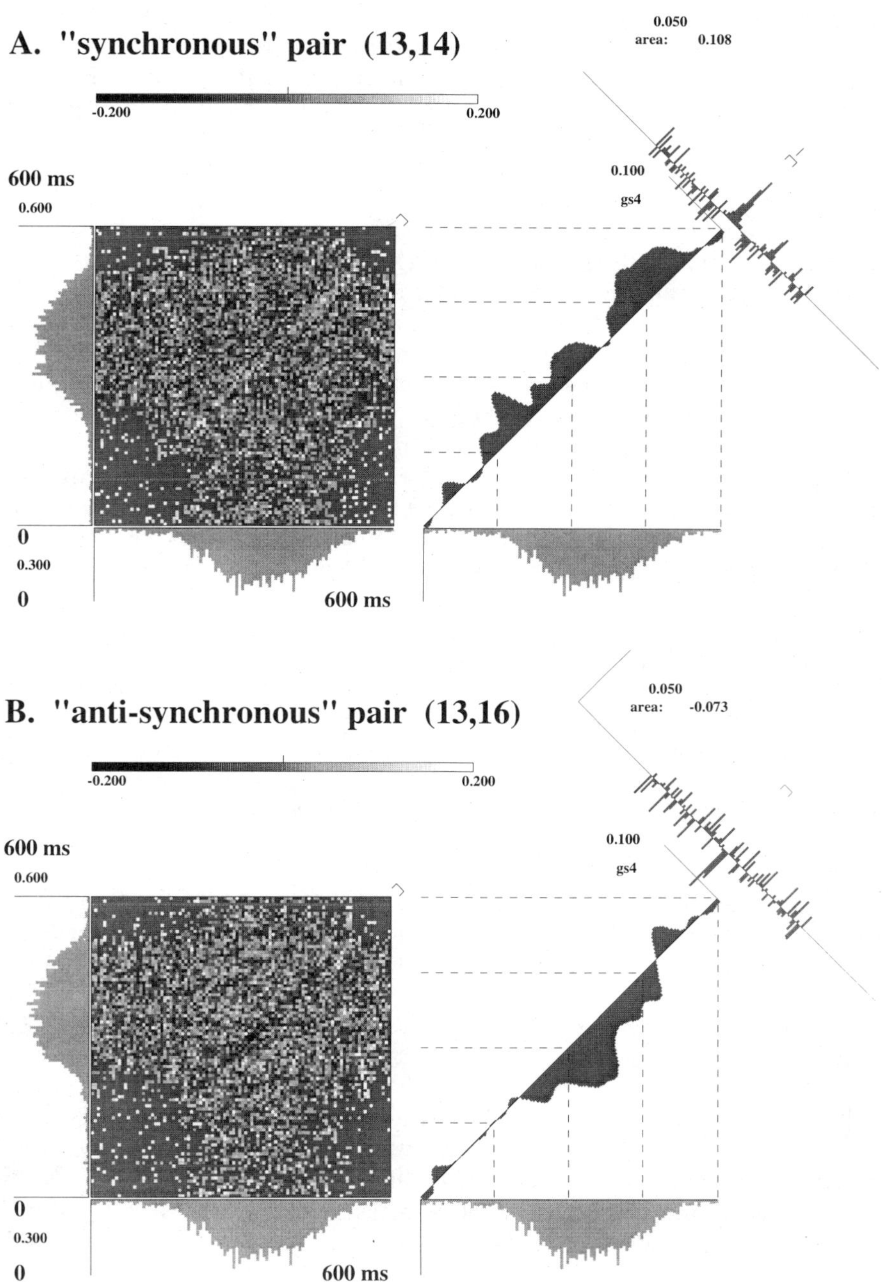

Figure 8. Similar firing rate patterns may be associated with very different correlation dynamics. Normalized Joint-PST histograms for two pairs of model neurons. **A:** Neurons (13,14) taken from two strongly overlapping model assemblies (groups 2a and 2b in Fig. 6). **B:** Neurons (13,16) taken from two 'competing' cell assemblies (groups 2 and 3 in Fig. 6). Notice the opposite correlation dynamics, in spite of the almost indistinguishable rate dynamics. Numbers of spikes: 2492 (neuron 13), 4152 (neuron 14), 4197 (neuron 16), recorded during 250 stimulus trials of 600 ms each. Further details as in Fig. 7.

The second example demonstrates that very similar firing rate patterns may go hand in hand with fully different correlation patterns. Figure 8 shows the JPSTHs of two pairs of neurons in response the same stimulus as used in Fig. 7. The first pair (Fig. 8A) was taken from two strongly overlapping model assemblies, the second pair (Fig. 8B) was taken from two different cell assemblies. The time course of the responses of all three neurons shows a very similar increase and decrease of the firing rate with time, as is evident from the PST histograms (neuron 13: x-axis, Figs. 8A,B; neuron 14: y-axis, Fig. 8A; neuron 16: y-axis, Fig. 8B). Particularly the neurons 14 and 16 can hardly be distinguished on the basis of their response rate profiles. Evidently, the stimulus succeeded to activate several different cell assemblies simultaneously, with very similar rate dynamics. In spite of these similarities, however, there is a dramatic difference in the firing correlations of the co-activated neurons. Whereas the first pair is engaged in synchronous firing during most of the activated period (Fig. 8A), the other, equally activated pair shows strong anti-synchronous firing, with distinctly fewer coincident spikes than expected on the basis of the firing rates (Fig. 8B). We found that this difference between synchronous and anti-synchronous firing distinguishes members of related assemblies from those of competing ones. Neurons that are involved in co-activated, related assemblies synchronize their activity and, thus, become temporarily bound into a single functional group. Neurons that are involved in co-activated, but competing cell assemblies fire in anti-synchrony, thereby effectively 'disconnecting' the different co-activated groups from each other. This 'contrast enhancement' is a direct consequence of the 'competition for resources', implemented by the activity regulation mechanism in our model.

A final interesting observation in these simulations concerns the shape and the width of the peak in the time-averaged crosscorrelograms. As can be seen in Figs. 7 and 8, these peaks were usually symmetric, several tens of milliseconds wide and straddling the origin. In none of the correlograms did we ever observe the typical narrow, asymmetric peaks that are usually associated with direct, monosynaptic connections (e.g., Perkel *et al.*, 1967; Moore *et al.*, 1970). In other words, although monosynaptic connections with short time constants were abundantly present in our model network (cf. Fig. 6), the firing correlations were generally dominated by loose synchrony. In fact, we could only elicit a typical 'monosynaptic' correlation if we made the corresponding synapse unphysiologically strong. Nevertheless, we did observe more precise forms of spike correlations in our model recordings. These fine temporal structures (not shown here) were, however, more complicated, and did not allow a straightforward interpretation in terms of local connectivities.

DISCUSSION

Our results demonstrate that cortical neurons may exhibit rapid modulations of discharge correlation in relation to behavioral events. These modulations may switch the neurons' firing behavior from being incoherent into a coherent state of joint synchrony. Alternatively, the modulation may enhance temporal segregation of the neurons' firing by a temporary increase of negative correlation. Each state may last from a few tens of milliseconds to several seconds. The observed modulations may be, but are not necessarily associated with changes in the individual neurons' firing rates. These findings support the notion that a single neuron can intermittently participate in different computations by rapidly changing its coupling to other neurons, thus switching its allegiance from one functional group to another (Vaadia *et al.*, 1995).

Our physiological results indicate that these groups are not randomly organized. Most pairs of neighboring neurons exhibited selective, rapid increases of positive correlation near behavioral events, whereas distant neuron pairs frequently showed compound patterns of correlation with enhancement of predominantly-negative correlation near the behavioral events. These findings support and extend several anatomical and physiological findings which indicated that functional groups are organized in clusters (Ts'o *et al.*, 1986; Eckhorn *et al.*, 1988; Selemon & Goldman Rakic, 1990). Our findings further suggest that neighboring neurons tend to share common inputs of the same sign (either inhibitory or excitatory) whereas the effects on more distant neurons (in our study: about 500-1000 μm apart) are mixed. Therefore, when the common drive is increased, neighboring neurons tend to be activated in unison, and can operate as a coherent functional group for a short while. On the other hand, the negative correlation between neurons in one group and those in another, more distant one can accentuate the separation among groups. Thus, the spatio-temporal organization of activity in the network allows for rapid association of neurons into a functional group, at the same time dissociating such a group from concurrently activated, competing groups.

Previous correlation studies concentrated on relatively precise coincidences with jitters of only a few milliseconds (e.g., Toyama *et al.*, 1981; Ts'o, 1986), the assumed jitters of direct synaptic interactions. In this paper we bring together physiological findings and results of model studies that focus on the phenomenon of loose synchrony. Precise coincidences do occur, however, in the present physiological data, as was discussed elsewhere (Abeles *et al.*, 1993). For example, JPST-analysis of the same data as in Fig. 3A, but using a narrower bin of 5 ms, revealed that the *modulation depth* of precise coincidences (with jitters of only ±2.5 ms) reached values up to 620%. Similarly, also our model data exhibited additional fine temporal structures in the spike correlations in. As in the physiological recordings, however, these more precise correlations were rather more complicated and, hence, not so much indicative of direct monosynaptic connections, but presumably caused by fine temporal details of the network interactions. These are, however, beyond the scope of our present discussion (see Abeles (1991, 1993a,b) and Aertsen *et al.* (1995b) for more elaborate discussions on this issue).

The wide peaks and troughs (tens to hundreds of milliseconds wide) in our cortical correlations indicate that the time-constraints of the processes evoking these correlations are loose. Such correlations (including precise coincidences) could emerge by repeated volleys of direct synaptic interactions between the two neurons or, more likely, by a change in the pattern of activity of a large number of neurons interacting in a correlated manner with the two sampled neurons. Also in our model network the firing correlations were generally dominated by loose synchrony, even though direct monosynaptic connections (with short time constants) were abundantly present. This predominance of loose synchrony for realistic synaptic connection strengths is in good accordance with our physiological findings, and presumably reflects a general property of feedback networks with distributed connectivity.

Regardless of the mechanism, the modifications of correlation between neurons in relation to stimulation and behavior most likely reflect changes in the organization of spike activity in larger groups of neurons. Indeed, our model studies demonstrate that similar dynamic organization can be accomplished in large networks, even without associated modifications of the synaptic weights (Aertsen & Preissl, 1991; Aertsen *et al.*, 1994; see also Kaneko, 1990; Hansel & Sompolinsky, 1992). Thus, considerable and rapid changes in functional coupling may occur, without the need for associated changes in local connectivity. Many of the features we encountered in our frontal cortex recordings were reproduced in the simulations in remarkable detail. For example, the opponent correlation dynamics within and across the model assemblies (Fig. 8) bear a striking resemblance to the different dynamics observed in the 'neighboring' versus the 'distant' cortical neurons (Fig. 4). We conclude that, even with the highly simplifying assumptions in our model, we have captured the essence of a global mechanism by which the flow of the activity in the network dynamically organizes the linking of neurons into functional groups. The structural properties of the network that enhance the manifestation of dynamic functional coupling are (Palm, 1993; Aertsen *et al.*, 1995a):

- sparse and weak connectivity in a very large network;
- low average activity;
- structured connectivity that favors excitatory connections within a number of subgroups of neurons (assemblies);
- a spike-generating mechanism that requires several (near-) coincident inputs to produce a spike.

These properties are well compatible with the anatomy and physiology of the cerebral cortex and with the idea of Hebbian learning of cell assemblies.

Dynamic modulation of firing correlation can facilitate the functional reorganization of the anatomical network according to the instantaneous computational demands. Repeated occurrences of dynamic correlation of the kind described here may have additional, more lasting implications. There is ample evidence that repeated co-activation of neurons may lead to enhancement of specific functional circuits by Hebb-like mechanisms for synaptic modification (Kelso *et al.*, 1986; Gustafsson *et al.*, 1987; Fregnac *et al.*, 1988; Bonhoeffer *et al.*, 1989; Ahissar *et al.*, 1992a). As a result, repeated ignition of functional neuron groups will leave traces in the connectivity of the neural network, thereby paving the way for rapid and reliable traversion on future occasions. Hence, dynamic modulation of firing coherence can facilitate learning by supporting the process of reorganizing the network, thereby enabling the organism to improve performance and acquire skill.

ACKNOWLEDGMENTS

This study was supported in part by grants from the Israeli Academy of Sciences and Humanities, the American-Israel Bi-National Science Foundation, the German "Bundesministerium für Forschung und Technologie" (BMFT), and the Human Frontier Science Program (HFSP).

REFERENCES

Abeles, M., 1982, *Local Cortical Circuits. An Electrophysiological Study*. Springer: Berlin.

Abeles, M., Vaadia, E., and Bergman, H., 1990, Firing patterns of single units in the prefrontal cortex and neural network models, *Network* 1:13-35.

Abeles, M., 1991, *Corticonics. Neural Circuits in the Cerebral Cortex*. Cambridge University Press: Cambridge, UK.

Abeles, M., Prut, Y., Bergman, H., Vaadia, E., and Aertsen, A., 1993a, Integration, synchronicity and periodicity. In: Aertsen A (ed), *Brain Theory: Spatio-Temporal Aspects of Brain Function*, Elsevier Science Publ.: Amsterdam, pp. 149-181.

Abeles, M., Bergman, H., Margalit, E., and Vaadia, E., 1993b, Spatio-temporal firing patterns in the frontal cortex of behaving monkeys, *J. Neurophysiol.* 70:1629-1643.

Aertsen, A. M. H. J., Gerstein, G. L., Habib, M. K., and Palm, G., 1989, Dynamics of neuronal firing correlation: modulation of "effective connectivity", *J. Neurophysiol.* 61:900-917.

Aertsen, A., and Gerstein, G., 1991, Dynamic aspects of neuronal cooperativity: Fast stimulus-locked modulations of 'effective connectivity'. In: Krüger J (ed), *Neuronal Cooperativity*, Springer: Berlin, pp. 52-67.

Aertsen, A., and Preissl, H., 1991, Dynamics of activity and connectivity in physiological neuronal networks. In: Schuster H (ed), *Nonlinear Dynamics and Neuronal Networks*, VCH Verlag: Weinheim, pp 281-301.

Aertsen, A., Erb, M., and Palm, G., 1994, Dynamics of functional coupling in the cerebral cortex: An attempt at a model-based interpretation, *Physica D* 75: 103-128.

Aertsen, A., Erb, M., Palm, G., and Schüz, A., 1995a, Coherent assembly dynamics in the cortex: Multi-neuron recordings, network simulations and anatomical considerations. In: Pantev C, Elbert T, Lütkenhoner B (eds), *Oscillatory Event-Related Brain Dynamics*, Plenum Press: New York, pp 59-84.

Aertsen, A., Diesmann, M., Grün, S., Arndt, M., Gewaltig, M-O., 1995b, Coupling dynamics and coincident spiking in cortical neural networks. In: Herrmann H, Pöppel E, Wolf DW (eds), *Supercomputers in Brain Research: From Tomography to Neural Networks*, World Scientific Publ.: Singapore, pp 213-223.

Ahissar M, Ahissar E, Bergman H, Vaadia E., 1992a, Encoding of sound-source location and movement: Activity of single neurons and interactions between adjacent neurons in the monkey auditory cortex, *J. Neurophysiol.* 67: 203-215.

Ahissar, E., Vaadia, E., Ahissar, M., Bergman, H., Arieli, A., and Abeles, M., 1992b, Dependence of cortical plasticity on correlated activity of single neurons and on behavioral context, *Science* 257:1412-1415.

Barlow, H. B., 1972, Single units and sensation: A neuron doctrine for perceptual psychology? *Perception* 1:371-394.

Barlow, H. B., 1992, Single cells versus neuronal assemblies. In: Aertsen, A., and Braitenberg, V. B. (eds), *Information Processing in the Cortex: Experiments and Theory*, Springer: Berlin, pp. 169-174.

Bonhoeffer, T., Staiger, V., and Aertsen, A., 1989, Synaptic plasticity in rat hippocampal slice cultures: Local 'Hebbian' conjunction of pre- and postsynaptic stimulation leads to distributed synaptic enhancement, *Proc. Natl. Acad. Sci* 86: 8113-8117.

Braitenberg, V., 1978, Cell assemblies in the cerebral cortex. In: Heim, R., and Palm, G. (eds), *Theoretical Approaches to Complex Systems. Lecture Notes in Biomathematics*, Vol. 21, Springer: Berlin, pp. 171-188.

Braitenberg, V., and Schüz, A., 1991, *Anatomy of the Cortex. Statistics and Geometry*. Springer: Berlin.

Eckhorn, R., Bauer, R., Jordan, W., Brosch, M., Kruse, W., Munk, M., and Reitboeck, H. J., 1988, Coherent oscillations: a mechanism of feature linking in the visual cortex? Multiple electrode and correlation analysis in the cat, *Biol. Cybern.* 60:121-130.

Eggermont, J. J., 1990, *The Correlative Brain. Theory and Experiment in Neural Interaction*. Springer: Berlin.

Eggermont, J. J., Smith, G. M., and Bowman, D., 1993, Spontaneous burst firing in cat primary auditory cortex: age and depth dependence and its effect on neural interaction measures, *J. Neurophysiol.* 69:1292-1313.

Engel, A. K., Kreiter, A. K., König, P., and Singer, W., 1991, Interhemispheric synchronization of oscillatory responses in cat visual cortex. *Science* 252:1177-1179.

Erb, M., and Aertsen, A., 1992, Dynamics of activity in biology-oriented neural network models: stability at low firing rates. In: Aertsen, A., and Braitenberg, V. (eds), *Information Processing in the Cortex: Experiments and Theory*, Springer: Berlin, pp. 201-223.

Fetz, E., Toyama, K., and Smith, W., 1991, Synaptic interactions between cortical neurons. In: Peters, A. (ed.), *Cerebral Cortex*, Vol. 9, Plenum Publ.: New York, pp. 1-47.

Frégnac, Y., Shulz, D., Thorpe, S., and Bienenstock, E., 1988, A cellular analogue of visual cortical plasticity, *Nature* 333:367-370.

Gerstein GL, Perkel DH., 1969, Simultaneously recorded trains of action potentials: Analysis and functional interpretation. *Science* 164:828-830.

Gerstein, G. L., and Perkel, D. H., 1972, Mutual temporal relationships among neuronal spike trains, *Biophys. J.* 12:453-473.

Gerstein, G. L., Bloom, M. J., Espinosa, I. E., Evanczuk, S., Turner, M. R., 1983, Design of a laboratory for multi-neuron studies, *IEEE Trans. Systems, Man and Cybernetics* SMC-13:668-676.

Gerstein, G. L., Bedenbaugh, P., AND Aertsen, A. M. H. J., 1989, Neuronal Assemblies, *IEEE Trans. Biomed. Eng.* 36: 4-14.

Gray, C. M., AND Singer, W., 1989, Stimulus-specific neuronal oscillations in orientation columns of cat visual cortex, *Proc. Natl. Acad. Sci.* USA 86:1698-1702.

Gray, C. M., Engel, A. K., König, P., AND Singer, W., 1992, Synchronization of oscillatory neuronal responses in cat striate cortex: temporal properties, *Vis. Neurosci.* 8:337-347.

Gustafsson, B., Wigström, H., Abraham, W. C., and Huang, Y. Y., 1987, Long term potentiation in the hippocampus using depolarizing current pulses as the conditioning stimulus to single volley synaptic potentials, *J. Neurosci.* 7:774-780.

Hansel, D., and Sompolinsky, H., 1992, Synchronization and computation in chaotic neural network, *Phys. Rev. Let.* 68:718-724.

Hebb, D., 1949, *The Organization of Behavior. A Neuropsychological Theory*. Wiley: New York.

James, W., 1890, Psychology (briefer course). In: Andersen, J. A., and Rosenfeld, E. (eds.), *Neurocomputing*, MIT Press: Cambridge. (1989).

Kaneko, K., 1990, Clustering, coding, switching, hierarchical ordering, and control in a network of chaotic elements, *Physica D* 41:137-172.

Kelso, S. R., Ganong, A. H., and Brown, T. H., 1986, Hebbian synapses in hippocampus, *Proc. Natl. Acad. Sci. USA* 83:5326-5330.

Krüger, J., 1983, Simultaneous individual recordings from many cerebral neurons: techniques and results, *Rev. Physiol. Biochem. Pharmacol.* 98:177-233.

Krüger, J., 1990, Multi-microelectrode investigation of monkey striate cortex: link between correlational and neuronal properties in the infragranular layers, *Vis. Neurosci.* 5:135-142.

Moore, G. P., Segundo, J. P., Perkel, D. H., and Levitan, H., 1970, Statistical signs of synaptic interaction in neurons, *Biophys. J.* 10:876-900.

Murthy, V. N., and Fetz, E. E., 1992, Coherent 25- to 35-Hz oscillations in the sensorimotor cortex of awake behaving monkeys, *Proc. Natl. Acad. Sci. USA* 89:5670-5674.

Newsome, W. T., Mikami, A., Wurtz, R. H., 1986, Motion selectivity in macaque visual cortex. III. Psychophysics and physiology of apparent motion, *J. Neurophysiol.* 55:1340-1351.

Palm, G., 1982, *Neural Assemblies. An Alternative Approach to Artificial Intelligence.* Springer: Berlin.

Palm, G., Aertsen, A. M. H. J., and Gerstein, G. L., 1988, On the significance of correlations among neuronal spike trains, *Biol. Cybern.* 59:1-11.

Palm. G., 1993, On the internal structure of cell assemblies In: Aertsen, A. (ed.), *Brain Theory: Spatio-Temporal Aspects of Brain Function*, Elsevier Science Publ: Amsterdam, pp. 261-270.

Perkel, D. H., Gerstein, G. L., and Moore, G. P., 1967, Neuronal spike trains and stochastic point processes. II. Simultaneous spike trains, *Biophys. J.* 7:419-440.

Sanes, J. N., and Donoghue, J. P., 1993, Oscillations in local field potentials of the primate motor cortex during voluntary movement, *Proc. Natl. Acad. Sci. USA* 90:4470-4474.

Selemon, L. D., and Goldman Rakic, P. S., 1990, Topographic intermingling of striatonigral and striatopallidal neurons in the rhesus monkey, *J. Comp. Neurol.* 297:359-376.

Toyama, K., Kimura, M., and Tanaka, K., 1981, Crosscorrelation analysis of interneuronal connectivity in cat visual cortex, *J. Neurophysiol.* 46:191-201.

Ts'o, D. Y., Gilbert, C. D., and Wiesel, T. N., 1986, Relationships between horizontal interactions and functional architecture in cat striate cortex as revealed by cross-correlation technique, *J. Neurosci.* 6:1160-1170.

Vaadia, E., and Aertsen, A., 1992, Coding and computation in the cortex: single-neuron activity and cooperative phenomena. In: Aertsen, A., and Braitenberg, V. (eds.), *Information Processing in the Cortex: Experiments and Theory*, Springer: Berlin, pp. 81-121.

Vaadia, E., Haalman, I., Abeles, M., Bergman, H., Prut, Y., Slovin, H., and Aertsen, A., 1995, Dynamics of neuronal interactions in the monkey cortex in relation to behavioral events, *Nature* 373:515-518.

von der Malsburg, C., 1981, The correlation theory of brain function, *Internal Report 81-2*, Max-Planck-Institute for Biophysical Chemistry, Göttingen (FRG).

von der Malsburg, C., 1986, Am I thinking assemblies? In: Palm, G., and Aertsen, A. (eds.), *Brain Theory*, Springer: Berlin, pp. 161-176.

Processing and Pattern Analysis of ECG in Health and Disease

16

ANALYSIS OF HEART RATE VARIABILITY
A Review

Otto Rompelman[1] and Ben J. TenVoorde[2]

[1] Delft University of Technology
Department of Electrical Engineering
Mekelweg 4, 2628 CD Delft, The Netherlands
[2] Medical Physics and Informatics Group
Academic Hospital, Free University
de Boelelaan 1118, 1081 HV Amsterdam, The Netherlands

ABSTRACT

Fluctuations in heart rate have long been the subject of investigation for almost as long as the electrocardiogram has been measured. Fluctuations with a period in the order of 2 to 100 seconds are usually referred to as heart rate variability (HRV) and are mainly of neuronal origin. This implies that the analysis of HRV may shed some light on the autonomous nervous system as it influences the neuro-cardiovascular system. Two main issues emerge if we want to analyze fluctuations in heart rate, viz. (a) how and with what accuracy can heart rate be assessed, and (b) in which way can variations in heart rate be quantified and analyzed? The lower bounds for the accuracy are discussed leading to the intrinsic signal-to-noise ratio of HRV. Consequently, the event series analysis of the cardiac event is reviewed and an interesting application of this approach is shown.

INTRODUCTION

The study of heart rate variability has gained a considerable increase in interest over the last 25 years. In his review paper Latson mentions that '*the current resurgence of interest in HRV is based upon the evidence that the analysis of HRV may (a) may help to elucidate the role of alterations in autonomic reflex activity in a number of pathophysiologic conditions and (b) it provides a noninvasive diagnostic tool for detecting alterations in autonomic reflex activity in specific patient populations*' (Latson, 1994). Furthermore the study of HRV is an essential tool in developing and verifying models of neuro-cardiovascular control.

It has been known for a long time that beat-to-beat fluctuations in heart rate can be observed. In 1847 Ludwig showed the existence of respiratory fluctuations in heart rate (Ludwig, 1847).

Advances in Processing and Pattern Analysis of Biological Signals, Edited by Isak Gath and Gideon F. Inbar
Plenum Press, New York, 1996

A more accurate investigation of variations in heart rate became possible after the invention of the electrocardiography. It was the famous Dutch physicist Willem Einthoven, who after having experimented with an electrometer (Einthoven, 1895), invented the string galvanometer, which was an instrument which, to put it in contemporary terms, 'outperformed the capillary electrometer in bandwidth, linearity and sensitivity' (Einthoven, 1903). Rapid fluctuations in heart rate were consequently described by Traube, Hering and Mayer as reviewed by Peñáz (1978) and, recently, by Latson (1994).

We now know that the variations in heart rate can be attributed to different origins viz. humoral and neural, the latter influence being composed of vagal and sympathetic activities.

In terms of frequency ranges humoral influences range from periods in the order of minutes, hours, days or even months or years, whereas neural influences cover periods between less than a second to about one hour. In more detail, both parasympathetic and sympathetic activity governs the lower frequencies (from 0.04 - 0.15 Hz) whereas parasympathetic activity is responsible for the fastest fluctuations in heart rate (> 0.15 Hz). The neurally induced variations are usually referred to as heart rate variability (HRV). Alternately, the analysis of fluctuations in heart rate may shed some light on the activity of the neural cardiovascular system.

Three main issues emerge if we want to apply the analysis of HRV into medical and physiological research and practice:

- How and with what accuracy can heart rate be assessed ?
- In which way can variations in heart rate be quantified and analyzed ?
- What are the underlying physiological hypotheses (models) for practical applications of HRV analysis ?

The rationale of the analysis of HRV is in the latter item. This means, that any application of HRV analysis is based on some hypothesis of the functioning of the neural cardiovascular system, which can be basically depicted as shown in Fig. 1.

The two important controlled variables are heart rate and arterial blood pressure. For a long time HR(V) has been the only measurable phenomenon of this system. Many studies were carried out relying completely on HRV. Since about 20 years, however, fluctuations in (arterial) blood pressure can be measured non-invasively, hence increasing the analysis power tremendously. This does not mean, that HRV analysis became less important. On the contrary: the ability of measuring both controlled variables of the neuro-cardiovascular system now enables us to gain much more insight into this system, hence the increased importance of both accurate measurement and sophisticated analysis of BPV and HRV.

It is the belief of the authors that a milestone in HRV-analysis was the special issue of ERGONOMICS in 1973 in which a keynote paper by Sayers from Imperial College, London, was published (Sayers, 1973). In this paper the relation between the analysis of HRV and a (non-linear) model for neuro-cardiovascular control was proposed. The interest spread from the U.K. (a.o. Hyndman, 1970) to the Netherlands (e.g. Rompelman *et al.*, 1977;

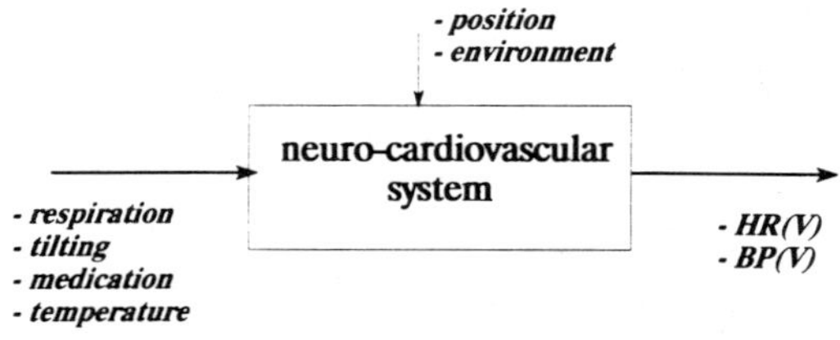

Figure 1. The neuro-cardiovascular control system with its inputs and outputs.

Kitney & Rompelman, 1987; de Boer *et al.*, 1983), the USA (e.g. Akselrod *et al.*, 1981; Saul *et al.*, 1989), Italy (e.g. Pagani *et al.*, 1986) and again the Netherlands (TenVoorde, 1992).

In this chapter we will discuss two important issues:

- How and with what accuracy heart rate can be assessed ?
- In which way variations in heart rate can be quantified and analyzed ?

A discussion of the models for neuro-cardiovascular control is beyond the scope of this chapter. The interested reader is referred to TenVoorde (1992).

CONVERSION OF THE ECG INTO AN EVENT SERIES

The Concept of the Waveform Occurrence Time (WOT)

The conversion of the ECG into the so called cardiac event series is in fact a two step procedure (Fig. 2).

The first step is the actual detection. Here it is decided whether or not a waveform of interest (either the P-wave or the QRS-complex) is present. The performance of this action is described in terms of reliability. This is usually done in probabilistic terms such as the detection probability and the false alarm probability. In the field of interest where HRV analysis is applied, the detection can usually be considered to be error free. The next step attributes a time instant to the detected waveform. This means that the time of occurrence of the waveform has to be estimated. This time instant is called the 'Waveform Occurrence Time (WOT)' (Rompelman, 1986). The problem now is how to give a measure for the accuracy of WOT estimation procedure.

Different ways of defining the WOT are used. In the literature the WOT-estimators are usually referred to as P-wave or QRS *detectors*. The most preferable definition will be that one which is least affected by disturbances of the ECG. When we observe an ECG recording it is apparent that the different waveforms are not time invariant. As an example, the amplitude is subject to fluctuations mainly related to respiration. Moreover the signal is disturbed by noise and mains interference. Since we want to establish the maximum achievable accuracy of the estimated occurrence time of the waveform it is important to quantify the mentioned effects. We will therefore introduce a hypothetical underlying ideal waveform *s(t)* (P-wave or QRS-complex), which is supposed to be time-invariant and not affected by additive noise or hum. The recorded waveform *x(t)* can now be described as:

$$\underline{x}(t) \;=\; (1 + \underline{r}).\underline{s}(t) + \underline{h}(t) + \underline{n}(t) \tag{1}$$

with: (underlined means random variable)

$\underline{x}(t)$: recorded waveform; $\underline{r}$: a factor depicting amplitude fluctuations; $s(t)$: underlying ideal waveform, but with slight random changes in morphology; $\underline{n}(t)$: additive noise (a.o. myopotentials, electrodes, amplifier, recording equipment); $\underline{h}(t)$: mains interference (hum).

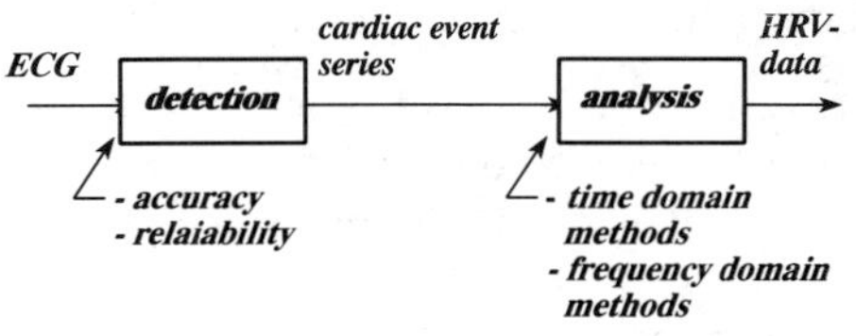

Figure 2. Derivation of the cardiac event series from the ECG.

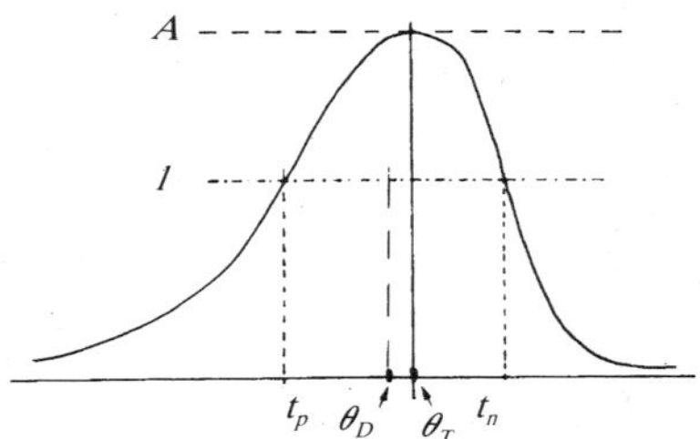

Figure 3. Two definitions of the WOT.

Changes in amplitude as well as morphology are due to changes in the relative position of the electrodes with respect to the heart. These changes are mainly caused by respiratory thoracic movements and rotations of the heart itself. It is assumed that the factor r doesn't affect $\underline{n}(t)$ nor $\underline{h}(t)$: possible amplitude variations can be encompassed by allowing $\underline{n}(t)$ and $\underline{h}(t)$ to be non-stationary. Moreover, $\underline{n}(t)$ is non-stationary due to the likewise non-stationary character of the myo-potentials. Finally, the phase relation between the mains interference and the waveform is different for each waveform, which explains the random character of $\underline{h}(t)$.

As mentioned before, the WOT can be defined in different ways. A few common methods are (Fig. 3): θ_T: time at which waveform reaches its maximum value; θ_D: midpoint between two level crossings (double level estimator); θ_M: time at which output of a matched filter reaches its maximum value.

There are two ways of investigating which WOT estimation procedure is least susceptible to disturbances: (a) experimentally with real data and (b) analytically with a model of the waveform. In short, the investigation is carried out along the following lines (for details we refer to the respective discussions in the literature):

(A) *WOT analysis with real data* (Rompelman *et al.*, 1986):

- create an estimate of the ideal waveform *x(t)* by means of coherent averaging of a number of recorded signals *s(t)*.
- add real noise and hum to this waveform (obtained from the iso-electric periods of the ECG-recordings), as well as slightly modify the shape of the waveform.
- calculate $\sigma = \sqrt{E[(\vartheta_s - \vartheta_x)^2]}$ and find out for which WOT estimate σ is minimal.

(B) *WOT analysis with analytical models of the signals* (Koeleman *et al.*, 1984):

- postulate an analytical model of the ideal waveform *x(t)* (e.g. a triangular waveform or a raised cosine).
- add 1/f-noise and 50 Hz to this signal, respectively, as well as slightly modify the shape of the waveform.
- compute (if possible analytically) $\sigma = \sqrt{E[(\vartheta_s - \vartheta_x)^2]}$ and find out for which WOT estimate σ is minimal.

Conclusions from detailed investigations are:

- the matched filter estimator (ϑ_M, which appeared to be the optimal estimator) is only slightly superior over the double level estimator (ϑ_D).
- the lower bound for the estimation error for the QRS-complex occurrence time: $\sigma_{QRS} \geq 0.1$ ms.
- the lower bound for the estimation error for the P-wave occurrence time: $\sigma_P \geq 1.1$ ms.

- the values for σ are roughly linearly related to the signal-to-noise ratio of the ECG-signal.

The Intrinsic Signal-to-Noise Ratio in HRV

HRV should reflect the fluctuations in the firing rate of the SA-node. Deviations (either or not random) of the generated event series from the SA-nodal firing moments will cause (random) errors in HRV, or in other words: noise. In Fig. 4 the ECG in relation to the SA-firing is shown, together with the relevant time shifts from the SA-nodal firing moment to the ECG-WOT's.

The relevant time shifts D are (underlined means: random variable):

$\underline{D}_{SA-P}$: SA-node → start P-wave
$\underline{D}_{d}$: time discretization ($\Delta t = 1/f_s$)
$\underline{D}_{P}$: estimation of P-WOT
$\underline{D}_{AV}$: P-WOT → QRS-WOT (AV-conduction)
$\underline{D}_{QRS}$: estimation of QRS-WOT.

For the expected values and the standard deviations we may find the following (approximate) values:

$E[\underline{D}_{SA-P}] \approx 70$ ms $\sigma_{SA-P} \approx$ few ms
$E[\underline{D}_{AV}] \approx 170$ ms $\sigma_{SA-P} \approx 1\text{-}5$ ms
$E[\underline{D}_{QRS}] = 0$ $\sigma_{QRS} \approx 10/(SNR_{ECG})$ $\approx 0,1$ ms
$E[\underline{D}_{P}] = 0$ $\sigma_{QRS} \approx 100/(SNR_{ECG})$ ≈ 1 ms
$E[\underline{D}_{d}] = 0$ $\sigma_{d} \approx 0.3\ \Delta t$ $\approx 1/(3)f_s)$

f_s is the sampling rate of the signal. This implies that each event will be shifted such that it coincides with a sampling instant. It is assumed that this shift is uniformly distributed. This shift is called synchronization and will be discussed later in more detail. If it is assumed (simplification), that the errors are mutually independent, the total estimation errors for the QRS-WOT and the P-WOT are respectively:

$$\sigma_{t,QRS} = (\sigma^2_{SA-P} + \sigma^2_{AV} + \sigma^2_{QRS} + \sigma^2_{d})^{1/2} \tag{2}$$

and

$$\sigma_{t,P} = (\sigma^2_{SA-P} + \sigma^2_{P} + \sigma^2_{d})^{1/2} \tag{3}$$

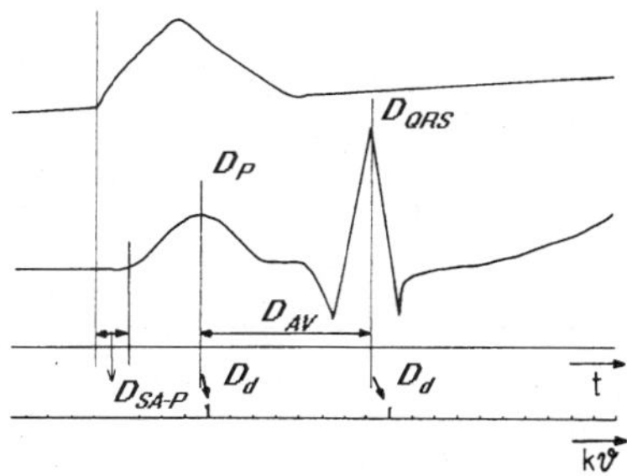

Figure 4. SA-nodal firing, the ECG and the relevant time shifts.

Taking the aforementioned values into account, we come to the conclusion, that (a) it is usually attractive to use the QRS-WOT (despite it is 'more distant' from the SA-firing Moment) and (b) the timing error for the cardiac event series (= series of QRS-WOT's) is approximately:

$$\sigma_{t,QRS} \sim (\sigma^2_{AV} + \sigma^2_{d})^{1/2} \sim [10 + 0.1\,(\Delta t)^2]^{1/2} \quad ms \tag{4}$$

Now we can relate this error to HRV. Assume that the variance of the RR-intervals is σ^2_{RR}. We may introduce the *intrinsic signal-to-noise ratio* η for HRV:

$$\eta = 20\log\frac{\sigma_{RR}}{\sigma_{t,QRS}} \quad dB \tag{5}$$

Two examples are given to indicate the importance of the sampling frequency:

Normal healthy subject	Autonomic neuropathy
$\sigma_{RR} = 50$ ms; $\sigma_{AV} = 3$ ms	$\sigma_{RR} = 10$ ms; $\sigma_{AV} = 1.7$ ms
fs $= 100$ Hz $\to \eta = 21$ dB	fs $= 100$ Hz $\to \eta = 9$ dB
fs $= 200$ Hz $\to \eta = 23$ dB	fs $= 100$ Hz $\to \eta = 13$ dB
fs $= 1000$ Hz $\to \eta = 24$ dB	fs $= 1000$ Hz $\to \eta = 15$ dB

ANALYSIS OF THE CARDIAC EVENT SERIES

In principle there are two ways of analyzing the cardiac event series. First we may derive a signal from the cardiac event series. Different methods are discussed in the literature such as the RR-interval tachogram, the instantaneous heart rate and the low pass filtered event series. For a comparative analysis of these methods we refer to e.g. Rompelman *et al.*, 1977.

A more suitable method is the event series analysis method, which has been proposed on a number of occasions (e.g. Rompelman, 1986). This approach is based on the differential counting process (Fig. 5).

Starting from the counting process $n(t)$ (t_i: event occurrence times):

$$n(t) = \sum_{\forall i} u(t - t_i) \tag{6}$$

The differential counting process $x(t)$ is consequently defined as:

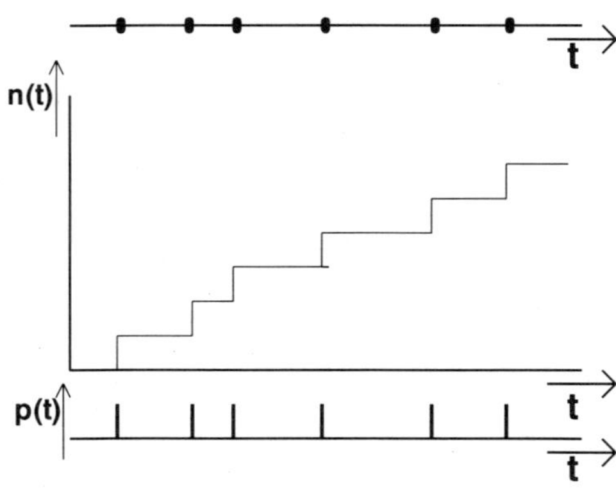

Figure 5. The derivation of the signal $x(t)$ from the cardiac event series.

$$x(t) = \frac{d}{dt}\, n(t) = \sum_{\forall i} \delta_D (t - t_i) \tag{7}$$

The *discrete time* differential counting process is given as:

$$x[k.\vartheta] = \sum_{\forall i} \delta_K [(k - k_i)\vartheta] \tag{8}$$

with ϑ the time discretization, or: sampling interval, and $k_i.\vartheta$: discrete i^{th} event occurrence time. Note the difference between δ_D (Dirac-δ) and δ_K (Kronecker-δ) in equation 8.

Linear filtering follows from the convolution of $x(t)$ with the impulse response $h(t)$ of the filter (continuous time):

$$y(t) = \int_{-\infty}^{\infty} x(t - \tau) h(\tau)\, d\tau =$$

$$= \int_{-\infty}^{\infty} \sum_{\forall i} \delta(t - t_i - \tau) h(\tau)\, d\tau$$

$$= \sum_{\forall i} h(t - t_i) \tag{9}$$

and discrete time $(t_i \rightarrow n_i.\vartheta)$:

$$y[k.\vartheta] = \sum_{\forall i} h[(k - n_i).\vartheta] \tag{10}$$

Note, that no multiplications are needed, only table look up instructions. The low pass filtered event series as an HRV-signal is based on this approach (Hyndman, 1973, Coenen *et al.*, 1977).

Fourier spectral analysis follows directly from applying the Fourier transform to the event series $x(t)$. For the complex spectrum we find:

$$\overline{X}(f) = \int_{-\infty}^{\infty} x(t)\, e^{-2j\pi f t}\, dt$$

$$= \int_{-\infty}^{\infty} \sum_{\forall i} \delta(t - t_i)\, e^{-2j\pi f t}\, dt$$

$$= \sum_{\forall i} e^{-2j\pi f t_i}\, dt \tag{11}$$

For the discrete time, discrete frequency spectrum $(t_i \rightarrow n_i\, \vartheta)$ we obtain:

$$X[m.\Delta f] = T_0 \sum_{\forall i} e^{-2j\pi\, m\Delta f\, n_i \vartheta} \tag{12}$$

where T_0 is the mean RR-interval length.

If $\Delta f = 1/T$ (T = segment length) and $N = T/\vartheta$:

$$X[m] = T_0 \sum_{\forall i} e^{-j\frac{2\pi m n_i}{N}} \tag{13}$$

Note, that no (complex) multiplications are needed: the calculation of the spectrum consists of table look up instructions and additions.

Finally we will give an interesting example of the spectral analysis. First we recall the spectrum of a linearly pulse frequency modulated event series as obtained by an Integral

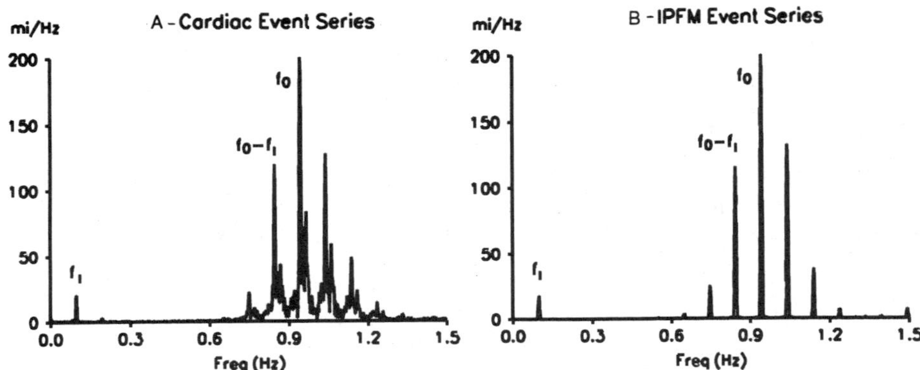

Figure 6. Left: cardiac event series spectrum of 0.1 Hz RSA. Right: event series spectrum of the IPFM if modulated with the same frequency.

Pulse Frequency Modulator. This modulator was shown to be a fair model of the neuronally influenced natural pacemaker, the SA-node (Hyndman, 1973). In Fig. 6 this spectrum is shown together with the HRV-spectrum of a subject breathing at 0.1 Hz.

Only one dominant peak is shown in the low frequency range of the spectrum, the range which is usually referred to as 'the HRV-spectrum'.

The spectrum consists of (Bayly, 1968):

- A component proportional to the modulating signal at f_1,
- components at multiples of the mean heart rate (unmodulated pulse frequency): $k.f_0$, and
- sum and difference components at $k.f_0 \pm m.f_1$.

The next example is obtained from a neonate, breathing at a rather high frequency. In fact the respiratory rate is larger than half the mean heart rate (mean heart rate $f_0 = 1.85$ Hz), which implies that (in the event series spectrum) the modulating (respiratory) component $f_1 = 1.15$ Hz has a frequency which is above the first side component $f_0 - f_1 = 0.7$ Hz. In Fig. 7 the simultaneously recorded respiratory waveform is shown with its spectrum, indicating the frequency of the modulating signal. The heart rate variability waveform (in this case the series of RR-intervals) shows some periodicity which could be identified as having a frequency equal to $f_0 - f_1$. If the event series spectrum is calculated up to 2.5 Hz both the modulating component and the side component are clearly visible. Applying a band pass filter to the event series (according to the linear filtering method as described above, the frequency characteristic is shown as a dashed line) reveals the actual respiratory arrhythmia. This is substantiated by comparing this band pass filtered event series to the respiratory waveform.

CONCLUSION

In the analysis of HRV the intrinsic accuracy of HRV information is limited due to mainly the sampling rate and the signal-to-noise ratio of the ECG. The smaller the fluctuations in heart rate, the more attention has to be paid to both aspects. In particular in neonatal, fetal and diabetic neuropathy studies the sampling rate should be at least 250 Hz.

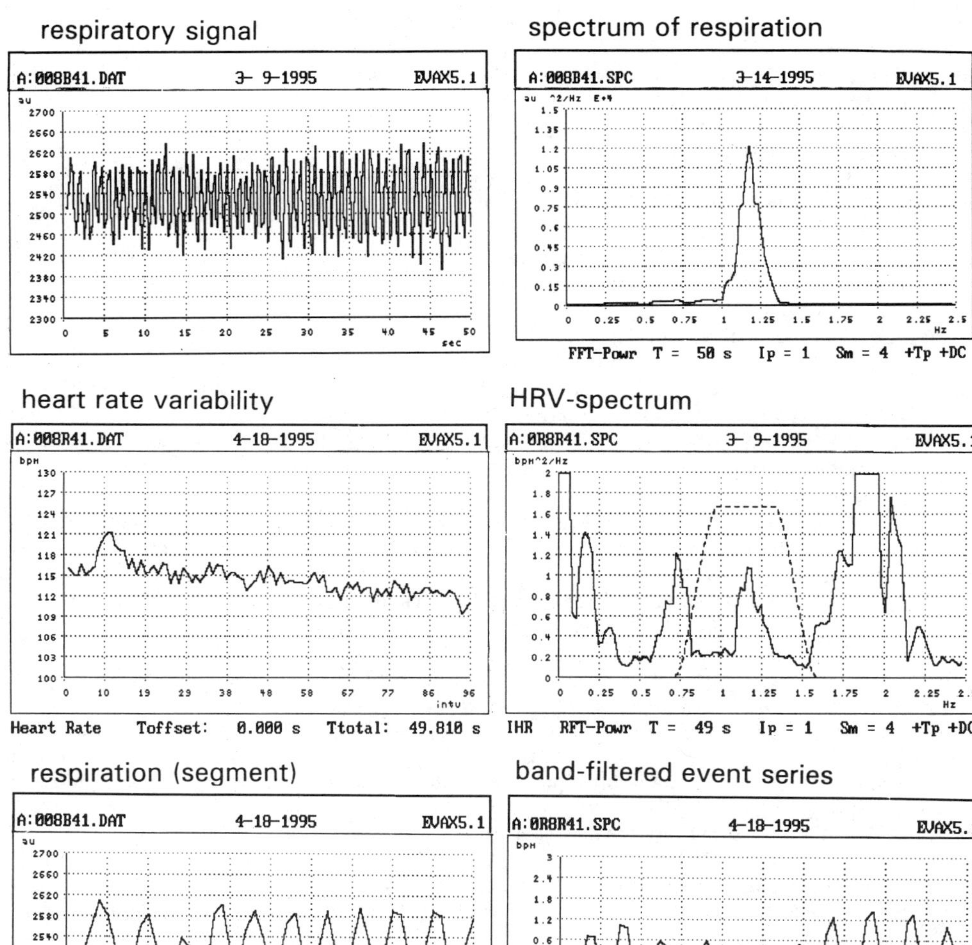

Figure 7. Analysis of neonatal RSA.

Spectral analysis of HRV can be advantageously carried out by event spectral analysis. In particular, if the modulating frequencies, such as respiration, get near or even surpass half the mean heart rate, this approach is obligatory in order to avoid erroneous interpretations.

REFERENCES

Akselrod, S., Gordon., D., Ubel, F. A., Shannon, D. C., Barger, A. C., and Cohen, R. J., 1981, Power spectrum analysis of heart rate fluctuations: a quantitative probe of beat-to-beat cardiovascular control, *Science* 213: 220-223.

Bayly, E. J., 1968, Spectral analysis of pulse frequency modulation in the nervous system, *IEEE Trans. Biomed. Eng.* BME-15:257-265.

DeBoer, R. W., Karemaker, J. M., and Strackee, J., 1983, Beat-to-Beat variability of heart interval and blood pressure, *Automedica* 4:217-222.

Coenen, A. J. R. M., Rompelman, O., and Kitney, R. I., 1977, Measurement of heart rate variability: Part II - Hardware digital device for the assessment of heart rate variability, *Med. & Biol. Eng. & Comp.* 15:423-430.

Einthoven, W., 1985, Über die Form des menschlichen Elektrocardiogramms, *Pflügers Arch. ges. Physiol.* 60:101-123.

Einthoven, W., 1903, Die galvanometrische Registrierung des menschlichen Elektrokardiogramms, zugleich eine Beurteilung der Anwendung des Capillar-Elektrometers in der Physiologie, *Pflügers Arch. ges. Physiol.* 99:472-480.

Hyndman, B. W., 1970, A digital simulation of the human cardiovascular system and its use in the study of sinus arrhythmia, *Ph.D. Thesis*, University of London.

Hyndman, B. W. and Mohn, R. K., 1973, A pulse modulator model of pacemaker activity, *Digest of the 10-th Int. Conf. on Med. & Biol. Eng., Dresden*, p. 223.

Koeleman, A. S. M., Van den Akker, T. J., Ros, H. H., Janssen, R. T., Rompelman, O., 1984, Estimation accuracy of P wave and QRS complex occurence times in the ECG: the accuracy for simplified theoretical and computer simulated waveforms, *Signal Processing*: 7: 389-405.

Kitney, R. I., Rompelman, O. (eds.), 1980, *The Study of Heart Rate Variability*, Clarendon Press: Oxford (UK).

Kitney, R. I., Rompelman, O. (eds.), 1987, *The Beat-by-beat Investigation of Cardiovascular Function*, Clarendon Press: Oxford (UK).

Latson, T. W., 1994, Principles and applications of heart rate variability analysis. In: Lynch C. III (ed.), *Clinical Cardiac Electrophysiology: Perioperative Considerations*, Lippincott Company: Philadelphia, pp. 307-348.

Ludwig, C., 1847, Beiträge zur Kenntnis des Einflusses der Respirationsbewegungen auf der Blutumlauf im Aortensystem, *Arch. Anat. Physiol. Wissenschaftl. Med.* p. 242-257.

Pagani, M., Lombardi, F., Guzetti, S., Rimoldi, O., Furlan, R., Pizzinelli, P., Sandrone, G., Malfatto, G., Dell'Orto, S., Picculagu, E., Turiel, M., Baselli, G., Cerutti, S., Malliani, A., 1986, Power spectral analysis of heart rate and arterial blood pressure variabilities as a marker of sympatho-vagal interaction in man and conscious dogs, *Circ. Res.* 59:178-193.

Peñáz, J., 1978, Mayer waves: history and methodology, *Automedica* 2:135-141.

Rompelman, O., Coenen, A. J. R. M., Kitney, R. I., 1977, Measurement of heart rate variability: Part I - Comparative study of heart rate variability analysis methods, *Med. & Biol. Eng. & Comp.* 15:233-239.

Rompelman, O., 1986, Tutorial review on the analysis of cardiac event series; a signal analysis approach, *Automedica* 7:191-212.

Rompelman, O., Janssen, R. J., Koeleman, A. S. M., van den Akker, T. J., and Ros, H. H., 1986, Practical limitations for the estimation of P-wave and QRS-complex occurrence times, *Automedica* 6:269-284.

Saul, J. P., Berger, R. D. Chen, M. H., and Cohen, R. J., 1989, Transferfunction analysis of autonomic regulation II. Respiratory sinus arrhythmia, *Am. J. Physiol.* 256 (*Heart Circ. Physiol.* 25):H153-H161.

Sayers, B. McA., 1973, Analysis of heart rate variability, *Ergonomics* 16:17-32.

TenVoorde, 1992, B. J., Modelling the Baroreflex, PhD-thesis, Free University, Amsterdam.

17

THE HEART RATE VARIABILITY SIGNAL

Among Rhythms, Noise and Chaos

Sergio Cerutti and Maria G. Signorini

Biomedical Engineering Department
Polytechnic University
via G.Ponzio 34/5, 20133
Milano, Italy

ABSTRACT

Heart rate variability (HRV) signal is supposed to be the effect of a variety of different controls acting through linear and non-linear mechanisms. In this paper, non-linear dynamic components of HRV signals are analyzed by the evaluation of invariant characteristics of the system attractor obtained from time series: correlation dimension, entropy, self-similarity parameter H, and Lyapunov exponents. The results confirm that nonlinear dynamics are involved in the HRV signal generating mechanism.

INTRODUCTION

Rhythmic phenomena in cardiovascular system have been extensively studied both in cardiovascular physiology and in clinical experiments. In fact, signals like depolarization and repolarization fronts propagating in the heart, ECG, blood pressure and pulse, as well as cardiac output, flux and others, do present a pseudoperiodic behavior synchronous to the heart period.

Traditionally, "periodicity" was considered a healthy sign, while "disclosure" from periodicity (or from rhythmic behavior) was considered a pathological sign. Normal cardiac rhythm has been assumed to be more regular (or more periodical) in respect to the pathological ones.

Not many years ago the famous physiologist A.M. Katz quoted that arrhythmias (which constitute an important issue of cardiovascular pathology) were characterized by "chaotic rhythm". Nowadays we are actually giving a different meaning to the same expression, as it will be remarked even in the present paper.

On the other hand, it is well known that the cardiovascular signals mentioned above (in particular ECG, arterial blood pressure and others directly related to them) are not strictly periodical and parameters measurable on them (such as RR intervals, systolic and diastolic

Advances in Processing and Pattern Analysis of Biological Signals, Edited by Isak Gath and Gideon F. Inbar
Plenum Press, New York, 1996

values, etc.) present more or less marked changes on a beat-to-beat basis: i.e., RR series in normal sinusal rhythm may generally have a variation of up to 10% in respect to their mean value. These changes are the indirect sign of the various controlling mechanisms of cardiovascular system (elicited through mechanical, neural, humoral effects and others) (Penaz, 1968; Sayers, 1973) (Malliani *et al.*, 1991). Such a spontaneous variability has often been correlated to healthy subjects, i.e., hearts which beat with too a synchronized rhythm may be associated to cases of diabetic neuropathy, heart failure, or to post-infarct patients, transplanted heart patients (Lombardi *et al.*,1987; Bianchi *et al.*, 1990; Mortara *et al.*, 1994). Such a variability also decreases with the age (Pagani *et al.*, 1986).

Kleiger and co-authors (Kleiger *et al.*, 1987) demonstrated that the variance (power) associated to heart rate variability is a strong prognostic index in post-infarct patients, thus suggesting that the characteristics of the analyzed signal strongly correlate with important clinical findings.

A great amount of literature has been made since the late 1960s, aiming at investigating the role of the autonomic nervous system (AN) in controlling heart rate, blood pressure and other related functions: beat-to-beat RR interval series, detected on ECG tracings, has become the most diffused and easy way to display heart rate variability (HRV) signal.

The analysis of such a signal in time and in frequency domains (Akselrod *et al.*,1981, Pagani *et al.*, 1986) has put into evidence that the complex regulating system acting on heart rate may be approached even through simple processing of a non-invasive and easy-to-develop examination like the traditional ECG recording. On the other hand, a careful analysis of HRV signal puts into evidence that the system involved in its generation is of high complexity and different behavior may correlate with pathophysiological states. In most of the normal cases we do not observe strictly periodical oscillations. Instead, an apparently erratic behavior in these circumstances may suggest that either pseudo-stochastic mechanisms are involved or non-linear dynamics in a deterministic system might give rise, under certain hypotheses, to chaotic behavior (through what are called the "routes towards chaos").

Undoubtedly, these experimental observations may build up the basis for a new concept of health and for the studying of models of pathologies which give more emphasis not to organ damage itself, but rather to controlling system failure: the so-called "dynamical diseases" (Glass *et al.*, 1979).

In the present paper, various algorithms to quantify the presence of non linear dynamics in HRV signal, will be illustrated in normal and pathological conditions, giving much emphasis to methods able to discriminate them from random noise components. A suggestive hypothesis is that the high complexity of the signal, and hence of the controlling system, may be associated to chaotic behavior. Pathological states were argued to be characterized by a lower degree of complexity (i.e., less chaotic behavior) (Goldberger *et al.*, 1990). The reported results seem to confirm these observations and all of the mentioned algorithms are coherent in suggesting that the normal cardiovascular control is characterized not by homeostatic behavior, but rather by etherostatic behavior with multiple equilibrium points (West, 1990).

HEART RATE VARIABILITY SIGNAL: PRESENCE OF OSCILLATORS, COLORED NOISE AND NON-LINEAR DYNAMICS

The HRV signal reflects the complex mechanisms of regulation of cardiovascular functioning. The traditional approach in the frequency domain (linear deterministic) is schematically reported in Fig. 1a. The signal is considered as a sum of sinusoids and the Fourier transform depicts the various frequency contributions. Through the use of Fast

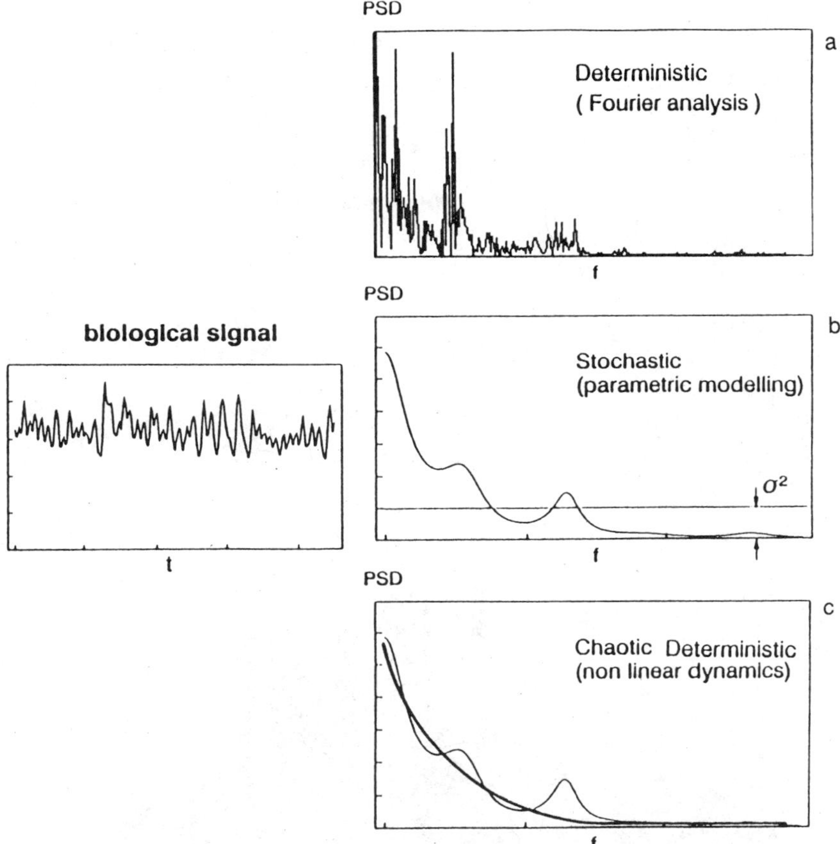

Figure 1. Approaches to the analysis of time series of biological origin. a) Linear deterministic Fourier approach by means of Fast Fourier Transform; b) parametric analysis by an Autoregressive (AR) linear model, c) non linear modeling by deterministic chaotic approach, here represented schematically by a fractal 1/f process.

Fourier Transform algorithm, the first power spectral densities (PSDs) of RR series have been obtained (Sayers, 1973), thus enlightening three main spectral components (rhythms) which were indicated by T, B and R. Rhythm T was considered relevant to thermoregulatory mechanisms, B to baroreceptor regulation (control of blood pressure signal) and R to the respiratory sinus arrhythmia.

Another approach, started from (Baselli *et al.*, 1986), yet considered linear the generation mechanism of the signal, but it was conceived to be pseudo-stochastic by nature: i.e., the signal was supposed as generated by a linear, time-invariant system driven by white Gaussian noise which constituted the stochastic (or non predictable) component superimposed to the signal. Therefore, a general ARMA (or more commonly an AR) model was conceptualized, whose coefficients were used to obtain a parametric PSD estimation, less than a white noise contribution of a certain variance (power). Figure 1b depicts schematically a parametric PSD with the white noise power superimposed. It is well known that a parametric PSD may be easily decomposed into its spectral components by means of the residual integration method (Zetterberg, 1969; Cerutti *et al.*, 1985). In this way, the three-peak pattern in RR series PSDs was confirmed. The peaks where labeled as LF (low-fre-

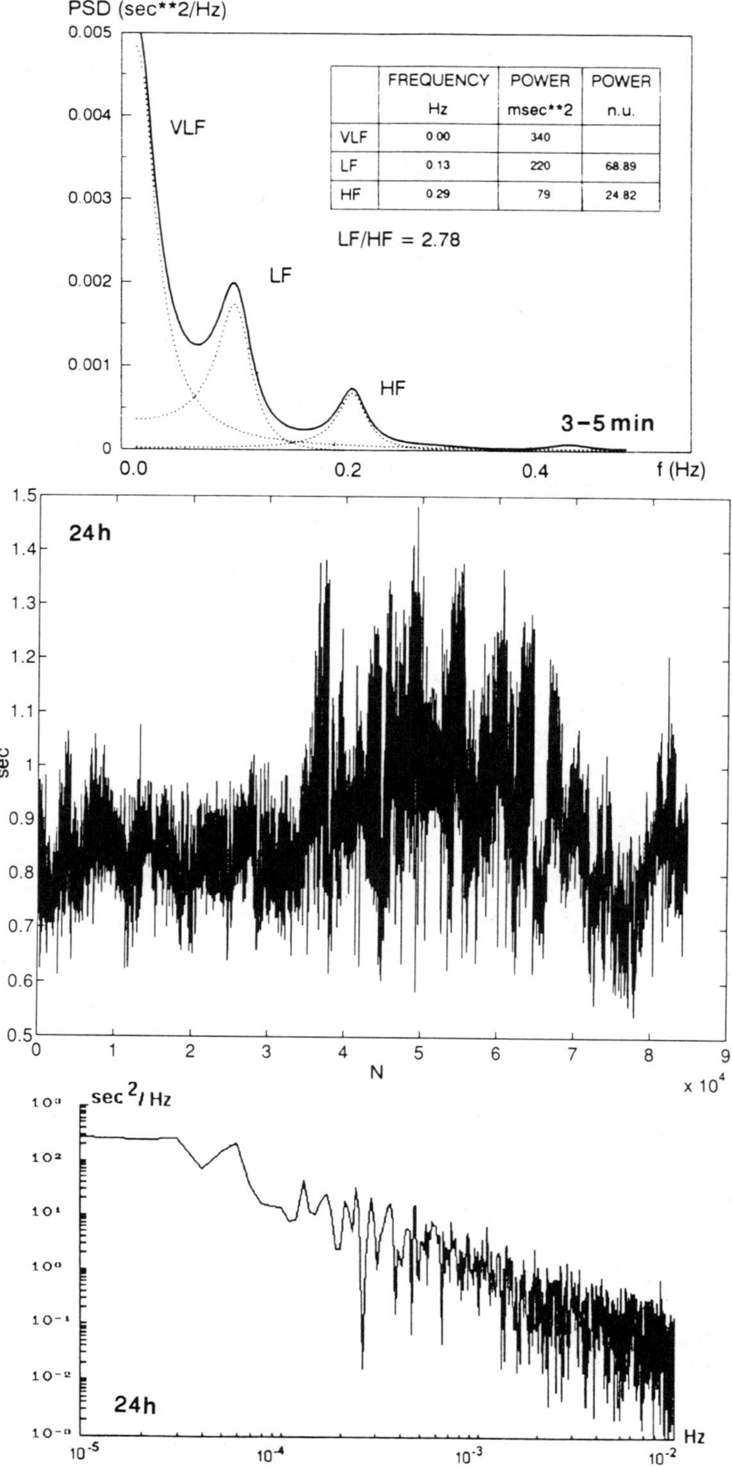

	FREQUENCY Hz	POWER msec**2	POWER n.u.
VLF	0.00	340	
LF	0.13	220	68.89
HF	0.29	79	24.82

Figure 2. HRV analysis can be performed in time windows of different lengths depending on the information one want to extract. Examples of short-term (200-500 beats) and long-term (24h) analysis are shown together with the relevant RR series measured in the 24 hours. For the deterministic chaotic approach, variables of the system have to be measured in a long time interval.

quency), HF (high frequency) and VLF (very low frequency) (Cerutti *et al.*, 1994). The associated powers, measured in absolute or normalized units, were the typical parameters which have been correlated mainly to sympathetic (LF power) and vagal (HF power) actions. The physiological correlate of VLF is still uncertain: possibly, long-term controlling mechanisms are involved, such as thermoregulation (as it was argued previously), humoral factors and others.

Parametric methods allow to reliably estimate rhythms in the spectra even on very short segments of data corrupted by random noise, in the cases in which stationarity does not maintain on longer time intervals. In this condition, it seems reasonable to think that HRV spectral peaks are corrupted by colored noise. From a physiological point of view, that can explain how these rhythms are often influenced by random frequency shifting and amplitude modulation in the spectrum.

Still, much information in the HRV signal spectra does not possess harmonic characteristics. In many conditions VLF may become predominant (as an example in 24h recordings), showing a *1/f* spectrum behavior if it is plotted in log-log scale (Kobayashi *et al.*,1982; Saul *et al.*,1987). In these cases, cardiovascular rhythms degenerate in a behavior which is apparently erratic but which can be thought as the consequence of a non-linear dynamic model characterized by deterministic chaos of low dimension. In fact, *1/f* spectra present a fractal (i.e., non integer) dimension which characterizes also chaotic strange attractors. Figure 1c depicts schematically such a behavior. As it will be illustrated in the following sections, there are various parameters which are thought to measure chaotic behavior: in most of these cases the number of RR values to be processed has to be significantly high in order to be able to capture efficiently the extreme complexity of the attractor trajectories (West, 1990).

Figure 2 reports the values of RR intervals measured in a normal subject during the 24h. We have two different kinds of analysis to be carried out on the data: i) a short-term analysis shown in fig.2, upper panel (carried out on $\cong$ 200-500 consecutive RR values), in order to capture the rhythmic components of the signal which maintains stationary conditions in the considered temporal segment; and ii) a medium/long-term analysis (from 20,000 up to 100,000 RR values shown in fig. 2, lower panel) in order to detect chaotic dynamics. In the present paper we shall investigate this last approach, while the first one is extensively described in (Cerutti *et al.*, 1991; Malliani *et al.*, 1991).

METHODS

Assessment of nonlinear dynamics in a biological system follows two principal directions. One can start from the identification of the model of the system under investigation through the solution of differential equations which describe the behavior of the model itself. In this case, the knowledge of the number of system variables is required together with system parameters themselves. In real cases it is not easy to obtain all this information: the system under analysis is often very complex, it is difficult to correctly estimate the initial conditions and both the number and the close relations between variables can adverse precise identification of the model. Further, there is a second approach starting from time series. In this way it is possible to determine the invariant characteristics of the non linear system, even if one has a measurement of only one state variable for a sufficiently long time interval. Non linear dynamics of the system may give rise either to equilibrium solutions (fixed point, limit-cycle, torus) or to chaotic behavior (characterized by strange attractors) (Eckmann *et al.*, 1986). The main geometrical and dynamic properties of the attractors identify the behavior of the system in the state-space. Strange attractors are characterized by non integer (fractal) dimension: the system trajectories moving in the state-space generate fold within

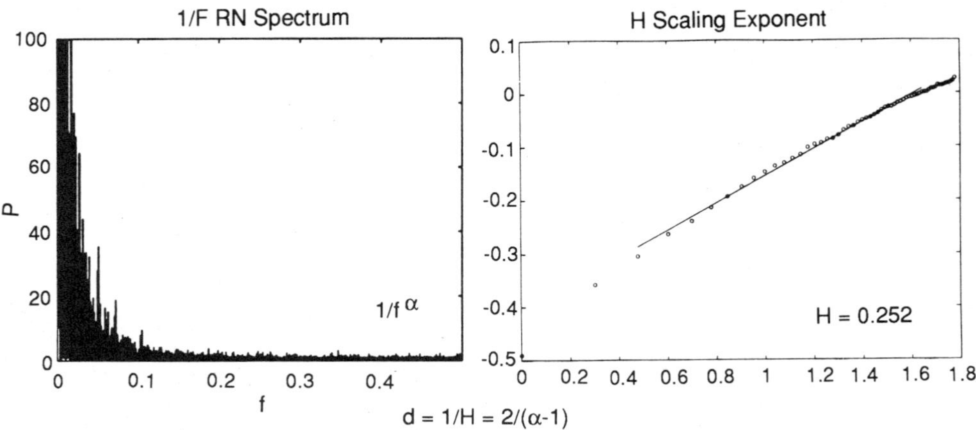

Figure 3. $1/f\alpha$ spectrum and *H* scaling exponent in a time series of a fractal random noise CRN. The link between the two indices (for $1 < \alpha < 3$) is underlined. d is the fractal dimension.

fold ad infinitum. In other words, the whole is better represented by an elementary structure which seems to repeat itself many different geometric scales (Mandelbrot,1981). From a dynamical point of view, strange attractors are characterized by an exponential divergence of trajectories. Nearby points of the attractor in the state-space tend to wander but, as the attractor has finite dimension, local divergence is compensated for by convergence of far points. Lyapunov exponents measure the rate of trajectories evolution: positive values of such exponents are characteristic of chaotic behavior.

Reconstruction procedure is the first step for the assessment of nonlinear properties of a time series. Conditions of Takens theorem (Takens, 1981) assure that the variable obtained by accordingly delaying the original *x(t)* series of an interval τ and reconstructing it in an *m*-dimensional space allows to compute the original attractor properties.

In the following, a few parameters will be introduced which are considered important quantitative indices for measuring chaotic behavior or for determining fractal geometrical (or dynamical) structures. All strange attractors possess a fractal geometry in the state-space; vice versa it is not sufficient to find fractal geometry for claiming that we are in presence of chaotic behavior. In fact, many fractal processes like Mandelbrot sets or fractional Brownian motions are by no way chaotic.

$1/f^{\alpha}$ Spectrum

Broadband pattern of the power spectrum constitutes a peculiar element of a process whose dynamics act over a large scale of values. Many apparently erratic processes show a power distribution decreasing as frequency values increase. Spectra which are expressed by $P(f) \cong 1/f^{\alpha}$ are called *power-law spectra* and α coefficient is the characteristic parameter. *P* is the power spectral density (PSD) and *f* the frequency.

This power-law spectral distribution indicates that a fractal process could be hypothesized to describe the observed phenomenon. In this way, the power spectral distribution over a long time interval can be used to separate broadband to narrowband distributions. In a biological context this means that the system showing a spectrum with a power-law distribution, possesses a variety of scales in which the system can operate. It is, in this sense, information rich and is characteristic of physiological conditions. The more or less tight correspondence of signals with the power-law distribution can be quantified by plotting in log-log scale the spectral power of signals vs. their frequency values and measuring the spectrum slope α. PSD's of heart rate variability signals, when calculated over long-term (from 8-12, up to 24 h), in normal subjects, present a value of $\alpha \cong 1$ This is a sign which is usually interpreted typical of healthy system with information richness, able to act in different frequency interacting scales.

Disclosures from $\alpha \cong 1$ values have been generally found in pathological cases. In fractal processes we have the following relation $d = 2 /(\alpha-1)$ for $1< \alpha<3$, where d is the fractal dimension measuring the degree of complexity of the fractal geometry (Osborne *et al.*,1989). The relation above puts into evidence that a precise link does exist between power-law spectra and fractal processes.

Self-Similarity

A quantity that contributes in the study of time series geometrical properties is the so called *structure function S(k)*. Its expression is given by:

$$S(k) = \sum_{i=1}^{N-k} \left[x(t+k\Delta t) - x(t)\right]^2$$

where $x(t)$ is the time series with N values, k is a scale factor and Δt is a time delay.

$S(k)$ measures the degree of "regularity" in a scalar time series and for fractal processes it has a scaling behavior, i.e.

$$S(k) \propto k^{2H}$$

for small k values. H is the scaling exponent: its value is included in $0 \div 1$ interval. By this expression one can establish a link between the power-law distribution characterizing fractal processes and the $S(k)$. As the power-law spectrum has the expression $P(f) \cong f^{-\alpha}$, we may obtain that $\alpha = 2H + 1$. This statement introduces a link between two different geometric properties of fractal curves.

This is pointed out in Fig. 3 where the link between α and H scaling exponent, for a time series, is illustrated.

Correlation Dimension ν

Grassberger and Procaccia (1983) introduced a method for the fractal dimension estimation by measuring the correlation of points of a time series in the space-state.

If the attractor has chaotic properties and its trajectories exponentially diverge, most of the pairs of points with coordinates $\mathbf{X}_j$, $\mathbf{X}_k$ with $j \neq k$ will be dynamically uncorrelated. Even if they appear as a random sequence, points that lie on the same attractor, are correlated in the state-space.

Correlation integral $C_m(\varepsilon)$, where m is the embedding dimension is then defined as:

$$C_m(\varepsilon) = \lim_{N \to \infty} \frac{1}{N(N-1)} \sum_{i=1}^{N} \sum_{j=1}^{N} \Theta\left(\varepsilon - \left|v_m(k_i) - v_m(k_j)\right|\right)$$

when $\mathbf{v}_m(k)$ is the state vector reconstructed in an m-dimensional space, ε is the correlation length, and $\Theta(\chi)$ is the Heaviside function. N is the number of points. If the attractor is chaotic, the following relation is established

$$C_m(\varepsilon) \approx \varepsilon^{\nu_m}$$

where

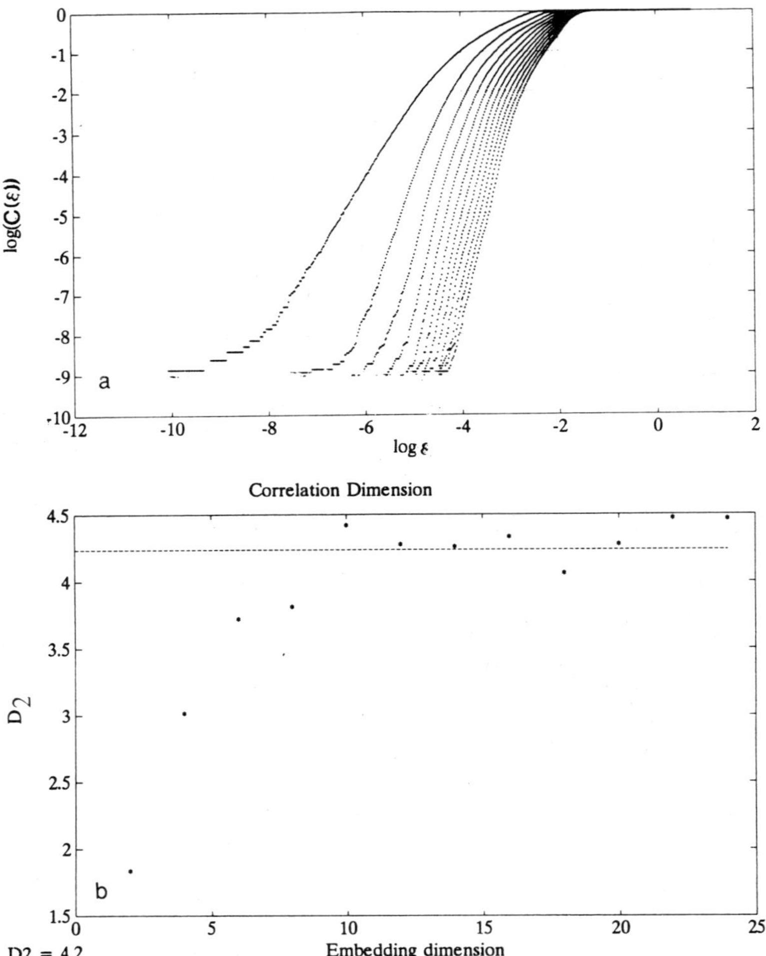

Figure 4. Correlation integral (a) and D_2 values (b) as a function of the embedding m in a transplanted subject. D_2 value is 4.2.

$$\lim_{m \to \infty} \nu_m = \nu$$

The parameter ν is the *correlation dimension* of the system attractor and is called D_2 in a multifractal way. D_2 value is obtained as the slope of the linear zone in the plot of $C(\varepsilon)$ vs. ε in log-log scale. It is an increasing function of the embedding value m until a saturation to the D_2 value is noticed. Figure 4 reports correlation integrals (Fig. 4a) and D_2 values (Fig. 4b) as a function of m. The method allows a quantification of the fractal dimension even for systems with more than three dimensions. This is often the case of experimental situations where typically high dimensional systems exist, with a large number of state variable involved in the generation of the process.

Considering independent realizations obtained as time-delayed versions of the same signal x_n (t) in an N-dimensional space, one can obtain a fractal trajectory which is parametrically represented by the set of x_n (t). The estimation of system fractal dimension one can obtain, for example, by the correlation dimension D_2, is related to the scaling exponent H by the expression $D_2 = 1/H$.

Spurious saturation could be found if a correlated noise is superimposed to the signal (Provenzale *et al.*,1992).

The calculation of D_2 for a time series obtained from a dissipative system through an embedding procedure, is supposed to provide finite values if the process is dominated by low-dimensional chaos governing the properties of a strange attractor. Measurement of this parameter implies a computation time that increases as N^2: this consideration, together with the requirement of N large enough to correctly perform the study of the attractor, suggests the use of proper computer machines with high parallelism.

Entropy K_2

In a time series the Entropy K quantifies the new information content that each new measured value adds to the process, whose prediction time is defined as $PT = 1/K$. Entropy values operate a classification within systems with different dynamic behavior. Theoretically, $K=0$ is for periodic signals and $K \to \infty$ for random signals. Finally K is assumed finite and > 0 for signals with chaotic dynamics.

The quantification of entropy value in a time series starts from the calculation of correlation integrals by the procedure proposed in (Grassberger et al,1985). If one evaluates $C_m(\varepsilon)$ and $C_{m+n}(\varepsilon)$ in the linear zone of the plot and for large m values, a good estimation of entropy K_2 (where $K_2 \leq K$) has the following expression:

$$K_2 = \frac{1}{n\tau} \ln \frac{C_m(\varepsilon)}{C_{m+n}(\varepsilon)}$$

n is the m increasing step and τ is the time delay.

A different definition of entropy will be introduced, in the following section, through the calculation of the entire spectrum of Lyapunov Exponents.

Lyapunov Exponents

Lyapunov exponents (LE) quantify the exponential rate of trajectories stretching. They represent a generalization of the eigenvalues in a regular attractor; for chaotic attractors

	L_1	L_2	L_3
Fixed point	(-)	(-)	(-)
Cycle	0	(-)	(-)
Torus	0	0	(-)
Strange attractor	(+)	0	(-)

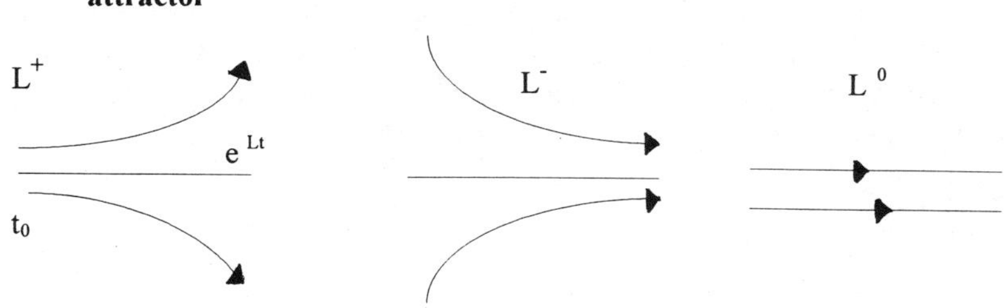

Figure 5. Lyapunov exponents in regular and chaotic attractors.

Table 1. Slope of the power-law spectrum of HRV signal in normal and pathological classes of patients (avg±std). The frequency range is $10e^{-04} - 10e^{-01}$

	Normal	Hypertensive	Severe Heart Failure	Transplanted
N	9	6	11	7
α $(1/f^a)$	1.12 ± 0.14	1.27 ± 0.11	1.29 ± 0.22	1.67 ± 0.29

only, there is at least one positive Lyapunov exponent as Fig. 5 points out. In this way, finding a positive exponent value means that the system is chaotic.

Starting from time series, we have two methods: i) calculation of the first positive exponent only (Wolf *et al.*,1985) or ii) calculation of the entire spectrum of LE (one exponent for each state-space dimension) (Eckmann *et al.*,1986). This second approach provides also the estimation of the entropy value. It is defined as the sum of positive exponent values, supposing that the stretching behavior does generate fractal structure of the attractor. Moreover, Kaplan and Yorke (Kaplan *et al.*,1974) conjectured that from LE's spectrum one could have also the information about the fractal dimension of the attractor. Lyapunov or Kaplan-Yorke (D_L or D_{KY}) dimension is defined as follows:

$$D_{KY} = j + \frac{L_1 + L_2 + \ldots + Lj}{\left| L_{j+1} \right|}$$

j is the progressive number of exponents obtained when adding the exponents themselves one at time (in decreasing value) until the sum remains positive. The $j+1$ exponent changes the sign of the sum (from positive to negative). The idea is that only the positive dynamical contributions, coming from the stretching of trajectories, are involved in the generation of the fractal structure of the attractor.

The conjecture provides a link between dynamic and geometric properties of the system attractor. The implementation of the algorithm used in this paper is from (Eckmann *et al.*, 1987).

Singular Value Decomposition

The representation of the attractor trajectories in the state-space is an useful tool to better evidence non linear dynamics. In particular, the singular value decomposition (SVD) approach is a very direct method to display the attractor structures (Broomhead *et al.*, 1985). It performs an extraction of the signal components which contain the information about its dynamics and separating them from the noise. This method is mainly applicable when one can suppose that periodic orbits of the attractor could be masked by a relevant noise contribution that fills some dimensions of the state-space. The procedure is based on a rotation of the trajectory matrix generated by an embedding procedure of the original time series in an m-dimensional state-space. The SVD operation generates a new diagonal matrix of trajectories where eigenvalues are ordered in decreasing values. Most important, eigenvalues identify the principal components acting in the generation of the nonlinear behavior of the system, whereas the smaller ones are a measurement of the noise superimposed to the signal.

RESULTS

The methods of non linear analysis introduced above, have been applied to HRV time series of normal and pathological subjects. In this way, we attempt at measuring nonlinear

Table 2. Parameters measuring characteristics of the system attractor

	Normal	Hypertensive	Severe Heart Failure	Transplanted
N	9	6	11	7
D_2	6.7 ± 1.4	5.8 ± 1.2	5.1 ± 1.8	4.5 ± 1.5
K_2	0.23 ± 0.04	0.21 ± 0.06	0.19 ± 0.07	0.13 ± 0.05
H	0.09 ± 0.05	0.12 ± 0.06	0.16 ± 0.11	0.34 ± 0.12
$D=1/H$	11.1 ± 4.3	8.3 ± 2.4	6.2 ± 3.5	2.9 ± 1.5

contributions in HRV signal regulation and assessing whether these parameters do evidence some differences between the normal subjects and the pathological ones.

As we have previously observed, a large amount of data is required in order to correctly estimate the invariant characteristics of the system attractor. For that reason it is useful to deal with long term recordings of ECG signals that can be easily obtained through an Holter system. By a traditional derivative and threshold algorithm, HRV series can be extracted through the recognition of R wave maxima in ECG signal. Series of RR intervals in the 24 h are constituted by about 100,000 values. When considering this signal we separated night and day periods to better preserve stationary characteristics of time series. The number of data points, for each epoch, is 20,000-30,000. Such numbers represent also a good compromise between the required large amount of data and the computational cost.

We considered HRV signals relative to 9 normal healthy subjects (N), 6 essential hypertensive (H), 11 severe heart failure (SHF) and 7 heart transplanted patients (T). All of the HRV time series extracted through the described procedure have been submitted to a test, performed by means of Fourier phases randomization, to verify the existence of determinism in system dynamics (Theiler, 1991) (Provenzale *et al.*, 1992). This is obtained by constructing surrogate data sets with the same amplitude (or power) characteristics of the original ones and where the phase relations are substituted by sequences of random numbers which should destroy the deterministic dynamics hidden in the signal itself.

The geometrical and dynamical properties of system attractor, from HRV time series, are properly described by a number of parameters: D_2, K_2, H and Lyapunov Exponents together with α slope of power-law spectrum identify the presence of fractal structure and chaotic dynamics in the cardiovascular regulation.

α slope is close to 1 when a healthy state characterizes the subject while it is different from 1 (α value generally increases) when a pathological condition is present. In particular, the PSD's of subjects N present a power-law distribution of frequency in the VLF band ($10e^{-4} - 10e^{-1}$ Hz). It seems that pathological subjects show a power distributions in this frequency band whose values of spectral α slope increase, and gradually the spectrum looses the power-law characteristic. This happens in hypertensive subjects and is more evidently in HF and T patients. Results are summarized in Table 1 as avg±std values This behavior may not be necessarily connected to a presence of chaotic dynamics in the signal generation mechanism but could reinforce the hypothesis that fractal characteristics of the time series might be the result of a chaotic system acting in cardiovascular regulation and control.

Effectively, quantitative parameters, giving information about attractor complexity, indicate a value of system attractor dimension which decreases when passing from normal to pathological subjects. Numerical results are summarized in Table 2 which reports, for each population, values of D_2, H, $D = 1/H$ and K_2. Values are reported as avg±std. Values of D_2 are computed for N, H, HF and T subjects on series of 20,000 values. Results in night and day epochs for each class of patients does not generally evidence significant differences.

A significant decrease is noticed among the different populations. As D_2 value approximately describes the number of variables that contribute in the generation of system dynamics, one can attribute this reduction to the action of pathological condition in the cardiovascular control. In the same way, other parameters follow the same trend of D_2 as well. K_2 parameter shows finite values, as expected for chaotic systems. Furthermore, $D=1/H$ parameter, starting from the measurement of different attractor characteristics, provides the same trend for the populations.

Finally, the computing of Lyapunov exponents provides for all the analyzed data at least one positive exponent, thus confirming, in a strong issue, that chaotic dynamics are

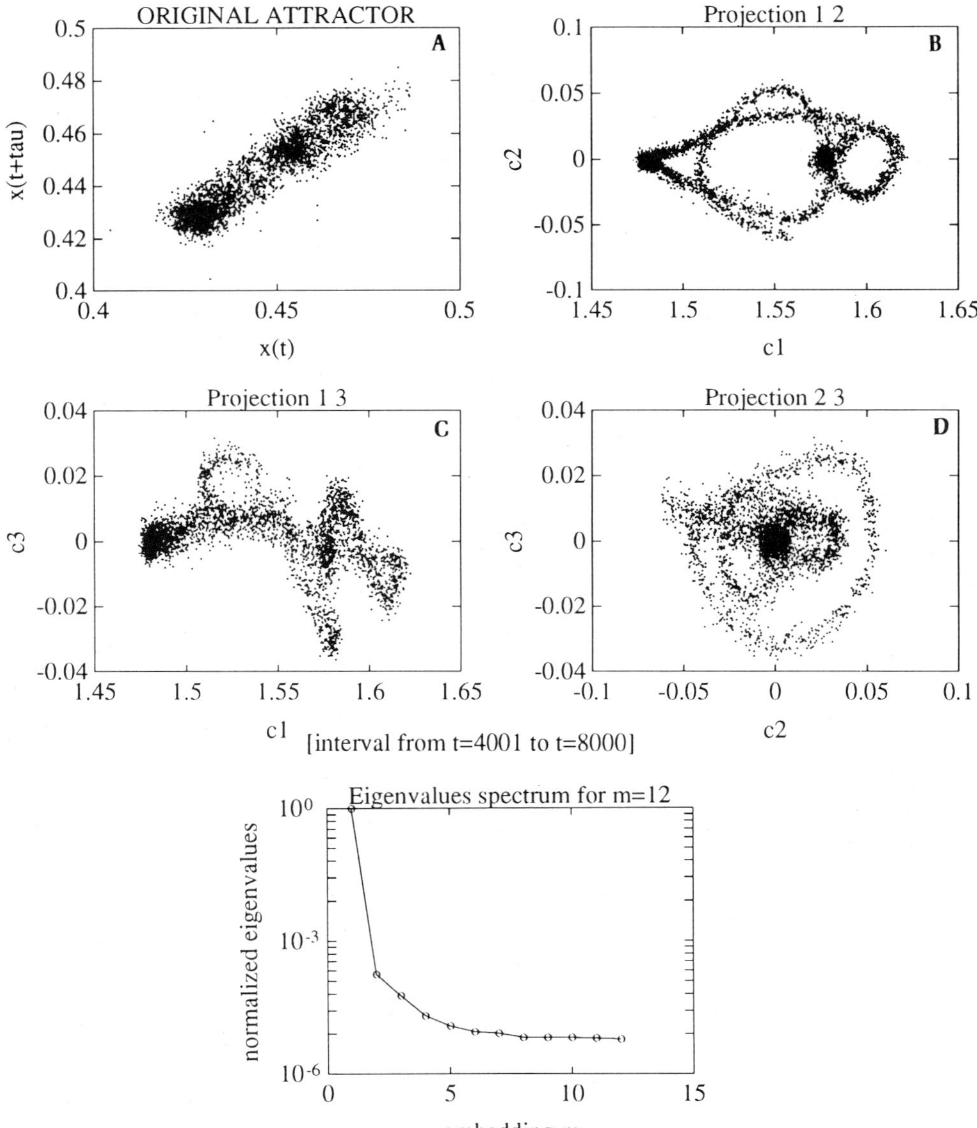

Figure 6. SVD of HRV series of a heart transplanted subject during night epoch. Plot (A) refers to the attractor representation without SVD procedure on trajectory matrix: no structures are evident. (B), (C) and (D) plots are the projections of the principal components of system trajectory obtained through the SVD procedure with the relevant eigenvalues.

Table 3. First three values of Lyapunov exponents of N, SHF, and T patients (avg±std). Low-pass FIR filtering performs a noise reduction without perturbation of nonlinear dynamics

Time Series Examined		N = 9	SHF = 11	T = 7
Lyapunov Exponents $[(sec^{-1})*10e^{-3}]$ after low-pass	Day interval	91.4 ± 16.2	61.3 ± 31.2	47.6 ± 33.9
		49.3 ± 4.6	36.2 ± 13.2	6.3 ± 1.7
		23.7 ± 3.8	-11.9 ± 14.9	-16.4 ± 20.4
filtering (0.1 Hz)	Night interval	85.4 ± 18.2	89.5 ± 18.2	50.5 ± 20.9
		40.6 ± 16.5	39.4 ± 13.6	19.3 ± 6.9
		21.1 ± 5.0	-7.3 ± 8.6	-3.2 ± 3.3

always involved in the signal generation mechanism. In our data we have often more than one positive exponent so indicating that the system possesses a higher degree of complexity (hyperchaos). Positive exponent number decreases from N to pathological classes of patients. It is proper to remark that LE's calculation may be critical due to the presence of noise and the finite precision of the algorithm employed. However, various simulations (Signorini, 1995) not shown in the paper, indicate a good correspondence to the results reported in literature.

In order to reduce the influence of the noise, HRV time series were filtered through a linear, low-pass, FIR filter. It does not change deterministic properties of the signal (Theiler et al, 1993). In this way high frequency components over 0.1 Hz are removed from the signal. Values of the first 3 Lyapunov Exponents, after FIR filtering, are then reported in Table 3, in day and night epochs.

A representation of the attractor characterizing HRV time series can be obtained using the SVD approach. In the example in Fig. 6, referring to a transplanted heart subject, the plot of the attractor in two dimensions, obtained by delaying of τ the original series, does not show any definite structure. When the SVD procedure is performed, the plots of the principal components of the trajectory matrix reveal a hidden and now discernible structure of data with close returns. The presence of the system attractor can be, in this way, pictorially shown, confirming the deterministic nature of the HRV signal control. On the other hand, when system complexity is higher as it happens for N subjects, it is not easy to represent the attractor in two-dimensional plots.

The SVD approach seems an useful tool even to identify the principal components generating the system dynamics and to perform a separation from the noise. In this context it is applied to increase the performance of algorithms for LE estimation or non linear noise reduction (Brown *et al.*, 1991) (Abarbanel *et al.*, 1993).

CONCLUSIONS

Results obtained on HRV time series in normal and pathological subjects, confirm the hypothesis, suggested by several observations, that a non linear model is involved in the generation and control of cardiovascular variability. Due to the difficulty in the analysis, the employment of different parameters is necessary for the reliability of results.

The presented method offers an interpretive tool that can be usefully integrated with other approaches like those ones based on linear models. In this way a better comprehension of the physio-pathological condition is obtained. Complexity parameters seem able to explain together the order and the variability of the rhythms interacting in cardiovascular system

REFERENCES

Abarbanel, H., Brown, R., Sidorowich, J. J., Tsimring, L., 1993, The analysis of observed chaotic data in physical systems. *Reviews of Modern Physics*, 65: 4.

Akselrod, S., Gordon, D., Ubel, F. A., Shannon, D. C., Barger, A. C., Cohen, R. J., 1981, Power spectrum analysis of heart rate fluctuations: A quantitative probe of beat-to-beat cardiovascular control, *Science* 213:220-222.

Baselli, G., Cerutti, S., Civardi, S., Liberati, D., Lombardi, F., Malliani, A., Pagani, M., 1986, Spectral and cross-spectral analysis of heart rate and blood varibility signals, *Comps. and Biomed. Res.* 19:520.

Bianchi, A., Bontempi, B., Cerutti, S., Gianoglio, P., Comi, G., and Natali Sora, M. G., 1990, Spectral analysis of heart rate variability signal and respiration in diabetic subjects, *Med. & Biol. Eng. & Comp.* 28:205-211.

Broomhead, D. S., and King, G. P., 1986, Extracting qualitative dynamics from experimental data, *Physica D* 20:217-236.

Brown, R., Bryan, P., and Abarbanel, H. D. I., 1991,Computing the Lyapunov spectrum of a dynamical system from an observed time series, *Phys. Rev.* A, 43,:2787-2805.

Cerutti, S., Bianchi, A., and Signorini, M. G., 1991, Spectral analysis of variability signals in cardiovascular system. In: Cerutti, S., and Minuco G. (eds.), *Spectral Analysis of Heart Rate Variability Signal: Methodological and Clinical Aspects*, La Goliardica Pavese: Pavia, pp. 31-43.

Cerutti, S., Bartoli, F, Baselli, G, 1985, AR identification and spectral estimate applied to the R-R interval measurements, *Int. J. Biomed. Comput.* 16:201-215.

Cerutti, S., Baselli, G., Bianchi, A., Mainardi, L., Signorini, M.G., and Malliani, A., 1994, Cardiovascular variability signals:from signal processing to modelling complex physiological interactions. *Automedica* 16:45-69.

Eckmann, J. P., Kamphorst, O. S., Ruelle, D., Ciliberto, S., 1986, Lyapunov exponents from time series, *Phys. Rev. A* 34:4971-4979.

Eckmann, J. P., Ruelle, D., 1985, Ergodic theory of chaos and strange attractors, *Rev. Mod. Phys.* 57:617-656.

Glass, L., Mackey, M. C., 1979, Pathological conditions resulting from instabilities in physiological control systems. *Ann. N.Y. Acad. Sci.*, 316:214-235.

Goldberger, A. L., Rigney, D. R., West, B. J., 1990, Chaos and fractals in human physiology, *Scientific Am.* 260:27-33.

Grassberger, P., Procaccia, I., 1984, Dimensions and entropies of strange attractors from a fluctuating dynamics approach, *Physica D* 13:34-54.

Grassberger, P., Procaccia, I, 1983, Measuring the strangeness of strange attractors, *Physica* 9D:189-208.

Hyndman, B.W., Kitney, R. I., Sayers, B. M. A., 1971, Spontaneous rhythms in physiological control systems *Nature* 233:339.

Kaplan, J. L., Yorke, J. A., 1979, Chaotic behaviour of multidimensional difference equations: *Lecture Notes in Mathematics*, vol. 730, (Springer, Berlin) pp. 228-287.

Keshner, M. S., 1982, 1/f noise, *IEEE Trans. Biomed. Eng.* 70:212-218.

Kleiger, R.E., Miller, J.P., Bigger, J.T., Moss, A.J., and the Multicenter Postinfarction Research Group, 1987, Decreased heart rate variability and its association with increased mortality after acute myocardial infarction, *Am. J. Cardiol.* 59:256-262.

Kobayashy, M., and Musha, T., 1982, 1/f fluctuations of heartbeat period, *IEEE Transaction* BME-29:456-464.

Lombardi, F., Sandrone, G., Perpuner, S., Sala, M, Garimoldi, M, Cerutti, S., Baselli, G., Pagani, M., and Malliani, A., 1987, Heart rate variability as an index of sympatho-vagal interaction after acute myocardial infarction, *Am. J. Cardiol.* 60:1239-1245.

Malliani, A., Pagani, M., Lombardi, F., and Cerutti, S., 1991, Cardiovascular neural regulation explored in the frequency domain, *Circulation* 84:482-1492.

Mandelbrot, B., 1983, *The Fractal Geometry of the Nature*, W.H. Freeman, New York.

Mortara, A., La Rovere, M.T., Signorini, M.G., Pantaleo, P., Pinna, G., Martinelli, L., Ceconi, C., Cerutti, S., and Tavazzi, L., 1994, Can power spectral analysis of heart rate variability identify a high risk subgroup of congestive heart failure patients with excessive sympathetic activation? A pilot study before and after heart transplantation, *Br. Heart J.* 71:422-430.

Osborne, A. R, and Provenzale, A., 1989, Finite correlation dimension for stochastic systems with power-law spectra, *Physica D* 35:357-381.

Pagani, M., Lombardi, F., Guzzetti, S., Rimoldi, O., Furlan, R., Pizzinelli, P., Sandrone,G., Malfatto,G., Dell'Orto,S., Piccaluga,E., Turiel,M., Baselli,G., Cerutti, S., and Malliani, A., 1986, Power spectral analysis of a beat-to-beat heart rate and blood pressure variability as a possible marker of sympatho vagal interaction in man and conscious dog, *Circ.Res* 59:178-193.

Pagani, M., Malfatto, G., Pierini, S., Casati, R., Masu, A. M., Poli, M., Guzzetti, S., Lombardi, F.,Cerutti, S., and Malliani, A., 1988, Spectral analysis of heart rate variability in the assessment of autonomic diabetic neuropathy, *J. Auton. Nerv. Syst.* 23:143-153.

Penaz, J., Roukenz, J., and Van der Waal, H. J., 1968, Spectral analysis of some spontaneous rhythms in the circulation, In: Crischel, H., and Tiedt, N. (eds.), *Biokybernetik*, Bd.I., Karl Marx Univ., Leipzig, pp. 233-241.

Provenzale, A., Smith, L.A., Vio, R., and Murante, G., 1992, Distinguishing between low dimensional dynamics and randomness in measured time series, *Physica D* 58:31-49.

Rompelman, O., Coenen, A. J. R. M., and Kitney, R. J., 1977, Measurement of heart rate variability: part 1. Comparative study of heart rate variability analysis methods, *Med. & Biol. Eng. & Comp.* 15:223.

Saul, J. P., Albrecht, P., Berger, R. D., and Cohen, R. J., 1988, Analysis of long term heart rate variability: methods, 1/f scaling and implications, *Proc. IEEE Computers in Cardiology. Conf.*, (IEEE Computer Society Press, Washington), pp. 419-422.

Sayers, B. Mc A., 1973, Analysis of heart rate variability, *Ergonomics* 16:17.

Signorini, M.G., 1995, Studio di Modelli di caos deterministico con applicazione al sistema cardiovascolare, *Ph.D Thesis*, Politecnico Univ. Milano, Italy.

Takens, F., 1981, Detecting strange attractors in turbulence, In: Rand, D. A., and Young, L. S. (eds.) *Dynamical Systems and Turbulence: Lecture Notes in Mathematics,* Springer, Berlin, vol. 898, pp. 366-381.

Theiler, J., and Eubank, S., 1993, Don't bleach chaotic data, *Chaos* 3:771-782.

Theiler, J., 1991, Some comments on the correlation dimension of noise, *Phys. Lett. A* 155:480-493.

West, B. J., Goldberger, A.L., Rovner, G., and Bhargava, V., 1985, Nonlinear dynamics of heartbeat (I), *Physica* 17D:198-206.

West, B. J., 1990, *Fractal Physiology and Chaos in Medicine*, World Scient. Publishing Co. Pte. Ltd.

Wolf. A., Swift, B. J., Swinney, H.L.,Vastano, J. A., 1985, Determining Lyapunov exponents from time series, *Physica* 16D:285-317.

Zetterberg, L. H., 1969, Estimation parameters for a linear difference equation with application to EEG analysis, *Math. Biosc.* 5:227-275.

18

ECG ARRYTHMIA ANALYSIS: DESIGN AND EVALUATION STRATEGIES

Roger G. Mark and George B. Moody

Massachusetts Institute of Technology
Harvard-MIT Division of Health Sciences and Technology
77 Massachusetts Avenue, Cambridge, Massachusetts 02139

ABSTRACT

Cardiac arrhythmias are a major cause of morbidity and mortality. Long-term ECG analysis is an important diagnostic technique for characterizing arrhythmias and documenting response to therapy. This paper reviews the technology of real-time automated ECG arrhythmia analysis, including principles of algorithm design, and the use of standard ECG databases in development and evaluation.

INTRODUCTION

Cardiac arrhythmias continue to be a major cause of morbidity and mortality, and have stimulated the investment of enormous effort from clinicians, biomedical engineers, physiologists and pharmacologists over the past three decades. Considerable progress has been made on all fronts, and yet the most devastating manifestation of cardiac arrhythmias - namely, sudden cardiac death, is still an unsolved problem. The inpatient management of acute myocardial infarction, with its associated increase in risk for cardiac arrhythmias, has improved dramatically over the past twenty-five years, largely as a result of the combined technologies of anti-arrhythmic pharmaceuticals and the development of the coronary care unit equipped with real-time cardiac arrythmia monitors. The problem of fatal arrhythmias in ambulatory patients has been considerably more difficult to solve, however. Perhaps the most vexing problem is that of identifying individuals who are at high risk for sudden cardiac death (SCD). The search for predictors has spurred the development of ambulatory ECG ("Holter") recorders, cardiac event recorders, anti-arrhythmic pacemakers, and automatic implantable cardioverter defibrillators (AICDs), and has motivated much of the research and development in automated arrhythmia analysis, and time series analysis of the ECG. Invasive electrophysiologic examinations and drug trials are used extensively in evaluating individuals known to be at particularly high risk for SCD based on major symptoms, including having survived one or more near-death episodes. Pharmacologic treatment has been generally

Advances in Processing and Pattern Analysis of Biological Signals, Edited by Isak Gath and Gideon F. Inbar
Plenum Press, New York, 1996

disappointing due to the high incidence of serious side effects and lack of convincing evidence of efficacy in preventing SCD. The most promising therapeutic intervention appears to be the AICD. This recently developed technology seems to have genuine promise in the treatment of individuals who are known to be at high risk for sudden cardiac death. However, the biggest unsolved problem facing cardiologists and biomedical engineers today is how to reliably detect individuals who are at high risk for SCD prior to their experiencing a life threatening event.

Arrhythmias are often transient, effervescent, and non-stationary. Their characterization requires the analysis of long periods of ECG data. Long term ambulatory ECG analysis is an important diagnostic modality for characterizing arrhythmias, and for documenting their response to therapy. Indeed, since some pharmaceutical interventions may be actually pro-arrhythmogenic, ECG monitoring is deemed essential as patients embark on certain new therapeutic regimens. Long term ECG data is also coming under intensive investigation through the use of time series analysis. The hope is to discover early warning signs which might identify patients at high risk for fatal arrhythmias.

Processing the enormous quantity of ECG data generated over hours or days requires automated methods. Indeed, at the present time, automated real-time arrythmia analysis plays a central role in both routine patient care (as in intensive care units), and in ambulatory patient monitoring. High quality arrythmia analysis software, capable of robust and stable QRS fiducial point assignment and accurate detection of ectopic beats is also essential for preparing ECG data for time series analysis.

This review will focus on the general topic of real-time arrythmia analysis using the ECG recorded from the body surface. Considerable recent attention has been devoted to intracardiac arrhythmia analysis which is of central importance in the technology of pacemakers and AICDs. However, the topic of arrhythmia analysis based on endocardial potentials is beyond the scope of this review.

DATABASES FOR ALGORITHM DEVELOPMENT AND EVALUATION

Carefully annotated collections of clinical ECG data are a critical resource for both the development and evaluation of automated arrythmia detectors. The diversity of waveform morphologies, underlying cardiac rhythms, and the presence of noise and artifact present continuing challenges to algorithm designers. Early researchers typically created their own collections of ECG data and used them in developing and evaluating their arrhythmia analysis programs. In the mid-1970's, however, it became apparent that the increasingly sophisticated designs for automated arrythmia detectors could be evaluated and compared objectively across laboratories only on the basis of their performance on standardized tests. During that period our group at MIT and Boston's Beth Israel Hospital developed the MIT-BIH arrythmia database (1), while nearly simultaneously K.L. Ripley, G.C. Oliver, and colleagues at Washington University and other cooperating institutions developed the American Heart Association Database for the Evaluation of Ventricular Arrhythmia Detectors (2). These databases are now widely available, and have contributed substantially to advancing the technology of arrhythmia analysis. Both databases consist of digitized excerpts from two-channel Holter ECG recordings, with beat-by beat QRS annotations provided by teams of human experts. The recording formats and beat labelling conventions are compatible.

The MIT-BIH arrythmia database was designed to provide representative samples from a wide variety of recordings which are observed in clinical practice . It consists of forty-eight 30-minute annotated ECG excerpts obtained from forty-seven subjects. About 60% of the records were obtained from inpatients. Twenty-three of the records were chosen at random from an extensive library of twenty-four hour Holter tapes, and the remaining twenty-five records were specifically selected to include a variety of rare but clinically important arrhythmias that would not be well represented by random sampling. Some records were chosen specifically because aspects of the rhythm, the QRS morphology, or the signal quality would be expected to present significant difficulty to arrythmia detectors. The MIT-BIH arrythmia database has considerable variability in signal quality with significant portions of noisy or unreadable data in at least one of the two leads. Since 1980, the MIT-BIH Arrhythmia Database has been distributed to more than two hundred sites world-wide; first on digital tapes, and since 1989 on a CD-ROM (3). The CD-ROM includes eight additional ECG databases including a collection of recordings of ventricular fibrillation assembled by Nolle and colleagues at Creighton University (4).

The AHA database was designed specifically to evaluate ventricular arrythmia detectors, and therefore emphasized ventricular ectopy. It is well structured, with particularly strong representation of clinically important arrhythmias, some of which occur only rarely. The signal quality is generally excellent, although a few recordings have small amounts of noise. The AHA database was developed in two parts, with eighty records released to the public and eighty kept "secret" for possible use by independent evaluation laboratories. Each record contains two leads of ECG, is 3 hours in duration, and has 30 minutes of annotated data. Only ventricular ectopic beats and rhythms are annotated. The AHA Database has been distributed since the mid-1980s on digital tapes, and in a shortened version (containing 35 minutes per record) on floppy disks.

More recently, an extensive annotated ECG database was released by the European Society of Cardiology to support the development and evaluation of ST-T analysis algorithms. The ESC ST-T Database contains 90 2-hour, 2-lead records containing annotations documenting individual beats, rhythm changes, noise episodes, and ST-T wave changes. Annotations and data formats are compatible with the MIT/BIH and AHA databases. This database is also distributed on CD-ROM (5).

The appendix lists distribution sources for the various databases. Table 1 lists a number of features of the MIT-BIH, AHA, and ESC ST-T Databases

Standardized databases have made it possible for investigators to compare quantitatively the performance of their algorithms with others in the field (6). The resulting competition has proven to be enormously beneficial in furthering the state-of-the-art over the past ten or fifteen years. Under the leadership of the Association for the Advancement of Medical Instrumentation (AAMI), standards have been recently promulgated which guide the proper use of the databases in evaluating arrythmia algorithms, and establish standards for reporting results (7).

The AAMI/ANSI standard for ambulatory electrocardiographic monitors specifies protocols for assessing monitor performance using these standardized databases, and requires that a variety of specific performance measurements be reported (8, 9). These protocols are designed to be strictly reproducible, so that fair comparisons of performance measurements can be made. In broad terms, conformance with the standard requires an evaluator to report detailed performance statistics for each record in the MIT-BIH and AHA Databases, and (depending on the claimed capabilities of the monitor) on the ESC ST-T Database and the CU Database. These performance statistics include the detector's sensitivity and positive predictivity for all QRS complexes, for ventricular ectopic beats (VEBs), for supraventricular ectopic beats (SVEBs), for couplets and runs of VEBs and of SVEBs, for ventricular and for atrial fibrillation and flutter, and for episodes of abnormal ST deviation. (Devices not designed to perform advanced functions, such as SVEB detection or ST analysis, are exempt from the reporting requirements aimed at assessing these functions.)

Table 1. Features of Standard ECG Databases.

	MIT–BIT Arrhythmia Database	AHA Database for Evaluation of Ventricular Arrhythmia Detectors	ESC ST–T Database
Available since	1980	1982	1991
# Records	48	80	90
Record duration	30 min	3 hrs (30 min ann.)	2 hrs
Leads	2	2	2
Resolution	360 Hz, 11 bits, 5µV	250 Hz, 12 bits, 5µV	250 Hz, 12 bits, 5µV
Anotations	110,000	170,000	800,000
Annotation types	QRS (16 types) Rhythm (15) Signal quality	QRS (5 types)	QRS (7 types) Signal quality ST,T changes

The reporting requirements also include results of noise stress testing (described below), a method for assessing the monitor's behavior as a function of input signal quality.

The performance of a detector is usually characterized in terms of four "raw" statistics (true positives, false positives, false negatives, and true negatives) and two normalized statistics (often sensitivity and specificity). Sensitivity is simply the percentage of positives that are detected (true positives), and specificity is the percentage of negatives that are rejected (true negatives). These statistics are appropriate for describing a discrete-event binary classification process such as a TB test. In contexts such as analysis of continuous signals, however, the number of true negatives (hence specificity) is indeterminate, and positive predictivity, the percentage of detections (both true and false positives) that are true positives, is used to characterize the degree to which a detector is prone to false detections.

In addition to the detailed statistics for each record in each of the relevant databases, the reporting requirements also include aggregate statistics of two types: "gross" statistics, in which the raw statistics (true positives, false positives, and false negatives) are summed over all records in a database to obtain the parameters required to calculate the derived statistics (sensitivity and positive predictivity); and "average" statistics, the means of the single-record sensitivities and positive predictivities. These aggregate statistics serve as a simplified basis of comparison among algorithms, although they tend to conceal record-to-record variability in performance. Depending on the intended clinical application, this variability may or may not be significant: in a life-support application, worst-case performance is far more important than in an off-line application such as interactive high-speed Holter analysis, for example, for which mean performance levels are more useful as figures of merit.

Although the AAMI committee charged with writing the ambulatory ECG standard considered including performance requirements in addition to these reporting requirements, doing so was deferred at least until the relationship between reported performance using the standard databases and "real world" performance in the clinical setting can be established. The databases present rigorous and reproducible challenges to a device under test. For this reason, they are highly valuable for exposing flaws in algorithm design, and for verifying that changes in algorithm design have the desired effects. All else being equal, an algorithm that performs well in a database test should be expected to perform better in the clinical setting than an algorithm that performs poorly in a database test. It is far from clear, however, what minimum levels of performance in a database test are sufficient to establish a reasonable expectation of satisfactory performance in the clinical setting (10). Such minimum levels may be possible to determine once the current generation of monitors has been tested according to the protocols specified by the standard.

Another approach to the problem of predicting real world performance is to use "bootstrap" estimation to obtain confidence limits for performance statistics (11, 12). The bootstrap is a procedure for obtaining an estimate of the distribution of a statistic by re-sampling the raw data from which the statistic is derived, randomly and with replacement (13). The statistic is recalculated from the re-sampled data set, and this procedure is repeated many times. The distribution of calculated statistics permits direct non-parametric estimation of confidence limits (thus, for example, the 95% confidence limits are given by the range that excludes the lowest 2.5% and the highest 2.5% of the calculated values). Use of the bootstrap requires only one (critical)assumption: that the original raw data are representative of the data for which the statistic is to be estimated. Since the databases have been selected in a way that virtually guarantees that their contents are not a random (i.e., representative) sample of the set of ECGs likely to be encountered in clinical use, it is still not possible to generalize performance estimates obtained using the bootstrap on database evaluations to "real-world" performance. Nevertheless, the bootstrap does permit us to determine if an observed difference in performance between two algorithms is significant.

For most well-designed arrhythmia analysis algorithms, noise is the major cause of errors. Noise can mask QRS complexes, or mimic them, and it can make accurate measurement of the QRS location (hence RR interval measurement) difficult or impossible. The effects of noise can thus be experienced at all levels of operation of an analysis algorithm, from QRS detection, timing classification, and characterization of morphology, to beat and rhythm labelling, ST analysis, and heart rate variability assessment. Although, as noted above, at least one of the standard databases includes records chosen specifically because of noise that might be expected to challenge an analysis algorithm, even these records do not exhibit the variety of noise characteristics, in the variety of ECG contexts, that may be expected to be encountered in the clinical setting. The noise stress test addresses this problem by providing a protocol for adding realistic noise in calibrated amounts to the standard annotated ECG recordings (14). The noise is obtained using standard ECG electrodes, cables, and Holter recorders,with the electrodes placed so that the ECG itself is not visible. By adding this noise to known ECG recordings at several signal-to-noise ratios,and measuring detector performance in each case, it is possible to determine how well a detector tolerates additive noise, such as electrode motion artifact (generally the most troublesome since it frequently mimics the ECG).

MAJOR SIGNAL PROCESSING TASKS

The objectives of real-time arrythmia analysis include: (1) accurate detection of all heart beats; (2) assignment of a physiologic label to each beat, based on its timing and morphology; and (3) characterization of the cardiac rhythm. To accomplish these objectives a number of specific signal processing tasks must be performed which are diagramed in Fig. 1. The details of implementation of the modules which perform these tasks vary considerably. In general, no optimal strategies are known. We review here the basic objectives of each signal processing module and provide a few examples of specific implementation strategies.

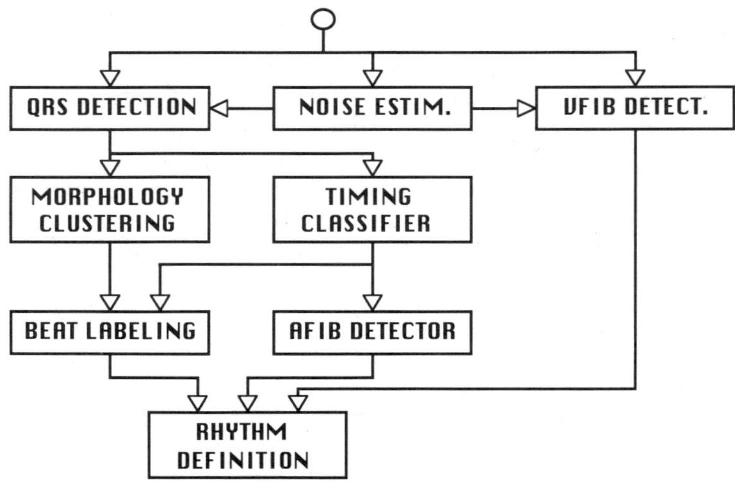

Figure 1. Major signal processing procedures in real-time arrhythmia analysis.

QRS Detection

Automatic arrythmia analyzers are data reduction systems, assigned with the task of deriving relatively small amounts of clinically crucial information from enormous volumes of input data. One or more ECG leads may be used for arrythmia analysis, and in fact most state-of-the-art detectors analyze two or more leads simultaneously. The ECG is amplified and filtered in the analog domain, with most modern systems performing additional filtering in the digital domain. Heart beats are detected by recognition of QRS complexes. (P-waves, because of their low amplitude, variable morphology, and variable relationship to QRS complexes are generally not reliably detectable, and are not processed in most real-time arrythmia analysis systems.) Accurate and robust QRS detection is a particularly critical part of any arrythmia analysis program and is the first stage of signal processing in which pattern recognition is important. Low QRS detector sensitivity will degrade system performance by increasing false negatives, and high false positive rates will unduly burden later algorithm modules which are responsible for artifact rejection.

The two generic steps involved in QRS detection are illustrated in Fig. 2. The ECG is first preprocessed to enhance the QRS complex while suppressing noise, artifact and non-QRS portions of the ECG. The output of this preprocessor stage is then presented to a decision rule which declares a QRS detection if the preprocessor output exceeds a threshold. The threshold may be fixed, but in general is adaptive to accommodate changes in signal amplitude which accompany shifts in lead axes, or changes in QRS morphology. Neural networks, which can be trained to selectively enhance QRS complexes, have also been used with reasonable success as QRS detectors. (15)

The decision threshold must be adjusted to achieve the best results for detector sensitivity and false positive rate. Figure 2 shows a hypothetical distribution of preprocessor output, V, for QRS complexes and for non-QRSs such as artifacts and noise. The position of the decision threshold defines the amplitude of V required in order to produce a detection. In general, some true events may be missed (false negative region) and some non-events may be falsely detected as QRS complexes (false positive region). As the decision threshold

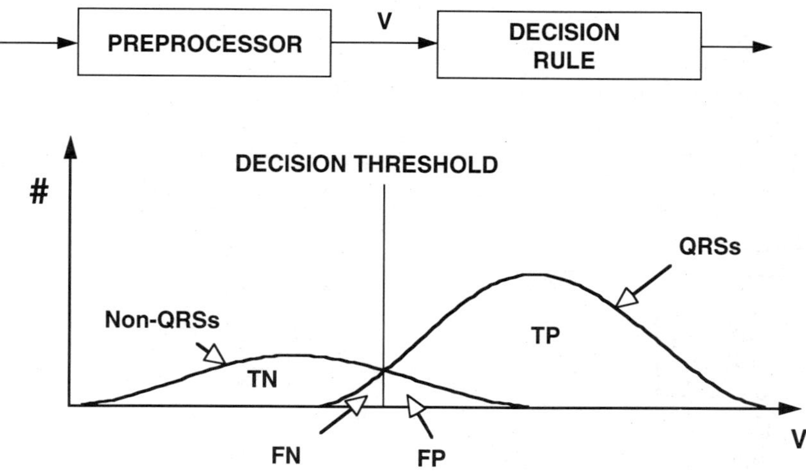

Figure 2. Generic event detection procedures. (From *Encyclopedia of Medica Devices and Instrumentation,* J.G. Webster, ed., copyright 1988 by John Wiley & Sons, Inc., Reprinted by permission of John Wiley & Sons Inc.)

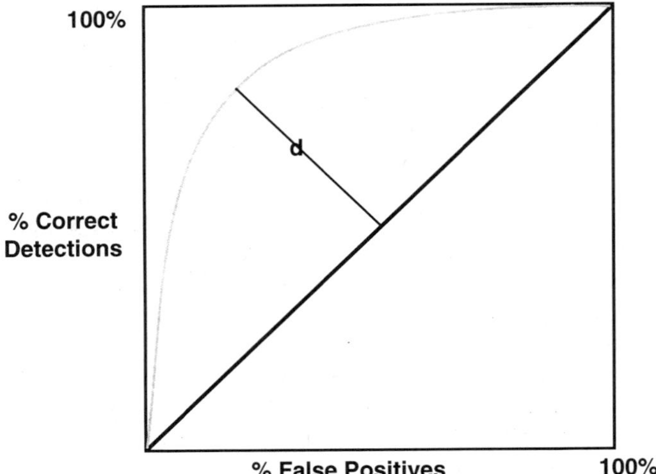

Figure 3. The receiver operating curve.

is moved, the false positive and false negative rates will vary, producing the "receiver operating curve" (ROC) shown in Fig. 3.

Algorithm designers must empirically adjust the decision threshold based on the intended application of their systems. The cost function which determines the optimal point on the ROC will vary considerably depending on system use. For example, a coronary care unit monitor would probably be tuned to produce a low false positive rate, while a human-assisted Holter tape scanner might be designed to emphasize sensitivity, while relying on the human operator to eliminate false positives.

Figure 4 illustrates the major processing steps which are employed by the QRS detector in ARISTOTLE (16), a multichannel arrhythmia analysis algorithm developed in our laboratory. Data from each ECG channel is passed through a digital matched filter (center

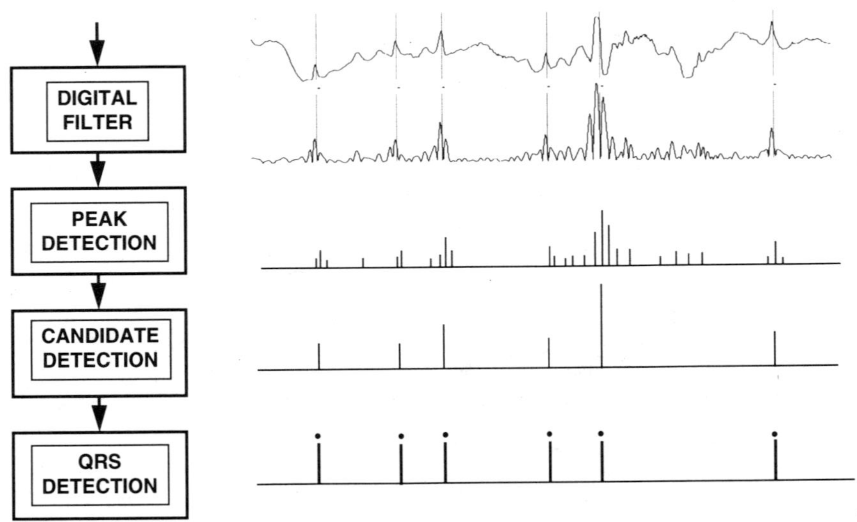

Figure 4. QRS detection strategy used in ARISTOTLE.

frequency 12 Hz, bandwidth 8 Hz.) The rectified output of the filter is examined to identify peaks of amplitude greater than 150 microvolts. Detected peaks from all analyzed ECG leads are then combined into a single data stream which is further processed.

The peaks are next examined by the module "candidate detector" which identifies most prominent peak in a sliding neighborhood of approximately 160 milliseconds. The resulting sequence of "candidates" is then subjected to an adaptive threshold procedure to detect QRS complexes. The QRS detector module in ARISTOTLE includes a "second pass" detection scheme in which a lower threshold is employed to reexamine data in regions suspected of containing a low amplitude event because of an unusually long RR interval.

The QRS detector module also assigns a fiducial mark to each detected QRS complex. Accurate assignment of fiducial marks is particularly important for systems in which QRS morphology is characterized by correlation or template matching, or is represented by orthonormal basis functions. Accurate fiducial assignment is also important in detecting slightly premature QRS complexes, and in providing robust temporal markers for time series analysis. The clinical definition for the onset of ventricular depolarization is the PQ junction, but this point cannot be located with confidence in the presence of noise, a task which is difficult even for careful human observers. The pragmatic approach of using a less ambiguous fiducial point is generally taken in automated analysis. Many techniques have been described in the literature, including identifying the point of maximum negative slope, the peak of the QRS complex, as well as the PQ junction. These are differential measures, however, and are quite sensitive to noise, to slight shape changes in the QRS complex and to sampling rates. Integral measures that take account of many samples, such as the peak of the output of a matched filter, or the "center of mass" of the QRS complex, are more robust measures for fiducial point assignment. ARISTOTLE uses the "center of mass" of the detected peaks from a given channel to assign the fiducial mark.

The ARISTOTLE QRS detector is designed to work with multiple ECG input leads. It is of key importance, however, that the multichannel QRS detector reject noisy channels in order to avoid excessive numbers of false positive detections. Accordingly, the "gain" of each input channel is a function of its estimated signal-to-noise ratio. The gain approaches 1 under conditions of high signal-to-noise ratio, and approaches 0 when signal quality is

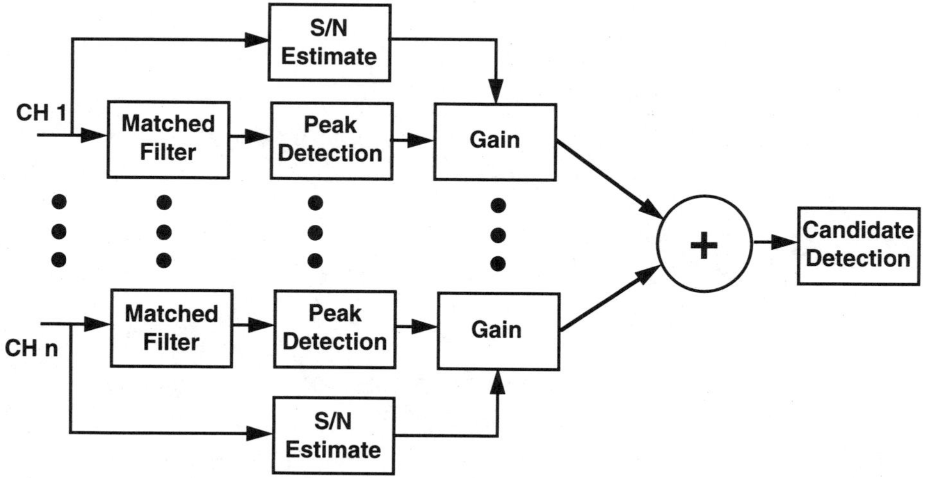

Figure 5. Multichannel QRS detection, showing individual channel gains controlled by S/N estimates. This technique reduces impact of noisy data.

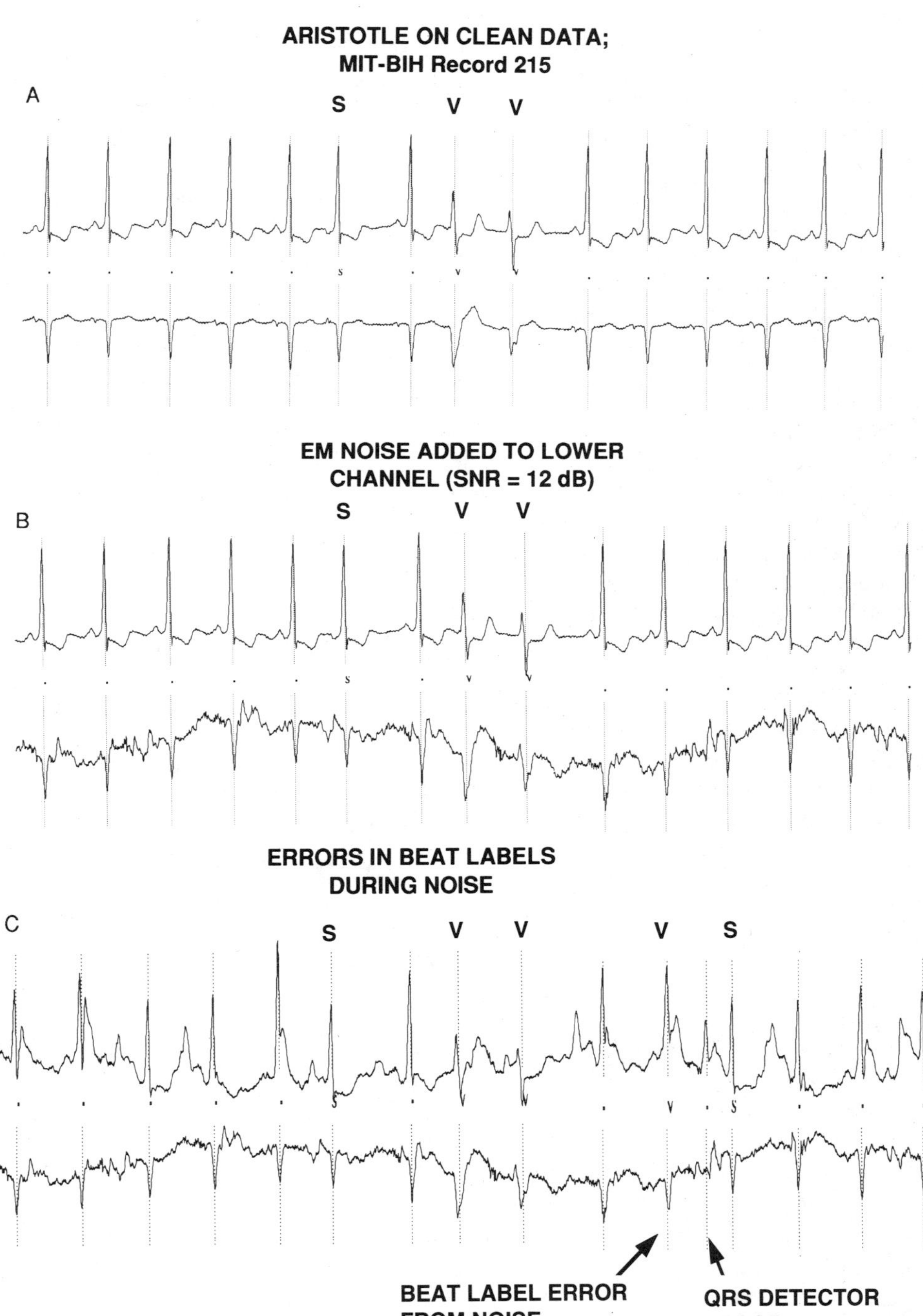

Figure 6. (A) ARISTOTLE's performance on clean data in both channels. (B) Performance with electrode motion artifact in one channel. (C) Performance with two noisy channels, showing both QRS detector error and two incorrect beat labels (one FP VPB, and one FP APB due to QRS detector error).

Table 2. Performance statistics: QRS detection (gross) of three algorithms using MIT/BIH and AHA databases. Algorithm I is ARISTOTLE. Algorithms II, III are state-of-the-art commercial systems

Algorithm	MIT/BIH		AHA	
	Sens.	PPA	Sens.	PPA
I (ARI)	99.71	99.77	99.90	99.77
II	99.65	99.89	99.73	99.82
III	99.64	99.77	99.40	99.70

severely degraded. (See Fig. 5) In the event that all leads experience severely degraded signal quality, ARISTOTLE QRS detection may cease.

Figure 6 demonstrates ARISTOTLE'S behavior in clean data, and in data with poor signal-to-noise ratio in one and in two channels. False positive QRS detection occurs only when artifact is present simultaneously in both channels.

The ARISTOTLE QRS detector has been thoroughly evaluated on both the MIT-BIH and the AHA databases. Performance statistics are shown in Table 2 and are compared with performance of two state-of-the-art commercial arrhythmia analysis algorithms. ARIS-TOTLE'S sensitivity of 99.71% implies that approximately thirteen beats per hour are missed. In addition, the positive predictive accuracy of 99.77% implies approximately ten false detections per hour. Notice that ARISTOTLE'S performance on the AHA database was slightly superior to that on the MIT-BIH database as expected given the superior overall data quality of the AHA database.

Timing Classification

In the presence of a reasonably regular cardiac rhythm, QRS complexes may be categorized as "on time," "premature," or "late." The observed RR interval is compared to an estimate of the expected RR interval. Many RR interval predictors have been discussed in the literature, and most use either a moving average or a first order digital filter. In either case, intervals that include abnormal beats are usually discarded or given low weight. Hence, RR interval estimation is generally dependent on beat morphology classification.

The moving average technique estimates the next RR interval as the average of the previous N intervals, where N may vary from 1-10.

The first order low pass digital filter estimates the next expected RR interval by updating the past estimate by a small fraction of the observed RR interval:

$$\hat{RR}_{n+1} = \alpha RR_n + (1-\alpha)\hat{RR}_n$$

where RR_{n+1} is the estimate of the next expected RR interval, RR_n and RR_n denote the observed and predicted values, respectively, of the present RR interval. As the parameter α becomes smaller, the effective averaging period increases.

In general, observed RR intervals are declared premature or "short" if they are less than some fraction (generally 0.85) of the predicted interval. Similarly an RR interval is declared "long" if it is greater than some fraction (often 1.15) of the predicted value.

Morphology Clustering

Detected QRS complexes must be subdivided on the basis of their morphology. (See Fig. 7.)

                                                      R. G. Mark and G. B. Moody

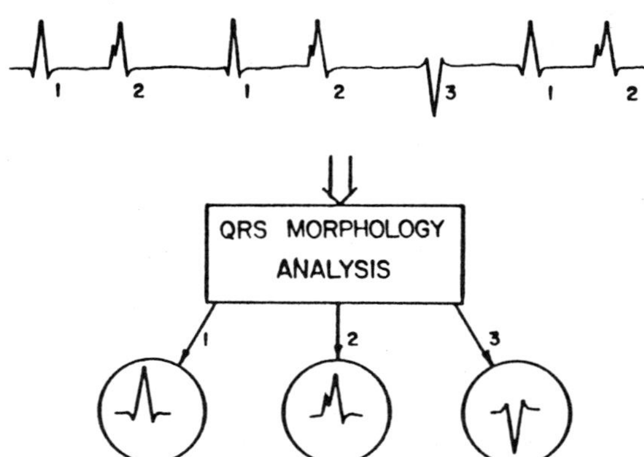

Figure 7. Morphology clustering of QRS complexes. (From *Encyclopedia of Medica Devices and Instrumentation*, J.G. Webster, ed., copyright 1988 by John Wiley & Sons, Inc., Reprinted by permission of John Wiley & Sons Inc.)

Many different techniques for analyzing QRS morphology have been described in the literature over the past several decades (17). All such algorithms may be described, however, in terms of a generalized process for QRS morphology analysis and clustering (Fig. 8).

The major steps are summarized as follows:

Data representation. The first step is to select a method by which to represent QRS shape using data from one or more ECG channels. Data representations that have been proposed in the literature include: (1) serial sample points which encompass the region of the QRS complex, (2) compressed sample points (every j^{th} sample, turning points, etc.), (3) heuristic features (QRS width, area, amplitude, etc.) or (4) formal features such as Fourier series, Karhunan-Loeve basis functions, etc. Each QRS complex is therefore represented by a set of numbers which may be considered to be the components of an n- dimensional "feature" vector. Typically, the vector elements will contain either QRS time sample values or derived QRS features. The ARISTOTLE algorithm developed in our laboratory utilizes the first five coefficients of the Karhunan-Loeve basis functions from each ECG lead. (Thus, for a two channel algorithm, each QRS feature vector has ten dimensions.) A simplified example of clustering using 2 Karhunan-Loeve coefficients from a single channel ECG is illustrated in Fig. 9.

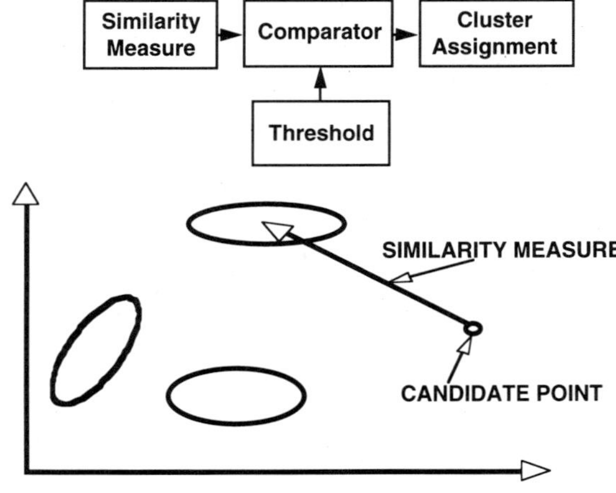

Figure 8. General procedure for QRS morphology analysis and clustering. (From *Encyclopedia of Medical Devices and Instrumentation*, J.G. Webster, ed., copyright 1988 by John Wiley & Sons, Inc., Reprinted by permission of John Wiley & Sons Inc.)

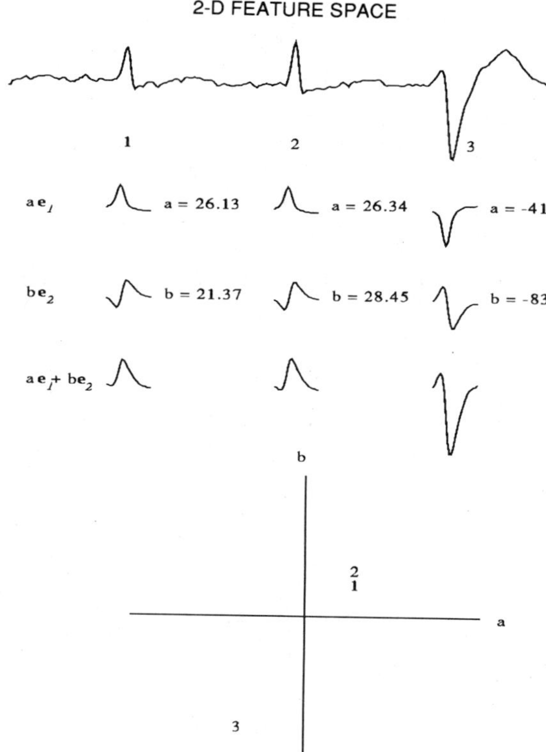

Figure 9. Representation of QRS waveforms by summation of KL basis functions. One channel of the original ECG is shown at the top of the figure. The first and second KL coefficients (a and b) and their contributions to the representation of the QRS complexes are shown in the center of the figure. In the lower part of the figure the three numbered QRS complexes are mapped into the a-b feature space, illustrating how the KLT preserves morphologic similarities and differences in preparation for clustering.

Similarity Measurement. Once QRS feature vectors have been defined, a clustering algorithm is used to group QRS complexes of similar shape. One of three types of similarity measures may be used to define the "distance" between the tips of two n-dimensional feature vectors. (See Fig. 10) The "city block" or L_1 distance is calculated by summing the component by component differences. The "straight line" or L_2 distance metric is calculated by measuring the magnitude of the vector difference between the tips of the feature vectors. A third method determines the n-dimensional cosine of the angle between the feature vectors (correlation coefficient).

Some form of normalization must be applied to the similarity measures in order to make them universal and independent of scaling factors. "Magnitude" normalizations divide the similarity metrics by the magnitude of the vector elements in one or both beats being compared. "Variance" normalization expresses similarity measures in terms of standard deviations of the vector elements.

Cluster Assignment. The normalized similarity metric is used to assign each candidate QRS complex to a particular morphologic cluster as shown in Fig. 8. The candidate is generally assigned to the nearest cluster unless the similarity metric is greater than some preset threshold. In that case, a new cluster is formed at the location of the candidate QRS. The threshold setting determines the closeness of fit required to match a candidate QRS complex to an existing cluster. Thus, the threshold setting will determine the number of clusters that are formed by an algorithm. In general, the management of a clustering routine requires that clusters are added and deleted dynamically as long term data analysis proceeds. Most algorithms for real-time arrhythmia analysis will maintain no more than ten to twenty

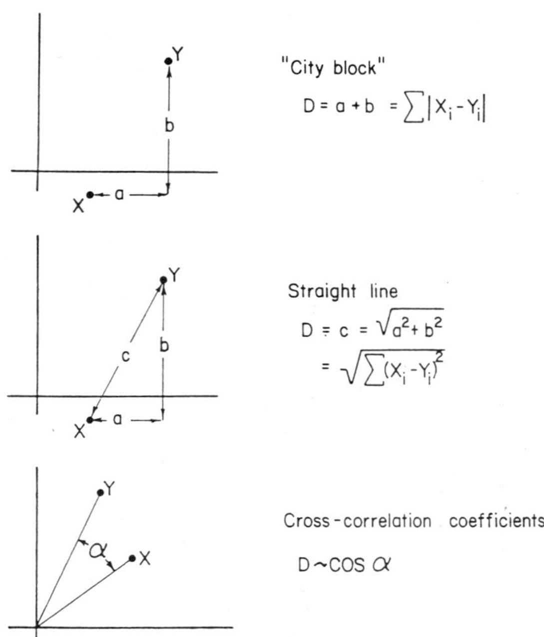

Figure 10. Three commonly used distance measures used in morphology clustering procedures.

clusters at any time in order to limit the amount of computation needed in assigning QRS candidates to clusters.

Beat Labelling

Having identified QRS complexes, determined their timing, and classified their morphology, the arrhythmia analysis algorithm must next attempt to assign a physiologic label to each QRS complex. The first task is to identify the "normal" QRS complex. This may be done using operator assistance, or automatically. If the latter strategy is used, it will usually select the most common QRS morphology occurring in a regular rhythm. (Clearly this approach may fail in situations such as continuous bigeminy, or when the frequency of VPBs exceeds that of normals.)

After the normal QRS cluster has been identified, other beats may be labelled using rules based on timing and morphology. Since P-waves generally are not analyzed, complete physiologic interpretation of the rhythm is impossible. Using QRS timing and morphology alone, the following beat classifications may be made: normal, supraventricular premature beat, VPB, ventricular fusion (on time, abnormal morphology), and ventricular escape (late, abnormal morphology).

The detailed logic which is employed to provide physiologic labels for QRS complexes may be quite intricate, and often includes subtle rules designed to minimize false positives. (e.g., premature beats that have been seen before or that have a compensatory pause are likely to be real VPBs, while premature events that have never been seen before are more likely to be false positives if they are interpolated.) Representative performance statistics for ARISTOTLE and two comparative commercial algorithms are shown in Table 3, which shows "gross" statistics on both MIT/BIH and AHA databases.

Table 3. Performance Statistics: VPB detection (Gross) of three algorithms using MIT/BIH and AHA Databases. Algorithm I is ARISTOTLE. Algorithms II, III are state-of-the-art commercial systems

Algorithm	MIT/BIH		AHA	
	Sens.	PPA	Sens.	PPA
I (ARI)	94.22	94.83	96.97	93.51
II	90.38	96.35	91.66	96.42
III	84.56	94.08	86.76	96.37

Artificial Neural Network Applications

QRS morphology classification may also be accomplished through the use of neural networks (18). Artificial neural networks (ANNs) are attractive for solving pattern recognition problems because they require very few assumptions about the underlying data. The development of successful ANN based pattern classifiers requires a well constructed, accurately characterized, and representative set of patterns for training; preprocessing and post processing algorithms; an appropriate network topology; and an evaluation database. We have had reasonable success using neural networks for detecting ventricular ectopic beats using two lead ECG data. The structure of the algorithm is shown in Fig. 11.

The input data to neural networks #1 and #2 consists of 16 normalized samples representing the QRS complex from each ECG lead. Each ANN is trained using prototypical VPBs, and normal QRSs chosen from the record under analysis. ANN output is near zero for normal beats and near one for VEB inputs. The third neural network is used to arbitrate discrepancies in decisions from the first two neural networks. ANN #3 receives information from each of the first two ANNs and also receives information concerning RR interval and noise level. Preliminary experiments in our laboratory indicate that VPB detection using neural networks is quite competitive with other techniques (See Table 4).

One of the major advantages of neural networks is the fact that their optimization as pattern classifiers is algorithmic and formalized. Rather good performance is achieved with a relatively short development time, as compared to the major and costly effort which typically must be invested in more traditional approaches.

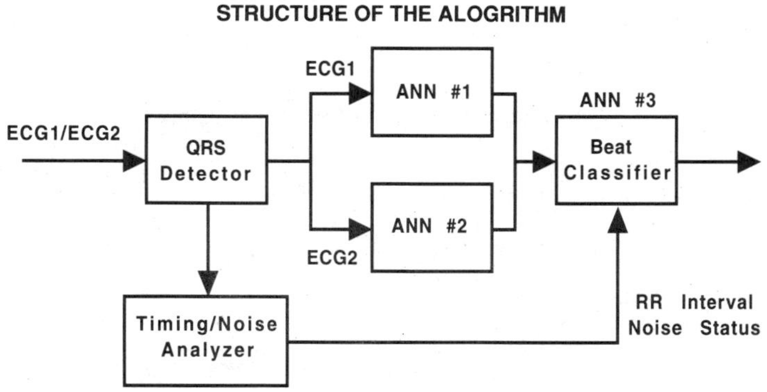

STRUCTURE OF THE ALOGRITHM

Figure 11. Structure of the ventricular ectopic beat detector using three ANNs.

Table 4. VPB detector performance on 69 AHA DB records. This table shows performance statistics of two conventional (ARISTOTLE and HSC) arrhythmia detectors relative to the neural network system)

Detector	Sens.	+Pred
Aristotle	98.6	94.2
HSC	94.8	92.7
ANN 2	96.9	92.3
ANN 3	97.4	93.6

Rhythm Definition

The final stage in arrhythmia analysis is rhythm definition based on sequences of QRS complexes. In the absence of P-wave detection, identifiable rhythms are restricted to those which can be defined by QRS type and timing only. Special strategies are generally employed to identify atrial fibrillation or ventricular fibrillation.

Atrial Fibrillation Detection

The timing sequence of QRS complexes may be used to detect abnormal rhythms such as atrial fibrillation. The detection of atrial fibrillation is important not only because of its clinical significance but also because of the need to modify algorithm rules for ventricular arrhythmia detection. In the presence of atrial fibrillation, for example, the concept of "prematurity" is no longer valid, and decision rules for identifying VPBs on the basis of shape and timing need to be changed.

One approach to detecting atrial fibrillation makes use of a Markov model of three states (19, 20). Each RR interval is classified as short, regular, or long (S, R, L). RR intervals bounded by VPBs are ignored. The transition probabilities between states can be determined by using a suitable ECG database containing many episodes of atrial fibrillation together with other rhythms (see Fig. 12).

The Markov model atrial fibrillation detector is applied to test data as follows: RR intervals are observed and classified as short, regular, or long as described previously. Next, the conditional probabilities of observing the test sequence are calculated using transition matrices for atrial fibrillation and for "other rhythms." The more probable rhythm is that for which the conditional probability is highest. Detectors based on Markov transition probabilities have been reasonably successful in detecting atrial fibrillation with reported sensitivities of over 90% and positive predictive accuracies over 85% for detecting atrial fibrillation beats (20).

Markov Afib Detector RR Interval Matrices

ATRIAL FIBRILLATION

FROM \ TO	S	R	L
S	.32	.44	.24
R	.18	.62	.21
L	.17	.49	.33

OTHER RHYTHMS

FROM \ TO	S	R	L
S	.25	.21	.54
R	.03	.95	.02
L	.25	.33	.42

Figure 12. Transition probability matrices used in Markov model detector of atrial fibrillation.

Table 5. Performance of several atrial fibrillation detectors in detecting heart beats within atrial fibrillation. Markov detectors are compared to artificial neural network (ANN) detectors

Detector	Sens.	+Pred.	FP
Markov	99.6	66.0	21.3
Mod. Markov	93.6	85.9	6.4
Pred. Array	75.8	91.9	2.8
ANN	92.9	92.3	3.0

Atrial fibrillation detection has also been accomplished using artificial neural networks (ANNs) (21). Sliding windows of 30 sequential RR intervals were characterized by 3 x 3 interval transition matrices in a manner analogous to that used in the Markov model technique. The statistical patterns represented by the transition matrices were presented to a neural network (9 input neurons, 12 nodes in a single hidden layer, and 1 output neuron) for pattern classification. The neural network was trained on a set of arrhythmias which included normal sinus rhythm (NSR), NSR with ventricular bigeminy, NSR with atrial tachycardia, NSR with frequent atrial premature beats, NSR with ventricular premature beats, and atrial fibrillation. The training data was extracted from records in the MIT-BIH Arrhythmia Database. The neural network approach was found to function rather well, and its performance compared favorably to atrial fibrillation detectors based on the Markov model (see Table 5).

DEALING WITH NOISE AND ARTIFACT

Despite the analog and digital filtering typically performed on the input ECG signal, some unwanted noise and artifact will remain since baseline wander, motion artifact, and muscle noise all have some energy that overlaps the ECG signal spectrum (see Fig. 13). Of particular concern is electrode motion artifact, since its frequency components overlap heavily with the spectrum of QRS complexes.

Automated ECG analysis algorithms generally perform quite well in clean data, but degrade substantially in the presence of noise and artifacts. It is therefore essential to deal effectively with the problem of noise and artifact in order to produce an arrhythmia analysis system which will have practical use in the real world. One approach to the problem of noise is to detect it, estimate a signal-to-noise ratio, and simply suspend arrhythmia analysis when the signal quality becomes excessively degraded. There is an obvious disadvantage to this "eye closing" approach, since significant arrhythmias may be missed during "shut down," and there may be a positive correlation between physical activity and clinically significant arrhythmias.

Multi-lead ECG processing has the potential to improve algorithm performance provided that at least one channel retains a favorable signal-to-noise ratio. Such systems, however, must estimate the signal-to-noise ratio on each ECG lead independently, and specifically attenuate the noisy channels prior to further processing of the data (see Fig. 5).

Signal-to-noise ratio is used elsewhere in arrhythmia analysis algorithms to adjust thresholds, to modify morphology clustering routines, and to modify alarm strategies. For example, in assigning a QRS complex to a particular morphology cluster, it would be advantageous to know whether that QRS complex is likely to be contaminated by noise. An approach which we have found useful in ARISTOTLE is to estimate noise content in the QRS complex by tracking the residual energy remaining after fitting the QRS complex with

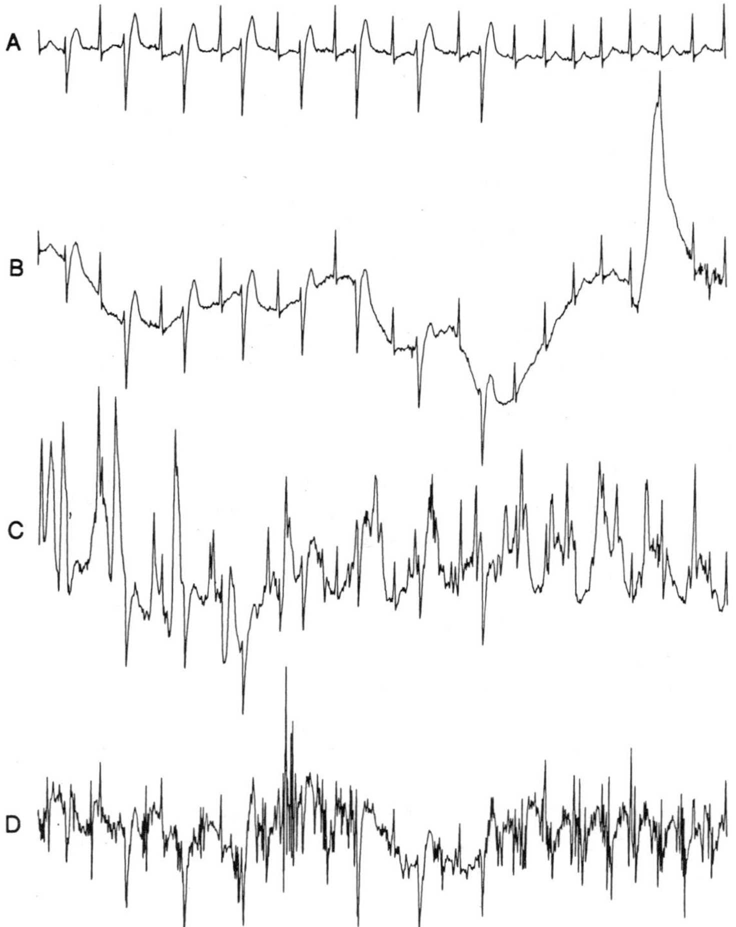

Figure 13. (A) Clean ECG signal; (B) Added baseline wander; (C) Added electrode motion artifact; (D) Added muscle noise. (From *Encyclopedia of Medical Devices and Instrumentation*, J.G. Webster, ed., copyright 1988 by John Wiley & Sons, Inc., Reprinted by permission of John Wiley & Sons Inc.)

the first five KL basis functions. These residual errors reflect components of the signal which are unlikely to be features of the ECG (22).

Since electrode motion artifact is particularly troublesome to arrhythmia detectors, it is quite helpful to obtain independent indicators of the probability of electrode motion. One approach, originally suggested by Feldman (23) makes use of two sets of electrodes which are placed in essentially the same recording positions, and therefore record the same shape QRS complex. (In fact, special concentric electrodes are sometimes used to insure precise alignment of the recording lead axes; see Fig. 14.) Any difference in potential between the two leads must be due to motion artifact one or both of the electrode sites. This "smart electrode" technique has been shown to substantially reduce false alarms in clinical monitoring settings.

A second approach to making independent assessments of electrode movement relies on the monitoring of the electrode-skin impedance (24, 25). Motion of the electrode with respect to the skin causes changes in the electrode-skin impedance as well as producing the

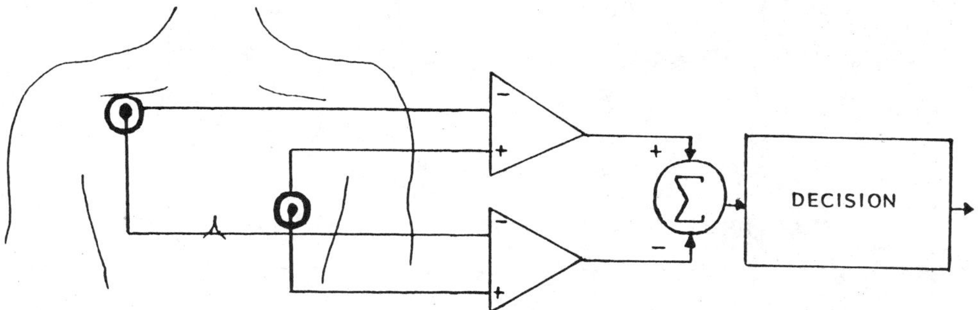

Figure 14. Detection of electrode motion artifact using Feldman's "smart electrode" concept. (From *Encyclopedia of Medical Devices and Instrumentation*, J.G. Webster, ed., copyright 1988 by John Wiley & Sons, Inc., Reprinted by permission of John Wiley & Sons Inc.)

electrical artifact (Fig. 15). The impedance signal therefore provides an independent measure of electrode motion and can be used to inform later arrhythmia analysis stages that motion artifact was present in the signal, and that routine signal processing may be suspect.

In the presence of extreme noise and artifact the performance of automated arrhythmia analysis systems deteriorates drastically. Under such conditions humans generally perform at a substantially higher level, primarily for two reasons. First, the human visual system can clearly identify beats using tiny, distinctive portions of non-corrupted QRS complexes. Second, during periods of clean data the human expert learns the patient's predominant heart rhythm, interbeat intervals, characteristic beat morphologies, and commonly observed abnormal beat sequences. During noisy data segments human experts make use of this knowledge to identify QRS complexes and differentiate them from noise and artifact. QRS complexes that are clearly recognizable in the noise are identified as "landmark" beats. Using these beats as points of reference, hypothetical beat sequences are then proposed and tested against the noisy data. If no reasonable hypothesis fits the data, human experts will generally skip ahead and restart their search procedures at a later landmark beat.

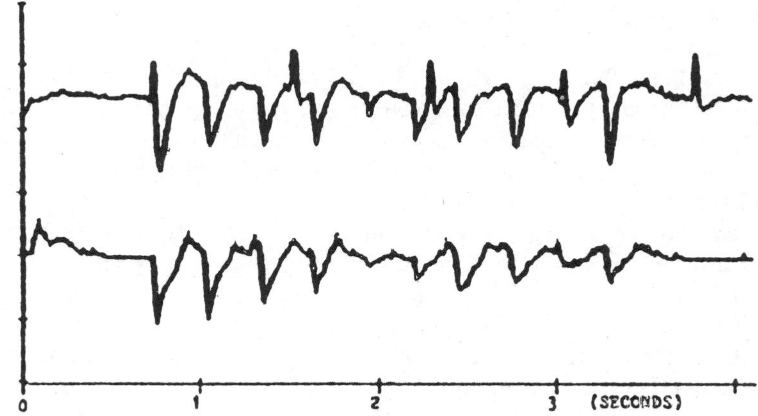

Figure 15. Top trace shows ECG contaminated with electrode motion artifact. The second trace shows the electrode-skin impedance signal (measured at 20 Khz).

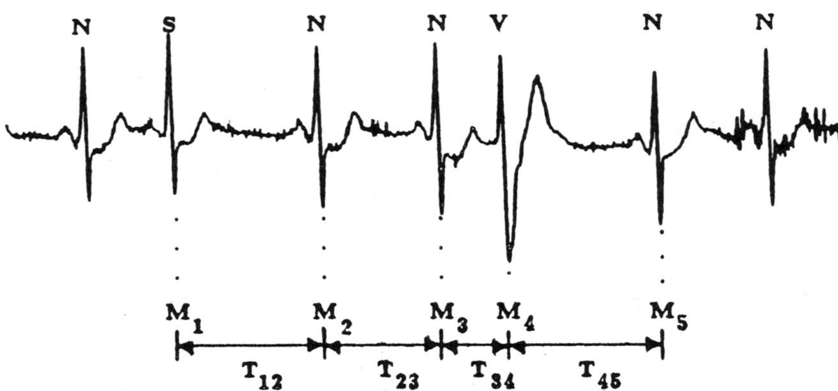

Each 5-beat sequence is mapped to a 9-dimensional feature vector.

$$V = \left[M_1, T_{12}, M_2, T_{23}, M_3, T_{34}, M_4, T_{45}, M_5 \right]$$

Figure 16. Contextual information is represented by HOBBES by describing five-beat sequences by nine morphology and timing features.

A similar approach may be automated by using intelligent or "expert" systems. One such program developed in our laboratory (HOBBES) [HOBBES = **H**ypothetical **OB**literated **BE**at **S**equences] was demonstrated to substantially enhance ARISTOTLE's performance in the presence of severe noise (26, 27, 28). HOBBES was designed to post-process the annotation stream created by ARISTOTLE and it does so in three distinct passes. In the first pass HOBBES learns from clean ECG data. The program identifies regions of noise-free data and analyzes the annotation stream to build a knowledge base. An estimate of heart rate is established and continually updated. HOBBES verifies whether the coupling intervals for isolated VPBs are constant and independent of heart rate, and establishes whether ventricular premature beats (VPBs) are followed by fully compensatory pauses. HOBBES also characterizes all five-beat contexts which are observed in clean data. Each five-beat sequence is represented by a nine-dimensional vector that includes information concerning the four interbeat intervals and five QRS morphologies (Fig. 16). The resultant clusters in nine-dimensional context "space" characterize the ECG.

In the second pass HOBBES attempts to find landmark beat sequences of normal beats in noisy data. In the third pass, HOBBES uses its knowledge base to generate and test hypothetical beat sequences to complete the analysis of noisy data. The previously identified "landmark" beats constrain the search. The hypothesis which best fits the observed data is adopted, and the original ARISTOTLE beat labels are modified as required. The technique generally results in discarding a number of ARISTOTLE beat labels as likely artifacts, thus improving the detector's positive predictive accuracy, while slightly decreasing sensitivity.

A test was developed to severely stress the ability of the expert system to discriminate between beats and artifact in noisy ECGs. Thirty-five half-hour ECG records were selected from the MIT/BIH and AHA databases, or from our extensive Holter library. The data contained a wide variety of types and levels of ectopic activity in the context of underlying normal sinus rhythm. For each record, the detector was allowed to "learn" on 16.5 minutes of clean data. Subsequently, severe electrode motion noise was added to the ECG data for one minute. Thereafter, three minutes of noise-free data was cycled with one minute of noisy data to give a total of 3 minutes of noise per record. Algorithm performance was evaluated

Table 6. Performance of several algorithms during noisy data

Arr. det.	QRS		N		VPB		SVPB	
	Sens	Pred	Sens	Pred	Sens	Pred	Sens	Pred
A	96.46	70.79	67.65	73.41	73.37	16.10	13.33	3.25
A/H	89.31	87.51	85.64	86.89	56.78	43.90	20.46	17.52
T/H	89.60	89.28	84.41	90.49	67.94	48.03	48.28	30.30
Total	8729		7667		627		435	

in only the noisy segments. (The test included 8,729 QRS complexes: 7,667 normals, 627 VPBs, and 435 SVPBs.) Table 6 shows performance results for ARISTOTLE (A), HOBBES as a post-processor to ARISTOTLE (A/H), and HOBBES as a post-processor to "truth" annotations (T/H).

The expert system improved ARISTOTLE's QRS and VPB predictive accuracy, while reducing sensitivity, because some true beats were rejected along with the majority of artifact. (HOBBES relabeled many noisy beats thereby enhancing the predictive accuracy of all events at the expense of mislabelling some VPB's as normals or SVPBs.)

The work suggests that expert systems are able to significantly enhance the performance of automated arrhythmia analyzers in very severe noise. It is likely, furthermore, that contextual expert reasoning would also enhance algorithm performance in clean data.

SUMMARY

Multi-lead real-time ECG arrhythmia analysis has achieved impressive progress over the past several decades, aided in no small measure by the remarkable advances in computational hardware which has permitted designers to employ sophisticated and complex software. The major remaining challenges are to develop improved P-wave detectors, and to continue to improve performance in the presence of noisy data. Available standardized databases are a major resource for both algorithm development and evaluation. Extrapolations from database trials to prediction of real world performance is difficult, but the "noise stress test" and bootstrap statistics may be helpful. Preliminary evidence suggests that artificial neural networks and expert systems are likely to be useful new techniques in further improving the technology of arrhythmia analysis.

REFERENCES

1. Mark, R.G., Schluter, P.S., Moody, G.B., Devlin, P.H., Chernoff, D., 1982, An annotated ECG database for evaluating arrhythmia detectors, pp. 205-210, in *Frontiers of Engineering in Health Care, Proc. 4th Ann. Conf. IEEE Engineering in Medicine and Biology Soc.*, IEEE Computer Society Press, Los Alamitos, CA.
2. Hermes, R.E., Geselowitz, D.B., Oliver, G.C., 1980, Development, distribution, and use of the American Heart Association database for ventricular arrhythmia detector evaluation, pp. 263-266, in *Computers in Cardiology 1980*, IEEE Computer Society Press, Los Alamitos, CA.
3. Moody, G.B., Mark, R.G., 1990, The MIT-BIH Arrhythmia Database on CD ROM and software for use with it, 1991, pp. 185-188, in *Computers in Cardiology 1990*, IEEE Computer Society Press, Los Alamitos, CA.
4. Nolle, F.M., Badura, F.K., Catlett, J.M., et al., 1986, CREI-GARD: a new concept in computerized arrhythmia monitoring systems, pp. 515-518, in *Computers in Cardiology 1986*, IEEE Computer Society Press, Los Alamitos, CA.

5. Taddei, A., Biagini, A., Distante, G., et al., 1990, The European ST-T Database: development, distribution, and use, pp. 177-180, in *Computers in Cardiology 1990*, IEEE Computer Society Press, Los Alamitos, CA.

6. Schluter, P.S., Mark, R.G., Moody, G.B., Olson, W.H., Peterson, S.K., 1980, Performance measures for arrhythmia detectors, pp. 267-270, in *Computers in Cardiology 1980*, IEEE Computer Society Press, Los Alamitos, CA.

7. [Arrhythmia Monitoring Subcommittee of the AAMI ECG Committee]. Testing and Reporting Performance Results of Ventricular Arrhythmia Detection Algorithms [AAMI ECAR]. Association for the Advancement of Medical Instrumentation, Arlington, VA, 1987.

8. Moody, G.B., Feldman, C.L., Bailey, J.J., 1993, Standards and applicable databases for long-term ECG monitoring, *J. Electrocardiology* 26 (Suppl): 151-155.

9. [Ambulatory ECG Subcommittee of the AAMI ECG Committee]. American National Standard: Ambulatory Electrocardiographs [ANSI/AAMI EC-38] Association for the Advancement of Medical Instrumentation, Arlington, VA, 1994.

10. Moody, G.B., Mark, R.G., 1983, How can we predict real-world performance of an arrhythmia detector?, pp. 71-76, in *Computers in Cardiology 1983*, IEEE Computer Society Press, Los Alamitos, CA.

11. Greenwald, S.D., Albrecht, P., Moody, G.B., Mark, R.G., 1985, Estimating confidence limits for arrhythmia detector performance, pp. 383-386, in *Computers in Cardiology 1985*, IEEE Computer Society Press, Los Alamitos, CA.

12. Albrecht, P., Moody, G.B., Mark, R.G., 1988, Use of the 'bootstrap' to assess the robustness of the performance statistics of an arrhythmia detector, *J. Ambulatory Monitoring* 1(2): 171-176.

13. Efron, B., 1979, Bootstrap methods: another look at the jackknife, *Annals of Statistics*, 7: 1-26.

14. Moody, G.B., Muldrow, W.K., Mark, R.G., 1984, A noise stress test for arrhythmia detectors, pp. 381-384, in *Computers in Cardiology 1984*, IEEE Computer Society Press, Los Alamitos, CA.

15. Hu, Y.H., Tompkins, W.J., Urrusti, M.S., and Afonso, V.X., 1993, Applications of artificial neural networks for ECG signal detection and classification, *J. Electrocardiology*, 26 (suppl): 66 - 73.

16. Moody, G.B., and Mark, R.G., 1982, Development and evaluation of a two-lead ECG analysis program, pp. 39 - 44, in *Computers in Cardiology 1982*, IEEE Computer Society Press, Los Alamitos, CA.

17. Rappaport, S.H., Gillick, L., Moody, G.B., and Mark, R.G., 1982, QRS morphology classification: quantitative evaluation of different strategies, pp. 33-38, in *Computers in Cardiology 1982*, IEEE Computer Society Press, Los Alamitos, CA.

18. Chow, H.S., Moody, G.B., and Mark, R.G., 1992, Detection of ventricular ectopic beats using neural networks, pp. 659-662, in *Computers in Cardiology 1992*, IEEE Computer Society Press, Los Alamitos, CA.

19. Gersh, W., Eddy, P., and Dong, E., 1970, Cardiac arrhythmia classification: a heart-beat interval Markov chain approach, *Comp. Biomed. Res.* 4: 385 - 392.

20. Moody, G.B., and Mark, R.G., 1983, A new method for detecting atrial fibrillation using RR intervals, pp. 227 - 230, in *Computers in Cardiology 1983*, IEEE Computer Society Press, Los Alamitos, CA.

21. Artis, S.G., Mark, R.G., and Moody, G.B., 1991, Detection of atrial fibrillation using artificial neural networks, pp. 173 - 176, in *Computers in Cardiology 1991*, IEEE Computer Society Press, Los Alamitos, CA.

22. Moody, G.B., and Mark, R.G., 1989, QRS morphology representation and noise estimation using the Karhunan-Loeve transform, pp 269 - 272, in *Computers in Cardiology 1989*, IEEE Computer Society Press, Los Alamitos, CA.

23. Feldman, C.L., and Hubelbank, M., 1978, Apparatus and method for ECG baseline shift detecting, U.S. Patent 41,112,930.

24. Devlin, P.H., Mark, R.G., and Moody, G.B., 1984, Detecting electrode motion noise in ECG signals by monitoring electrode impedance, pp. 51 - 56, in *Computers in Cardiology 1984*, IEEE Computer Society Press, Los Alamitos, CA.

25. Ferguson, P.F., and Mark, R.G., 1986, An algorithm to reduce false positive alarms in arrhythmia detectors using dynamic electrode impedance monitoring, pp. 511-514, in *Computers in Cardiology 1986*, IEEE Computer Society Press, Los Alamitos, CA.

26. Muldrow, W.K., Mark, R.G., Long, W.J., and Moody, G.B., 1986, CALVIN: A rule-based expert system for improving arrhythmia detector performance during noisy ECGs, pp. 21 - 26, in *Computers in Cardiology 1986*, IEEE Computer Society Press, Los Alamitos, CA.

27. Greenwald, S.D., Patil, R.S., and Mark, R.G., 1992, Improved detection and classification of arrhythmias in noise-corrupted electrocardiograms using contextual information within an expert system, Biomed. Instrum. Technol. 26 (2): 124 - 132.

28. Greenwald, S.D., 1990, Improved Detection and Classification of Arrhythmias in Noise-Corrupted Electrocardiograms using Contextual Information. Doctoral Thesis, Massachusetts Institute of Technology, June, 1990.

19

FUNDAMENTAL ANALYSES OF VENTRICULAR FIBRILLATION SIGNALS BY PARAMETRIC, NONPARAMETRIC, AND DYNAMICAL METHODS

Nitish V. Thakor,[1] Ahmet Baykal,[1] and Aldo Casaleggio[2]

[1] The Johns Hopkins University
Biomedical Engineering Departmant
720 Rutland Ave., Baltimore, Maryland 21205
[2] ICE-CNE
Genova, Italy

ABSTRACT

Ventricular fibrillation (VF) is the malignant electrical rhythm of the heart. Fundamental understanding of this rhythm can only be obtained by considering signals generated by single heart cells, isolated heart tissue, and the whole organ. Interpretation of these complex phenomena also requires that we employ the modern signal processing methods that consider the temporal, spectral and dynamical features of this rhythm. We recorded action potentials (AP) from single cells in isolated fibrillating hearts with the aid of a floating microelectrode technique, and in cardiac tissue using optical fluorescence imaging. 1) Time-frequency analysis of these signals reveals different characteristics of VF signal during early and late stages of fibrillation. Perfusion of the heart maximized the short-term, high-frequency events, while without perfusion, the time-frequency distributions showed dispersion. Time-frequency analysis was thus shown to be helpful in characterizing AP during various stages of VF. 2) Parametric modeling was next considered to determine the evolution of fibrillation with extended periods of time, a situation that would arise during resuscitation procedures. Autoregressive modeling of VF signals was carried out to identify changes in the dominant poles of the VF signals with time course of evolution of VF. Parametric modeling of VF was thus shown to be useful in predicting the duration of cardiac arrest. 3) Finally, we sought to determine how dynamics of VF signals change with time, and in particular whether fibrillation can be considered chaotic at the cellular levels. Dynamical analysis, carried out by the methods of dimensional analysis and Lyapunov exponents revealed that VF has a relatively low dimensional attractor at the single cell level even though VF on the heart or the body surface may be a high dimensional process. Such analyses may help suggest methods to track and modify low dimensional chaotic VF in its

Advances in Processing and Pattern Analysis of Biological Signals, Edited by Isak Gath and Gideon F. Inbar
Plenum Press, New York, 1996

early stages of evolution. 4) Finally, algorithms were developed for application in a clinical device such as the implantable cardioverter-defibrillator. Here, the emphasis is on the reducing false positive and false negative rates. This objective is accomplished using a sequential hypothesis testing algorithm that trades off accuracy for detection time and vice versa. In summary, non-parametric, parametric, and dynamical methods together provide quantitative insights into the fibrillation phenomenon at the cellular and whole heart levels, and may help in discrimination of various stages and forms of VF for the purposes of possible therapy.

INTRODUCTION

The electrocardiogram (ECG) signal serves as a signature at the body surface of the normal or abnormal electrical function of the heart. The surface ECG is, however, a direct representation of the electrophysiologic function of the cardiac cells and tissues. Thus, specific patterns of the normal ECG signals correlate with the electrophysiologic signals of the cardiac cells. The QRS complex is generated by the electrical depolarization of the cardiac cells, with the depolarization indicated by the upstroke of the cardiac AP. Analogously, the ST and the T waves are the result of repolarization in the cardiac AP. The P wave is the result of depolarization of the atrial cells. Thus, ECG signal provides a direct window into the electrophysiologic activity at the cellular level. Signal analysis of the ECG potentially offers an insight into the electrophysiological aspects of the cardiac system, and vice versa.

Abnormal function of the heart usually manifests itself as characteristic changes of the ECG signal. Abnormal rhythms or arrhythmias show up as morphological as well as rhythmic changes in the ECG. Arrhythmias range from relatively benign, such as certain atrial rhythms and occasional premature ventricular contractions, to malignant, such as ventricular tachycardia (VT) and ventricular fibrillation (VF). Abnormal ventricular signals have their corresponding basis in the cellular electrophysiology (1). VT and VF rhythms result from self-sustaining but aberrant activity. VT may result from a very rapid rhythm of the ventricular tissue, and also from abnormal pattern of activation that maintains itself in the heart through reexcitation of the tissue along the propagation pathway. This is called "reentry," implying that a rhythm reenters or repeats itself along the same pathway (2-4). VF is a fatal arrhythmia that may arise from reentry or from multiple reentrant or irregular pathways (or wavelets (5)) that simultaneously persist, causing irregular and uncoordinated beating of the heart and making it unable to support circulation. ECG signals show a fairly regular beating pattern but abnormal morphology, such as wider complexes without apparent atrial activity, in VT. VF, on the other hand, fails to show any characteristic pattern at all. Irregular and uncoordinated activity of the cardiac tissue manifests itself at the body surface as an apparently random signal without any visualizable temporal or spectral features. Indeed, VF is often called a "chaotic" rhythm (6, 7). However, at the cellular level, the cells show well defined, periodic pattern of AP in VT and, even in VF, the cells show characteristic features of APs (such as a rapid upstroke and a broad repolarization period). It would appear surprising that AP integrity is maintained at all, at least in the early stages of the VF signal, but fundamental electrophysiologic studies reveal that some electrophysiologic processes are still intact (8, 9). Based on these cellular studies we are encouraged to find recognizable patterns in the surface VF signals. To find a measurable pattern in VF would be desirable and beneficial. We hope to find the patterns during the early stages to be regular, and even instances throughout VF we hope to find evidence of underlying regularity in the VF rhythm. Such observations would indicate electrophysiologic processes that may be healthy enough to salvage the heart's function. Thus, a fundamental study of the VF signals is warranted.

Early stage identification of the signal character may help identify strategies for low energy cardioversion or defibrillation (10). Late stage VF may also be important to characterize, since it may offer directions for cardiac resuscitation therapy, such as how cardiopulmonary resuscitation is carried out (11). Analysis of intervening periods may facilitate successful defibrillation, for example if such periods are found to be regular, representing a well synchronized electrophysiological activity (10). Underlying regularity in the irregular or chaotic rhythm may offer the potential for interdiction or management. Thus, fundamental analyses of VF signals may provide an insight into the underlying electrophysiological processes and offer possible therapeutic solutions to terminating malignant arrhythmias.

VF ELECTROPHYSIOLOGY

At the cellular level, a cardiac AP represents changes in the potential across cell membrane as a result of time-varying changes in various ionic currents. Fast sodium (Na) channels conduct currents during the AP upstroke for the first few ms. Slow channels carrying mainly the calcium (Ca) currents maintain the plateau for the next 100 ms or so. Potassium (K) channels are mainly responsible for the repolarization period which comprises the final phase of the AP. Many other types of channels carrying these ions in somewhat different voltage and time dependent manner have been discovered using the modern experimental techniques of voltage and patch clamp, and these channels are the contributors of a slight variability in AP in different parts of the heart as well as among different species. The role of all these channels under pathological conditions is less well understood, at least in part because of difficulty in rigorously simulating pathological conditions at the cellular level during voltage clamp or patch clamp experiments. Another difficulty arising in studying the mechanisms of cardiac arrhythmias is that they result from pathologies that affect not only the single cells but also the collective behavior of the cardiac tissue, such as resulting from abnormal conduction in the myocardium.

VF is thought to arise due to irregular, or uncoordinated activity of the cells in the entire myocardium. VF may be induced in the laboratory by electrical shock to the heart. In clinical situations, VF may occur as a result of ischemic disease. Aberrant beats falling in the vulnerable period of the cardiac cycle (R-on-T) can cause VF. Often a rapid self-sustaining but regular rhythm such as VT degenerates into the irregular rhythm VF. During these processes, cardiac APs show corresponding behaviors - slower than normal excitation or depolarization, repeated excitation resulting in an incomplete repolarization, and irregular inter-beat intervals. These observations can partially be explained by the following sequalae of events. Aberrant beat or a reentrant rhythm induced by stimulation, reperfusion or drugs initiates VF (12). The consequence is that there is no regular pattern of electrical excitation in the myocardium. Mapping of electrical activation events has been able to identify in some parts of the heart and during early stages of the rhythm, a reentrant pattern such as figure-of-8 (13). These arise when a region of the heart which was previously excited is re-excited almost immediately upon repolarization of the tissue by a returning wavefront (14). This process can occur because of the passage of the wavefront around an anatomical obstacle such as caused by ischemic or infarcted tissue, or a functional block simply created by a conduction block between two excitable pieces of tissue. However, this reentrant pattern soon degenerates into possibly multiple fragmented waves or what we may describe more descriptively as wavelets (5). Multiple wavelets may coexist transiently, break up, new ones form, and all together no coherent activity is seen when observing the entire heart.

At the cellular level, the APs also show characteristic changes in VF. These may be due to ischemic changes resulting from VF which may then go on to affect the metabolic processes and consequently the activity of the ion channels and pumps responsible for maintaining the cardiac AP (9). These may also be due to the abnormal patterns of excitation. As excitation occurs rapidly, it has the immediate effect of reducing the AP duration due to incomplete recovery of long time-dependent currents such as the K currents. Further, the excitation is quite irregular, and therefore a given cell may get depolarized at various times during the cardiac AP. This depolarization naturally curtails the recovery process of the premature slow channels, and causes irregular shapes (mainly the duration) of the AP. Since a given cell is very likely to be depolarized from an elevated voltage plateau, the upstroke resulting from voltage dependent Na currents is also abnormal. Mainly, one sees slower AP upstroke possibly due to reduced activity of the Na channels (8, 9). Since even neighboring cells are expected to be stimulated quite irregularly, potential differences exist that cause currents to flow in the extracellular resistive space. This interaction between neighboring cells shows up as electrotonic interactions (15). Thus, the overall observation is that of APs with slower upstrokes, irregular repolarization periods, varying cycle lengths, generally elevated voltage plateau during repolarization or recovery, and considerable subthreshold activity which may be in part due to electrotonic interaction among neighboring cells. VF leads to an immediate cessation of circulation, and therefore deprivation of metabolic and energy delivery mechanisms. An immediate consequence is a failure of energy dependent processes, such as active pumps and exchange mechanisms. Thus, the Na-K pump and Na-Ca exchange mechanisms are affected. A consequence is accumulation of intracellular Na, failure of Ca homeostasis leading to Ca dependent oscillations, accumulation of extracellular

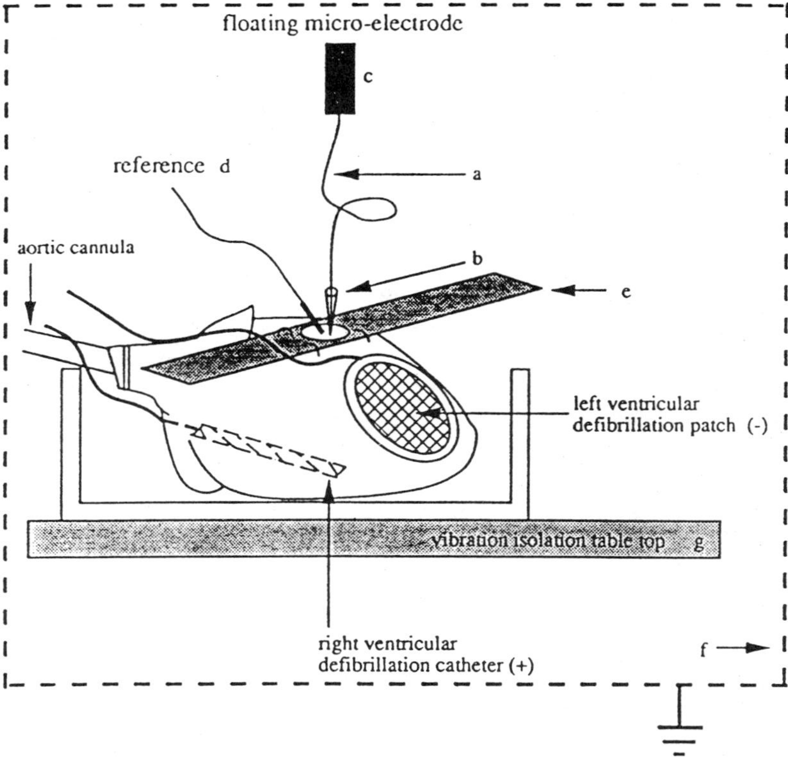

Figure 1. The experimental set-up for recording AP from single cells during VF. Coiled wire (a) suspending a floating microelectrode (b), immobilizing plate and recording chamber (c) with a reference electrode (d), a pair of defibrillation electrodes (+ catheter and - patch), Faraday cage for shielding (f), and vibration isolation table (g).

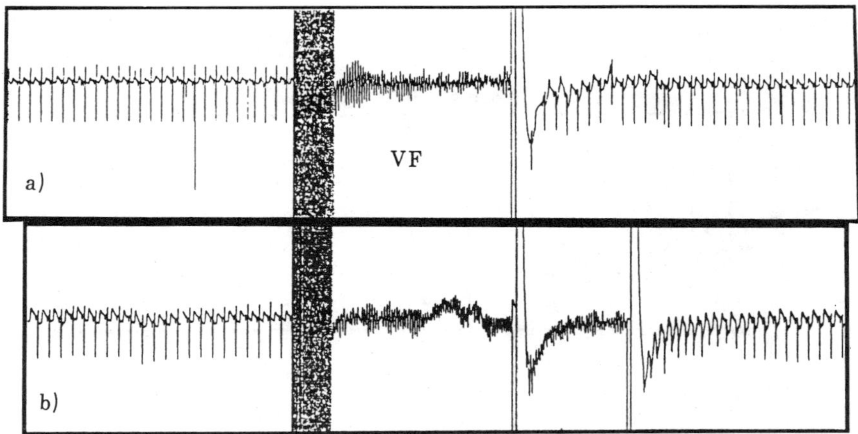

Figure 2. Induction of VF by ac stimulation. The initial high level interference is from 60 Hz stimulation which causes VF. A first shock is unsuccessful, while a later shock at the same energy is successful, suggesting that VF patterns are dynamic and there are desirable instances when low energy shocks may be successful.

K and so on. This metabolic failure only gets more severe with time. Therefore, not surprisingly, the electrical activity in the heart cells shows continuing but more severe changes indicative of this general failure of the active processes as time for which VF persists increases. Events occurring with increasing VF duration include further reductions in AP upstroke, considerable variability in repolarization intervals, possibly increased oscillations, very irregular repolarization patterns with reduced re-excitation and increased cycle lengths (8, 9). These patterns continue to degenerate further as the duration of VF increases without any therapy (such as defibrillation of CPR) (16). Even after therapy, some residual changes may be expected to persist, and may be indicative of the pathological conditions existent at that time.

EXPERIMENTAL STUDIES

Isolated Heart Experiments

Electrophysiologic studies of VF can be done only in the whole large animal hearts; a heart tissue of a critical large mass only supports VF. To this end we have developed an isolated heart preparation in which we obtain electrophysiologic recordings while the heart is fibrillating (Fig. 1). Often these studies were done when certain interventions, such as changes in perfusion or administration of drugs, were carried out. Large New Zealand rabbits (4-6 kg) were anesthetized and their hearts were rapidly excised via median sternotomy and perfused by a Langendorff-style preparation (8). Retrograde perfusion at a peak perfusion pressure of 80 mmHg was maintained during normal sinus rhythm. An electrode pair attached to the epicardium was used for pacing the heart or for inducing fibrillation by 60 Hz ac stimulation. A separate pair, consisting of a large area patch electrode and an intraventricular catheter, was used for delivering defibrillation shocks. Additional bipolar electrodes with 2 mm separation were placed in a 1 cm square grid surrounding the pacing electrode. For intracellular recordings, a floating microelectrode technique was developed (17). A glass pipette was pulled to a resistance of the order of 10-20 Mohms and then its front tip along with about 1 cm of the shank was separated and suspended using a coiled Ag-AgCl wire

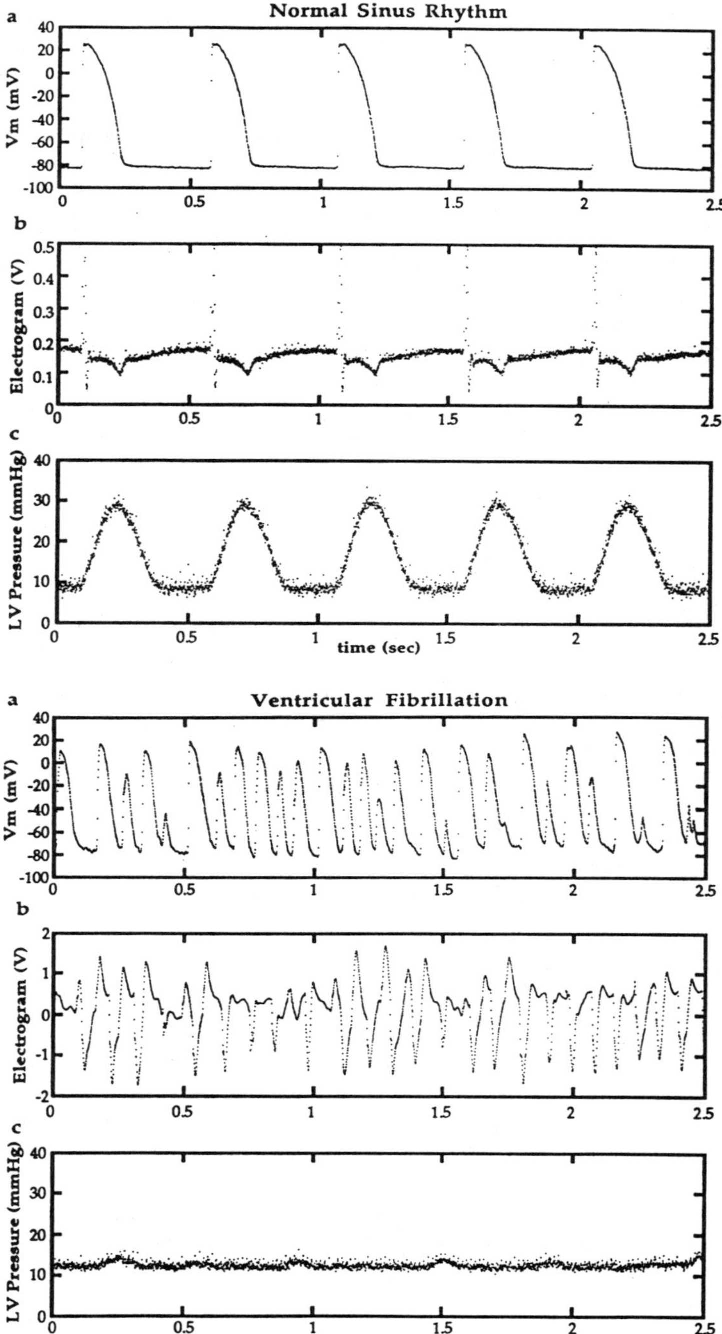

Figure 3. Bipolar and microelectrode (AP) recordings from isolated heart during A) normal sinus rhythm, and B) ventricular fibrillation.

which also acted as the sensing element. The coiled wire provided the compliance needed to hold the electrode in the cell in spite of the heart motion. The microelectrode was lowered onto the heart surface using a micromanipulator until it achieved cell penetration due to natural gravitational force exerted over the microscopic tip diameter. Through the use of careful experimental techniques impalement was often maintained and AP recordings acquired for several minutes while fibrillation was induced, and interventions such as drug infusion or defibrillation were carried out (9, 11).

Experimental studies involved induction of VF by ac shocks (or using certain drugs, such as oubain) and eventual defibrillation by delivery of 2 to 20 J shocks (Fig. 2). VF was allowed to persist in some experiments for several minutes while bipolar and microelectrode recordings were obtained (Fig. 3). In other experiments perfusion was either completely stopped or perfusion pressure was reduced to simulate realistic conditions existing during a cardiac arrest, while in other experiments, drugs such as Lidocaine or TTX were infused (9, 11). Experimental data were either collected on a wide band FM tape recorder such as Model MR30, Teac, Japan, or directly digitized on a Macintosh personal computer using National Instruments' Lab View data-acquisition software. Data recorded on the FM tape were digitized using a waveform scrolling software and hardware board (Model WFS200PC from Dataq instruments, Akron, OH) operating on an IBM compatible personal computer. AP characteristics such as upstroke, duration, and cycle lengths during normal sinus rhythm and VF were analyzed using a custom software.

Isolated Tissue Experiments

VF in the whole heart is a very complex system to study. Not only is the tissue structure three dimensional, but geometry of the fiber bundles and anisotropy make the study of activations and propagations very complicated to interpret (18). Since VF has been shown

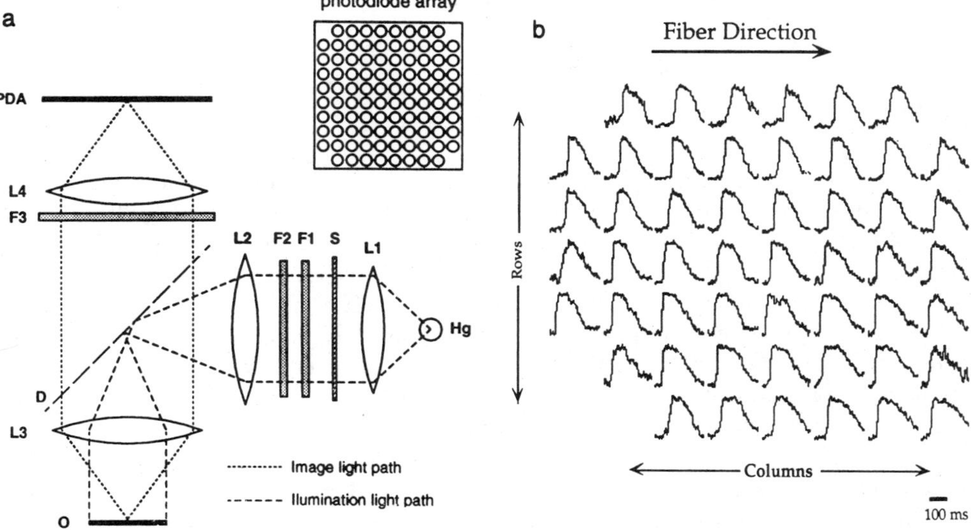

Figure 4. (a) An optical set up for measurement of optical fluorescence in the cardiac tissue. Light from a Tungsten Halogen (Hg) source is passed through filters (F) and lenses (L) and reflected onto the tissue with the aid of a dichroic mirror (D). Collated fluorescence light from lens L3 is partially transmitted by to filter and lens on to the photodetector array (PDA). (b) Optical fluorescence measurements of action potential are sensed by the photo-detector array.

to be produced or sustained by mechanisms such reentry, an experimental preparation that provides a simpler simulation of such a behavior would be desirable. A simplification of the experiment and VF-like rhythm is achieved by doing studies on isolated tissue slice, cut thin enough to provide essentially a 2-dimensional surface over which the rhythm can persist. Such a persistent rhythm is produced by generating "spiral waves," which are self-sustaining rotational patterns that can persist because the tissue is self-excited immediately after repolarization by a returning wavefront from the same rhythm (18, 19). Experimental measurements are done using optical techniques; a voltage sensitive dye, di-4-Anepps (Molecular Probes, Eugene, OR) is used in the perfusate to bind to the cell membrane. When excited by a green light source, the tissue gives out fluorescence in proportion to the transmembrane voltage (20-22). Changes in the fluorescence level are measured using sensitive photodetectors. The optical set up is shown in Fig. 4a and AP recordings from the array are shown in Fig. 4b. An array of 96 photodetectors is used to measure and reconstruct activity over a 2x2 cm patch of the heart tissue. Patterns such as rotors and spiral-waves on the heart surface are imaged in this manner.

Clinical Studies

Clinical studies were carried out typically during independently scheduled cardiac catheterization or implantation of pacemakers or defibrillators. During many of these studies, arrhythmias were induced by electrical stimulation of the heart to evaluate the sensitivity or performance of antiarrhythmic drugs, or threshold for pacing or defibrillation. In course of these studies, various ventricular and atrial arrhythmias were conveniently generated and made available for subsequent analysis and algorithm development. Signals were recorded via endocardial catheters situated in atrial as well as ventricular chambers. Signals were amplified using clinical grade instrumentation and directly digitized on a personal computer using the data-acquisition and display system described earlier. Signals were digitized at up to 1000 samples/s and digitized data were archived on an optical WORM drive. Rhythms induced and recorded in these studies included normal sinus rhythm, atrial and ventricular tachycardia, and VF.

VF SIGNAL PROCESSING

VF signal processing algorithms are necessary because rapid and accurate identification of this rhythm is of paramount importance. Other than this crucial application, the study of underlying mechanisms and pathology may also be facilitated by signal analysis. Clinical application of VF signal analysis are many. In cardiac patient monitoring, especially in the intensive care unit, immediate identification of the VF signal is the basis for instituting the life-recovery therapy of the patient (such as CPR and defibrillation). Need for automated algorithms is further justified when monitoring and therapeutic devices are used in homes or emergency vehicles where expert clinical aid may not be available. These applications for many years have been the driving forces behind the development of signal analysis methods for identification of VF. Further applications to emergency care situations may be possible. If we can learn how long the heart has remained in VF (11, 23), we can possibly improve the resuscitation therapy which may include use of drugs, CPR and increased shock energy (or in extreme cases, consideration for withholding therapy altogether). An emerging application is the detection and termination of VF by implantable devices. The automatic implantable cardioverter-defibrillator first detects VT or VF rhythm, and then delivers an electrical shock directly to the heart (24). Of course, an extremely high level of accuracy is essential: a missed detection can result in death, and a false detection can result in an unnecessary shock which can be hazardous too. Similarly, any delay in identifying this rhythm can also be severely detrimental to successful therapy. An intriguing possibility is to improve the success of defibrillation by timing the defibrillation shock. Since there exist somewhat regular patterns of activation waves in the heart during VF, at least in its early stage, there possibly exist patterns in the body surface or endocardial VF recordings suggestive of this regularity. Therefore, we should explore the possibility that defibrillation shock can be timed to improve the probability of a successful outcome. At a more fundamental level, signal analyses of the cellular AP, epicardial and endocardial electrogram, and body surface ECG can be helpful in quantitating the character of the evolving VF signal and relating it to the electrophysiologic basis (25, 26, 27, 28). This can improve our understanding of the fundamental mechanisms and processes existent in the initiation, maintenance and termination of VF.

These applications have greatly spurred the development of algorithms for identification the VF rhythm and quantification of its characteristics. The methods employed can be grouped into three categories: parametric, nonparametric, and dynamical. Parametric methods study the VF signal by a signal model, assuming that some underlying deterministic system model generates the measured pattern. The methods include autocorrelation and autoregressive modeling of the VF signal. The nonparametric method is most often the spectral analysis (29, 30). Use of FFT helps identify the characteristic frequencies at various stages in VF (31). Time-frequency analysis is helpful in defining the characteristics of the AP during VF. VF is variously called random or chaotic (6, 32). Indeed, now we have the possibility of providing mathematical rigor to this assertion. Dynamical analysis includes measurement of correlation dimension and Lypunov exponent to show whether VF is indeed chaotic (33). Applications of these methods to VF signal analysis at the cellular as well as at the whole heart level are considered below.

Action Potentials during VF

Cardiac AP during normal sinus rhythm has a characteristic shape with features that include a rapid upstroke during cell depolarization, a plateau phase lasting about 200 ms, a repolarization phase, and a diastolic interval. During VF, however, the pattern of excitation changes. AP recorded during VF show widely varying levels of upstroke, repolarization, and cycle lengths (34, 35). Further, there is substantial evidence of electrotronic interactions with the neighboring cells (15, 36). An analysis of AP can be done in conventional manner, that is the upstroke (V max), AP duration (APD), and cycle length (CL) typically measured . The variety of waveform morphologies encountered during VF can possibly be better interpreted using spectral analysis (31, 37). However, time-varying patterns of an AP indicates that a joint time-frequency analysis approach be used to characterize the AP (Fig. 5) (38, 39). One method to do so is the short-term Fourier transform (STFT). By defining the sampling rate and window size, spectra can be obtained over short or long time-varying segments. By proper windowing and sliding of the analysis window, a continuous spectral measurement throughout the AP is obtained. STFT of a normal AP shows a sharp peak during upstroke, with instantaneous frequencies of several thousand Hz. Following this upstroke, the repolarization period is characterized by a period of several hundred ms over which relatively low signal frequencies are seen. Throughout the AP, a characteristic contour diagram is generated giving the quantitative character of the AP during its depolarization and repolarization. This method is particularly useful in distinguishing the AP recorded during VF. The upstroke is greatly reduced, as characterized by significantly lower instantaneous frequencies, while even the repolarization frequencies are lowered and often distributed over a greater time-interval. APs recorded in VF when perfusion is stopped show more of these changes (further reduced depolarization frequencies and repolarization frequencies and

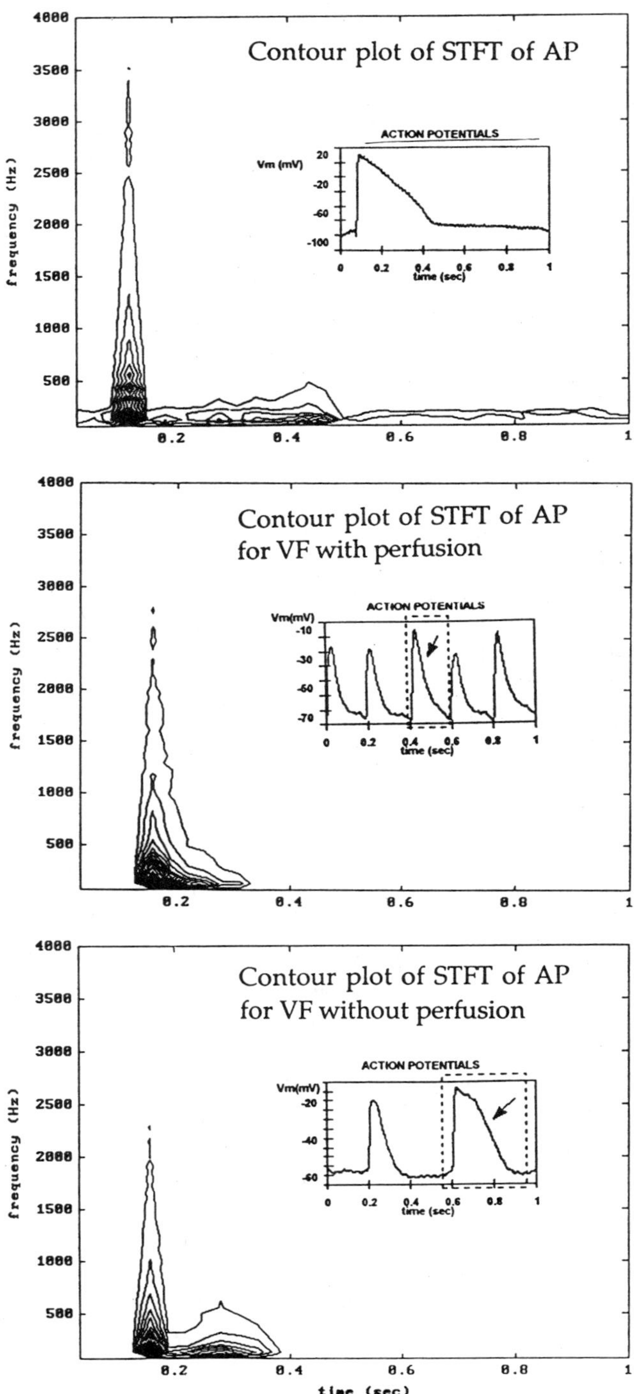

Figure 5. Short-term Fourier transform of AP recorded in an isolated heart during normal rhythm and VF (with and without perfusion). AP recordings are shown in the insets of each plot.

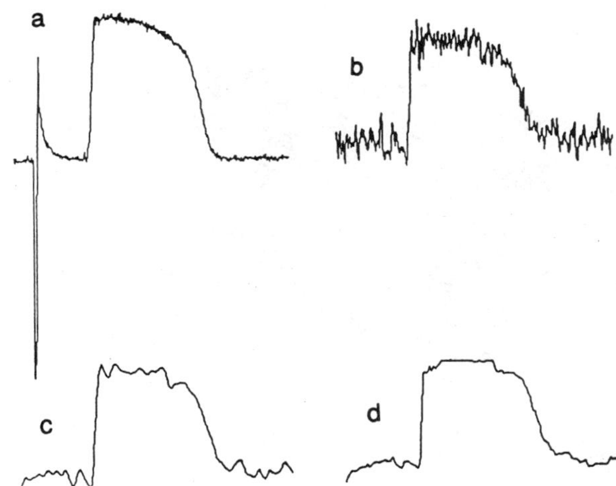

Figure 6. (a) AP recorded by an optical fluorescence detector, and the results obtained using various filters; (b) lowpass filter; (c) adaptive filter; (d) variable convergence (adaptation rate) adaptive filter (adapted from (40)).

interval). Thus, time-frequency analysis of AP is a novel quantitative way of studying cardiac AP during various arrhythmias, and in particular, helpful in discriminating instantaneous features during various phases of AP occurring during fibrillation.

An alternative method of obtaining AP from the optical fluorescence signal poses further signal processing challenges (20-22). The signal is now corrupted by a broad band

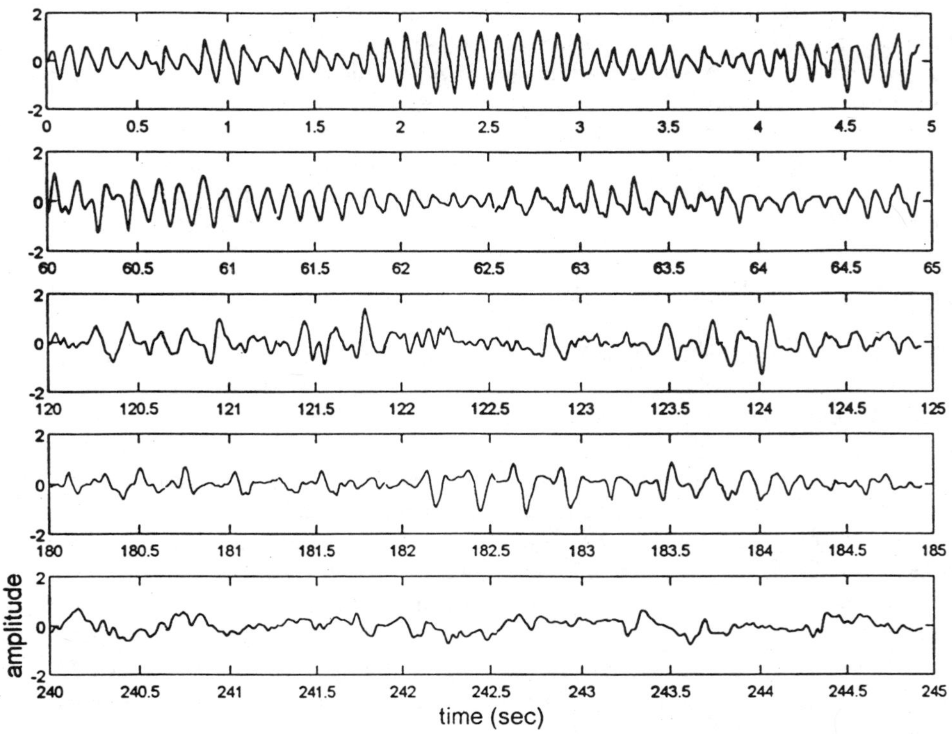

Figure 7. A long-term recording of bipolar VF signal in an isolated heart, with perfusion stopped to simulate cardiac arrest.

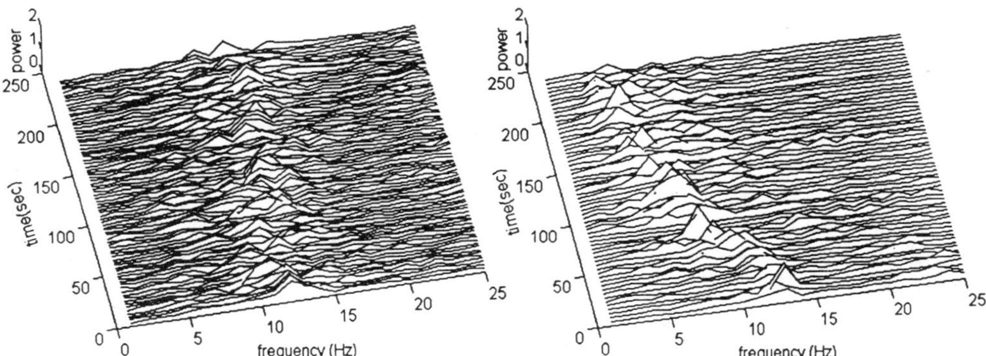

Figure 8. Power spectra of VF signals calculated using the short-term Fourier transform. (left) When perfusion is maintained, the spectra remain relatively stable, although there are fluctuations from epoch to epoch. (right) When perfusion is stopped, the center frequency of the spectrum shows a gradual decline towards lower values.

noise, possibly contributed by the optical system (background light, dark current of the detector) and the electronics (shot and thermal noise of the detector, resistors, and the amplifier). Thus, the problem is to filter the AP but at the same time not distort the characteristic such as the upstroke (40). Hence, instead of using a conventional lowpass filter which would reduce or smooth out the estimate of the AP upstroke, we use an adaptive filtering algorithm which dynamically adapts the filter characteristics (essentially the cutoff frequencies) (Fig. 6). The adaptive filter utilizes the principle of mean-squared error (MSE) minimization to estimate the signal from noise (41, 42). The adaptation rate is however continuously adjusted, so that during the repolarization and diastolic phases, the adaptation rate is slow allowing a fairly low misadjustment error. The adaptation rate is quite fast during the upstroke allowing a faster tracking of the potential change, albeit at a lowered misadjustment error. By adaptively striking a compromise between the extent of filtering and reduced misadjustment or distortion, as needed during various phases of AP, a noise-free signal with a high quality estimate of the AP parameters (in particular V max) is obtained.

Bipolar Signals during VF

In experiments with isolated hearts, we observed that the bipolar ECG (or more accurately electrogram) signals display none of the characteristic features of the AP signal. The bipolar electrodes record local extracellular electrical activity that electrophysiologically relates to the AP of the underlying tissue (43). However, simple time-domain analysis is complicated by the diversity of all the AP, with their widely fluctuating features (upstroke, duration, etc.). These AP characteristics manifest themselves as undulating and apparently random or chaotic patterns of the VF signal (9, 17). Nevertheless, there is an evolutionary change in the character of the signal. This is all the more apparent when VF is observed over extended time periods or when some interventions are carried out. With the passage of time, the character of the signal changes to a more fine and possibly slowing pattern (Fig. 7) (11).

To interpret these VF signals, we first obtained the spectra of overlapping segments. These evolutionary spectra show that the range of frequencies during VF gradually but persistently shift towards lower frequencies at reduced powers (Fig. 8) (44). This trend is significant when VF is allowed to continue but without perfusion of the heart (as in complete cardiac arrest). In absence of metabolic support, the underlying electrophysiologic changes

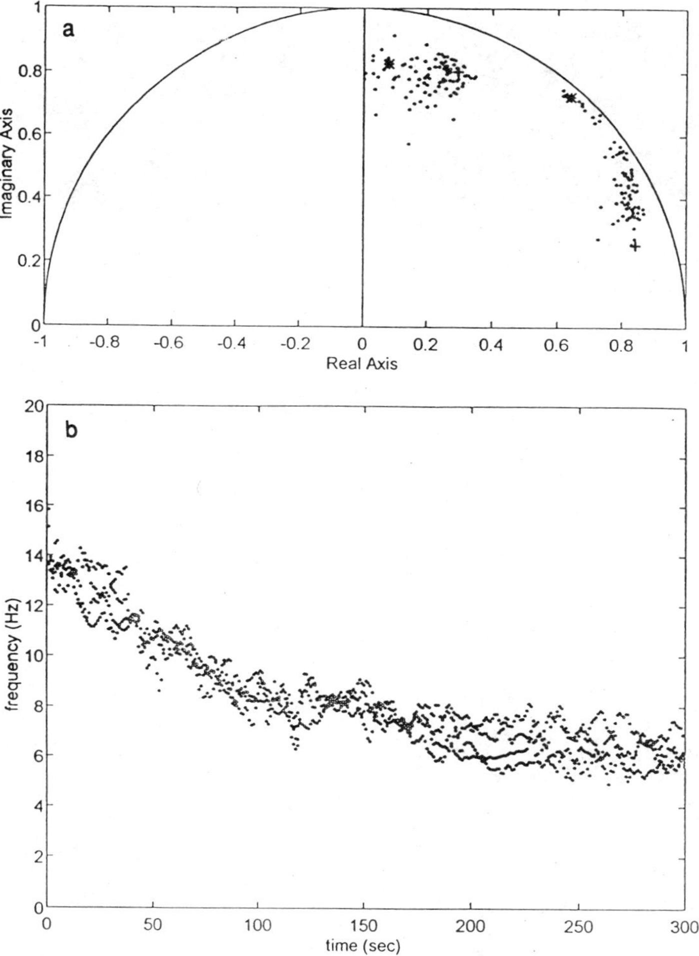

Figure 9. (a) Changes in the location of poles with time for a VF signal recorded over 100 seconds. (* time=0 s and + time=100 s). A fourth order autoregressive model is fit to the data. (b) Change in the dominant frequency, that is the frequency of the first pole, with time. The frequency values can be fit to a first order exponential.

manifest themselves as changes in frequency spectra over time. This observation suggested to us that signal analysis may yield a measure of how long the heart might have been in fibrillation.

To estimate the time-dependent changes in the spectra of VF signal, we sought to model the signal using parametric methods. VF signals were modeled by an autoregressive (AR) process whose order was determined using conventional approaches (Akaike information criterion and standard error) (44). A fourth order AR model of the VF signal was found to be optimal for our data, and accordingly this model displays two dominant poles or frequencies. The lower pole and its characteristic frequency are tracked in Fig. 9a. From the initial phases of VF to its later phases, the pole shows a shift towards lower frequencies. Further analysis showed that this trend could be fit to an exponential curve (Fig. 9b) (11). A trend showing changing VF signal frequencies with time may be indicative of a general slowing of the electrophysiologic processes, and in particular of changes in longer cycle lengths and reduced upstroke. A quantitative measurement of such a trend would have useful

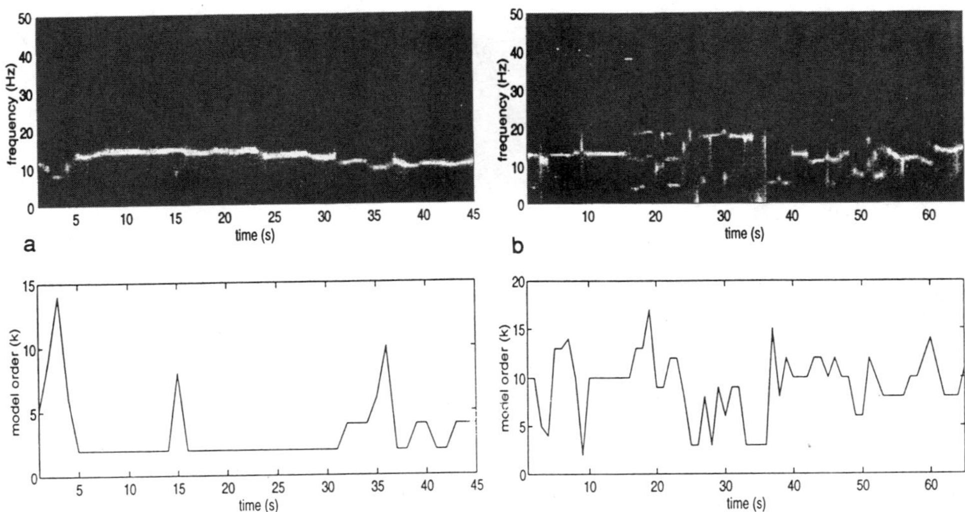

Figure 10. Dynamic tracking of the model order and the autoregressive model based power spectra: (a) initial VF, (b) reinduced VF.

clinical application - determination of the time over which heart may have been in fibrillation (23). An observation of the VF signal frequencies at a given time would be interpreted to determine the condition of the arrest and may be useful in guiding the therapy. Not unexpectedly, we find the trend to be quite stable when the heart is perfused during VF: in our Langendorff preparation this is done artificially by a pump, while in a clinical situation this would be done by means of chest compression. This signal analysis presents the evidence that artificial perfusion sustains the character of VF signal and possibly the electrophysiologic behavior of the underlying cells.

While VF signals do follow a long-term trend as described above, they also show a great deal of moment-to-moment variation which is clearly indicative of the diversity of AP signals described earlier. Therefore, a fixed order AR process may not be optimum to model the signal. When adaptive model order estimation is attempted, we see that over the short term there is a noticeable fluctuation in the model order and the power at the corresponding frequencies (Fig. 10). What is interesting is that the character of VF spectra estimated from AR modeling show striking differences under different circumstances (44). For example, a VF signal in Fig. 9 has a conventional evolution, with an exponential decline in the power at the dominant pole (11). However, there are occasions when the model order and the signal power slightly fluctuate. The same VF record spontaneously converted to normal rhythm, suggesting that stability of the VF signal as seen in its model and spectra may be beneficial. The same heart was subsequently electrically fibrillated and this time showed very high levels of fluctuations in the model order and the power at various dominant frequencies. We may speculate that this more complex rhythm as interpreted from the corresponding spectrum indicates a VF that may not be easy to convert to a normal rhythm.

Dynamical Analysis of VF

Conventional linear systems analysis or parametric analysis of VF may not be optimum in view of nonstationarities and possible nonlinearities present in our data (45). Indeed VF has often been called chaotic (6). Technically this interpretation of VF suggests

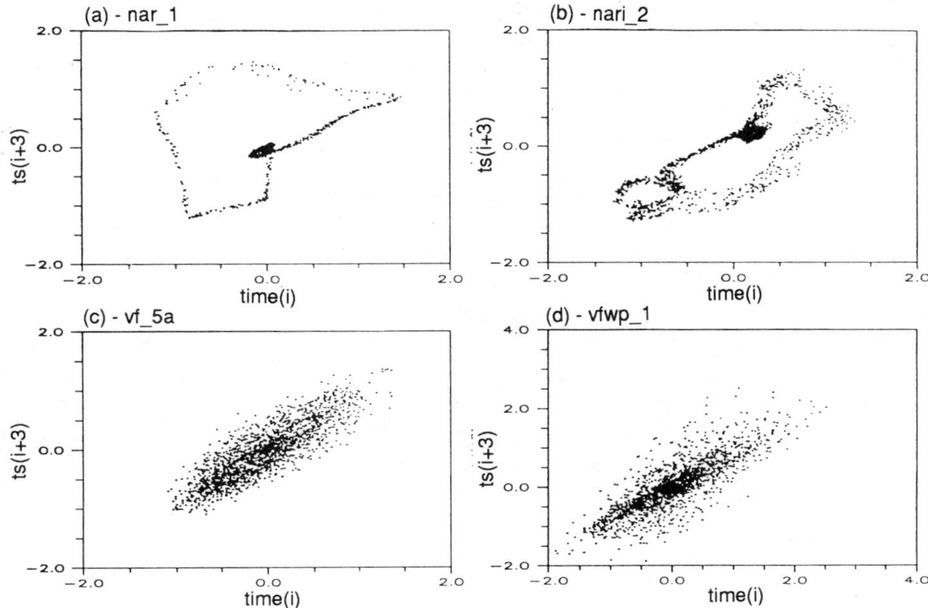

Figure 11. Phase plane plots of various signals: a) normal sinus rhythm, b) ischemic ECG, c) ventricular fibrillation, d) ventricular fibrillation with perfusion.

that there is an underlying deterministic system, which may however be nonlinear. As a consequence of the system nonlinearities or sensitivity to initial conditions, a chaotic rhythm that is quite unpredictable in its evolution may arise (46). That there may exist an underlying deterministic cause to generation of VF is physiologically plausible and experimentally or clinically attractive; such a formulation would supply a theoretical framework and improved insight into the management of this rhythm. Chaotic signals are evaluated using the methods of correlation dimension and Lypunov exponents (46). This is done so as to investigate whether different dynamics underlie various arrhythmias, and if a low dimensional attractor can be associated with them (a low dimensional attractor would indicate a dynamical system that is more amenable to identification and control). Signals obtained during experiments described earlier are analyzed using dynamical analysis techniques. Correlation dimension is calculated for normal and ischemic signals and for VF signals obtained during experiments in which perfusion was maintained or stopped. Phase plane plots of data recorded in these experiments clearly reveal a pattern of organization that deteriorates with ischemia and fibrillation (Fig. 11) (47).

Not unexpectedly, the correlation dimension of the normal sinus rhythm is seen to be close to 1. This result differs from previous attempts at measuring the dimension of surface ECG (32), but can be explained by the fact that in our isolated heart experiment, the rhythm is quite stable in the absence of autonomic regulation. The correlation dimension of the ischemic heart signal is increased as would be expected due to changes in the ECG pattern brought about by ischemia. The correlation dimension of the bipolar VF recordings ranged in values from 5 to 8, and a similar range was seen in conditions with or without perfusion (47). At an initial glance this result is not unexpected, because VF was indeed seen as a high dimensional process in previous studies too. However, further analysis of the same experimental data reveals that the correlation dimension of the microelectrode recordings made during VF shows a low dimensional attractor (33). That is, at the cellular level, VF is possibly a low dimensional system with a strong deterministic basis and implied regularity in the rhythm. This observation, if further validated, may prove to be useful in applying dynamical analysis approaches to interdiction and termination of VF.

Clinical Interpretation of VF

At the clinical level, the problem is to accurately identify the VF rhythm. First of all, malignant ventricular arrhythmias such as ventricular tachycardia (VT) and VF must be discriminated from the normal sinus rhythm(NSR) and atrial arrhythmias (24, 48-51). This is because NSR and atrial arrhythmias do not require a strong therapeutic solution such as electrical shock. Next, VT must be discriminated from VF, since VT can be cardioverted at lower energies while VF must be immediately defibrillated (51). Detection of VF is thus important in clinically used defibrillators, and in particular in implantable defibrillators which must make an autonomous and independent decision. In these applications, it is essential that false positive (FP) and false negative (FN) rates are minimized to reduce the possibilities of false shocks or missed shocks, respectively.

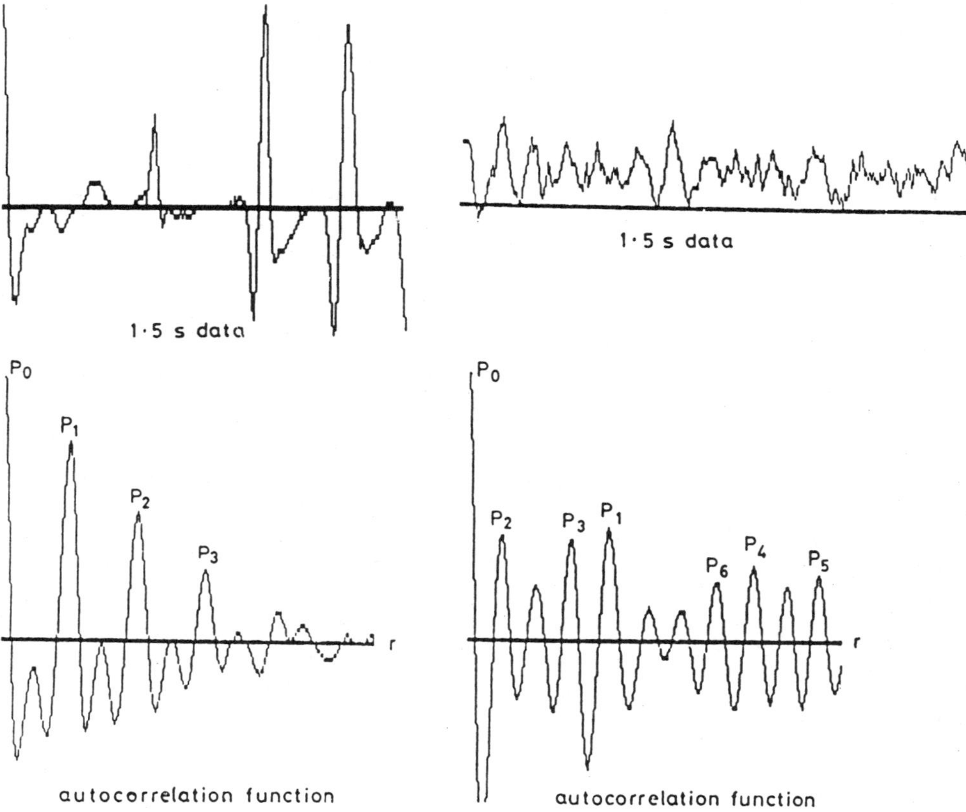

Figure 12. Short-term autocorrelation analysis of a non-fibrillatory (normal rhythm with an aberrant beat) and fibrillatory rhythm. Peaks for the normal rhythm show an ordered decline in magnitude, while for fibrillation the peaks are completely disordered (P0, P1, ... indicate the peaks numbered according to their magnitude) (adapted from (53)).

Autocorrelation Analysis. Time domain analysis of VF is particularly convenient, because beat-by-beat patterns of correlation can readily be seen in NSR and atrial, and to some extent, ventricular tachycardia (37, 52). Thus, autocorrelation, or for short experimental recordings short-term autocorrelation, shows obvious peaks at intervals corresponding to the heart-beats (30). Autocorrelation analysis thus readily reveals the periodicity and the inter-beat interval information with which NSR and VT can be discriminated from VF (Fig. 12) (53). The VF signal does not show any apparent periodicity, and hence no regular and repeatable peaks in the autocorrelation are seen. For short data segments, it can be shown that autocorrelation peaks follow a linear declining trend, and this seems to be the case for NSR and VT. However, no such trend, as quantitated by linear regression, can be discerned for VF. In a large data set of VT and VF signals, this discrimination criterion was shown to yield an excellent classification of the rhythms. Since VT signals show occasional irregularities and VF signals show occasional regular patterns, repeat testing of three or more segments is recommended to improve accuracy (53).

Sequential Hypothesis Testing. Most classical methods do not directly address the problem of accuracy of the algorithm. That is, it is not feasible to define or specify *a priori* the FP and FN rates. A direct approach to rhythm classification was taken in a technique called sequential hypothesis testing (51, 54, 55). We note that the inter-beat-interval (or simply zero or threshold crossing interval in case of VF) has distinctively different distributions for various arrhythmias (Fig. 13a). That is, the mean and standard deviation values (or more generally, probability distributions) of inter-beat intervals differ for these rhythms. These distributions, however, overlap, which is why when a fixed criterion such as a value of inter-beat interval is used as a threshold, either FP or FN results. To achieve discrimination of the rhythms based on inter-beat intervals and *a priori* knowledge of distributions for various rhythms, algorithms can be developed to test the hypotheses that the observed data belong to: H(NSR or atrial rhythm) vs. H(Ventricular rhythm); analogously hypotheses can be tested to determine whether observed data lead to a positive test of H(VT) or H(VF) (in general, multiple simultaneous hypotheses can also be tested) (51). Based on the probability distributions, a compact likelihood test is conducted at each successive beat to determine whether one of the hypotheses is true at the desired confidence level or, more specifically, at the specified FP or FN level. A conventional test would discriminate the rhythm based on

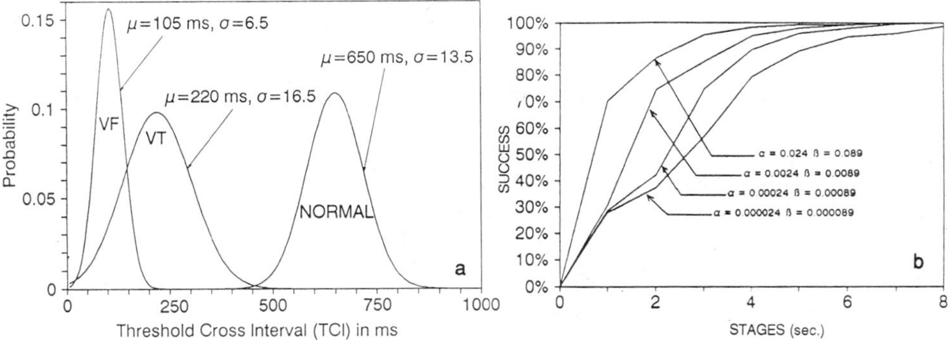

Figure 13. (a) Probability distributions of threshold crossing intervals (equivalently inter-beat intervals for NSR and VT) for three rhythms. While the distribution means are distinct, they overlap and hence account for error probabilities α and β. (b) A plot of percentage success with time (each stage represents test done after acquiring one additional second of data). Error probabilities α and β represent the two types of errors. Higher the tolerated error, faster the test result, and vice versa (reproduced from (51)).

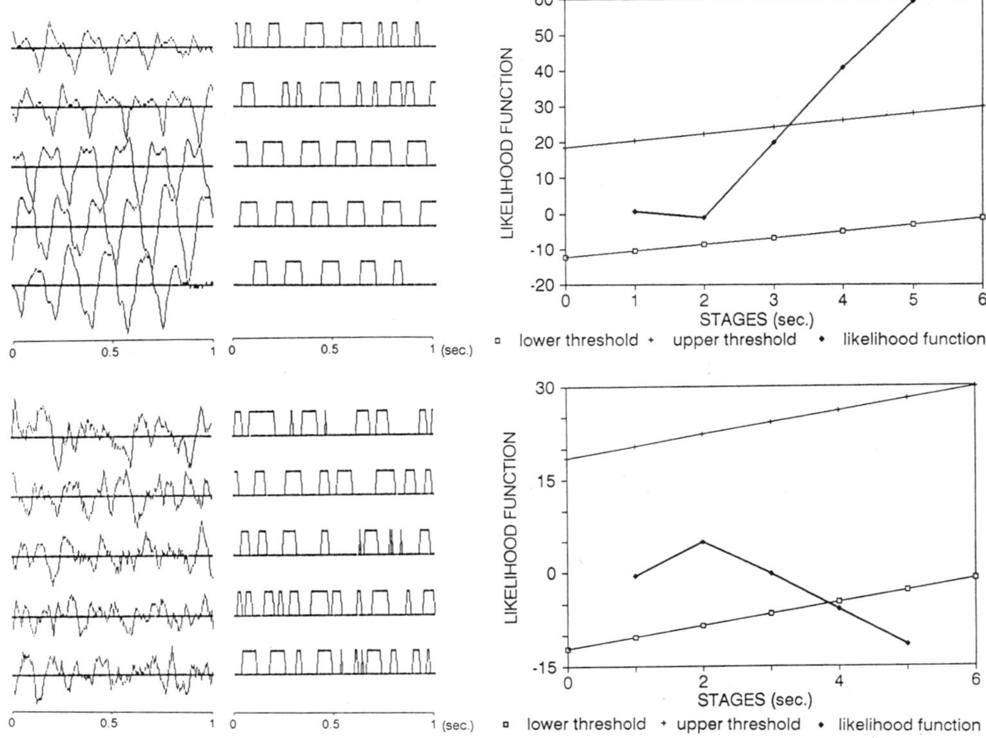

Figure 14. Plots of likelihood function versus time (each stage represents tests done every second) for two rhythms. The test threshold is explicitly dependent on the specified error probabilities; when the threshold is crossed, the rhythm (a) VT or (b) VF is identified (reproduced from (51)).

a single fixed threshold. In our formulation, this discrimination threshold is explicitly replaced by a test conditioned on the specified error probabilities (dependence of error probabilities α and β on threshold is readily seen in Fig. 13b).

This test is next reformulated as a sequential hypothesis test: at each successive data sample various hypotheses are tested. If the hypothesis concerning any of the rhythms is satisfied (that is, conditions of expected error probabilities are met), the test is terminated and the rhythm is identified, otherwise the test is continued. The benefit of this approach is that testing is continued until the accuracy criteria are met. Consequently, for relatively high values of desired FP and FN rates (equivalently, larger error probabilities α and β), as we might wish for the discrimination of NSR and atrial arrhythmias, the test is terminated quickly (55). For relatively low desired values of α and β, as we might wish for the discrimination of VT and VF, the test takes longer to terminate. Thus, a direct specification of FP and FN is achieved, and these desired rates are obtained by trading off time for accuracy, and vice versa. In studies of clinical data acquired from patients undergoing cardiac catheterization or device implantation procedures, we have shown that excellent discrimination of various rhythms can be attained. In one study, 85 cases each of VT and VF could be discriminated accurately in less than 8 seconds (Fig. 14) (51). In another study, NSR, SVT, and VT could similarly be discriminated accurately within 3 to 8 seconds (55). These performances and detection times are compatible with application in clinical devices such as automatic implantable defibrillators.

DISCUSSION

VF is a very complex rhythm. Complexity of this rhythm can be traced to activities both at the cellular and the whole heart level. At the cellular level, cardiac AP show repetitive but irregular excitation (11, 17). Since most of the time the cells are excited while in partially repolarized states, the subsequent AP also show reduced upstroke and irregular durations. This behavior can be traced at the channel level, where in reduced activity of Na channel can be verified, while Ca channel activity is shown to be necessary for maintenance of VF (8, 9). Because adjacent cells are in various stages of repolarization, the potential difference between cells contributes electrotonic interactions. These rapid but irregular excitations, variable upstrokes repolarization intervals and cycle lengths, along with electrotonic interactions, give VF even at the cellular level a fairly complex character. At the whole heart level, when measurements are made extracellularly, the character of the VF signals is also strikingly complex. There is no apparent periodic activity and, with visual inspection, the signals look random or chaotic. Electrophysiologic analyses, such as by activation mapping or computer modeling, verify this complexity (56). At the same time, these studies also reveal patterns of regularity or perodicity. In some experiments these patterns are attributed to the existence of organized wavefronts and rotors. Fundamental electrophysiologic investigations have thus revealed VF to be a complex phenomenon with cellular as well as organ level substrates. For these reasons, VF signals pose considerable challenges for both analysis and interpretation. Signal processing problems at the cellular level are to obtain high quality, noise-free data (such as by optical fluorescence imaging) or obtain better interpretation of the AP (such as by time-frequency analysis (39)). Further, recordings from an array of sensors can be used to generate propagating patterns on the heart surface. This problem poses the challenge of simultaneous analyses of data from many sensors, such as calculating time delays and propagation velocities, for the purpose of reconstructing wavefronts. At the whole heart level, similarly, we can obtain recordings of single bipolar detector or an array of epicardial or endocardial detectors. The challenge here is once again to interpret single electrogram for electrophysiologic insights and analyze measurements from large arrays for reconstruction of activation wavefronts and interpreting the mechanism of generation and termination of the arrhythmia.

On a practical level, analysis of VF is essential for cardiac arrhythmia recognition devices used in patient monitoring (57, 58). Such instrumentation exist especially in cardiac intensive care units where VF rhythm when detected can be immediately treated using a cardioverter-defibrillator (59). However, this is not the most common scenario. Sudden cardiac death, presumably brought on by sudden onset of VF, affects half a million people in the USA alone. Therefore, a device that is permanently available on a stand-by basis could be particularly useful. Such a device is the implantable cardioverter-defibrillator (57, 59). The implantable device works completely autonomously and delivers a low energy shock directly to the heart when VF is recognized. Therefore, an extremely high level of accuracy is required (51, 54). An FP results when a rhythm other than VF is incorrectly identified and a shock may unnecessarily be given. An FN results when VF detection is missed, ultimately resulting in death. To achieve the desired very high accuracy, both FP and FN must be minimized, but if necessary, they can be traded off. Rather than trade-off FP for FN and vice versa, another strategy is to trade off time for accuracy. Allowing a longer time for detection of arrhythmia may improve accuracy (55). Thus, current development of algorithms should lead to improved performance of the therapeutic devices.

At the therapeutic level, the consideration is interpretation of VF so that diagnostic and therapeutic devices can help with its interdiction or termination. VF rhythm is fatal and hence immediate detection and application of therapy to terminate this rhythm are strongly desired (57, 59). Immediate and conventional therapy is to use a cardioverter-defibrillator which delivers a high energy shock to terminate the arrhythmia. However, further improvements would be desirable. For example, delivery of an early shock may reduce the energy required for defibrillation, perhaps because the rhythms are more regular and heart is metabolically stronger. Similarly, periods of apparent regularity in an otherwise irregular rhythm raises the possibility that methods may be developed to control such a rhythm. This raises the possibility that defibrillation shock can be timed to achieve defibrillation at a higher success rate and at a lower energy (10). Indeed, if VF is not completely random, it may have nonstationary and perhaps nonlinear components. Some ongoing studies suggest that it may lend itself to interpretation with the help of dynamical analysis methods. Dimensional analysis shows that VF on the heart or the body surface is a high order process (high correlation dimension) (47). At the cellular level, however, our analysis has found it to be a low order process (60). If VF is chaotic, in particular with low order attractors, then it may lend itself to an active control that stabilizes the rhythm. Finally, when heart has been in VF for extended periods of time, different therapies and strategies might be called for, which may include cardiopulmonary resuscitation, use of drugs, and so on (11).

Future work may lead to surprising and refreshingly different approaches to diagnosis and therapy. Thus, electrophysiological studies of VF provide important, provocative insights in to possibilities for regulating and terminating the arrhythmia. This might be accomplished in two different ways: interdiction while the rhythm is deterministic chaotic or control through intelligent stimulation therapy. The classical approach is to detect VF and terminate it by delivery of high energy shocks. New approaches are needed that not only detect but even anticipate. For example, a rhythm preceding VF is identified, preemptive therapy may be instituted. Another approach may be to manage the rhythm or interdict the evolving rhythm. The possibility here is that patterns of regularity or electrophysiologic synchrony and stability may lend themselves to interdiction of the evolving rhythm by pacing therapy. In the field of chaotic signals, this is the strategy of chaos control (46), although this strategy needs adaptation (which may be non-trivial) to the cardiac arrhythmia problem. To achieve such goals, improved quantitative analysis of VF at local as well as global levels (i.e., at the single cell and micro-electrode versus multi-electrode and activation map) may be needed. Signal analysis may need to decipher the dynamics or transients of the rhythm. To do this, a combination of parametric, nonparametric, and dynamical analysis approaches will be needed. The result will be intelligent algorithms for detection, interdiction, and termination of life-threatening arrhythmias.

CONCLUSIONS

VF is a fatal cardiac arrhythmia; hence its immediate detection and termination is essential in critical care settings. Fundamental electrophysiologic investigations have shown that VF is a complex rhythm based on irregular, if not chaotic, patterns at the cellular and whole heart levels, although patterns of regularity may exist in the form of wavefronts and rotors. At the cellular level, the APs show irregular cycle lengths, upstrokes, and durations along with probable electrotonic interactions. As a result, the interpretation of VF necessitates analysis in time and spectral domains. The apparent irregularity, randomness or chaotic nature of VF signal has led to the development of parametric (11, 44), non-parametric (55), and dynamical analysis (11, 33, 44, 47) algorithms. Autocorrelation analysis of VF shows that very poor correlations exist in the short-term (53). Spectral analysis shows that while there is a dominant frequency band around 8-12 Hz where the signal power is concentrated, the magnitude of the power and its frequency distribution fluctuates moment to moment and over extended periods of time (31, 61). The trends in the dominant frequencies, analyzed using autoregressive modeling technique show an exponential decline in the value of the frequencies over the first several minutes of VF (11). However, if perfusion is maintained in the heart, the power in the same dominant frequencies is stabilized over the first few minutes. Changes in the signals correlate well with the cycle lengths and other changes in the APs of the cardiac cells. Using this method, it is possible to predict how long heart may have been VF; this knowledge may prove to be useful in guiding the emergency resuscitation procedures (11). Finally, the chaotic or random nature of VF is evaluated using dynamical analysis techniques (correlation dimension and Lypunov exponent). VF at the heart surface is shown to have a high dimension, although at the single cell level a low dimensional system is seen (33, 47). Should further evidence of low dimensional processes in VF become available, it may then be possible to control the chaotic phenomenon. Together, these data suggest a new understanding of the mechanism of VF electrophysiology and how VF signals are generated and how they can be interpreted. For clinical applications of implantable devices, extremely accurate algorithms are needed for VF detection without false positives or false negatives; this is accomplished with the help of a sequential hypothesis testing algorithm (51, 62). Automatic interpretation of VF is directly useful in therapeutic devices such as external and internal pacemakers and defibrillators (57, 58).

ACKNOWLEDGMENTS

The authors thank many students and collaborators who contributed to the past research on which this review is based. In particular, we thank Shoupu Chen, Ananth Natarajan, Kong-yan Pan, Jan Heber, Ravi Ranjan, Denise Hodgson, Matthew Fishler, Gordon Tomaselli and Morton Mower. This research was supported in part at various times by grants from the National Institutes of Health, National Science Foundation, and industry.

REFERENCES

1. Arnsdorf, M. F., 1990, The cellular basis of cardiac arrhythmias. A matrical perspective, *Annals N. Y. Acad. Sci.*, 601: 263-280.
2. de Bakker, J. M. T., van Capelle, F. J. L., Janse, M. J., Wilde, A. A. M., Coronel, R., Becker, A. E., Dingemans, K. P., van Hemel, N. M., and R. N. W. Hauer, 1988, Reentry as a cause of ventricular tachycardia in patients with chronic ischemic heart disease: Electrophysiologic and anatomic correlation, *Circulation* 77: 589-606.
3. Bernstein, R. C., and Frame, L. H., Ventricular reentry around a fixed barrier. Resetting with advancement in an *in vitro* model, Circulation 81: 267-280.
4. Panfilov, A. V., and Keener, J. P., 1993, Generation of reentry in anisotropic myocardium, *J. Cardiovasc. Electrophysiol.* 4: 412-421.
5. Moe, G. K., 1962, On the multiple wavelet hypothesis of atrial fibrillation, Archives Int. Pharmacodynamics 140: 183-188.
6. Kaplan, D. T., and Cohen, R. J., 1990, Is fibrillation chaos?, *Circ. Res.* 67: 886-892.
7. Goldberger, A. L., and Rigney, D. R., 1988, Sudden death is not chaos, In: *Dynamic Patterns in Complex Systems*, Kelso, J. A. S. , Mandell , A. J., Shlusinger, M. F. , (ed.), World Scientific Publ: Singapore, pp. 248-264.
8. Hodgson, D., 1991, Electrophysiology of ventricular fibrillation studied by the microelectrode method, *M.S.E. Dissertation*, Johns Hopkins University, Baltimore, MD.
9. Hodgson, D., Fishler, M. G., Chan, R., and Thakor, N. V., 1992, Intracellular recordings during VF: role of ion channels from experiments and comptuter models, *Proc. Comput. Cardiol.*, IEEE Comput. Soc. Press, pp. 537-540.

10. Carlisle, E. J. F., Allen, J. D., Bailey, A., et. al., 1988, Fourier analysis of ventricular fibrillation and synchronization of DC countershocks in defibrillation, J. Electrocardiol., 21: 337.

11. Baykal, A., Ranjan, R., and Thakor, N. V., 1996, Estimation of ventricular fibrillation duration by autoregressive modeling, *IEEE Trans. Biomed. Eng.*, in press.

12. Ideker, R. E., Klein, G. J., Harrison, L., Smith, W. M., Kassel, J., Reimer, K. A., Wallace, A. G., and Gallagher, J. J., 1981, The transition of ventricular fibrillation induced by reperfusion following acute ischemia in the dog, *Circulation* 63: 1371-1379.

13. El-Sherif, N., 1985, The Figure 8 model of reentrant excitation in the canine postinfarction model, In: *Cardiac Electrophysiology and Arrhythmias*, Zipes D.P., Jalife J, (eds.), Grune & Stratton: Orlando, pp. 363-378.

14. Allessie, M. A., Bonke, F. I. M., and Schopman, F. J. G., 1977, Interactions across an inexcitable region as a cause of ectopic activity in acute regional myocardial ischemia. A study in intact porcine and canine hearts and computer models, *Circ. Res.* 50: 527-537.

16. Dzwonczyk, R., Brown, C. G., and Werma, H. A., 1990, The median frequency of the ECG during ventricular fibrillation: Its use in an algorithm for estimating the duration of cardiac arrest, *IEE Trans. Biomed. Eng.* 37:640-645.

17. Akiyama, T. 1981, Intracellular recording of in situ ventricular cells during ventricular fibrillation, *Am. J. Physiol.* 240: H465-H471.

18. Allessie, M. A., Schalij, M. J., Kirchhof, C. J., Boersma, L., Huyberts, M.and Hollen, J., 1990, Electrophysiology of spiral waves in two dimensions: the role of anisotropy, *Annals N.Y. Acad. Sci.* 591: 247-256.

19. Davidenko, J. M., Pertsov, A. V., Salomonsz, R., Baxter, W., and Jalife, J., 1992, Stationary and drifting spiral waves of excitation in isolated cardiac muscle, *Nature* 355: 349-351.

20. Dillon, S., and Morad, M., 1981, A new laser scanning system for measuring action potential propagation in the heart, *Science* 214: 453-456.

21. Efimov, I. R., Huang, D. T., and Salama, G., 1994, Optical mapping of repolarization and refractoriness from intact hearts, *Circulation* 90: 1469-1480.

22. Knisley, S. B., Blitchington, T. F., Hill, B. C., Grant, A. O., Smith, W. M., Pilkington, T. C., and Ideker, R. E., 1993, Optical measurements of transmembrane potential changes during electric field stimulation of ventricular cells, *Circ. Res.* 72: 255-270.

23. Dzwonczyk, R., Brown, C. G., and Verma, H. A., 1990, The median frequency of the ECG during ventricular fibrillation: its use in an algorithm for estimating the duration of cardiac arrest, *IEEE Trans. Biomed. Eng.* 37: 640-646.

24. Camm, A. J., Davies, D. W., and Ward, D. E., 1987, Tachycardia recognition by implantable electronic devices, *PACE Pacing Clin. Electrophysiol.* 10: 1175-1190.

25. Geselowitz, D. B., Smith, S., Mowrey, K., and Berbari, E. J., 1991, Model studies of extracellular electrograms arising from an excitation wave propagating in a thin layer, *IEEE Trans. Biomed. Eng.* 38: 526-531.

26. Greenhut, S. E., Dicarlo, L. A., Jenkins, J. M., Throne, R. D. and Winston, S. A., 1991, Identification of ventricular tachycardia using intracardiac electrograms: a comparison of unipolar versus bipolar waveform analysis, *PACE Pacing Clin. Electrophysiol.* 14: 427-433.

27. Lin, D., DiCarlo, L. A. and Jenkins, J. M. 1988, Identification of ventricular tachycardia using intracavitary electrograms: analysis of time and frequency domain patterns, *PACE* 11: 41-249.

28. Throne, R. D., Jenkins, J. M., and DiCarlo, L. A.. 1990, Intraventricular electrogram analysis for ventricular tachycardia detection: statistical validation, *PACE Pacing Clin. Electrophysiol.* 13: 1596-1601.

29. Chen, S., Thakor, N. V., and Mower, M. M., 1987, Ventricular fibrillation detection by a regression test on the autocorrelation function, *Med. Biol. Eng. Comput.* 25: 241-249.

30. Steinhaus, B. M., Wells, R. T., Greenhut, S. E., Maas, S. M., Nappholz, T. A., Jenkins, J. M., and DiCarlo, L. A., 1990, Detection of ventricular tachycardia using scanning correlation analysis, PACE. Pacing Clin. Electrophysiol. 13: 1930-1936.

31. Nygards, M., and Hulting, J., 1977, Recognition of ventricular fibrillation utilizing the power spectrum of the ECG, *Proc. Comput. Cardiol.*, IEEE Comput. Soc. Press, pp. 393-397.

32. Babloyantz, A. and Destexhe, A., 1988, Is the normal heart a periodic oscillator? *Biol. Cybernetics* 58: 203-211.

33. Casaleggio, A., Ranjan, R., and Thakor, N. V., 1994, Correlation dimension analysis of epicardial cell action potentials during different cardiac arrhythmias, *Proc. Comput. Cardiol.*, IEEE Comput. Soc. Press, pp. 697-700.

34. Russel, D. C., Smith, H. J., and Oliver, M. F., 1979, Transmembrane potential changes and ventricular fibrillation during repetitive myocardial ischemia in the dog, *Br. Heart J.* 42: 88-96.

35. Hogancamp, C. E., Kardech, M., Danforth, W. H., and Bing, R. J., 1959, Transmembrane electrical potentials in ventricular tachycardia and fibrillation, *Am. Heart J.* 57: 214-222.

36. Joyner, R. W., Picone, J., Veenstra, R., and Rawling, D., 1983, Propagation through electrically coupled cells. Effects of regional changes in membrane properties, *Circ. Res.* 53: 526-534.

37. Morkrid, L., Ohm, O. J., and Engedal, H., 1984, Time domain and spectral analysis of electrograms in man during regular ventricular activity and ventricular fibrillation, *IEEE Trans. Biomed. Eng.* 31: 350-355.

38. Cohen, L., 1989, Time-frequency distributions: a review, *Proc. IEEE* 77: 941-981.

39. Boashash, B., 1992, *Time-Frequency Signal Analysis: Methods and Applications*, Australia: Longman Cheshire.

40. Yi-Sheng, Z., Bin, Z., and Thakor, N. V., 1996, Variable convergence adaptive filter and its application to cardiac action potential, *Med. Biol. Eng. Comput.*, in press.

41. Thakor, N. V., and Yi-Sheng, Z., 1991, Some applications of adaptive filtering to ECG analysis: noise cancellation and arrhythmia detection, *IEEE Trans. Biomed. Eng.* 38: 785-794.

42. Laguna, P., Jane, R., Meste, O., Poon, P., Caminal, P., Rix, H., and Thakor, N. V., 1992, Adaptive filter for event-related bioelectric signals using an impulse correlated reference input: comparison with signal averaging techniques, *IEEE Trans. Biomed. Eng.* 39: 1032-1044.

43. Spach, M. S., Miller III, W. T., MillerPJones, E., Warren, R. B., 1979, Extracellular potentials related to intracellular action potentials during impulse conduction in anisotropic canine cardiac muscle, *Circ. Res.* 45: 188-204.

44. Baykal, A., Ranjan, R., and Thakor, N. V., 1994, Model based analysis of ECG during early stages of ventricular fibrillation, *J. Electrocardiol.* 27 (suppl.): 84-90.

45. Kaplan, D., Smith, J., Saxberg, B., and Cohen, R., 1988, Nonlinear dynamics in cardiac conduction, *Math. Biosci.* 90: 19-48.

46. Parker, T. S., and Chua, L. O. 1987, Chaos: a tutorial for engineers, *Proc. IEEE* 75: 982-1008.

47. Casaleggio, A., Ranjan, R., and Thakor, N. V., 1996, Dimensional analysis of the electrical activity at the epicardium of isolated hearts, *Int. J. Bifurc. Chaos,* to be published.

48. Jenkins, J., Bump, T., and Mukenbeck, F., 1984, Tachycardia detection in implantable antitachycardia devices, *PACE Pacing Clin. Electrophysiol.* 7: 1273-1277.

49. Khoury, D. S., and Wilkoff, B. L., 1990, Tachycardia recognition algorithms for implantable systems, *IEEE Eng. Med. Biol. Mag.* 9: 40-42.

50. Ropella, K. M., Baerman, J. M., Sahakian, A. V., and Swiryn, S., 1990, Differentiation of ventricular tachyarrhythmias, *Circulation* 82: 2035-2043.

51. Thakor, N. V., Zhu, Y. S., and Pan, K. Y., 1990, Ventricular tachycardia and fibrillation detection by a sequential hypothesis testing algorithm, *IEEE Trans. Biomed. Eng.* 37: 837-843.

52. DiCarlo, L. A., Throne, R. D., and Jenkins, J. M., 1991, A time-domain analysis of intracardiac electrograms for arrhythmia detection, *PACE Pacing Clin. Electrophysiol.* 14: 329-336.

53. Shoupu, C., Thakor, N. V., and Mower, M. M., 1987, Ventricular fibrillation detection by regression test of autocorrelation function, *Med. Biol. Eng. Comput.* 25: 241-249.

54. Thakor, N. V. and Pan, K., 1990, Tachycardia and Fibrillation Detection by Automatic Implantable Cardioverter-Defibrillators: Sequential Testing in the Time Domain, *IEEE Eng. Med. Biol. Soc. Mag.* 9: 21-24.

55. Thakor, N. V., Natarajan, A., and Tomaselli, G., 1994, Multiway sequential hypothesis testing for tachyarrhythmia discrimination, *IEEE Trans. Biomed. Eng.* 41: 480-487.

56. Fishler, M. G. and Thakor, N. V., 1991, massively parallel computer model of propagation through two dimensional cardiac syncytium, PACE Pacing Clin. Electrophysiol. 14 (Part II): 1694-1699.

57. Mirowski, M., and Mower, M. M., 1985, The automatic implantable defibrillator, *Am. Heart J.* 100: 996-1004.

58. Thakor, N. V., 1984, From Holter monitors to automatic defibrillators: developments in ambulatory arrhythmia monitoring, *IEEE Trans. Biomed. Eng.* 31: 700-708.

59. Fisher, J. D., Kim, S. G., and Mercando, A. D., 1988, Electrical devices for treatment of arrhythmias, *Am. J. Cardiol.* 61: 45a-57a.

60. Casaleggio, A., 1993, Differences on the correlation dimension of MIT-BIH ECG database recordings, *Proc. Comput. Cardiol.,* IEEE Comput. Soc. Press, pp. 539-542.

62 Ropella, K. M., A. V., M., Swiryn, S., 1989, The coherence spectrum. A quantitative discriminator of fibrillatory and nonfibrillatory cardiac rhythms, *Circulation* 80: 112-119.

20

FETAL ECG DETECTION AND APPLICATIONS

Solange Akselrod, Jacob Karin, and Michael Hirsch

Center for Medical Physics
School of Physics
Tel Aviv University, Israel

ABSTRACT

A new fetal electronic monitoring system (FEMO) has been developed, based on the automatic detection of fetal heart beats from the combined maternal abdominal ECG signal. It uses a single lead abdominal signal and submits it to an elaborate algorithm for the on-line automatic detection of fetal heart rate and the computation of an average fetal ECG complex.

The ability to noninvasively measure the fetal ECG complex and to detect accurate beat-to-beat fetal heart rate from abdominal signal, can become a diagnostic tool of particular importance. The average fetal ECG complex provides direct information about the fetal heart (its intrinsic electrical activity, possible congenital malformations or fetal arrhythmia), from early stages of pregnancy on, and not only at birth. By means of the "true" beat-to-beat fetal heart rate, our system can provide the "classical" direct clinical insight in the well-being of the baby. In addition, in conjunction with tools for spectral analysis of beat-to-beat HR fluctuations, it permits the investigation of the development and maturation of autonomic control in the fetus. It allows estimation of fetal behavioral stages, in particular the indirect detection of fetal breathing movements from its HR variability, instead of requiring cumbersome ultrasonic imaging.

INTRODUCTION

ECG is one of the most widely used monitoring methods available. It is well known and is used in a variety of clinical situations. ECG is available today for adults, children and newborns. Unfortunately, it is not available for fetuses, excluding a very short time during labor, when an invasive scalp electrode can be attached in order to obtain a fetal ECG trace.

Currently fetal monitoring is performed by Doppler ultrasound devices, measuring the fetal heart rate. The clinical evaluation of these heart rate traces is not sensitive enough to allow for clear-cut decisions, thus leading to a high rate of operational deliveries. An accurate and reliable instantaneous fetal HR might provide a much better insight in the

Advances in Processing and Pattern Analysis of Biological Signals, Edited by Isak Gath and Gideon F. Inbar
Plenum Press, New York, 1996

behavioral state, the autonomic functioning and well-being of the fetus. Extensive clinical studies, performed with scalp electrode fetal ECG during delivery, have shown that the additional data available from the fetal ECG complex may reduce the number of unnecessary operational deliveries by 46% (1).

Nowadays, the non-invasive fetal ECG measurement during pregnancy is not included in any available commercial monitoring system. Using the FEMO system, developed in our laboratory, it is possible to record fetal ECG complex as early as 20 weeks of gestation and the clinical implications of this new data remain to be determined.

The core of such a system is the need for a reliable separation of the fetal signal from the maternal contribution. The task of the fetal ECG detection is complicated by the variability of fetal and maternal heart rate, by the overlap in frequency content of their respective ECG's, by the small amplitude of the fetal complex, often buried in the noise or hidden behind the "huge" maternal complex. The algorithm is based on a combined use of first and second derivatives of the abdominal signal, using a derivative procedure which is also a low-pass filter to enhance either fetal or maternal contribution, depending on parameters of the derivative. Maternal templates are computed for these derivatives and correspondingly subtracted. Algebraic computations with the residuals and adequate thresholding allow localization of the fetal ECG complex. Reliable detection is achieved, with high sensitivity and specificity. Validation has been performed against the "gold standard" scalp ECG signal, which can be measured during labor. Excellent agreement is obtained both in the fetal heart rate and in the shape of the fetal ECG complex.

The single most difficult problem in the development of a reliable fetal ECG monitoring system is the detection of the fetal heart beat from the combined maternal abdominal ECG signal. The small fetal heart signal is often hidden by the much larger maternal complexes, or by muscle noise and is usually not visible in the recorded trace.

The FEMO system described below performs accurate detection of fetal heart beats from abdominal ECG. The core of this system is an elaborated detection algorithm, developed at the Center for Medical Physics at Tel-Aviv University, and now implemented in a monitoring system for clinical research (Medco Electronic Systems Ltd.).

The on-line detection of fetal R-waves results in the creation of RR intervals time series which are then transformed into an instantaneous fetal heart rate chart.

Using an off-line utility, the fetal average ECG complex is computed and drawn. The averaging utility allows on-screen measurements of the ECG complex and comparison of up to four different fetal ECG complexes from different patients or from sub-traces of a single examination.

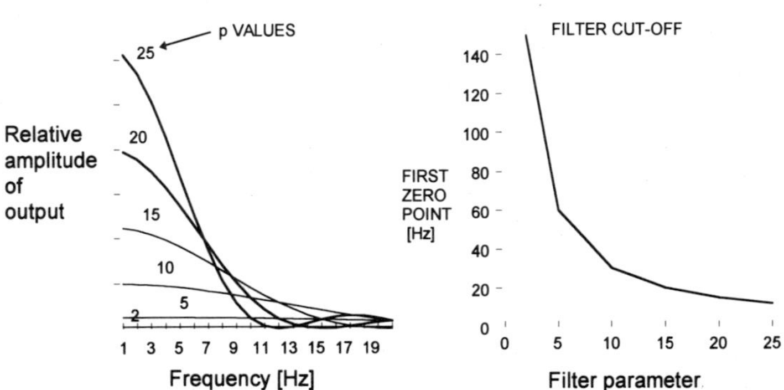

Figure 1. The response curves of some filters and the first zero point of the filtration curves as a function of the parameter p.

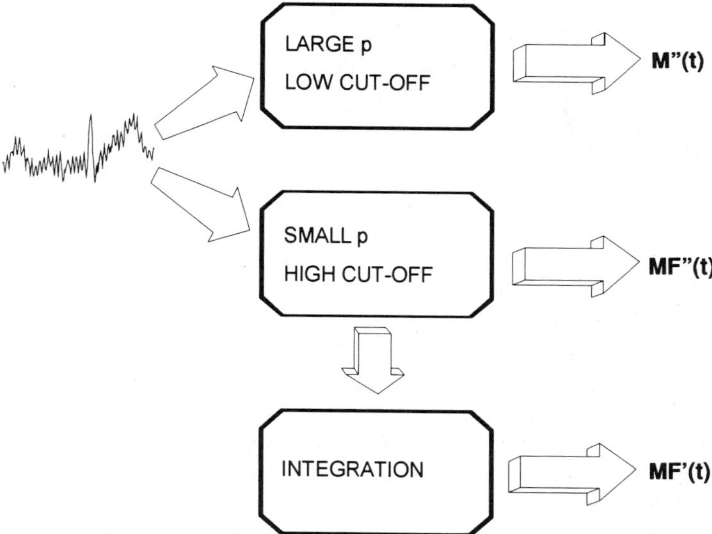

Figure 2. Splitting the abdominal ECG into three channels: A second derivative with maternal contribution only (M″), and a second and first derivatives with combined fetal-maternal (MF″, MF′) contributions.

METHODS

The core of the detection algorithm is the use of first and second derivatives of the abdominal ECG to detect fetal R-waves. The derivatives eliminate the influence of baseline

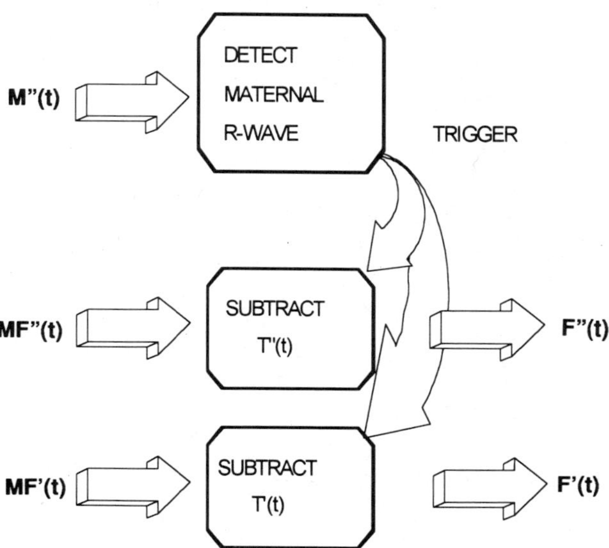

Figure 3. Subtraction of previously made templates (T′ and T″) of the maternal contribution from the combined signal. Each channel is processed separately to increase detection accuracy. The outputs (F′,F″) are the pure contribution of the fetus to the derivatives.

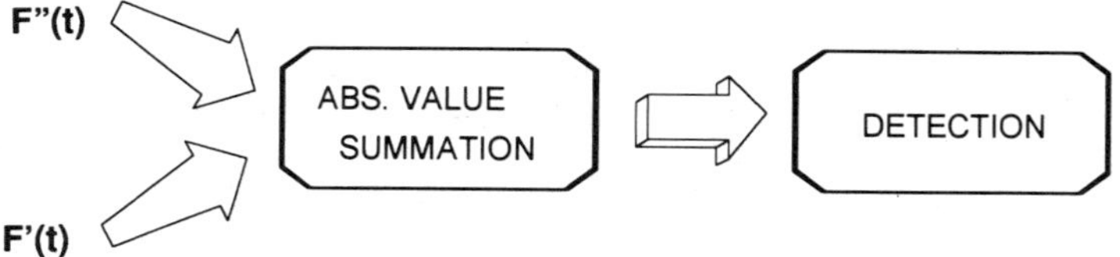

Figure 4. Combining the fetal-only derivatives into a single wave suitable for threshold detection of fetal R-waves.

wandering, while the specific derivation procedure we use, applies additional filtering to the signal, thus helping to differentiate the maternal contribution from the fetal contribution.

The derivation process applied is simple, time saving and effective. It incorporates a low-pass filtration described in Fig 1. The parameter p determines the cut off point of the filter and we use two different p parameters in the algorithm. Being the product of a much smaller heart, the fetal R-wave is narrower than the maternal R-wave, reaching higher frequencies. By using a large value of p we can eliminate the fetal contribution to the derivatives, producing a maternal-only signal M″ (see Fig 2). By using a small value of p, we only filter out the very low frequencies and the resulting signal MF″ and MF′ (see Fig 2) is a combined maternal fetal signal.

The maternal-only signal (M″) is used to detect maternal R-waves and to trigger a subtraction procedure for the elimination of the maternal contribution from the combined derivatives (MF′ and MF″). For this purpose, maternal templates (T′ and T″) are created and subtracted from each derivative signal MF′ and MF″.

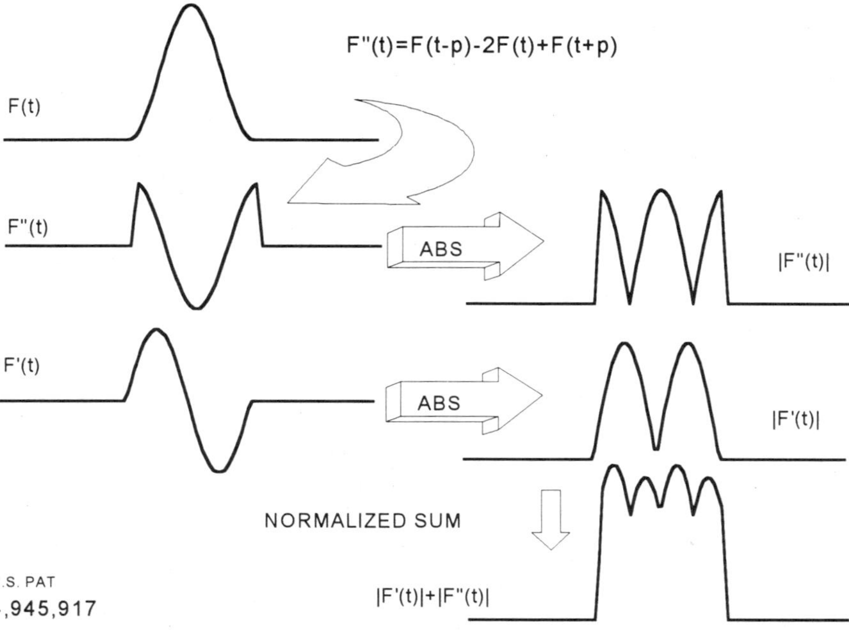

Figure 5. Illustration of the derivation process, describing the advantage of using the normalized sum of the first and second derivatives.

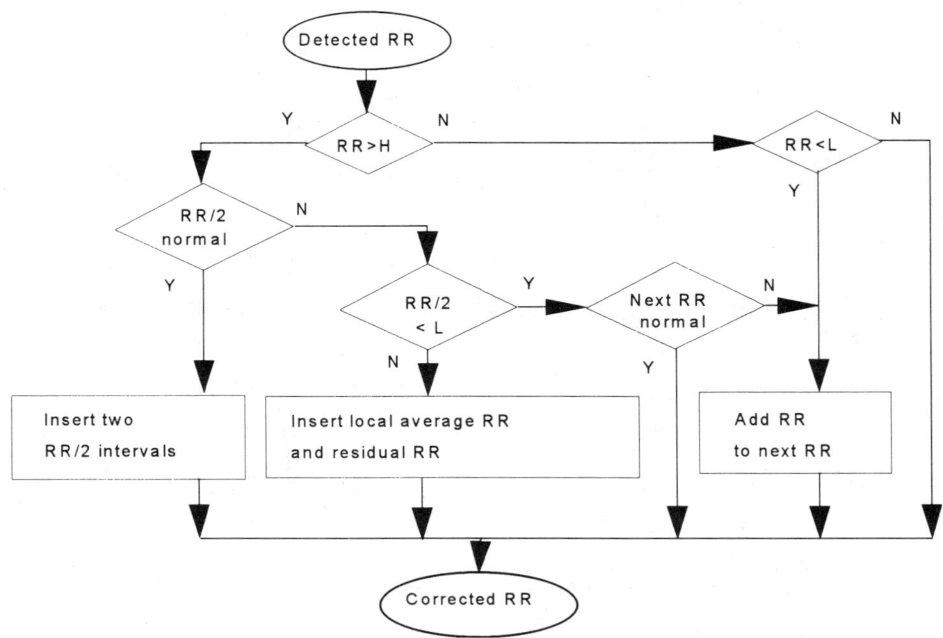

Figure 6. Principles of the automatic RR correction procedure. The actual procedure is more complicated and uses a large set of rules to correct for various situations.

The subtraction process described in Fig. 3 is most vulnerable to computational errors. If the subtraction is not performed at the correct moment, the residual value of the maternal contribution is larger than the fetal contribution.

In addition to solving the baseline wandering problem, processing the two fetal-maternal derivatives (the first and the second derivatives of the abdominal ECG) as two independent data channels in parallel, reduces computational errors. We found out that computational errors did not occur in the two channels simultaneously. Thus the double computation and the combination of the resulting functions reduced the number of detection errors. Fig. 5 illustrates the combination of two derivatives and the logic behind it.

Combining the first and second derivatives of the signal is effective for signals of triangular morphology. As illustrated in Fig.5, the Gaussian like R-wave produces two derivatives that are complementing upon summation of their absolute normalized value, producing an almost square wave for each R wave. The square wave function produced is more robust than the original R-waves and thus easier to detect.

The RR intervals detected at this stage are scanned automatically by a correction procedure, and misplaced R-waves are fixed. The correction principles are presented in Fig.6, but the actual procedure is more complicated, employing a large set of rules in order to correct accurately in many situations. Once the location of each fetal and maternal R-wave is determined, we can return to the original abdominal ECG trace, use the fetal R-waves that are not overlapping maternal R-waves and compute an average fetal ECG complex.

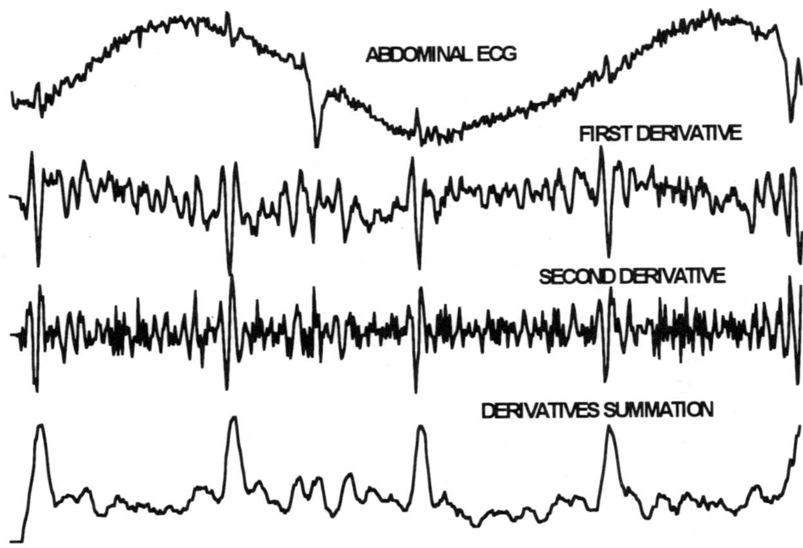

Figure 7. An example of the elimination of baseline wandering. Although the abdominal ECG is wandering, the derivatives are not influenced by it, and the resulting derivatives summation is of good quality.

RESULTS

In the examples presented (Figs 7 and 8), the effects of baseline wandering and large signal to noise ratio are demonstrated. One of the common problems is an overlap between maternal and fetal R-waves. Figure 9 illustrates this kind of overlap, and demonstrates the

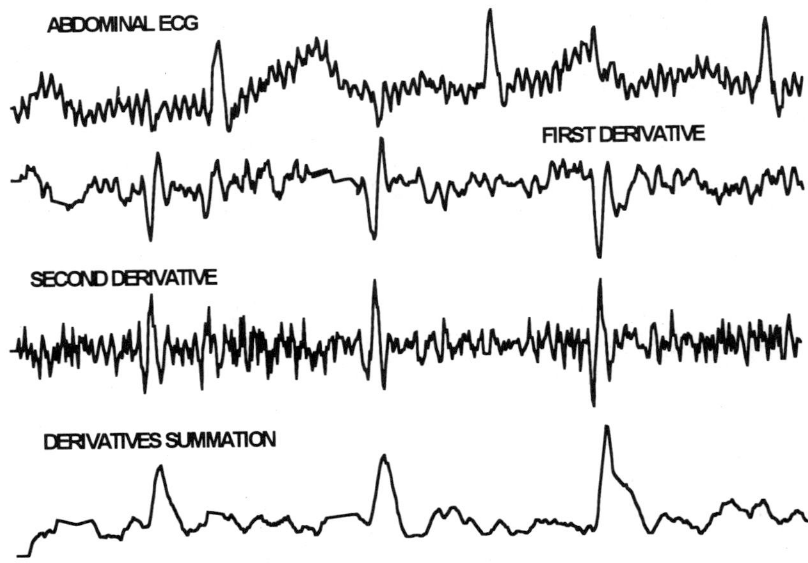

Figure 8. Detecting fetal R-waves from a noisy abdominal ECG trace. It is easy to see the added value of the combined derivatives relative to each derivative on its own.

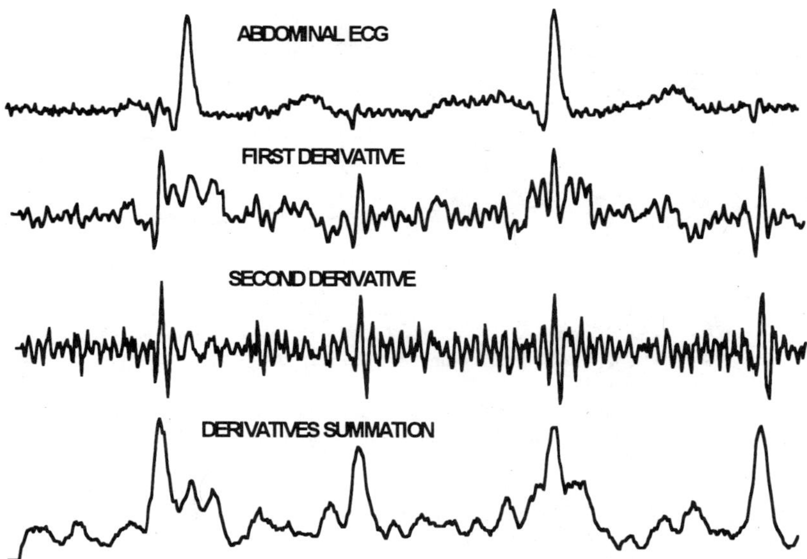

Figure 9. An example of detection of fetal R-wave overlapping maternal R-waves. Using the derivatives of the signal it is possible to remove the maternal contribution without the removal of the fetal contribution, therefore the detection is accurate even when the R-waves are overlapping.

ability of the FEMO system to detect accurately fetal R-waves even when overlapped by maternal R-waves.

The FEMO system achieves a very accurate detection of fetal R waves. In a controlled study, 40 patients were examined (gestational age 36 to 41 weeks, average 39.3) validating the detection accuracy beat by beat using a special viewing utility.

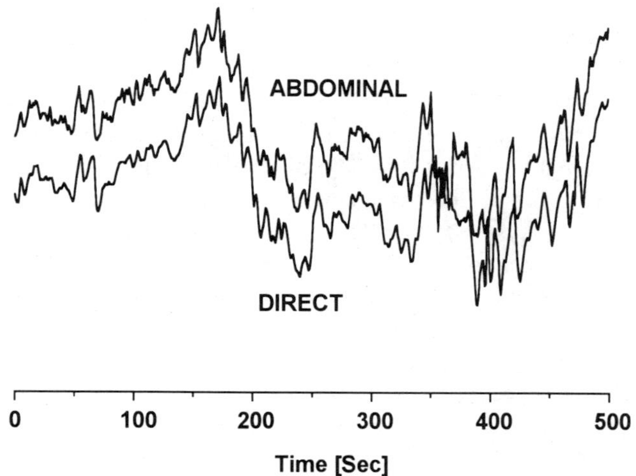

Figure 10. Fetal heart rate charts from abdominal ECG (top trace) and from direct scalp electrode measurement (bottom trace) measured simultaneously during labor.

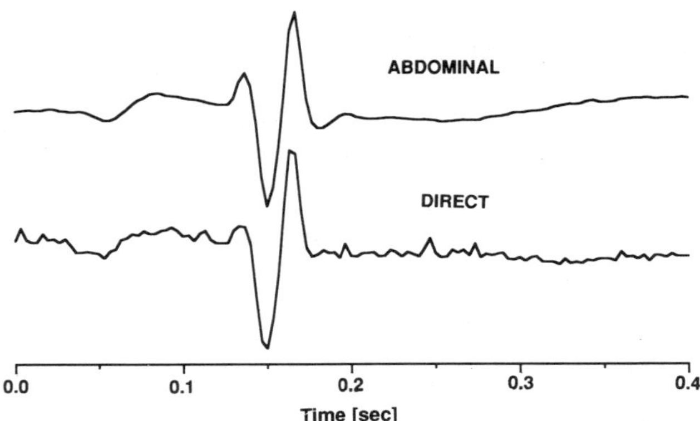

Figure 11. An example of comparison between the average abdominal ECG complex and a single ECG complex of the corresponding direct fetal ECG signal recorded by means of scalp electrode.

In 6 patients the system failed to detect the fetal R-wave, and in two patients detection accuracy was below 80%. The population success percentage is therefore 80% (32 out of 40). Though the system works for only 80% of the near-term population, the detection accuracy is excellent.

Indeed, the average accuracy of R wave detection per successful patient is 95.2% for false negative and 97.1% for false positive. Instantaneous fetal HR detected by the FEMO system, simultaneously with fetal HR obtained from scalp electrode from the same fetus during labor is displayed in Fig. 10, indicating the excellent agreement between our detection algorithm and the scalp electrode 'gold standard'.

Average fetal ECG derived from the FEMO system was compared with the direct scalp electrode ECG (see Fig. 11). Based on the measured QRS widths of eight patients tested, the correlation between the non-invasive abdominal ECG and the invasive scalp electrode ECG is 0.896 and is significant at $p<0.0001$.

CONCLUSIONS

The fetal R-wave detection algorithm presented has been developed to a point that enables trained personnel to operate it as a bedside fetal monitor. The fetal heart rate chart (Fig. 10) produced by the system is equivalent and more accurate than the fetal HR produced by the Doppler ultrasound systems. It can thus be used as a sensitive tool for fetal well-being assessment. The measurement of an average fetal ECG complex (Fig. 11) opens new fields for physicians to explore and new diagnostic tools to apply. Recording fetal ECG at early stages of pregnancy can detect congenital malfunction of the heart. Towards the end of pregnancy and during labor, fetal ECG can provide an early warning to fetal distress, which may be more specific than the fetal HR chart by itself. The accurate fetal HR can be used to apply the well-known techniques of spectral analysis of heart rate fluctuations in order to obtain information about the state of the fetal central nervous system (CNS) (2) and in particular about the existence of fetal breathing movements (3, 4). Being a low cost system and easy to implement, the proposed system can be used as a massive screening test for all pregnant populations.

REFFERENCES

1. Westgate, J., Harris, M., Curnow, J. S., and Greene, K.R., 1993, Plymouth randomized trial of cardiotocogram only versus ST waveform plus cardiotocogram for intrapartum monitoring in 2400 cases, *Am. J. Obstet. Gynecol.* 169(5):1151-1160.
2. Karin, J., Hirsch, M., and Akselrod, S., 1993, An estimate of fetal autonomic state by spectral analysis of fetal heart rate fluctuations, *Pediatric Res* 34(2):134-138.
3. Myers, M. M., Fifer, W., Haiken, J., and Stark, R. I., 1990, Relationship between breathing activity and heart rate in fetal baboons, *Am J Physiol.* 258:R1485-1497.
4. Hirsch, M., Karin, J., Shechter, B., Segal, O., Merlob, P., Kaplan, B., and Akselrod, S., 1993, Detection of fetal breathing activity in real-time by means of spectral analysis of fetal heart rate fluctuations, *2nd World Congress of Perinatal Medicine* pp. 535-539.

21

PROCESSING, FEATURE EXTRACTION AND CLASSIFICATION OF BODY SURFACE POTENTIAL MAPS

Dan R. Adam

The Cardiac Research Center
Department of Biomedical Engineering
Technion, I.I.T., Haifa 32000, Israel

ABSTRACT

Body surface potential mapping (BSPM) is an electrocardiographic technique significantly better than standard ECG, yet its technical complexity and high cost have limited its use. A simplifying approach is suggested where the electrodes, spread out over the whole thorax, do not measure the whole ECG signal but only detect the time instants when the potentials at each electrode location cross some given threshold level. An algorithm for reconstruction of the complete BSPM information from these level-crossings has been utilized, based on the Karhunen-Loeve (K-L) expansion method. Complete BSPM recordings of a training set from subjects produces the set of basis functions for the reconstruction procedure. The reconstructions from level-crossings of a set of subjects with similar pathologies produce average RMS reconstruction error of about 140 uV, and correlation between original and estimated BSPMs of around 0.91. The simplified approach to BSPM measurement makes it a possible replacement to clinical ECG; here features of BSPM sequences are defined, and tested for classification and easiness of display. Three dimensional luci of the centers of mass of the BSPMs have been found to be a feature easy for display and serve as a good classifier.

INTRODUCTION

The process of zero-crossing (ZC) has been extensively used as a means for data compression. For one dimensional (1D) data (e.g., a dependent variable as a function of time), the process can be simply described as registering the time instants when the function crosses a voltage level (zero volts in the case of ZC). The compression ratio is significant, but usually destructive, as in many cases reconstruction of the original signal is difficult. There is a need to identify conditions for proper signal reconstruction, guaranteeing stable and unique solutions. Since a real function is well defined by its ZC if all its zeros are real

Advances in Processing and Pattern Analysis of Biological Signals, Edited by Isak Gath and Gideon F. Inbar
Plenum Press, New York, 1996

(1), and a band-limited function is complete (analytical everywhere) and therefore is defined by its zeros (complex or real) up to a scaling and exponential factor (2), the task has been to define conditions under which a signal is guaranteed to have only zeros. A less demanding condition has been defined, where the rate of real zeros is higher in some respect than the information content provided by these signals. A class of band-limited signals was defined (3), which are represented uniquely, within a scale factor, by their ZC. Most signals, unfortunately, though real and band-limited, do not comply with the restrictions, and cannot be uniquely reconstructed from their ZC.

When multidimensional signals are considered, two dimensional (2D) and above, significantly better results have been achieved. While ZC of 1D signals produce a discrete and usually limited number of points, 2D data create 2D contours, and ZC of three dimensional (3D) data produce surfaces. ZC of multidimensional data produce, in theory, an infinite number of zero-crossing points. Though in reality the number of points is finite, in many cases the number is large enough to satisfy the requirements of the Nyiquist theorem. For those cases, two different approaches have been suggested: A direct application of 2D Fourier Series Expansion (4, 5), and an iterative algorithm (5-7). In the first approach, the expansion coefficients appear as the set of unknowns of a linear system of equations; their solution provides an estimate of the expansion coefficients from the measured ZC instants. This approach assumes the data to be a periodic, band limited image, which provides a finite number of unknowns, and requires precise measurements of the ZC. The reports cited above describe reasonable reconstruction when the number of ZC is 4 times the number of Fourier coefficients. The second approach requires imposing restrictions on the reconstructed signal, alternatively in the signal (e.g., time) and frequency domains, for each iteration step. The restriction imposed on the signal in the frequency domain might be a low-pass filter (e.g., limited the reconstructed signal to the frequencies contained in the original signal), while in the signal domain the restriction might impose the correct sign/polarity on the reconstructed signal, similar to that of the original signal. The transition between the domains is done by DFT.

These two approaches are applicable to certain signals only, and are not useful for body surface potential maps (BSPM) compression/reconstruction: for 2D BSPM (one map), the finite, though large number of electrodes, allows only rough approximation of the ZC; for 3D (a time sequence of maps), the spectral information in the maps requires a relatively large number of coefficients. There is a need for a different approach, and the one suggested here utilizes intrinsic properties of the BSPM, thus providing an additional constraint and limiting the reconstructed signal to a specific class of signals.

Body surface potential maps (BSPM) provide the required high spatial sampling of potential distribution over the torso, that conventional 12 lead electrocardiogram (ECG) does not provide. The current ECG measuring procedure provides only spatially averaged information, thus losing fine details of the local movement and intensity of the electrical sources within the muscle. These fine details were reported to have a great significance in expressing the electrophysiological condition of the heart (8, 9). Due to this poor quality of the ECG, only some correlation exists between abnormal features of the ECG wave form and cardiac diseases such as myocardial infarction. A somewhat better correlation is found with major conduction pathologies like arrhythmia, Wolff-Parkinson-White syndrome (WPW) and bundle brunch block. Though in many cases the existence of the disease is recognized through the ECG, its severity and location cannot usually be defined. The BSPM exhibit sensitivity to local cardiac electrical events (10, 11), and are capable of detecting physiological and diagnostic information which is not obtained by conventional ECG. The majority of published reports on BSPM describe and analyze surface maps qualitatively. An earlier work done by Taccardi (12) describes a sequence of maps during normal beat, while Mirvis *et al.* (13) explore the maps in dogs with myocardial infarction. In special cases of myocardial infarction it has been shown that although the electrocardiograms exhibit normal wave form, the surface maps behave abnormal (14, 15). Similar results have been reported for abnormalities provoked by exercise in cases of ischemia (16, 17). Contrary to the low correlation between ECG and myocardial electrical processes, a more direct correlation was found between the negative and positive area in the BSPMs and epicardial events, and even with major activation sequences within the myocardium (18). The maps have been reported to be also advantageous as a research tool. Potential maps calculated on the body surface as part of the forward problem, are sensitive to the time dependency of the electrical generators and their number, and therefore useful in evaluating mathematical models of the cardiac electrical activity. These models are used today only in research (solving the inverse/forward problem) with continuous effort towards their use in clinical applications (19, 20).

Quantitative analysis of BSPM was only lately addressed. It is an issue of major importance, since it determines whether the technique becomes widely used by the clinicians. The maps contain clinically important and useful information regarding the electrical status of the heart, but visual inspection can not detect and comprehend the relevant details. Hence it is desirable to have an objective and sensitive tool that can extract and express in a quantitative way the desirable information. Few quantitative diagnostic methods have been suggested for discriminating between normal and abnormal maps. A method for detecting infarction was proposed (21), in which the movements of minima and maxima of the BSPM peaks are traced during the cardiac cycle. The Karhunen - Loeve expansion coefficients (22, 23) were used to characterize BSPM of normal and abnormal patients and to detect myocardial infarction and hypertrophic cardiomyopathy (24). Singular value decomposition method was utilized to localize the ventricular pre-excitation area in WPW (10). Isointegral maps were calculated in order to detect areas of infarction in the heart muscle (25, 26). Montague *et al.* (25) characterized normal subjects by an averaged isointegral map calculated for each period in the ECG wave form: P, Q, QRS, ST-T and T. Difference maps between normals and each MI patient were computed by subtracting the corresponding isointegral maps. The difference areas were then checked for a threshold value in an attempt to produce discrimination. These methods do not solve the problems of data compression and quantitative characterization simultaneously in a way that would make the BSPM clinically popular. A different approach was used by Nikias (27) to quantitatively investigate the effect of an induced ischemia (in dogs), on the myocardial electrical wave front activation. Estimating a frequency wave number spectrum, a single map was created from the whole BSPM sequence. This map was further used to estimate the number, orientation and bandwidth of wave fronts inside the heart muscle.

This study presents a new technique for reconstructing the complete sequence of BSPM from their measured ZC; thus allowing the use of a significantly simplified measurement system in which each electrode just measures the time instants of the ZC. This study also presents a method for characterizing body surface potential maps in a compressed and compact way. It preserves the uniqueness of the maps without losing the fine but significant details, which play a major role in diagnosing the various kinds of abnormalities. By thresholding the 3D isopotential maps at different levels, annuli are formed. The annuli may then be characterized by a number of features, which are used to discern between normal and abnormal patterns. This approach leads towards a practical and useful diagnostic method, which is suitable for laboratories and clinical use.

METHODS

Sampling and Compression of Data from the BSPM

An assumption inherent to the approach outlined below is that the distribution of potentials on the body surface is a 3D process, varying in time and the torso space. It is described as a continuous surface. The electrodes, spaced at specific locations around the torso, sample the process in its spacial coordinates, and the measurement of the ZC sample the process in time, at varying intervals. These samples establish a sample set, which represents the BSPM function.

Technical considerations dictate the spacing between the electrodes so that at some locations there are just few electrodes (e.g., no electrodes under the armpits), and spacing is uneven. Additionally, the potential level set for the comparator for performing the ZC measurement, is usually not set to zero. Setting the threshold above the normal noise level minimizes the number of crossings measured to those related to the signal (+noise), and leaving out those related to noise alone. Selecting various potential levels for subsets of electrodes improves somewhat the performance of the procedure (termed level crossings (LC)). Each electrode is exposed to the continuous potential changes on the skin's surface. The electronic circuitry does not use amplification, filtration, signal conditioning and sampling at a rate of several hundreds of samples (if not thousands) per sec, but instead it measures only several LC instants per heart beat.

Reconstruction of BSPM

Each electrode measures several LC instants per heart beat, vs. several hundreds of samples (if not thousands) during regular ECG measurement. The sampled set of the BSPM is very small. This produces a significant data compression, bur also requires that reconstruction of the complete BSPM from their LC should be based on a very compact representation of the BSPM. Since the BSPM are viewed here as samples of a random process, random signal processing transforms/expansions can be used, e.g., the K-L transform. Using such an approach, the autocorrelation kernel of the random process is decomposed to its eigenfunction basis. From such a basis, and allowing a specified RMS error, reconstruction is possible with a very small set of eigenfunctions: they are arranged in decreasing order, and only the largest ones are used to produce a reconstructed signal similar to the original within the allowed error.

The Eigenfunction Basis. Reconstruction of the 3D BSPM from their LC requires a small set of eigenfunctions, which serve as a basis that characterizes a large group of BSPMs, recorded from a population of normal subjects and patients. Such a basis is constructed by measuring a full set of BSPMs, over at least one representative cardiac cycle, from various subjects and patients that describe (from the statistically point of view) the normals and the different pathologies. Each measurement is transformed into a vector, which comprises all the time samples from all electrodes; a matrix is constructed, in which each vector represents a subject. The covariance matrix is estimated and then decomposed so that the eigenfunction (orthonormal) basis is calculated. The calculation of the eigenfunction basis is performed by the use of the Singular Value Decomposition (SVD), instead of the K-L method, since the latter requires calculations with huge matrices. The SVD procedure produces the required matrix which is diagonal, and contains the ordered singular values. A threshold is set, depending on the chosen RMS error value, and all singular values below it are set to zero, thus reducing the matrix/vector rank to a very low one.

Once the expansion matrix was calculated for a representative population, measurements of additional subjects are made according to the procedure described in this paper - i.e., by measuring and storing only the LC instants instead of the full BSPMs. When the subjects' BSPMs are required for analysis or display - they can be rapidly reconstructed as described below.

Reconstruction of a BSPM from the LC Measurements. According to the procedure presented in this report, the standard measurement of BSPM is reduced to registration of the LC time instants only. A set of such time instants is all that is stored. Once there is a need to observe, display or diagnose the subject's BSPM, they must be reconstructed from their LC instants. The reconstruction procedure is an iterative estimation method, in which the vector which describes the data collected from a subject, which is supposed to comprise all the time samples from all electrodes, initially contains just a few values different from zero - the value of the threshold voltage at the specific electrode and specific time sample at which it was crossed. Using the least-squares method, the decomposition coefficients of this particular subject are estimated, by solving a set of linear equations. Once the set of coefficients is estimated, the vector mentioned above can be calculated, to include the full set of all time samples in all electrodes. The vector can easily be converted to any mode of display, and diagnosis can be performed from the full maps, or features can be extracted from it for quantitative analysis.

Annuli Formation from BSPM

Quantitative evaluation and simple display of the information contained in the BSPM requires further processing. This report offers a two step procedure - first a set of annuli is calculated from a series of maps, which adequately preserves their shape and movement as a function of time, then features are extracted from this set of annuli.

Image processing techniques are used to calculate the shape of the annuli. Amplitude thresholding is performed and an edge detection procedure is taken place. The annuli are calculated in the following way:

- Each 2D map is thresholded twice at two different potentials.
- Each thresholded image is converted to a binary format. Thus two binary images are created from each map.
- The annula is formed by subtracting the two images. More than one annula may be generated by this procedure, depending on the potential distribution.

This procedure is performed separately for the positive and negative voltages of each map. The data reduction procedure outlined above is significant in saving CPU time and in occupying minimal storage area.

The thresholding procedure requires selection of two parameters: The height of the basic thresholding level and the difference between the two levels. The basic level is determined by the noise level, which is computed as the average RMS voltage as measured by four electrodes at the inferior right midaxillary line. These electrodes are far from the myocardial electrical resources and can hardly detect any ECG information. The highest level found in the whole study group is taken as the noise level. The basic thresholding level is then chosen as twice the noise level, in order to minimize the effects of noise, while not losing the fine details. This produces a threshold level which is 14% of the maximum normalized positive peak potential (the R wave). This lowest possible potential was found to be valuable in discriminating between normal and abnormal patterns in maps (8, 9). The choice of low thresholding level is also important for obtaining an annula during the major part of the heart cycle. The difference between the basic thresholding level and the second one, is the other parameter of choice. It is selected empirically in order to enable further processing of the annuli. The requirement is for an accurate estimation of the gradient near the basic level, and therefore the difference between the two levels should be small as possible. This requirement is limited by the spatial resolution and by the requirement of a suitable width of the annuli, (at least one pixel). A difference of 6% is chosen between the two threshold levels and the second threshold level is set at 20% of the positive peak potential. These values of the two threshold levels are used for the whole study group.

Threshold Sensitivity Analysis. Sensitivity of the method to the thresholding levels is examined by estimating the gradient using Robert's operator. For a given function f(x,y) which has partial and continuous derivatives, one may define the operator as the sum of the partial derivatives calculated for the two directions. The amplitude resolution of f(x,y) is dictated by the quantization of the measured BSPM, which is here 12 bit, and results in 10.0 microvolt resolution. The spatial resolution that determines the increments at the two directions, is the distance between the points in the 44x72 grid. Accurate estimation of the gradient is achieved through small increment values, while the increments should be greater than one pixel, so as to enable reasonable results. High increment values above a certain limit, will cause small changes in the potential distribution to pass unnoticed.

Extraction of Features from the Annuli

As mentioned earlier, annuli are calculated from the BSPM (either directly or from those reconstructed from the LC measurements), and can easily be displayed for monitoring and diagnosis. Quantitative analysis of the shape and movement of the annuli series (or of the respective potential distributions in the BSPMs) is made possible by defining features to be extracted from the data. Several features are described below, and are calculated as a function of time, so as to represent the dynamic nature of the annuli (and the BSPMs):

- The loci, in 3D space (the two spatial coordinates of the torso, and time), of the center-of-mass of the annuli during the cardiac cycle. The loci represent the advancement of the activation front in the myocardium. It should differ, therefore, from the common measure of the potential peak (and its movement with time), which has physiological meaning and is sensitive to noise. This is an important feature for discriminating between normal and abnormal myocardial activation, since the sequence of activation changes in abnormal tissue and may result in different time trajectories.
- Two-Dimensional cross-correlation coefficient series estimator. It is a well known measure of similarity between images, and is calculated for each consecutive pair of images, to form a sequence of coefficients. Since it is evaluated here for annuli (instead of full, 3D images) the calculation of the cross-correlation coefficient series estimators is extremely fast. It is expected to express the changes in annuli shape and size, specifically around the QRS; low values of the coefficients are expected during these periods of rapid changes.
- Time of appearance and time of disappearance (relative to the P-QRS-T complex) of the annuli, and also the annuli duration. These indices are dependent on the selected threshold level.
- The distance between the centers of mass of consecutive maps. This index is a measure of the velocity of movement of the annuli, representing the velocity of displacement of the activation front in the myocardium.

Data Acquisition

Records were taken by a BSPM system developed at the BioMedical Engineering Department of Case Western Reserve University as described by Liebman *et al.* (28). The measurements were made at the Rainbow Babies and Children's Hospital, Cleveland, Ohio. The system utilizes 180 uniformly spaced passive dry electrodes which are organized in a grid and mounted in a vest. (11 rows and 18 columns of electrodes, of these 105 electrodes are located on the anterior side and 75 electrodes are on the posterior side of the torso). There are number of vests to match torso size. The ECG from each electrode was band passed between 0.01Hz - 500Hz and sampled at a rate of 1KHz. Each signal was modified by baseline adjustment, gain normalization noise filtering and artifact detection. ECG recordings with excessive noise or with artifact were rejected.

Study Population. This report is based on a one second recording made from 281 subjects: 200 of these were grouped to form the training group, and the rest (81 subjects) formed the test group. The training group included subjects who have been diagnosed earlier, by other conventional means, as follows: 65 normals, 65 patients with WPW, 10 cases of different CAD, 40 of aortic stenosis and 20 VSD patients. The test group included similar distribution of pathologies.

Preprocessing. Several preprocessing steps are required before performing the data reduction procedure. Two dimensional weighted average of nearest neighbors interpolation is carried out to generate a map of 44x72 pixels. Eight neighbors amplitude values are weighted according to 1/(square of D), where D is the distance between the desired and each neighbor points. The analysis (e.g., thresholding) is performed over sequences of BSPM recorded over a full cardiac cycle. Since measurements are done continuously over a few heart cycles, a fiducial point must be chosen, so that a cycle from each subject can be aligned. The fiducial point (in time) is selected as the beginning of the QRS wave, (the first negative deflection of the Q wave in the v2 precordial lead). Amplitude normalization is performed in order to decrease intersubjects variability. All the measurements are divided therefore by the maximum positive R wave value in a whole sequence of maps.

RESULTS

When comparing the original ECG wave forms to the reconstructed ones, either at a specific electrode or at the whole array of electrodes, they match quite well. It is quite obvious that the amplitude is not always accurately reconstructed, but the differences are small in nearly all cases studied. It was found empirically that the number of eigenfunctions used should be approximately the number of unknowns, the number of electrodes (traces needed to be reconstructed). When a smaller set of eigenfunctions was used, the errors were much larger. Increasing the number of eigenfunctions above the number of unknowns did not reduce the error but added to the complexity of the solution.

The main parameter of choice for modifying the data analysis/reconstruction procedures is the voltage value of the threshold for the LC. When using ZC instead of LC (i.e., the level=0 V.), The number of crossings is the largest, due to the noise present in the recordings, but also the RMS error is large because of that noise (RMS error is defined as the square root of the sum, normalized by the number of electrodes, of the squared difference between the 3D discretized original BSPM sequence and the reconstructed one). When increasing the threshold level above (or below) zero, the number of crossings goes down, and so does the number of equations of the linear system that can be defined; the description of the system becomes less over deterministic, and the determination of the decomposition coefficients becomes less robust. There is, however, a range of voltage levels that produces a minimal value of the RMS error. This range starts at values higher than zero (>50uV) up to about 200uV when the RMS error starts to increase. Selecting a crossing-level of +- 100uV produces about 350 sampling points, which is about 4 times the number of unknowns. This allows a large value of overdeterminicity. Choosing the negative value (-100uV) usually produces a larger number of LC. The appropriate RMS error is around 250uV, which is quite large. (See discussion for ways of improving these results).

Once the BSPM's have been compressed/reconstructed, the second stage of quantifying the maps have been attempted: annuli generation and extraction of features for display purposes and classification. The generation of the annuli required performing the threshold procedure (i.e., thresholding twice, at two voltage levels, so as to obtain two isopotential lines). It utilized only a subset of the whole grid of electrodes mounted around the thorax. In each map the above subset produces the annula, while the rest of the electrodes are not used. In a sequence of maps of a particular subject, in each map, a different subset of electrodes may be utilized. It has been found that up to 62% of the whole grid of electrodes mounted in the vest, (an average of 41%) measure the required potentials. (An average of 71 electrodes out of 172). The distribution of the subset of electrodes which measure potentials higher than the higher threshold level, has been found to be as follows: 59% of these electrodes are located on the left side of the thorax above the heart (from the midsternal line to the midaxillary line); 33% of these electrodes are located on the back, and various numbers of electrodes (from 1 to 19) are found on the right side of the thorax. The subset of active electrodes has been always limited to the electrodes mounted in between the third to the eleventh row and between the fourth to the fourteenth column. (Where electrode (1,1), i.e., the first row and first column of the grid, is placed over the right shoulder).

In order to justify the processing described in this report, the spectral 2D contents of the maps has been calculated. It has been found that most of the energy is concentrated at low frequencies. The horizontal spatial sampling rate is 0.28(1/cm.) and the vertical 0.22(1/cm.). In all subjects 99.8% of the energy is bounded by a spatial frequency band of 0.07(1/cm.) (in the vertical direction) and 0.027(1/cm.) (in the horizontal direction).

The application of the threshold procedure to sequences of 3D maps produces sequences of 2D annuli. The maps are measured 2 ms apart, and are calculated from potentials measured during the QRS by a grid of 11x18 electrodes equally spaced on the torso. The annuli are calculated for the thresholding potential levels of 14% and 20% of the maximum normalized positive peak. A similar procedure is performed for the negative part of the maps, near the time of the breakthrough (the time in which the activation front in the right ventricle reaches the epicardium). It is manifested in the BSPM by a negative peak. The normalization and the fiducial point are kept the same as for the positive part. The annuli sequences preserve the differences between normal and abnormal patterns. Even though normalization of the time of activity is done, the differences between the annuli sequences of a normal subject and that of a WPW patient are quite evident: during the QRS period the WPW annuli sequence is shifted leftwards and the widths of annuli are significantly greater.

The features extracted from the annuli: In all the cases studied, a sequence of annuli calculated for the positive potentials section of the maps exist during the QRS period. In normal subjects often more than one annula exist in the same map, in different spatial locations; the largest annula, which lasts for the longest period is treated here. Since the cardiac cycle includes periods of low activity and correspondingly low level potentials, annuli appear only during part of the heart cycle.

The measurements of the time of appearance, time of disappearance, and duration of the annuli sequence during the cardiac cycle, are of major importance, as they are correlated with elector- physiological characteristics and conditions. The values obtained for the normal group differ from all other groups in both starting and terminating instances during the QRS wave. During the T wave these values differ from WPW and MI groups, while being similar (only in annuli appearance), to the AS group.

The spatial behavior of the annuli and their movement in time are characterized by the 3D locus of the centers of mass. The 3D locus is constructed by calculating the spatial location of the centers of mass as a function of time. When comparing typical examples of the normals with examples taken from patients with different cardiac pathologies, during the QRS, significant differences are apparent: Normal loci begin at the upper front thorax at the vicinity of the midsternal line; then the center of mass moves inferiorly and to the left towards the left midaxillary line; the loci end at the lower thorax, penetrating the dorsal area. The overall projection of the loci on the chest surface is of a concave shape. AS patients have a similar conceived path, but shifted up towards the left shoulder. The loci of the AS patients differ from that of the normal group by a sharp turning at the end of the conceived line, pointing towards the upper back. The loci of the WPW patients originate lower than the normals, near the xiphoid, then continue towards the left midaxillary line with a less steep slope, ending in the same area as the normals but forming an acute turn upwards (as in the AS cases). The MI patients draw a different curve on the chest, moving mostly on the left thorax from the frontal to the dorsal areas.

An estimate of the resemblance between successive annul is given by the cross correlation coefficients series. When studying the coefficients series as a function of time, plots of the normal group (during the QRS) are characterized by having two main peaks, one between 14 msec and 18 msec (with one exception of 22 msec), and the second between 32 msec and 44 msec. The second main peak, in most cases, contains a small depression of 4 msec - 6 msec in length. The cross correlation coefficients reach values between 0.78 and 0.94 in the peaks with low values between 0.15 - 0.53 at the minima. WPW patients are characterized by consistent high values of coefficients, with no distinct peaks. MI patients do exhibit two peaks but with the first peak appearing at 34 msec. These peaks are delayed with respect to the normal cases. AS patients exhibit various patterns which differ from each other, although in two cases two main peaks are observed, similar to the normals. The graphs of these two patients are characterized by high coefficient values at the longer time intervals.

The fourth extracted feature is the distances between centers of mass of successive annuli. In the normal group the curve usually consists of one main peak with values between 3.0 and 7.0 pixels' per map. In all patients, except the AS, the curves do not display any significant peak, but even in these cases the curves look different from the normal group curves. When observing the behavior of this feature simultaneously with the cross correlation coefficients series curve, additional and valuable information is supplied about the dynamic changes in the annuli sequence. 1) High cross correlation coefficient value accompanied by high distance value implies change in location of the annulus without changing of shape. 2) High cross correlation coefficient value accompanied by low distance value is a result of a marginal changes in both location and shape. 3) Low cross correlation coefficient value accompanied by high distance value is a measure of major variations in both location and shape. 4) Low cross correlation value accompanied by low distance value implies variation only in shape (only increase on decrease of the amplitude of the potentials distribution). Large values (greater than 10 pixels per map) in the distance curve, which are usually accompanied by low cross correlation coefficients values, imply a creation or disappearance of a second annulus in a map.

DISCUSSION

The error obtained with LC of either +100uV or -100uV was the lowest, and most of it was due to the difficulty of reconstructing the amplitudes. The errors in reconstructing the amplitudes of the BSPM are due to the intrinsic property of the method, i.e., that amplitude information is ignored by the LC method. There are two ways of reducing this error: setting groups of electrodes to various voltage levels; each electrode still detects crossing of a specific level, but since different electrodes are responsive to different voltages, the whole array becomes more sensitive to the ECG amplitudes. The second approach requires measuring the whole ECG signal by one or just a few electrodes. The standard ECG electrode configuration may be used for this purpose. This approach produces excellent results, by both reducing the total RMS error, and allowing presentation of the standard ECG trace with no errors at all.

The generation of a good eigenfunction basis is extremely important for being able to reconstruct the maps, and do that well. In order to create a robust eigenfunction basis, the training set must include as much a-priory measured data as possible. Since there is a physical limit to the amount of data that can be stored, as well as to the size of matrices that can be used, it is advisable to include in the training set similar sized groups of the different pathologies and normals.

The BSPM contains undoubtfully more information concerning the behavior of regional electrophysiological cardiac sources than the standard ECG (10, 11). Unfortunately the lack of suitable and simple data compression techniques, prevented practical usage of the BSPM in clinics. BSPM low level potentials have been found to contain fine details of high significance for diagnostic purposes. The BSPM which include the low level potentials, usually produce more complex maps with fine spatial details. The main requirement therefore from data compression technique is a preservation of the fine spatial details. The LC technique, together with the reconstruction method presented, and the annuli thresholding procedure, are both data compression techniques which preserve these details, and may serve as the proper base for extracting clinically significant features. Although features can be extracted from the original data, the implementation is not practical, due to the requirement of handling an overwhelming amount of data.

When comparing the method suggested here to earlier reports which utilize orthogonal expansion (22-24), the need to preserve the fine details requires the utilization of a very high number of eigen values components, thus consuming a lot of cpu time and storage capacity, preventing the utilization as a real-time, clinical system. The present method consumes only small cpu time and storage capacity. The amount of data reduced hardly affects the preservation of fine details. In this method the choice of a proper thresholding level, at the basic low potential is important. When the right level is selected the relevant information is preserved for classification and discrimination between normal and abnormal patterns in the maps.

The attempts of reducing the amount of recorded data by preselecting a small limited number of best electrodes, using orthogonal expansion (29, 30), or specifically for diagnosing myocardial infarction (14), limit the spatial resolution gained by the BSPM. The present study selects a flexible (map to map), rather than rigid, best electrodes configuration, suitable for both normal and pathological BSPM. Since the potential distribution has a spatially dynamic nature and is not confined to a limited location on the thorax, the measurement of the whole array is required, and not of a smaller set of electrodes, according to the distribution of the potentials in each map. It was found here that 41% of the electrodes mounted in the vest are valuable to the creation of the annula sequence. As expected 59% of these electrodes are located over the left side of the thorax, but also that valuable information is recorded over the back. Information recorded from the right side of the thorax has been found to be of value only in a few cases. It is concluded that electrodes located over the midright axiallary line and over the right side of the back are redundant.

Isointegral methods (25, 26), have been used to identify patients with myocardial infarction. The integration procedure is evaluated for different time intervals during the cardiac cycle, to produce one map for a sequence of maps, thus compression is achieved. By its definition, this method is insensitive to the dynamic changes of the maps and therefore also to the time dependency of the electrocardiac process. The isointegral method has been found to be influenced only by the repolarization process (which is slow) and not by the fast activation process (26). The descriptions of the maps by the locations of the maximum and minimum potentials in maps, at various instances during the heart cycle (21), are highly susceptible to noise. It describes very poorly the overall distribution of potentials on the thorax, as it has no value of a statistical descriptor of the distribution. The location of the extrema may be completely unrelated to the general shape of the 3D potentials distribution. The feature used here, the center of mass calculated over the whole cycle, is found to be a reliable parameter, which takes into account the 3D potentials distribution and its movement in time.

The procedures described here, the data compression by LC and the compatible reconstruction process, the data reduction by LC and the features extracted from them, present a complete new method of processing and quantifying the BSPMs, in a technically feasible and inexpensive way. The new method preserves the dynamic changes present in the sequence of the BSPM, and therefore describes well the dynamics of the underlying process. The method is also sensitive to both the depolarization and repolarization periods and is not limited to a specific kind of disease. Due to the enormous amount of information produced by the measurement of BSPM, until now there was no technical way of measuring and displaying the data in real time or in a sequence like a movie. Using the LC allows measuring long sequences in real-time, while calculating the annuli sequences instead of the whole BSPM sequence. It allows display and simplified visual inspection of the maps. The dynamic features characterize the annuli and the BSPM, and therefore may serve as a quantitative method for classification and discrimination of various abnormalities.

ACKNOWLEDGMENT

The work presented in this review was done together with Dr. I. Vitsnudel and Dr. S. Gilat. Their major contributions and laborious studies enabled accomplishing the task of developing the new technique. The clinical measurements of the BSPM were made at the Biomedical Engineering Department of Case Western Reserve University and the Rainbow Babies and Children's Hospital, Cleveland, Ohio. The authors wish to thank Dr. Rudy, Dr. Liebman and their colleagues for their generosity in supplying the clinical data. This research was supported by the Fund for the Promotion of Research at the Technion.

REFERENCES

1. Requicha, A. A., 1980, Zeros of entire functions: Theory and engineering applications, *Proc. IEEE* 68:308-328.

2. Levine, B. J., 1980, *Distribution of Zeroes of Entire Functions*, American Math. Soc., Providence, RI.

3. Logan, B. F., 1977, Information in the zero-crossings of bandpass signals, *Bell Tech. J.* 56:487-510.

4. Curtis, S. R., and Oppenheim, A. V., 1987, Reconstruction of multi-dimensional signals from zero-crossings, *J. Opt. Soc. Am.* 4, 221-231.

5. Curtis, S. R., Oppenheim, A. V., and Lim, J. S., 1985, Signal reconstruction from Fourier transform sign information, *IEEE Trans. ASSP* 33:543-657.

6. Papoulis, A., 1975, A new algorithm in spectral analysis and band-limited extrapolation, *IEEE Tran. Circuits and Systems* 22:735-743.

7. Youla, D. C., and Webb, H., 1982, Image restoration by the method of convex projections. Part I. Theory, *IEEE Tran. Med. Imaging* MI-1:81-94.

8. Green, L. S., Lux, R. L., Stilli, D., Haws, C. W., and Taccardi, B., 1987, Fine detail in body surface maps: accuracy of maps using limited lead array and spatial and temporal data representation, *J. Electrocardiol.* 20:21-26.

9. Spach, M. S., Barr, R. C., Benson, W., Walston, A., Wonen, R. B., and Edwards, S., 1979, Body surface low-level potentials during ventricular repolarization with analysis of the ST segment: Variability in normal subjects, *Circulation* 59:822-836.

10. Uijen, G. J. H., Heringa, A., and Van Oosterom, A., 1984, Data reduction of body surface potential maps by means of orthogonal expansion, *IEEE Trans. Biomed. Eng.* 31:706-714.

11. Kornreich, F., and Rautaharju, P.M., 1981, The missing waveform and diagnostic information in the standard 12-lead electrocardiogram, *J. Electrocardiol.* 14:341-350.

12. Taccardi, B., 1963, Distribution of heart potentials on the thoracic surface of normal human subjects, *Circ. Res.* 12:341-352.

13. Mirvis, D. M., 1985, Ability of standard E.C.G parameters to detect the body surface isopotential abnormalities of pacing induced myocardial ischemia in the dog, *J. Electrocardiology* 18:77-85.

14. DeAmbroggi, L., Bertoni, T., Rabbia, C., and Landolina, M., 1986, Body surface potential maps in old inferior myocardial infarction: assessment of diagnostic criteria, *J. Electrocardiol.* 19:225-234.

15. Osugi, J., Ohta, T., Toyama, J., Takatsu, F., Nagaya, T., and Yamada, K., 1984, Body surface isopotential maps in old inferior myocardial infarction undetectable by 12 lead electrocardiogram, *J. Electrocardiol.* 17:55-62.

16. Kubota, I., Ikeda, K., Ohyama, T., Yamaki, M., Kawashima, S., Igarashi, A., Tsuiki, K., and Yasui, S., 1985, Body surface distributions of ST segment changes after exercise in effort angina pectoris without myocardial infarction, *Am. Heart J.* 110:949-955.

17. Simoons, M. L., and Block, P., 1981, Towards the optimal lead system and optimal criteria for exercise electrocardiography, *Am. J. Cardiol.* 47:1366-1374.

18. Spach, M. S., and Barr, R. C., 1971 Physiologic correlates and clinical application of isopotential surface maps, in (Hoffman ,I., Hamby, R. I., Glassman, E., editors) *Vectorcardiography 2.* North Holland Publishing: Amsterdam, pp. 131 - 141.

19. MacLeod, R., Johnson, C., Gardner, M., and Horacek, B. M., 1992, Localization of ischemia during coronary angioplasty using body surface potential mapping and an electrocardiographic inverse solution, *Computers in Cardiology* 92:251-254.

20. Rudy, Y., and Plonsey, R., 1980, A comparison of volume conductor and source geometry effects on body surface and epicardial potentials, *Circ. Res.* 46:283-291.

21. Pan huy, H., Gulrajani, R. M., Roberge, F. A., Nadeu, R. A., Mailloux, G. E., and Savard, P. S., 1981, A comparative evaluation of three different approaches for detecting body surface isopotential map abnormalities in patients with myocardial infarction, *J. Electrocardiol.* 14:43-56.

22. Lux, R. L., Evans, A. K., Burgess, M. J., Wyatt, R. F., and Abildskov, J. A., 1981, Redundancy reduction for improved display and analysis of body surface potential maps I. Spatial compression. *Circ. Res.* 49:186-196.

23. Evans, A. K., Lux, R. L., Burgess, M. J., Wyatt, R. F., and Abildskov, J. A., 1981, Redundancy reduction for improved display and analysis of body surface potential maps II. Temporal compression, *Circ. Res.* 49:197-203.

24. Lux, R. L., and Green, L. S., 1983, Surface potential mapping: a problem in statistical imaging of the heart, *IEEE Frontiers of Eng. and Comput. in Health Care* 83:37-40.

25. Montague, T. J., Smith, E. R., Johnstone, D. E., Spencer, C. A., Lalond, L. D., Bessoudo, R. M., Gardner, M. J., Anderson, R. M., and Horacek, B. M., 1984,Temporal evolution of body surface maps pattern following acute inferior myocardial infarction, *J. Electrocardiol.* 17:319-328.

26. Tonooka, I., Kubota, I., Watanabe, Y., Tsuiki, K., and Yasui, S., 1983, Isointegral analysis of body surface maps for the assessment of location and size of myocardial infarction, *Am. J. Cardiol.* 52:1174-1180.

27. Nikias, C. L., Raghuveer, M. R., Siegel, J. H., and Fabian, M., 1986, The zero delay wavenumber spectrum estimation for the analysis of array ECG signals - an alternative to isopotential mapping, *IEEE Trans. on Biomed. Eng.* 33:435-451.

28. Liebman, J., Rudy, Y., Thomas, C., Ko, W., Plonsey, R., and Diaz, P. J., 1984, *Body Surface Potential Mapping System Reference Manual*, Department of Biomedical Eng., Case Western Reserve University Cleveland Ohio.

29. Kornreich, F., Holt, J., Rijlant, P., Barnard, A. C. L., Tiberghien, J., Kramer, J., and Snoeck, J., 1976, New ECG techniques in the diagnosis of infarction and hypertrophy, in (Hoffman, I., and Hamby R. I., editros) *Vectorcardiography* 3rd. ed., Elsevier North Holland: Amsterdam p. 171.

30. Lux, R. L., Smith, C. R., Wyatt, R. F., and Abildskov, J. A., 1978, Limited lead selection for estimation of body surface potential maps in electrocardiography. *IEEE Trans. Biomed. Eng.* 25:270-276.

Processing and Pattern Analysis of EMG and Human Movement

22

SOURCE CHARACTERISTICS FROM INVERSE MODELING OF EMG SIGNALS

Herman B. K. Boom and Willemien Wallinga

University of Twente
Faculty of Electrical Engineering
and Institute for Biomedical Technology, P.O. Box 217
7500AE Enschede, The Netherlands

ABSTRACT

If one tries to solve the inverse problem in single fiber electromyography (SFEMG), the question is which signal to consider as the proper source signal. Often the intracellular action potential (IAP) is taken as such. Measuring shape parameters and distances of Single Fiber Action Potentials and active fibers in muscle and comparing them with muscle structure model predictions reveals that it is transmembrane current which is the true source of the SFEMG.

INTRODUCTION

The surface electromyogram (EMG), as is clinically well-known, is a noise-like, non repetitive signal (Hermens *et al.*, 1992; Stegeman & Linssen, 1992). Its source has a complexity somewhere in between that of the electroencephalogram (EEG) and the electro-cardiogram (ECG). While EEG signals are summed from the voltages generated by millions of nerve cells, the ECG arises from predictable membrane processes in one, tightly coupled multicellular muscular organ. The EMG also is generated by muscle cells, which are, however, not directly electrically coupled. The timing of cell depolarization is determined by motor axon branches connecting to them by way of the motor end-plates. Thus, membranes of skeletal muscle cells are the localizations of the sources of the EMG. The membrane phenomena are complex and influenced by electrical properties of the cell interior as well of the muscle tissue surrounding the active cell. Phenomena are complexly and non-linearly interrelated and regeneratively coupled, making it difficult to decide which signal is causal and which is result.

Transmembrane potential is a well studied physiological variable. If it is non uniform over a part of the cell, external volume currents will ensue, and transmembrane potential could be envisioned as the source of the EMG (Plonsey, 1969). However, transmembrane

Advances in Processing and Pattern Analysis of Biological Signals, Edited by Isak Gath and Gideon F. Inbar
Plenum Press, New York, 1996

potential is the difference of intracellular potential and extra(peri)cellular potential. A question is: which of these two, most typically causes transmembrane potential to act as the EMG source. Although extracellular potential might seem most closely linked with the muscular volume current field and thus with the EMG, it is quite small with respect to intracellular potential, which is therefore often considered the actual EMG source. Complicating things further, nonuniform transmembrane potential is also associated with transmembrane *current*. Since current density is proportional to intramuscular voltage gradient, and thus with the forward conduction of the bioelectric signal, membrane current could be considered as the source of the EMG. Again, it might be argued that membrane current and transmembrane potential are linked. However, as will be shown in this chapter, this relationship is more complicated than the simpler models suggest. It is, therefore relevant to EMG analysis to question the nature of the EMG source. Knowledge of this nature is necessary to fully solve the inverse model for electromyographic single fiber potentials.

INVERSE AND FORWARD MODELING OF THE SINGLE FIBER POTENTIAL

Inverse modeling of a bioelectrical signal is the determination of sources or source parameters from registrations such as EMG, EEG, ECG, or, more exotically, magnetic myographic data. Although there are special inverse modeling techniques available, which directly estimated source parameters from measured data, the usual way is repetitive forward modeling followed by tentative adaptation of source parameters. Figure 1 depicts the flow of thoughts in solving inverse problems in single fiber electromyography, where the electric output of one single skeletal muscle fiber, the single fiber action potential, (SFAP) is studied. There are several points of interest shown by this picture:

- A model of both the signal source and of the passive volume conductor must be available.
- The quality of the prediction given by the forward model is also dependent on the correctness of the constituent models. For instance, if the most correct model for the source is not known a priori, several models can be tried. The source model

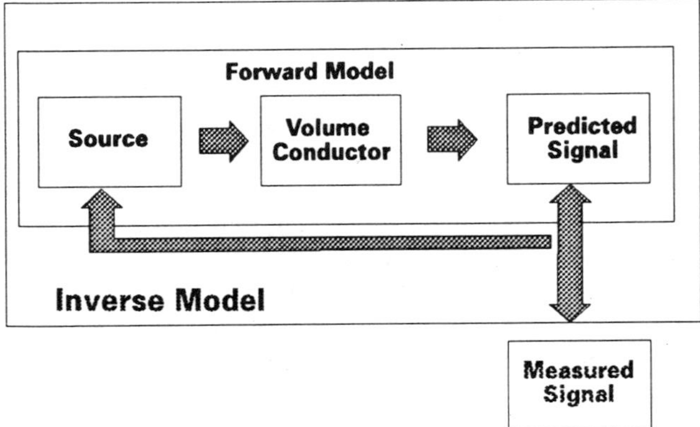

Figure 1. Forward and Inverse modeling of the single fiber action potential (SFAP). A model of the source generates a signal, which is conducted by the modeled passive volume conductor to a remote measurement site, also modeled. Signals, thus forwardly computed, are compared with measured values, and their difference minimized by adopting the forward model.

which, after a complete iterated identification gives the best prediction, then, is the most correct model.

- Comparison between predicted and measured signals can be extended to the prediction of relations between them. For instance, if one source model predicts a linear relationship between two signals parameters as contrasted to an alternative model which suggests exponentiality, the latter model will be adopted if measured relations also tend to an exponential behavior. This method can be advantageous if it is difficult to obtain numbers predicted and measured under precisely the same conditions.

- Both source models and volume conductor models contain many parameters unknown at the outset, of which only a limited number are identifiable by the above process. Thus, measurements have to be done to further determine the properties of the models. For instance, in the volume conductor models amplitude-distance relations can play a role. Such relations require knowledge which could be complemented by measuring field quantities as a function of the distance to the active source.

THE ACTIVE MUSCLE FIBER AS THE SOURCE OF THE EMG

Measurement Techniques for EMG Source Signals

A major advantage in considering the above mentioned variables, intracellular potential, transmembrane potential, extracellular potential and membrane current, as potential source signals for the EMG is the fact that they can be measured.

Intracellular Potential. Measurement of intracellular potential is a classic technique. A micropipette, drawn from a glass tube down to a diameter of less than 1 μm and filled with saline, is inserted into the muscle fiber. Intracellular potential is measured via the high intraluminal resistance in the pipet, with respect to a reference electrode. The positioning of this reference electrode is of some concern, because no cross-talk of the EMG signals from the muscle must have access to it. The special application to skeletal muscle fiber (as contrasted, for instance to nerve axons) may call attention to the possibility of mechanical effects due to the fibers contraction. Special very flexible pipettes have been designed for this purpose by Fedida *et al.* (1990). A common problem with most cellular signal techniques is that, also when measuring on small objects like cells there can be much difference between local values and global, averaged ones. Micropipettes measure on a scale which is small with respect to cell length, but perhaps not with respect to finer structures like, for instance, the T system. Local modulations of intracellular voltages, during action potentials may then be difficult to catch.

Extracellular Potential. It could seem easier to measure voltages extracellularly, but that is only partially true. The problem is that, unlike intracellular potential, extracellular potential is ill defined, if the distance to the fiber surface is not well known. Strictly speaking, extracellular potential, the amplitude of which decreases rapidly in close proximity of the fiber, has to be measured at negligible distance from the fiber surface, which is not practicable. Van Veen *et al.* (1994) used needle multielectrodes (Fig. 2). If such needles are inserted into the muscle, one of the electrodes may happen to be near an active fiber and a SFAP as depicted in Fig. 3 is recorded. This SFAP is triphasic, as expected on theoretical grounds (e.g., Clark & Plonsey, 1966). Since the shape of the SFAP is of interest in single fiber electromyography as well for the analysis in the current paper, it will be described with

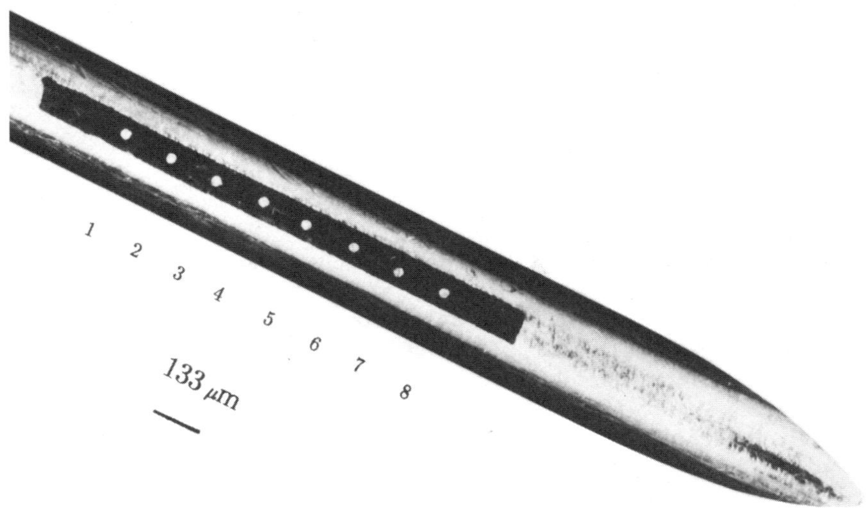

Figure 2. Needle multi-electrodes for measuring extra cellular single fiber action potentials. One of the electrode tips must be manoeuvred closed to the fiber surface by trial and error (Van Veen *et al.* 1994, Fig. 1).

the usual characteristic parameters V_{tt}: amplitude, V_1: amplitude of the first deflection, and Δt: falling time (for definition see Fig. 3).

Membrane Current. In contrast to extracellular potential, membrane current can be measured in a relatively defined way. Hamill *et al.* (1981) showed that, if a fire polished pipette is pressed to the extracellular surface, the "patch" of muscle cell membrane within

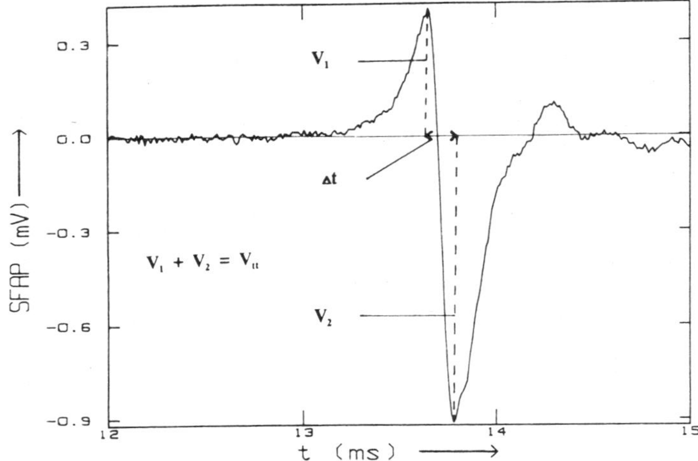

Figure 3. Single fiber action potential measured with the multielectrode of Fig. 2. It shows a characteristic triphasic shape, justifying the assumption that it has been recorded near the active fiber surface. SFAPs are usually characterized by peak-peak time V_{tt}, first phase amplitude and falling time Δt (Van Veen *et al.* 1994, Fig. 2).

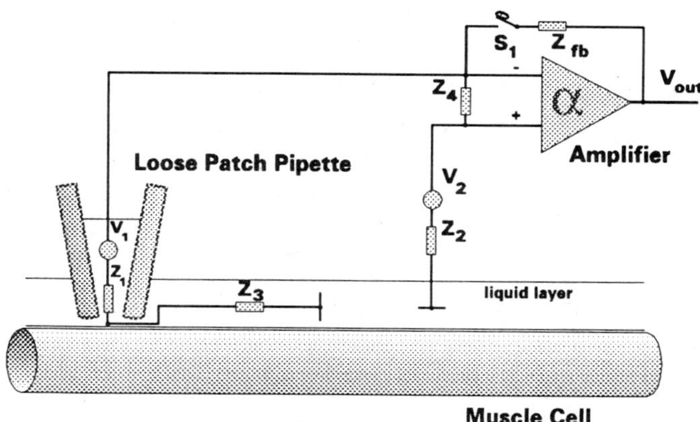

Figure 4. Equivalent circuit for the loose-patch clamp measuring situation. Pipette current flows through Z_1, Z_4, and Z_2, seal current through seal resistance Z_3.

the pipette is electrically sealed off from its surroundings. Currents measured from such a pipette have been interpreted as transmembrane currents since the time that "patch clamping" became a standard technique in cellular muscle physiology. Patch pipettes come in several sizes. Narrow pipettes, 1-2 μm are used to measure as little ionic channels as possible, typically 1 to a few. Wider pipettes, up to 20 μm less effectively shield off surrounding membrane and extracellular fluid (loose patch clamp technique). However, with such clamps it is possible to measure spatially averaged cell behavior and, by moving the electrode, channel density distributions over the cell membrane (Milton *et al.*, 1992). The incomplete sealing of the pipette from the cell membrane causes the fluid between membrane surface and pipette rim to act as a parasitic conductance. This is effectively switched in parallel to current path from the membrane patch to the input stage of the measuring amplifier (Fig. 4).

Membrane current passing the membrane patch divides into two branches, one flowing through the pipette into the amplifier input, and, passing Z_4 and Z_2 to the extracellular fluid (considered at zero potential). The other branch is the seal passage, of which impedance is assumed to be represented by impedance Z_3. V_1 and V_2 are electrode potentials assumed constant and equal, thus canceling each other. If switch S_1 is closed, input impedance of the amplifier is low and all patch current flows into Z_{fb}. The output of the amplifier then is proportional to patch current i_{patch}, and the seal current is negligible, because Z_3 is much greater than Z_1 and the input impedance of the amplifier (voltage clamp mode). Opening S_1 causes input impedance to become high and virtually all patch current is flowing through Z_4 (current clamp mode). Thus, in the voltage clamp mode:

$$V_1 = V_{out} = Z_{fb} \cdot i_{patch} \cdot \frac{Z_3}{Z_1 + Z_3}$$

and in the current clamp mode:

$$V_2 = V_{out} = \alpha \cdot i_{patch} \cdot Z_3$$

thus:

$$\frac{V_2}{V_1} = \frac{\alpha}{Z_{fb}} (Z_1 + Z_3) \tag{1}$$

Now α/Z_{fb} is a known equipment parameter, and thus $Z_1 + Z_3$ is also known. Since Z_1 can be measured separately, the seal impedance then will be known. Thus, by comparing

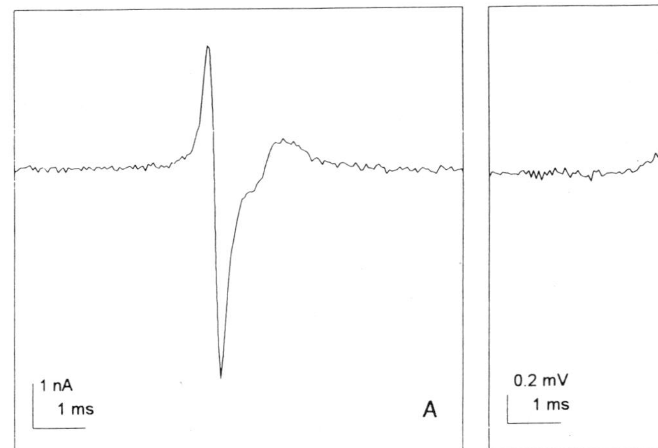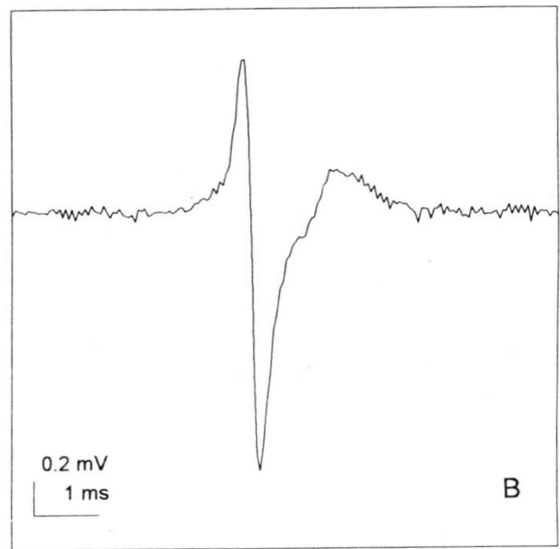

Figure 5. Loose patch clamp recordings of adult mice EDL single fiber action potentials. A: Voltage clamp mode measuring transmembrane current. B: Current clamp mode output now showing pipette tip potential. Note good proportionality between current and voltage indicating linear seal (resistance 310KΩ). Pipette resistance was 350 KΩ (Wolters *et al.*, 1994, Fig. 1).

voltage clamp mode and current clamp mode at various conditions, that is, different phases of the membrane current, seal properties can be studied, for instance its linearity.

Figure 5 shows voltage clamp mode and current clamp mode signals obtained from mice EDL fibers. The two registrations, membrane current (voltage clamped) and the extracellular potential (voltage clamped) are virtually identical. Their interrelation shows a regression coefficient of >>0.98. Thus Z_3 also must be linear. This fact is far from trivial, since it not only demonstrates a well determined and well behaved seal resistance, but it also indicates that the membrane patch has current source properties. Any internal source impedance is closely associated with alinear membrane channel kinetics and would show up as alinear behavior of the dependence of patch voltage on patch current. Figure 5 clearly shows that this is not the case.

The similarity of current clamped and voltage clamped signals might also suggest that part of the problem discussed in this chapter is fictitious. Of the three possible EMG sources, intracellular potential, extracellular potential and membrane current, the latter two seem identical. This, however, is not the case. Current clamped extracellular patch voltage is obtained by electrically isolating the extracellular microvolume in the patch pipette from the surrounding volume conductor. This inhibits volume currents flowing from adjacent peri-membrane areas to that of the patch, allowing patch current and patch voltage to become proportional.

The Relation between Transmembrane Potential and Transmembrane Current

Transmembrane potential is the potential difference between the interior of the cell and the thin extracellular layer directly adjacent to the outer side of the membrane. Classically, the relation between transmembrane potential and transmembrane current is given by the core conductor model first formulated by Lorente de Nó (1947). Current passing the cell membrane result from the gradient of axial intracellular current, which, in turn, only depends on axial potential gradient. Thus, transmembrane current i_m must be proportional to the second axial space derivative of intracellular potential. If a constant propagation velocity U of the action potential along the muscle fiber is assumed (i.e.,):

$$\frac{\partial^2 V}{\partial t^2} = \frac{2U^2 R_i}{a} i_m$$

(2)

with R_i the intracellular resistance. Equation (2) looks like another simple relationship between two of the candidate source variables. There is, however, considerable uncertainty how membrane current i_m is composed. Adrian and Peachey (1973) give

$$i_m = i_i + i_T + C_m \frac{dV}{dt}$$

(3)

where i_i represent the ionic channels, and i_T is the current delivered by the transversal (tubular) system. The existence of a tubular system has consequences which are only poorly understood. Equation (3) may stay valid, but the tubular capacitance, for instance, needs to be included in C_m. The way to do this is unknown. There is evidence that the tubular orifices contain an entrance resistance. Furthermore, there will be an effect on propagation velocity c, because of the slowing down of the upstroke of the propagated action potential. Even the validity of Eq. (2) might be questioned, because it is unclear whether the part of i_m passing the tubular membrane contributes to the intracellular axial current. Actually, it will be shown in the last section of this paper that this doubt is experimentally justified. The provisional conclusion is that intracellular potential (transmembrane potential) and transmembrane current give complementary information on skeletal muscle cell membrane behavior.

FORWARD TRANSFER FUNCTIONS FOR THE SINGLE FIBER ACTION POTENTIAL

Models of Volume Conduction in Skeletal Muscle

As explained in the Introduction, forward models of volume conduction of SFAP's can be used for solving inverse problems concerning EMG sources. For this, forward models have to be realistic and accurate. Forward models tend to filter out higher frequencies from the source signal being conducted (Albers *et al.*, 1986). In order to reconstruct as much as possible of these frequencies, significant realism should be contained by the models. Realistic models of volume conduction by skeletal muscle tissue should take into account geometry (fiber location, muscle size and shape), correct conductivities including anisotropy, structure (represented by the existence of a fibrous tissue), inhomogeneities like blood vessels and micro deposits of fat. In addition, realistic values for all parameters involved in representing these structural refinements should be known. Many parameter values have been measured under conditions different from their occurrence in the muscle conduction model. For instance, classical parameters like conductance, g, being well defined as the ratio of current density and voltage gradient, each of these variables being well measurable, cause problems in anisotropic media. Under such conditions measured voltage current ratios have to be transformed to those defined by the conductance tensor σ_{ij}, for which a model is required which is not readily available (Gielen *et al.*, 1984, 1986).

In addition, available realistic models tend to be very complex, requiring huge amounts of computer time. This, in contrast to the above, calls for simpler models, so in the end compromise cannot be avoided.

Model realism and complexity can also vary depending on the part of the muscle the model is applied to. Since EMG sources are microscopic, at least in some of their dimensions, models describing volume conduction in the direct proximity of such a source call for greater detail than would be required at greater distances. Combining a complex model applied to a limited part of space with a more global one, valid for the greater, more remote areas, offers a way of economizing (Van Veen *et al.*, 1992).

One aspect, much debated, is whether a model should preferably have a closed analytical form rather than use numerical computational techniques. Although this also is a trade-off, the difference is less fundamental then it would seem. Some of the 'analytical' models discussed in the following require such an elaborate numerical assessment that the difference from a numerical simulation is only gradual.

Analytical Models of Volume Conduction in Skeletal Muscle

The simplest geometry taking into account the more or less circular cross-section of some skeletal muscles is the anisotropic cylinder. Such models can calculate effects of changing conductivities and other electrical parameters at muscle and muscle fiber boundaries. A muscle fiber does not need to be situated along the muscle cylinder axis (Rosenfalck, 1969), in order to keep the model analytically solvable (Fig. 6). Only some essential features of the mathematics will be discussed here. Details can be found elsewhere (e.g. Ganapathy *et al.*, 1987; Meier *et al.*, 1992). Expressed in Cartesian coordinates, potential in an anisotropic conducting continuum is subjected to:

$$\sigma_x \frac{\partial^2 \Phi}{\partial x^2} + \sigma_y \frac{\partial^2 \Phi}{\partial y^2} + \sigma_z \frac{\partial^2 \Phi}{\partial z^2} = \nabla I_i \tag{4}$$

with I_i impressed current, the strength of source current density at that location. For the geometry involved, transformation to cylindrical, 'corrected' coordinates, r^*, z^*, φ is convenient

$$r^* = \frac{r}{\sigma_x} = \frac{r}{\sigma_y} \qquad\qquad z^* = \frac{z}{\sigma_z} \tag{5}$$

thus:

$$\frac{\partial^2 \Phi}{\partial r^{*2}} + \frac{1}{r^*}\frac{\partial \Phi}{\partial r^*} + \frac{\partial^2 \Phi}{\partial z^{*2}} + \frac{1}{r^{*2}}\frac{\partial^2 \Phi}{\partial \varphi^2} = \nabla I_i \tag{6}$$

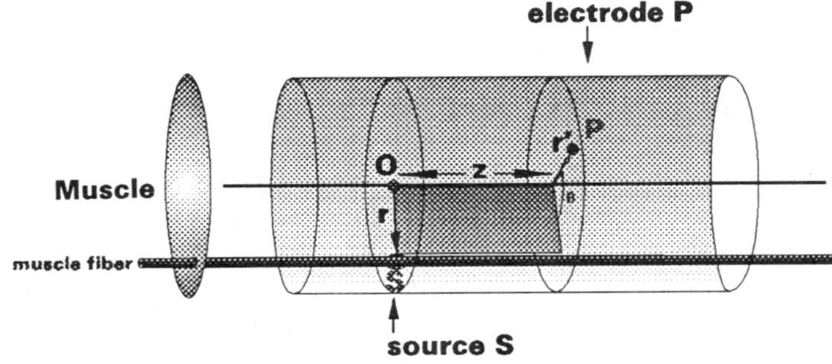

Figure 6. Cylindrical model of skeletal muscle. By placing the source muscle fiber excentrical the influence of its distance to the muscle boundary can be evaluated. This model can be solved analytically. $\vec{r}$ is radius vector of source. $\vec{r}'$ is radius vector of any volume conduction field point.

The detailed solution of Eq. (6) depends on the special case at hand. Usually, a primary source field $\Phi_{primary}$, which is associated with the inhomogeneity term on the right hand side of Eq. (6) in a Coulomb-like way, is treated separately from a secondary field $\Phi_{secondary}$ arising at surfaces separating areas with different conductivities, and has to found by specifying boundary conditions, and for which the right hand side of Eq. (6) vanishes:

$$\Phi = \Phi_{primary} + \Phi_{secondary}$$

while $\Phi_{primary}$, if the source current distribution is known, can be calculated straightforwardly, $\Phi_{secondary}$ can be found by solving the homogeneous version of Eq. (6), greatly helped by the fact that it is separable in its independent variables. It can be written as:

$$\Phi = R(r^*) \cdot P(\varphi) \cdot Z(z^*) \tag{7}$$

For $P(\varphi)$ and $Z(z^*)$ exponential solutions are found. The exponential for $P(\varphi)$ has a discrete spectrum of exponential coefficient because of the required circumferential equivalence. For $Z(z^*)$, no such condition exists. Its exponent thus has a continuous spectrum. $R(r^*)$ is the solution of Bessel's differential equation and can be written as a series of Bessel functions, the orders of which corresponds to the coefficient of the associated $P(\varphi)$ exponential (a discrete number). To obtain the general solution, all these special ones are to be summed, causing the $Z(z^*)$ coefficients to act as integration variables because of their continuity. Likewise, there will be a summation over the discrete $P(\varphi)$. Thus

$$\Phi(r^*, \varphi, z^*) = \Phi_{prim}(r^*, \varphi, z^*) + \sum_{n=-\infty}^{\infty} e^{-jn\varphi} \int_{-\infty}^{\infty} A_n(k) I_n\left(|k| \cdot r^*\right) \cdot e^{-jkz^*} dk \tag{8}$$

where n is the discrete $Z(z^*)$ parameter, k the discrete one in $P(\varphi)$. Formally, k represents a spatial frequency, and n a comparable frequency referring to the angle φ. The A_n must be found from boundary conditions imposed on the outer boundaries of the muscle cylinder. $\Phi_{primary}$ can likewise be written in terms of cylindrical harmonics and mathematically combined with the secondary field. This yields an expression solely consisting of a summation and an integration of Bessel's function with exponentials as weighting factors. The mathematical equivalence with the retransformation of a Fourier transform must now be evident.

Further elaboration of Eq. (8) also depends on the source assumed. The Fourier transform character of Eq. (8) makes representation of also the source in a spatial, z directed frequency domain attractive (Ganapathy et al., 1987). However, this would obscure a direct evaluation of various source configurations as is the aim of this chapter. Therefore the approach of Meier (1992) is taken, assuming a point current source I at S(r,0,0), for which a practicable Fourier representation can also be obtained:

$$\Phi_{prim}(r^*, \varphi, z^*) = \frac{I}{4\pi R} = \frac{I}{4\pi\sqrt{r^{*2} + z^{*2}}} = \frac{1}{\pi} \cdot \int_{-\infty}^{\infty} K_0(k \cdot r^*) \cdot e^{ikz^*} dk \tag{9}$$

For an electrode site near the surface of the muscle (at greater distance from the axis than the active fiber) the result is found to be (Meier et al. 1992)

$$\Phi(r^*, \varphi, z^*) = \int_{-\infty}^{\infty} \left[\sum_{n=-0}^{\infty} \left\{ A_n(k) I_n\left(|k| \cdot r^*\right) + \frac{I}{4\pi^2}(2 - \delta_{n0}) I_n(kR) \cdot K_n\left(|k| \cdot r^*\right) \right\} \cdot \cos(n\varphi) \right] dk \tag{10}$$

where the exponential in φ has been replaced by the cosine because of symmetry, and δ_{n0} is zero if $n=0$ and otherwise 1.

After A_n has been found by imposing the boundary condition for the muscle surface (depending on experimental conditions) the potential field can actually be calculated. This is performed by digitizing Eq. (10) and using a fast Fourier Transform.

This case demonstrates that even in an analytically solvable model, real understanding by inspecting the equations is not much more easily done than in the case of a completely numerical model, in which fields are calculated by numerically solving the necessary partial differential equations.

Anisotropy is the only structural aspect in the above cylindrical, homogeneous model that refers to the fibrous structure of muscle. Anisotropy is brought about by the muscle's fibers conducting current parallel to the fibers differently than transverse to them. However, anisotropy is but one consequence of the fibrous structure. Another is the fact that current can be distributed between the intracellular spaces and the extracellular spaces in a variable way, depending on the global current distribution. There are several ways to account for this. One of them is the bidomain approach.

Bidomain Models of Volume Conduction in Skeletal Muscle

Plonsey and Barr (1982) pointed out that it is relatively simple to extend the continuum model to one incorporating more explicit intracellular and extracellular current distributions. The main addition is a second continuum, coexisting with the first. So, instead of one potential, $\phi_o(r^*, \varphi, z^*)$, extracellular potential, there is intracellular potential ϕ_i (r^*, φ, z^*) at the same point $P(r^*, \varphi, *)$, The potentials are connected by an interdomain volume current I_m:

$$\nabla i_m = \frac{\beta}{R_m} (\Phi_i - \Phi_o) \tag{11}$$

with β the surface-to-volume ratio of the tissue. Of course, I_m is the current flowing from the intracellular space to the extracellular. Plonsey and Barr (1982) applied this model approach to cardiac tissue, which has a moderate anisotropy. Roth *et al.* (1987) elaborated the case of skeletal muscle, which is more extreme, since it has to be assumed that in the intracellular domain, current is only flowing in the z direction. Taking the relatively simple case of the cylindrical muscle and a fiber along the axis, in the summation of Eq. (10), only the term with n=0 will not vanish. Equation (6) now has to be applied twice with zero φ derivatives:

$$\frac{\partial_2 \Phi_o}{\partial r^{*2}} + \frac{1}{r^*} \frac{\partial \Phi_o}{\partial r^*} + \frac{\partial^2 \Phi_o}{\partial z^{*2}} = -\frac{\beta}{R_m} (\Phi_i - \Phi_o) \tag{12a}$$

$$\frac{\partial^2 \Phi_i}{\partial r^{*2}} + \frac{1}{r^*} \frac{\partial \Phi_i}{\partial r^*} + \frac{\partial^2 \Phi_i}{\partial z^{*2}} = \frac{\beta}{R_m} (\Phi_i - \Phi_o) \tag{12b}$$

A problem with bidomain models is that one has to assume how the different anisotropic conductivities relate to known macroscopic electrical properties and structure. Macroscopically, conductivities along the fiber and transverse to it have been measured by Gielen *et al.* (1984), with a four electrode method. (Even these values can be questioned because they are measured at the surface of the muscle). However these conductivities are not equivalent to those implied by Eqs. (12a) and (12b). The usual assumption that in skeletal muscle intracellular current is only longitudinal, makes the r-derivative terms on the left hand of (12b) zero. Again, it seems practical to transform to the spatial, z related frequency domain. If this is done with Eq. (12b) an algebraic relation between ϕ_o and ϕ_i,

$$\Phi_i = \cfrac{1}{\cfrac{Z_m}{\beta}\sigma_i^z k^2 + 1}\,\Phi_0 \tag{13}$$

results, which can be substituted in Eq. (12a). From Eq. (13), which represents a spatial-frequency low-pass filter, a characteristic length can be defined by putting

$$\frac{Z_m}{\beta}\sigma_i^z k^2 = 1 \qquad \text{or} \qquad \gamma = \frac{1}{k} = \sqrt{\frac{z_m}{\beta}\sigma_i^z}$$

Estimating Z_m/β as $10^{-6}\,\Omega\text{m}^3$ (Plonsey & Bar, 1982), and σ_i at 0.4 Siemens/m (Albers *et al.*, 1988), the result for the characteristic length λ is 0.63 mm. It would imply that an unbalanced intracellular-extracellular current distribution resulting from the extracellular injection of current on any location, would be redistributed to equilibrium over approximately that distance This could be verified experimentally, but is yet to be done.

Relation (13) can be solved similarly to the procedure for Eq. (6). The result has to be Fast Fourier Transformed to the spatial domain. This again makes this method more laborious than its elegance suggests.

Boundary conditions in this problem are equal to those valid for the monodomain described by Eq. (10). This is because intracellular radial current is zero. Thus current entering the bidomain from or to an active fiber or from or to the muscle surface is injected into the external domain, just as in the monodomain case. In the more general case, where a fiber is situated non-axially (such cases are important because of the possibility of computing fields by motor units) relation (12b) cannot be solved algebraically and numerical methods are necessary.

A weakness of the bidomain concept is that some choices seem arbitrary. External conductivities to be chosen will be different from those valid for extracellular fluid since extracellular currents will follow curved paths, avoiding cell membranes. The precise relationship could only be calculated with precise models, a problem already at hand. To proceed, one needs models more based on known muscle structure than the continuous bidomain.

Fiber Structure Models of Volume Conduction in Skeletal Muscle

Volume current generated by the EMG source is injected by the active fiber into the surrounding extracellular space. Most of this current is conducted further through this space to be measured by measuring electrodes on a distance. Some of the conduction current will pass adjacent cell membranes and enter intracellular space, to be conducted along the length of a muscle fiber. The way this conduction takes place primarily depends on microscopic conduction parameters, like σ_e, extracellular space conductivity, σ_i, intracellular space conductivity, C_m, cell membrane capacitance, G_m, membrane conductivity, and, A, fiber radius. Gielen *et al.* (1986) have introduced a model taking some of these microscopic structural elements into account (Fig. 7).

The above electrical parameters are treated as lumped quantities like resistors and capacitors interconnected in such a way that they define (electrically) extracellular and intracellular spaces, with their interconnections. The muscle is divided into slices, and each muscle fiber membrane segment in a slice is represented by six impedances. Each impedance consists of one resistor (membrane Resistance = $1/G_m$) and one capacitor (membrane capacitance = C_m) in parallel. In transversal and longitudinal directions, extracellular space is accounted for by connecting these membrane impedances, at their extracellular ends, by

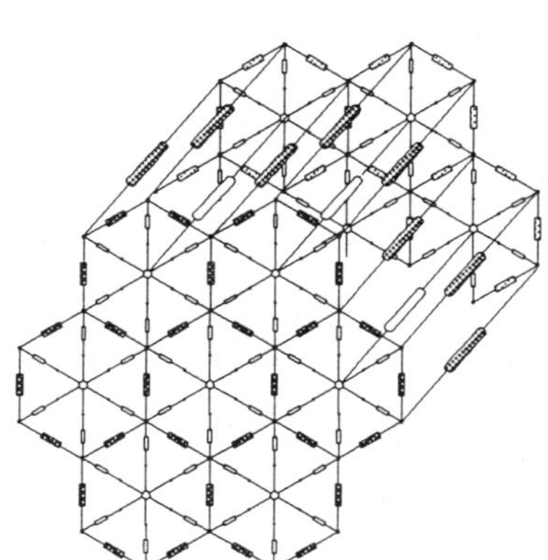
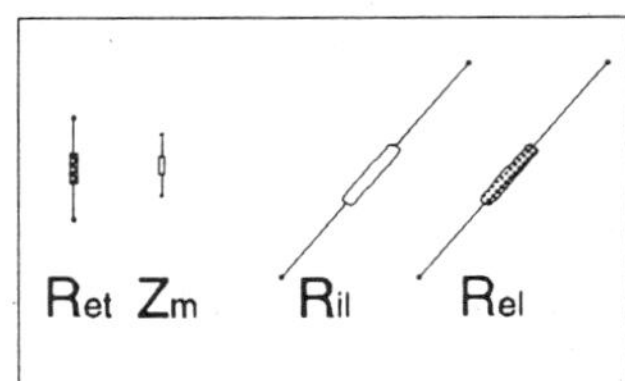

Figure 7. Electrical network representing intracellular, extracellular and membrane electrical parameters. The values of the components can directly be compared with their physiological counterparts in real skeletal muscle. R_{et}, R_{el}, and R_{il} are ohmic resistances representing extracellular and intracellular resistances along and transverse to the fiber. Z_m is membrane impedance consisting of a resistance and a (membrane) capacitance in parallel.

resistors, assuming thus extracellular space to be resistive. Any anisotropy can then be incorporated by selecting these extracellular resistances.

Current can be impressed on the muscle simulating network for two reasons: a, to simulate EMG sources, b, to represent boundary conditions to the network.

For point sources, assuming one fiber to be active the six extracellular resistance nodes surrounding the fiber in one slice were supplied with one sixth each of the total current to be impressed. For longitudinally more extended EMG sources two (for a dipole) or more slices were likewise connected to current sources. Boundary conditions were imposed in the same way.

Since, in this model each network element is linear and connects two points, with two values for the potential, the potential distribution in the network follows from

$$\mathbf{Y} \cdot \mathbf{V} = \mathbf{I} \tag{14a}$$

where $\mathbf{V}$ and $\mathbf{I}$ are signal vectors containing the discrete extracellular node voltages and the source currents impressed respectively. $\mathbf{Y}$ is a conductance matrix related to the network elements depicted in Fig. 7. Since each extracellular node is connected to the adjacent nodes only, the matrix $\mathbf{Y}$ is typically sparse. This makes it suitable for being solved by Gausz-Seidel iteration (Gerald & Wheatley, 1994), finding its reciprocal:

$$\mathbf{V} = \mathbf{Y}^{-1} \cdot \mathbf{I} = \mathbf{H} \cdot \mathbf{I} \tag{14b}$$

and with it, the potential distributions looked for. However, with a fiber radius of 25 μm and muscle dimensions of 1 mm diameter and 10 mm length the number of nodes would be in the order of 40 x 40 x 10 = 16000, which is computationally unmanageable.

There are several ways to approach this problem. Decreasing the number of nodes is inadmissible since the grid is coupled to the muscle structure. Albers *et al.* (1986) placed the active fiber along the axis of a cylindrical muscle model, using cylindrical symmetry, thus eliminating one dimension. They could actually calculate SFAPs at various distances. The same authors (Albers *et al.*, 1988) showed that those SFAP amplitudes were in partial

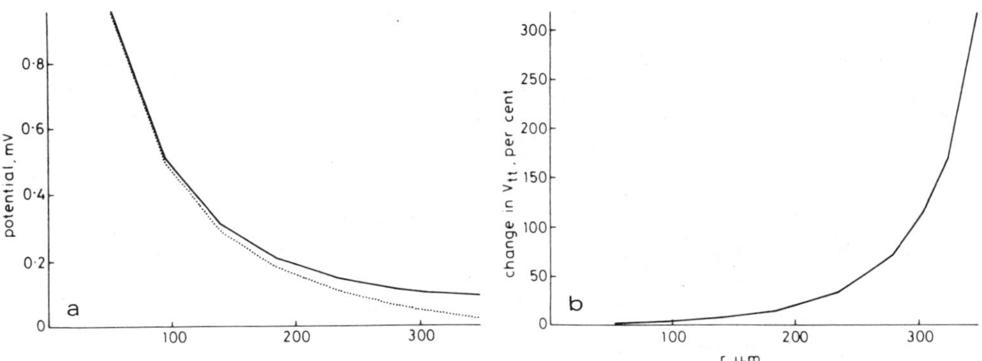

Figure 8. The influence of fiber position in a muscle on muscle boundary effects predicted by a hybrid network-homogeneous analytical model. Radius of cylindrical muscle: 1. 5 mm. **a:** SFAP amplitude at various distances from the active fiber. Solid line: active fiber positioned in the centre of the model. Dotted line: active fiber 500 μm beneath muscle surface. **b:** Relative change resulting from the fiber shift from axial to more peripheral. At a shift of 300 μm the increase in amplitude 100% (from van Veen *et al.*, 1992, Fig. 8).

accordance with experimental results reported by Gath and Stålberg (1979). At a distance of 53 mm SFAP amplitudes calculated by the network model is approx. 60% from the value obtained by homogeneous, analytical models. This difference decreased to only a few percent at a distance of 300 μm. So, at those distances the fibrous structure of muscle has ceased to be of influence on SFAP shape and amplitude. This figure gives an estimation of the minimal radial size a network model should have. More generally, under those distances the capacitive properties of the muscle cell membrane perceptibly increase SFAP amplitude and should not be neglected in models of EMG conduction.

Cylindrical symmetry in a EMG conduction model is a serious limitation. Many interesting questions, then, cannot be asked, because they implicitly refer to non-symmetry. For instance, the influence of the muscle boundary. For an axial fiber muscle boundary is too far away to study its influence on potential distribution closer to the fiber. Also, noncylindrical muscle shape, as most skeletal muscles have, cannot be investigated by the above models.

Van Veen *et al.* (1992) computed potential distributions by a completely three-dimensional network model, and matched this to a surrounding analytical homogeneous model at its outer boundary. The matching was performed by also calculating an entirely analytical model, and using current densities found at the location where the transition to the network model would have been as boundary conditions to the network part of the hybrid network-analytical model. It was subsequently shown that for cylindrically symmetric situations (still possible, then) both models calculated the same potential fields at greater distances from the active fiber. Figure 8 gives an application of this model. It compares the effects of fibers at two locations in the muscle: central and peripheral at two thirds of the distance to the muscle boundary. Extrapolating to this boundary, the model predicts SFAP amplitude by the peripheral to be several times greater than by the central fiber.

Frequency Domain Modeling of EMG Conduction

It has long been recognized that skeletal muscle as most bioelectrical volume conductors are effectively frequency dependent. (Clark *et al.*, 1978.) This fact, together with the linearity of the conduction process makes skeletal muscle tissue an attractive goal for

Fourier analysis. Conduction parameters can be studied as a function of frequency, and the response of the volume conductor can be found in the time domain by summing source Fourier components at the output.

Two phenomena cause frequency dependent conduction: a) The reactive properties of the muscle cell membranes, containing both resistive and capacitive elements; b) The finite conduction velocity of the EMG source (approximately 4 m/s).

Network models are exceptionally suited to study the frequency dependent conduction caused by membrane reactances, because of the direct link such models have with measurable membrane parameters like capacitance. In general, the net effect of membrane capacitance may be expected to be an increasing transfer by the network at the higher frequency. Actually it must be the cause of the marked higher SFAP amplitude shown in Fig. 8. However, that this model does predict this amplitude increase is not trivial since the second frequency dependent effect mentioned above can be shown to act in the opposite direction. Qualitatively it can be reasoned that a finite propagation velocity of the EMG source results in a low-pass character of the volume conductor. At any time during this propagation different parts of the active fiber will be in different phases of the action potential. This spatial extension makes the volume conductor difficult to consider as an electrical circuit with one input and one output and thus as a frequency dependent filter. Albers *et al.* (1988) have circumvented this problem by making use of the symmetry in the network response. If at one longitudinal coordinate

$$V_{z_i'r_n}(\omega) = \sum_{j=-j_{max}}^{j=j_{max}} H_{z_i'z_jr_n}(\omega) \cdot I_{z_j}(\omega) \tag{15}$$

where $V_{z_ir_n}$ is the Fourier transforms of the measured EMG voltage at longitudinal coordinate z_i and radial coordinate r_n, $I_{z_j}(\omega)$ that of source current at longitudinal coordinate z_j, and $H(\omega)$ that of the H matrix in Eq. (14b). Or, in view of the constant propagation velocity of the source:

$$V_{z_i'r_n}(\omega) = \sum_{k=-k_{max}}^{k=k_{max}} H_{z_i'z_jr_n}(\omega) \cdot I\cos\left(k\omega\frac{\Delta z}{U}\right) \tag{16}$$

In a sufficiently long volume conductor and for a sinusoidal extending source current density. thus:

$$V_{z_j'r_n}(\omega) = \sum_{k=-k_{max}}^{k=k_{max}} H_{z_m'z_jr_n}(\omega) \cdot \cos\left(k\omega\frac{\Delta Z}{U}\right) \cdot I(\omega) = \sum_{k=-k_{max}}^{k=k_{max}} H_{z_k'r_n}'(\omega) \cdot I(\omega) \tag{17}$$

showing that propagation effects in the transfer function can be accounted for by adapting the transfer matrix H, i.e., only by the volume conductor structure. The cosine in Eq. (17) demonstrates the low pass character of the transfer H'. For frequencies for which $k\omega \cdot \Delta z < \pi \cdot U/2$, the cosine is < 1.

Representation of volume conduction transfer in term of Fourier transformations offers the possibility of predicting various source functions by inverse transformation of their Fourier transform. Complementarily, such a representation gives a more defined way of typifying the properties of the conducting medium. Figure 9 exemplifies isopotential lines

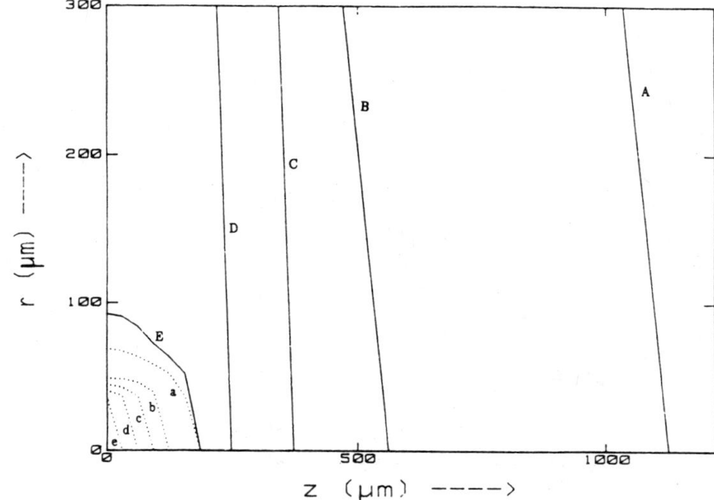

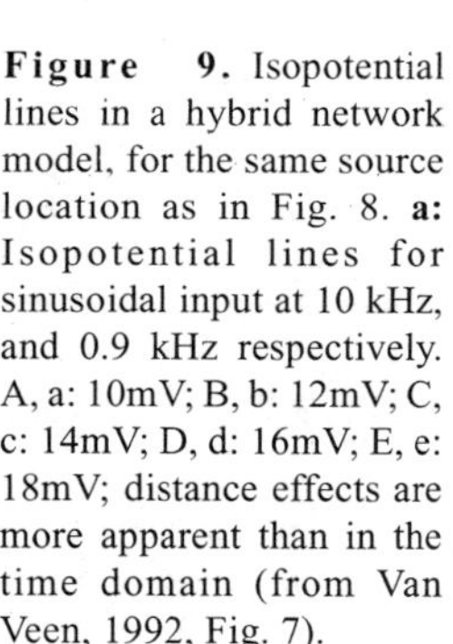

Figure 9. Isopotential lines in a hybrid network model, for the same source location as in Fig. 8. **a:** Isopotential lines for sinusoidal input at 10 kHz, and 0.9 kHz respectively. A, a: 10mV; B, b: 12mV; C, c: 14mV; D, d: 16mV; E, e: 18mV; distance effects are more apparent than in the time domain (from Van Veen, 1992, Fig. 7).

in the hybrid network model. Isopotential lines for higher frequency (10 kHz) keep closer to the source than those for lower frequency (0.9 kHz), corresponding to low pass behavior. In a time domain representation, this would be more ambiguous because of the shape distortion of the SFAPs to be used.

THE INVERSE PROBLEM IN SINGLE FIBER ELECTROMYOGRAPHY

Approaches of the Inverse Problem

As stated in the Introduction, characterization of the source of the SFAP will now be treated. According to Fig. 1, various sources can be tried and the one best fitting the experimental results considered the optimal solution. A major difficulty is that available volume conduction models predict SFAPs as a function of intramuscular coordinates and distances, which experimentally are difficult to obtain. A somewhat more indirect way, avoiding this problem, makes use of the fact that SFAP shape describing parameters, because of frequency dependent volume conduction, will vary with distance to the source, and, thus, are also mutually interdependent. This relationship can be obtained for each measured SFAP and compared with volume conduction model predictions. This approach (Van Veen, 1993) will be described first. It will be found that, of the sources discussed, directly measured transmembrane current shows the best behavior.

Although this is a result, a more direct way would be welcome. This will be described next. A limited set of data on actual source distance could be obtained by histological techniques and substituted into the volume conduction model.

The SFAP Shape at Different Places in the Muscle

The SFAP shapes were defined by two relationships (see Fig. 3): a) first positive phase amplitude (V_1) versus total amplitude (V_{tt}): the V_1-V_{tt} relation (Fig. 10a). This would be a straight line under 45° in a frequency independent, linear volume conductor. b) Negative

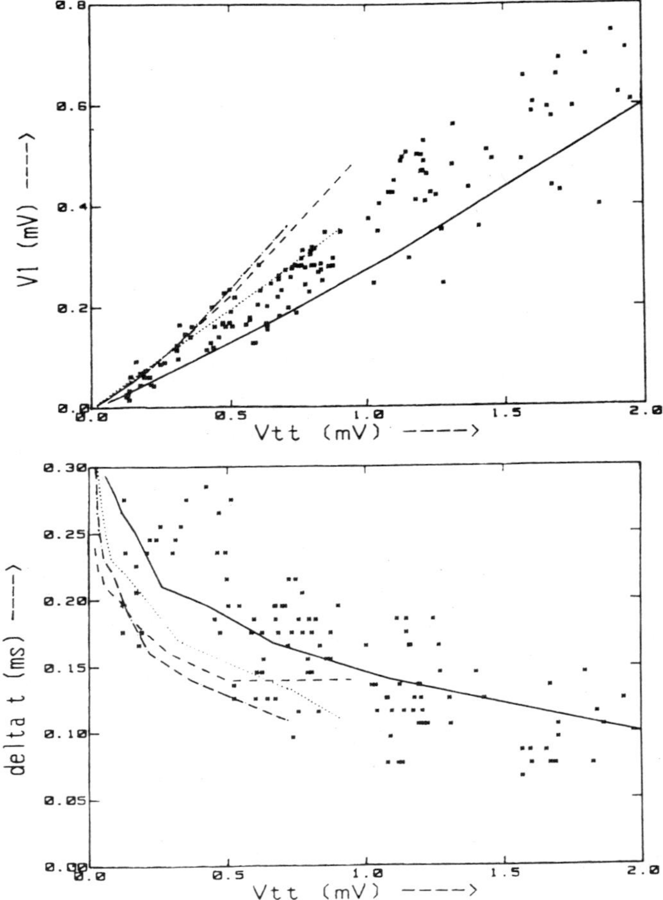

Figure 10. Relations of SFAP shape defining parameters at various, unmeasured locations in the muscle. Comparison with theory and experiment. a. V1 versus V_{tt}. b. Δt versus V_{tt}. Symbols: Experimental data. Dashed line: Source is IAP according to Rosenfalck (1969). Dashed-Dotted line: Source is measured IAP by Wallinga *et al.* (1985). Dotted line: Source is the SFAP measured with the greatest $V_{tt}/\Delta t$ found during the experiments with that muscle. Solid line: Source is Transmembrane Current actually measured with loose patch-clamp technique (Wolters *et al.*, 1992; Almers *et al.*, 1983). Direct measured transmembrane current gives the best correspondence between theoretical and experimental relationship of both V_1 and Δt with V_{tt} (from: Van Veen, 1993, Figs. 6, 7).

deflection time (Δt), also versus V_{tt}. In a resistive volume conductor this would stay constant. The EMG source candidates introduced at the beginning of this paper were tried for their correctness by: computing V_1-V_{tt} and Δt-V_{tt} relations for the SFAPs they generated (the four line styles in Fig. 10b and c) by being an input to the hybrid network model depicted in Fig. 7. These theoretical relations were compared with those calculated from SFAPs. Intracellular action potentials were input to the network model in two ways, as IAPs analytically described (Rosenfalck, 1969) and as experimental ones (Wallinga *et al.*, 1985). These signals were transformed to transmembrane current equivalents by substituting into Eq. (2). A SFAP as a source should be one measured in the closest proximity of the source. Thus the SFAP with the greatest amplitude and the shortest duration measured during the experiments was taken and also scaled to transmembrane current (proportionally). Transmembrane current was measured with the loose patch-clamp technique, as described above (Wolters *et al.*, 1992). Since these measurements were *in vitro* and on mice, instead of *in vivo* and in rat, these differences had to be corrected as much as possible.

Figure 10 clearly suggests that directly measured transmembrane current is the best source. The V_1-V_{tt} relation for the IAPs as a source stays higher than the cloud of experimental points. SFAP and Membrane current do comparably better. However, for the Δt-V_{tt} relation membrane current stands out as the best. Since all sources were scaled to an equivalent transmembrane current with the same amplitude and duration, the differences

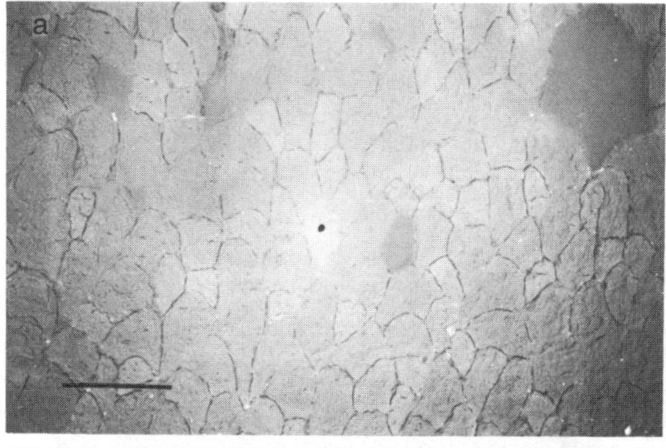
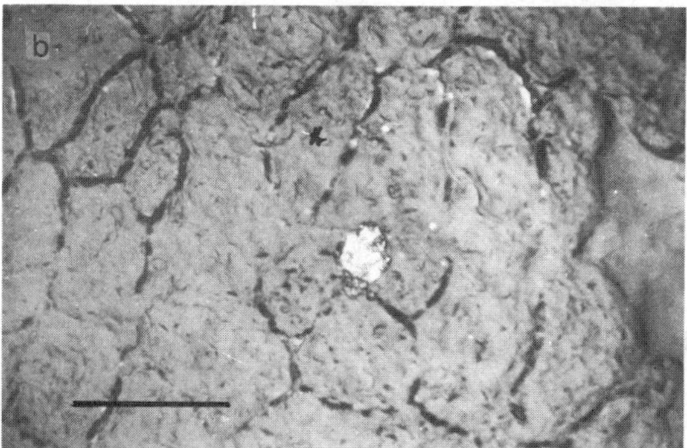

Figure 11. Muscle slices showing a: active fiber made visible with Lucifer Yellow, and b: deposited silver dot, originating from electrode at that position at SFAP recording time. Slices like a and b usually could be associated in such a way that the distance between active fiber and electrode site was established (from Van Veen 1994, Fig. 6).

must be ascribed to difference in shape. Since in the IAP cases as well as in the SFAP case in Fig. 10, Eq. (2), had to be used for calculating equivalent membrane current, the outcome of this investigation clearly suggests this relation to be incorrect, a conclusion that was drawn earlier in this chapter.

The SFAP Amplitudes at Measured Distances

Direct comparison of theoretical and experimental SFAP-distance relations can only be achieved if precise knowledge can be obtained where the active fiber was and where an electrode measured its SFAP. Van Veen *et al.* (1994) show data on this. SFAPs were observed with a 14 channel array wire electrode arrangement. To make one fiber active, a micropipette was inserted as in a fiber whose SFAPs might be detectable by the wire electrodes. With the micropipette the fiber was stimulated by hyperpolarizing break excitation.

As a volume conductor model, again, the hybrid network model was used (with the active fiber located axially). Source electrode distance, to be specified in the elements of the matrix $\mathbf{Y}$ in Eq. (14) were found experimentally as follows. The position of the radial position of the stimulated fiber was found by electrophoretically labeling it with the fluorescent dye Lucifer Yellow CH, supplied through the stimulating pipette, after SFAPs could be recorded successfully. When the muscle had been sliced, the fluorescent cross-section of the active fiber could be retraced. The electrode position was recorded well. At the conclusion of a successful experiment silver was deposited by DC current pulses at the top of those wire electrodes which actually had recorded SFAPs. Then the wires were removed in order to enhance the slice quality. Slicing was after the muscle had been frozen and stored at a temperature of 193 K. Figure 11 gives examples of active fiber and electrode showing slices, and Fig. 12 the result of comparing experimental SFAP amplitudes (symbols) with model

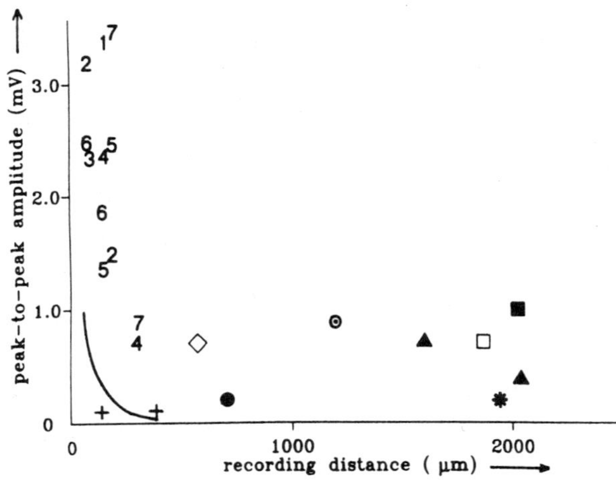

Figure 12. Measured SFAP amplitude (symbols) and network model computed SFAP amplitude, using histological SFAP-source data . Model predictions of SFAP amplitude are systematically too low (from Van Veen, 1994, Fig. 7).

computations using the histological data (solid line). The most obvious finding is that experimental SFAP are significantly greater in amplitude than predicted. Although the experimental scatter is also significant, the finding is consistent: all experimental SFAPs follow the same pattern.

It is difficult to explain the above results. It seems clear that the relation between SFAP amplitude and recording distance is more complicated than assumed. In the literature, direct comparison of theory and experiment is very scarce in this area (Gath and Stålberg, 1978). To the knowledge of the authors, the current approach is the only attempt to relate experimental SFAPs to muscle tissue distances. Recordings of one active fiber are rarely encountered.

The theoretical SFAP amplitudes in this investigation were based on intracellular potential as a source, patch clamp measurements not yet being available at the time. Comparison with Fig. 10 shows that measuring transmembrane current directly can improve the situation, because SFAP amplitudes would be computed twice as large, which, however, is still not enough. Only a significantly better conduction in skeletal muscle tissue than assumed can bring theory in accord with the practice of Fig. 12.

CONCLUSION

We are still far from fully understanding the relationships between SFAPs, *a fortiori* clinical SFAPs and the characteristics of the active fiber. As the optimal EMG source, Transmembrane Current can be measured, by the loose patch-clamp technique, a method not yet applicable in really *in situ* situations. The results in this paper show significant differences between what can be measured and what is known about muscle. Particularly intriguing is the question, where the higher conductivity, suggested by opposing theory and experiment, will be found to reside. For this, a more refined model of muscle and muscle fiber, and more measurements are required.

REFERENCES

Adrian, R. H., and Peachey, L. D., 1973, Reconstruction of the action potential of frog sartorius muscle, *J. Physiol.* 235:103-131.

Albers, B. A., Put, J. H. M., Wallinga, W., and Wirtz, P., 1989, Quantitative analysis of single muscle fibre action potentials recorded at known distances, *Electroencephalogr. Clin. Neurophysiol.* 73:245-253.

Albers, B. A., Rutten, W. L. C., Wallinga, W., and Boom, H. B. K., 1988, Microscopic and macroscopic volume conduction in skeletal muscle tissue, applied to simulation of single-fiber action potentials, *Med. & Biol. Eng. & Comp.* 26:605-610.

Albers, B. A., Rutten, W. L. C., Wallinga, W., and Boom, H. B. K., 1986, A model Study on the influence of structure and membrane capacitance on volume conduction in skeletal muscle tissue, *IEEE Trans. Biomed. Eng.* 33:681-689.

Almers, W., Stanfield, P. R., Stühmer, W., 1983, Lateral distribution of sodium and potassium channels in frog skeletal: measurements with a patch clamp technique, *J. Physiol.* 336:261-284.

Clark, J. W. Jr., Greco, E. C., Harman, T. L., 1978, Experience with a Fourier method for determining the extracellular potential fields of excitable cells with cylindrical geometry, *CRC Crit. Rev. in Bioeng.* 3: 1-22.

Clark, J., and Plonsey, R., 1966, A mathematical Evaluation of the Core Conductor Model, *Biophys. J.* 6:95.

Fedida, D., Sethi, S., Mulder, B. J. M., and Ter Keurs, H. E. D. J., 1990, An ultracompliant glass microelectrode for intracellular recording, *Am. J. Physiol. 258(Cell Physiol. 27)*: C164-C170.

Ganapathy, N., Clark, J. W. Jr., and Wilson, O. B., 1987, Extracellular potentials from skeletal muscle, *Math. Biosc.* 83:61-96.

Gath, I., and Stålberg, E., 1978, The calculated radial decline of the extracellular action potential compared with in situ measurements in the human biceps, *Electroencephalogr. Clin. Neurophysiol.* 44:547-552.

Gath, I., and Stålberg, E., 1979, Measurement of the uptake area of small size electromyographic electrodes, *IEEE Trans Biomed. Eng.* 26:374-376.

Gerald, C. F., Wheatley, P. O., 1989, *Applied Numerical Analysis*, 4th ed., Adison-Wesley Pub. Co., Reading, Ma.

Gielen, F. L. H., Wallinga, W., Boon, K. L., 1984, Electrical conductivity of skeletal muscle tissue: experimental results from different muscles in vivo, *Med. & Biol. Eng & Comput.* 22:569-577.

Gielen, F. L., Cruts, H. E. P., Albers, B. A., Boon, K. L., Wallinga, W., and Boom, H. B. K., 1986, Model of the electrical conductivity of skeletal muscle based on tissue structure, *Med. & Biol. Eng & Comput.* 24:34-40.

Hamill O. P., Marty, A., Neher, E., Sakmann, B., and Sigworth, F. J., 1981, Improved patch-clamp techniques for high-resolution current recording from cells and cell-free membrane patches, *Pfluegers Arch.* 391: 85-100.

Hermens, H. J., van Bruggen, T. A. M., Baten, C. T. M., Rutten, W. L. C., and Boom, H. B. K., 1992, The median frequency of the surface EMG power spectrum in relation to motor unit firing and action potential properties, *J. Electromyography Kin.* 2: 15-25.

Lorente de Nó, R., 1947, Analysis of the distribution of action currents of nerve fiber in volume conductors, *Stud. Rockefeller Inst. Med. Res.* 132:384.

Meier, J. H., Rutten, W. M. L., Zoutman, A. E., Boom, H. B. K., and Bergveld, P., 1992, Simulation of multipolar fiber selective neural stimulation using intrafascicular electrodes, *IEEE Trans. Biomed. Eng.* 39: 122-134.

Milton, R. L., Lupa, M. T., and Caldwell, J. H., 1992, Fast and slow twitch muscle fibers differ in their distributions of Na channels near the endplate, *Neurosci. Lett.* 135:41-44.

Plonsey, R., 1969, *Bioelectric Phenomena.* New York: McGraw-Hill.

Plonsey, R., and Barr, R., 1982, The four-electrode resistivity technique as applied to cardiac muscle, *IEEE Trans. Biomed. Eng.* 29:541-544.

Rosenfalck, P., 1969, Intra- and extracellular potential fields of active nerve and muscle fibers. A physico-mathematical analysis of different models, *Akademisk Forlag*, Copenhagen, Denmark.

Roth, B. J., and Gielen, L. H., 1987, A comparison of two models for calculating the electrical potential in skeletal muscle, *Ann. Biomed. Eng.* 15:591-602.

Stegeman, D. F., and Linssen, W. H. J. P., 1992, Muscle fiber action potential changes and surface EMG: a simulation study, *J. Electromyography Kin.* 2:130-140.

Van Veen, B. K., Wolters, H., Wallinga, W., Rutten, W. L. C., and Boom, H. B. K., 1993, The bioelectrical source in computing single muscle fiber action potentials, *Biophys. J.* 64:1492-1498.

Van Veen, B. K., Rijkhoff, N. J. M., Rutten, W. L. C., Wallinga, W., and Boom, H. B. K., 1992, Potential distribution and single-fiber action potentials in a radially bounded muscle model, *Med. & Biol. Eng & Comput.* 30:303-310.

Van Veen, B. K., Mast, E., Busschers, R., Verloop, A. J., Wallinga, W., Rutten, W. L. C., Gerrits, P. O., and Boom, H. B. K., 1994, Single fibre action potentials in skeletal muscle related to recording distances, *J. Electromyography Kinesioly*, 4:37-46.

Wallinga, W., Gielen, F. L. H., Wirtz, P., de Jong, P., and Broenink, J., 1985, The different intracellular action potentials of fast and slow muscle fibers, *Electroencephalogr. Clin. Neurophysiol.* 60:539-547.

Wallinga W., Albers, B. A., Put, J. M. H., Rutten, W. L. C., and Wirtz, P., 1988, Activity of single muscle fibtres recorded at known distances. In: *Electrophysiological Kinesiology*, Wallinga, W., Boom, H. B. K., and de Vries, J. (eds.). Elsevier Science Publishers: Amsterdam, pp. 221-224.

Wolters, H. W., Wallinga, W., and Ypey, D. L., 1991, Recording of membrane current and action potential on the same spot in mammalian skeletal muscle fibers, *Pfluegers Arch.* 418:R152.

Wolters, H., Wallinga, W., Ypey, D. L., and Boom, H. B. K., 1994, Ionic currents during action potentials in mammalian skeletal muscle fibers analyzed with loose patch clamp, *Am. J. Physiol. (Cell Physiol. 36):* C1699-C1706.

23

THE EMG AS A WINDOW TO THE BRAIN: SIGNAL PROCESSING TOOLS TO ENHANCE THE VIEW

Werner M. Wolf

Institut für Mathematik und Datenverarbeitung
Universität der Bundeswehr München
85577 Neubiberg, Germany

ABSTRACT

The paper discusses processing tools for electromyographic signals (EMG) with particular consideration of needle EMG and its decomposition in spike trains for single motor units (MU). Examples are given for combined application of surface and needle EMG, and possibilities for further developments of EMG signal processing tools are critically commented.

ELECTROPHYSIOLOGY - A LOOK INTO THE BODY

The basic foundation for the modern biocybernetic view which considers the living being a machinery consisting of sensors, information processing units, actuators and a power supply was set by Descartes (1596 - 1650) and his doctrine of the dualism of body and mind. Analyzing the human's information processing system, experimental psychology observes the behavior and performance of subjects solving specific tasks (e.g., reaction time experiments) with the aim to establish a black box description; those investigations were totally noninvasive. Electrophysiological measurements like evoked EEG potential recording and skin resistance reading supplement those experiments by permitting some blurred access to internal processes: even if they are considered noninvasive methods their principle includes physiological structure already. But a clear view is provided by invasive techniques which allow to place the receiving antenna near to the "broadcasting station" in the body. Invasive electromyography (EMG) allows to monitor muscle activity and was developed by Adrian and Bronk (1929).

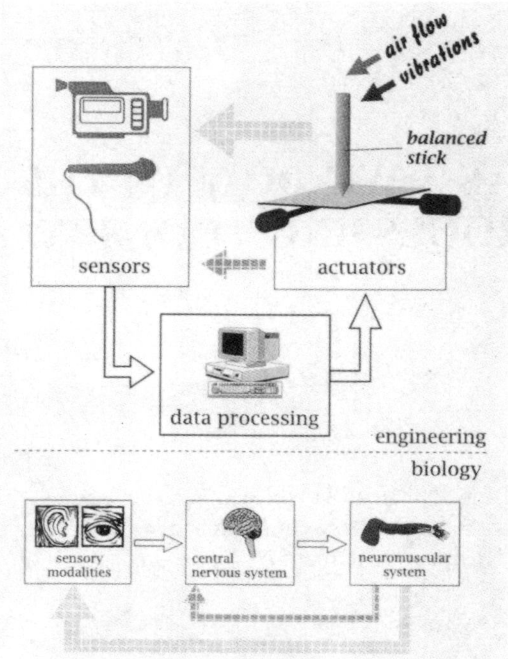

Figure 1. Structure of sensorimotor systems.

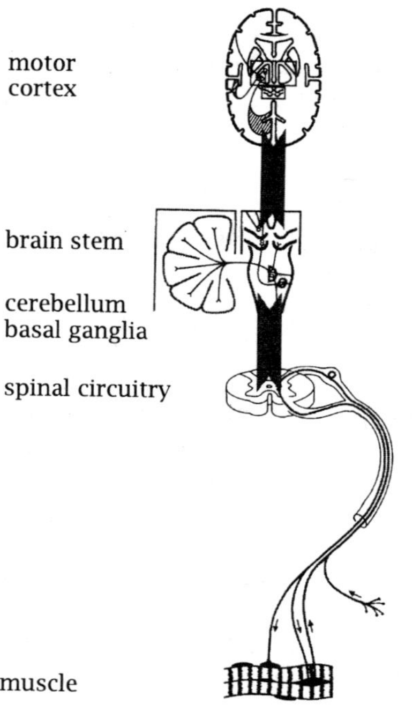

Figure 2. Organization of the neuromuscular system.

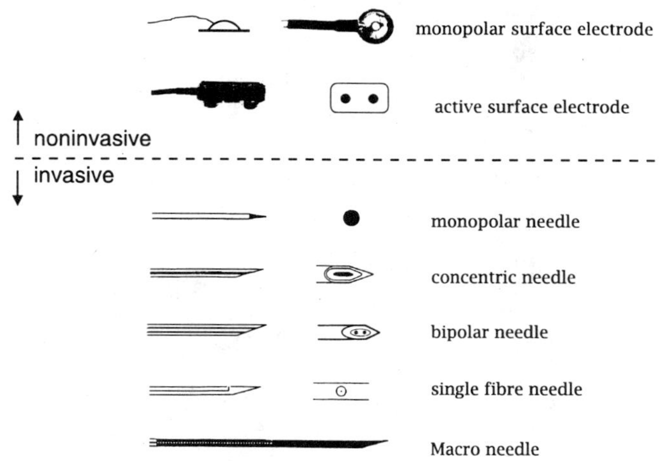

Figure 3. EMG electrodes.

ELECTROMYOGRAPHY - OBSERVING THE MUSCLE MACHINERY

Starting out from the engineering model of an autonomous system acting in an unpredictable environment (Fig. 1) (e.g., a robot) we structure such a black box system into a sensory front end (i.e., the different modalities), followed by a data analyzing and processing unit (i.e., the Central Nervous System, CNS) and, finally, an actuator machinery (i.e., the neuromuscular system). The latter to which we will look more closely again can be structured into 4 layers (Fig. 2): (i) motor cortex, (ii) brain stem and other subcortical areas, (iii) spinal circuitry, and the (iiii) muscle. The sequence given indicates the hierarchy within the system. In humans, we can observe the system only at the front ends by common electrophysiological measures: (i) cortical motor potentials (MP) pickup from the scalp, and (ii) the electromyogram (EMG) recorded from the muscle.

The **MP** is a very global signal and it is difficult to extract specific information from it. Like in evoked EEG potentials, averaging of a lot of trials is necessary to raise the MPs over the noise given by the background EEG. But averaging needs a trigger event: either the stimulus onset for a go cue can serve for that (in case of reaction time or reflex experiments) or - more common - the observed movement itself. Unfortunately, both events are not tightly coupled to the MP because the motor cortex level is distant to the early sensory input stage as well as to the last element in the sensorimotor chain, the muscle. Thus the MP latencies will show some stochastic delay variations with respect to either trigger event, which results in a smearing of MP by the averaging process.

In contrast, the **EMG** recorded from the muscle site does not suffer from low signal to noise ratios (S/N) like the cortical MP. Depending on electrode type and configuration, we can obtain EMG records of different selectivity (Fig. 3) spanning from global muscle activity (surface electrode) to single muscle fibre monitoring (needle electrode). Simplistically assuming the spinal circuitry a simple relay stage, the muscle activity and, consequently, the EMG reflects the task related activity of the CNS structures - it provides *a window to the brain*.

What do we measure by the EMG? The basic element of a muscle is the motor unit (MU) consisting of an alpha motor neuron and the muscle fibers connected to it (Fig. 4). When the alpha motor neuron discharges, each fibre of the MU produces an extracellular

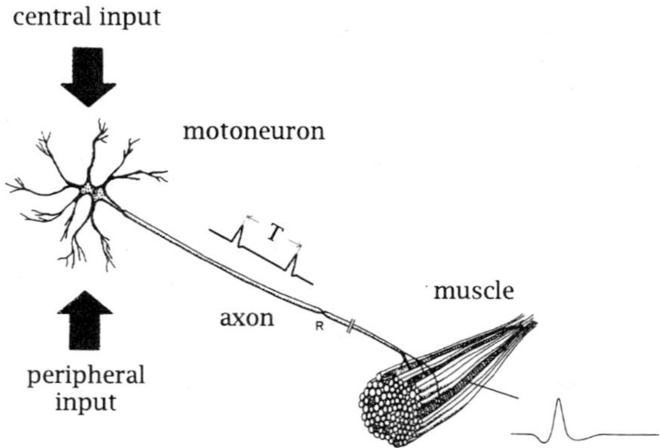

Figure 4. The Motor Unit (MU).

spike potential. With the very selective single fibre needle this single action potential (AP) can be recorded, while the concentric needle collects the signals of several fibers of this MU, which together form the motor unit action potential (MUAP). A surface electrode, however, sums up the MUAPs of all active MUs composing a measure of the total activity of the muscle. The shape of the MUAP is governed by muscle determinants, while the MUAP rate directly reflects the central drive. Therefore, we talk about **peripheral** EMG when analyzing the shape, and about **central** EMG when observing the discharge rate of a MU. Since muscles consist of several hundreds (e.g., hand muscle) up to several thousands of MUs (eye muscle) which all act in a go/nogo manner, a sophisticated schedule for MU activation is required to achieve a continuous smooth movement. This proper timing is disturbed by several central motor disorders and the central EMG allows analysis of those central deficits.

Clinical evaluation of needle EMG was mainly introduced by the Buchthal group (Buchthal *et al.* 1954 a, b). They processed the EMG recordings manually and they established reference values still widely used today. Intelligibly that people invested efforts to have the computer performing this tedious work automatically. These research activities started in the late sixties (e.g., Kopec and Hausmanova-Petrusewisz, 1969; Bergmans, 1971; Prochazka *et al.*, 1972, 1973; Lee and White, 1973), peaked in the eighties (e.g., Andreassen,

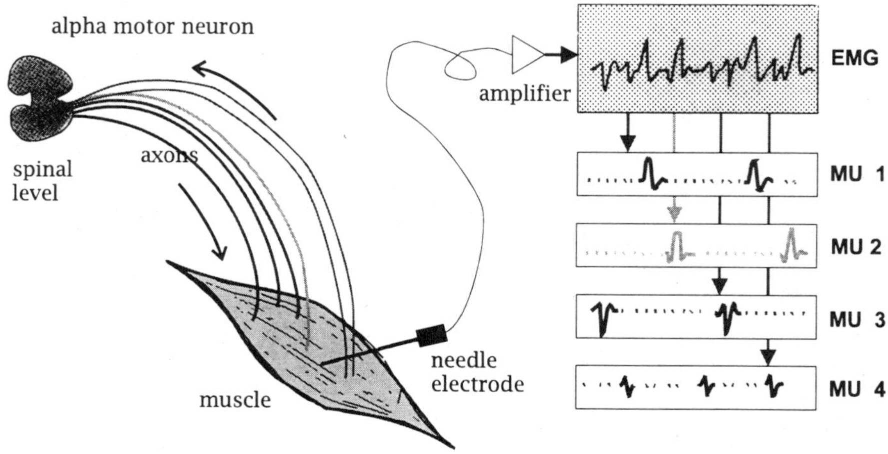

Figure 5. EMG decomposition in MU specific MUAP trains.

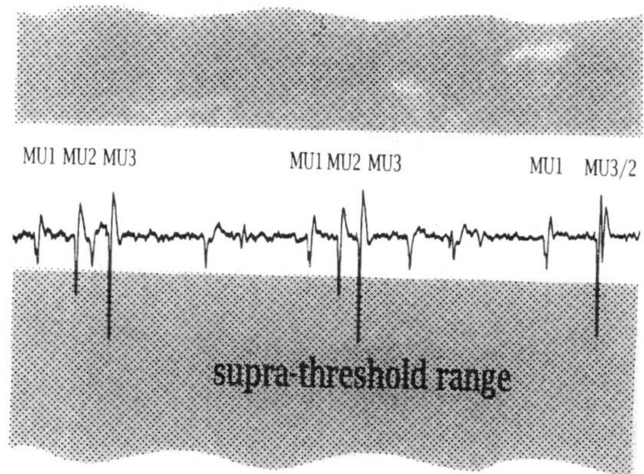

Figure 6. Segmentation by amplitude threshold.

1983; DeLuca *et al.*, 1982a, b; Gerber *et al.*, 1984; Guiheneuc, 1985; McGill and Dorfman, 1985; Stalberg and Antoni, 1983; Stashuk and DeBruin, 1988; Studer *et al.*, 1984; Wolf and Dengler, 1982, 1985, 1987; Wolf *et al.*, 1985, 1987), and converged to a low steady state level in the early nineties (e.g., Pattichis *et al..*, 1992; Stashuk *et al.*, 1992).

EMG DECOMPOSITION - THE SALIENT POINT IN NEEDLE EMG

Except single fibre recordings, all other EMG data comprise MUAPs of several MUs, which calls for their decomposition in MU-specific spike trains (Fig. 5) before clinical

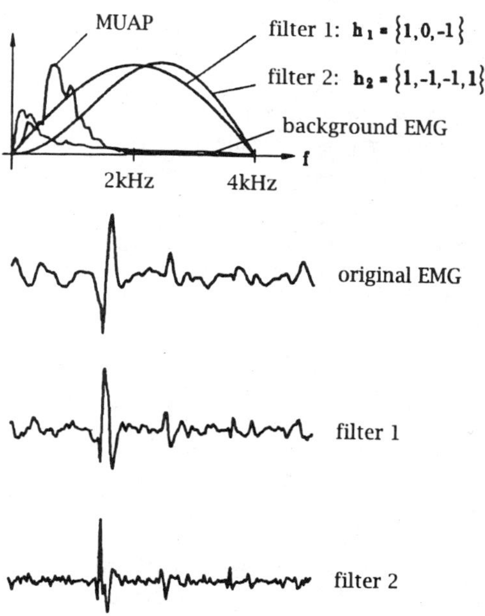

Figure 7. Low-pass differentiation to enhance segmentation.

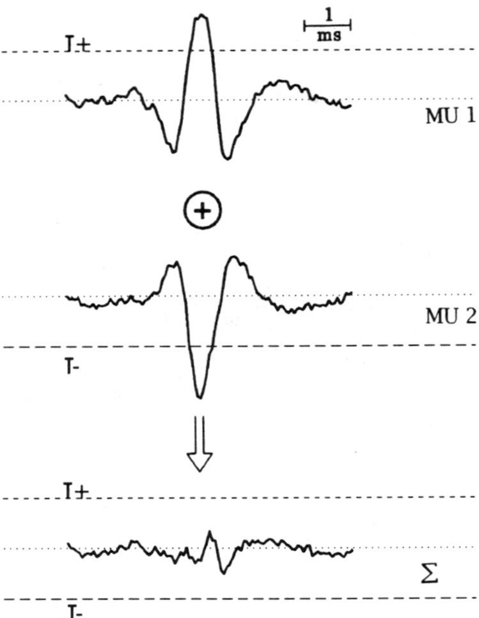

Figure 8. Adverse superimposition of two MUAPs of different polarity.

evaluation. Opposing intentions during examination are to capture as many MUs as possible simultaneously in a record (in order to obtain information on coordination and interaction of MUs) but to have decomposition as simple as possible through clearly separated and distinguishable MUAPs. Selection of electrode type, its placement and the contraction level of the muscle must be matched for achieving a compromise. There also are some promising attempts to isolate single MU from surface EMG (Reucher *et al.*, 1987; Dorfman *et al.*, 1989), but in routine work the concentric needle and the rather selective bipolar needle are commonly used. Needle recordings show a very large frequency bandwidth up to 10 kHz, thus sampling rate usually should be set to about 20 kHz (e.g., Stalberg and Stalberg, 1989; Wolf and Dengler, 1982). Lower rates (e.g., Guiheneuc *et al.*, 1983) cause some loss of high frequency information, higher rates (e.g., LeFever and DeLuca, 1982) lead to an increase of storage requirement and computational load.

EMG decomposition needs two steps: (i) segmentation of signal to determine **where** in the time series MUAPs are located, and (ii) classification of these MUAPs to define to **which** MU class they belong.

Segmentation

Usually, segmentation does not attract attention considerably; signals of a high S/N can be processed by a simple amplitude threshold or an amplitude window in case of components with different polarities (Fig. 6).This approach requires modest computations but small clearly distinguishable MUAPs like MU 1 in Fig. 6 cannot be captured. In some cases, a preceding signal conditioning by a low-pass differentiation (e.g., Usui and Amidor, 1982; Wolf, 1991) can improve the performance; it enhances the sharp transient in the MUAP without accentuating the high frequency noise (Fig. 7), and computational costs are reasonable. But nevertheless, there are some tricky cases in needle EMG which bypass our detection

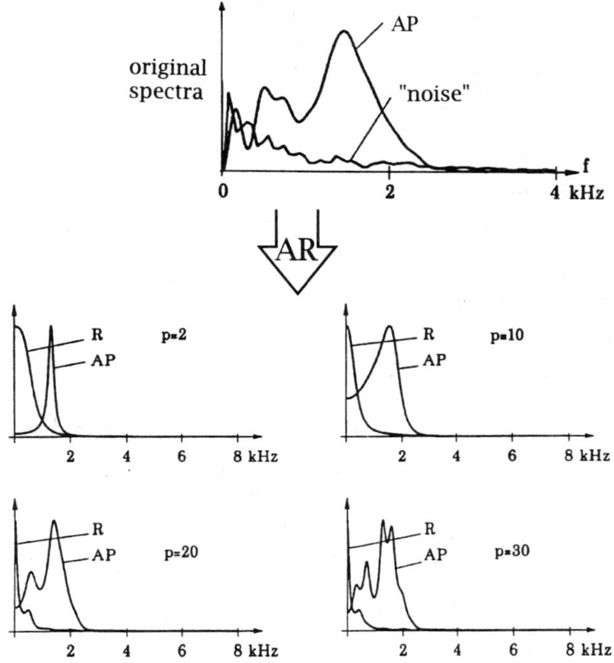

Figure 9. AR modeling of needle EMG.

procedure (Fig. 8): as polarities of MUAPs in a record can be opposite it happens sometimes that they superimpose each other and the amplitude of the resulting residue gets very small.

Since the frequency plots of signal and "noise" differ significantly (Fig. 7) , the use of a spectral criterium instead of an amplitude threshold seems appropriate. AR modeling of EMG signal x describes

$$x_k = -\sum_{i=1}^{p} a_i * x_{k-i} + e \tag{1}$$

the signal in the frequency domain by p coefficients a_i (e.g., Paiss and Inbar, 1987) where e represents the residual error. As shown in Fig. 9, the higher the model order p the more detailed is the approximation of the spectra. But with increasing p also the amount of calculations necessary to set up coefficients a_i increases, so a value between 6 and 10 is recommended. The detection of MUAPs (Fig. 10) is then achieved by determining first the coefficients $a_i^{(r)}$ for a so-called reference window which should not contain any MUAP, and then do so for a test window with coefficients $a_i^{(t)}$; the test window will be shifted sample by sample over the signal trace. If the used distance measure D, the "Generalized Likeli-hood-Ratio" (Appel and Brandt, 1983), detects significant deviations between both sets of coefficients, the alarm is given and a MUAP detected. This algorithm works very stable even in case of very noisy EMG signals. It should be noted that the threshold should be better underestimated since a false alarm can be corrected during classification, but a missing alarm will be lost.

With the exception of the very early studies, most decomposition methods reported use a cascade of a filter followed a threshold element. Also, the AR-approach proposed here can be interpreted as a filter but with an adaptive behavior. This causes a significant higher computational load, which must be taken into account, in particular where on-line operation is aspired. On the other hand, segmentation and classification are separated processes and can be distributed on different CPUs allowing their pipe-lined parallel execution.

Classification

Every detected MUAP must be assigned to the class it belongs to by the classification procedure. In principle, classification of a MUAP can be taken on the basis of a comparison between the new MUAP and those already known, the so-called templates. Unfortunately, the shape of MUAPs depend on so many factors, thus no a priori templates can be used from some memory (e.g., as in recognition of printed characters) but they must be collected from the actual record itself. The easiest solution is to display some EMG traces and select the templates interactively by cursor settings (mouse). More sophisticated, an unsupervised cluster program builds MUAP groups from the initial phase of the recording. Groups of high correlation between members and, in addition, comprising a reasonable amount of members are regarded as the MUAP classes, those groups of low correlation between members or rare occurrence are put to garbage. Finally, the templates are build up by averaging the members of each group. In many applications, template selection and classification are combined to a single process because of overlapping tasks. Especially in off-line EMG evaluation, template selection and classification could be done iteratively, which means that every reliably classified MUAP will be included in the template estimation of its group. By this mean, also a continuous update of templates can be performed in order to adapt them to slowly changing MUAP shapes, e.g., caused by some needle slipping.

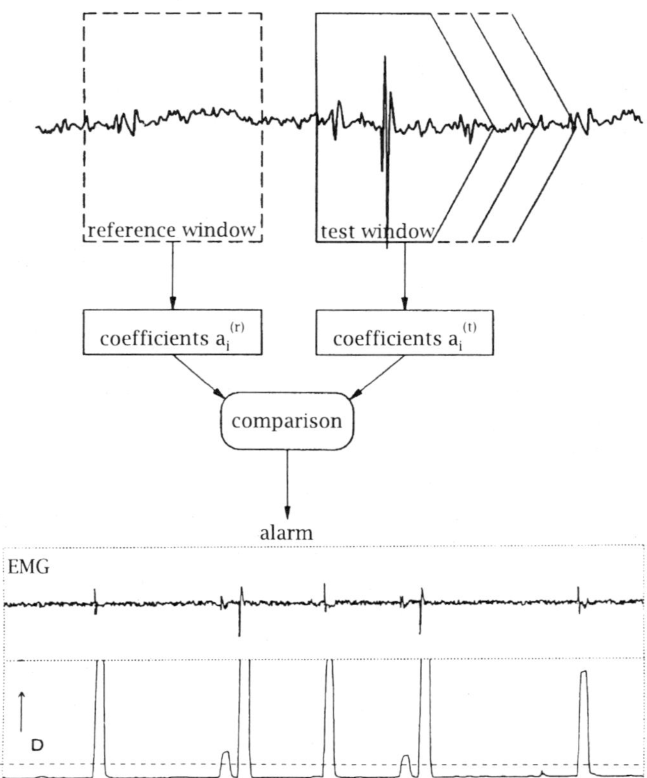

Figure 10. AR detection schema.

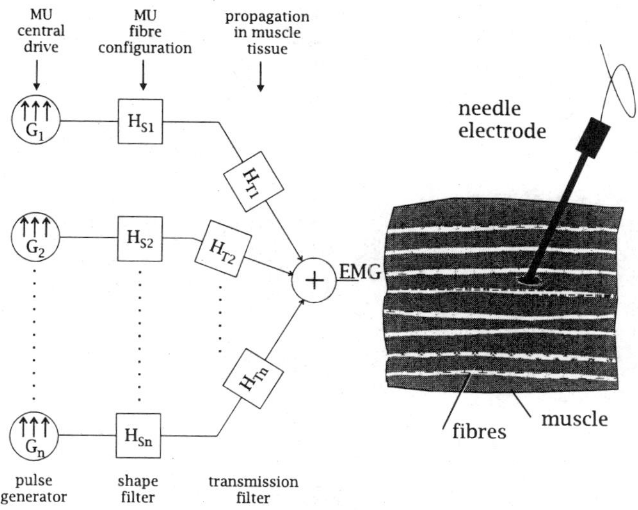

Figure 11. MUAP signal model.

Classification itself is achieved by some correlation techniques, linear parametric and nonparametric as well as nonlinear approaches like Neural Nets and other AI concepts. Most work is based on linear concepts since the problem is basically a linear problem (Fig. 11): a superimposition of different field components which are generated by the muscle fibers of the individual MUs and filtered on their way through the muscle tissue to the electrode tip. (Certainly, the questions whether the propagation medium is homogeneous and whether some interaction between field components occurs must be posed.) Separation of these components can be achieved by a multi-channel matched filter set. Impulse functions h_j of the M channels (one for each MU) can be obtained through templates t_j by fitting

$$\sum_k \sum_u t_j(k-u) \cdot h_i(u) \;\Rightarrow\; \min \qquad 1 \leq i,j \leq M, \quad i \neq j \tag{2}$$

$$\sum_k \sum_u t_j(k-u) \cdot h_j(u) \;\Rightarrow\; \max \qquad 1 \leq j \leq M \tag{3}$$

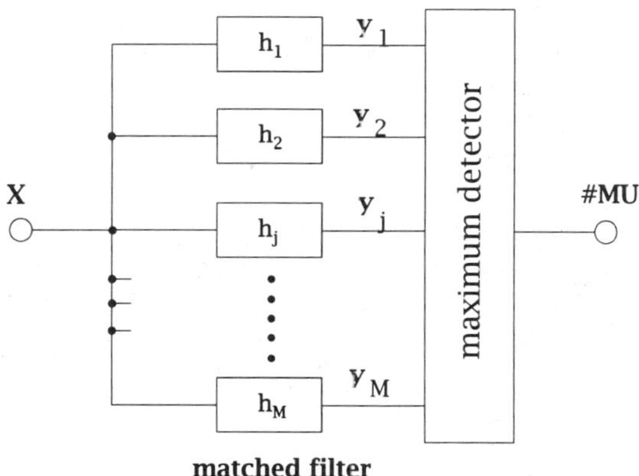

Figure 12. Decomposition filter.

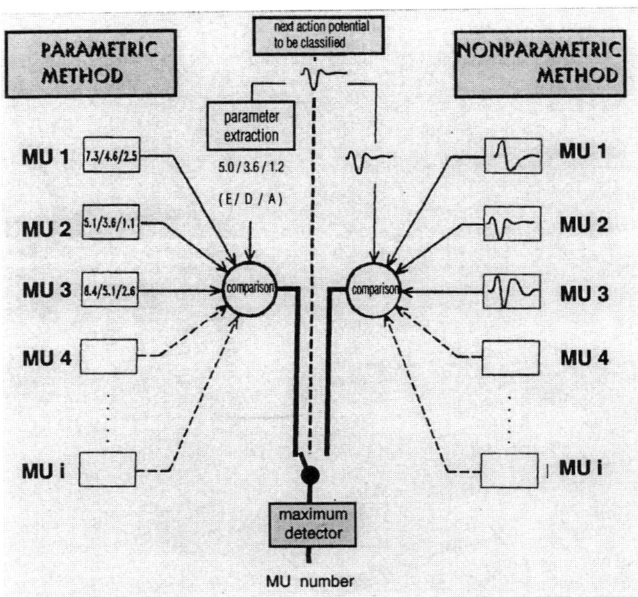

Figure 13. Classification methods.

i.e., to minimize cross sensitivity by (2) and to maximize responses to MUAPs of the corresponding group by (3). The input signal vector x provided by the segmentation procedure will be filtered by all these channel impulse functions h_j to obtain the outputs y_j

$$y_j(k) = \sum_u x(k-u) \cdot h_j(u) \qquad 1 \leq j \leq M \tag{4}$$

Assignment of the input MUAP vector x to the specific MU class j will be performed by a simple maximum detector which evaluates these filter channel outputs (Fig. 12). The optimization of h_j is the crucial point because its success depends on the criteria (Studer, 1984) and those are different between methods (and often not detailed).

In principle, this design concept described is common to nonparametric and parametric methods as well. As shown in Fig. 13, parametric methods uses not the input signal vector x itself for classification, but a parameter vector F determined from the input signal vector. Usually, the parameters used are chosen so that they are almost independent from each other

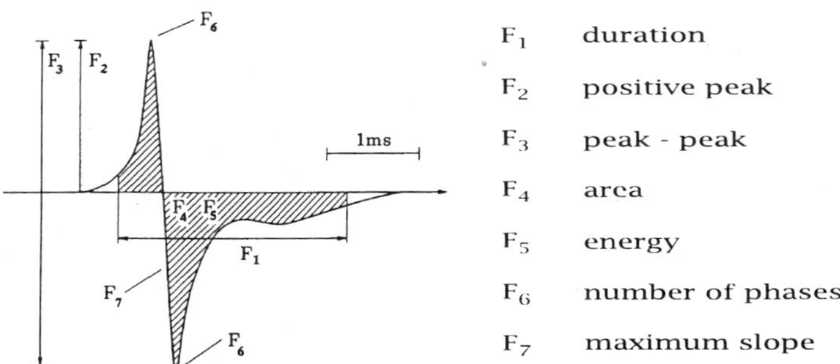

Figure 14. MUAP features.

to avoid redundancy. Andreassen (1977) in his pioneering work, however, adapted his selection to the parameters set up by Buchthal *et al.* (1954a, b) and widely used by the clinicians, as shown in Fig. 14. Their redundancy becomes obvious by the fact that the 3 features "positive peak amplitude", "peak-to-peak amplitude" and "area" already provide a rate of 97, 5% correct classification of isolated MUAPs; considering more features do not improve the result significantly. Some efforts were made to introduce more effective features like orthonormal basic functions (Nandedkar and Sanders, 1984) and wavelets (Pattichis and Pattichis, 1993), but further research is required.

In the late 80s, Neural Nets (NN) were promoted for pattern recognition tasks. In principle, MUAP classification can be done by this artificial intelligence (AI) method (Schizzas *et al.*, 1990; Pattichis, 1992), but the problem is that a training set of (manually) classified MUAPs is required. Therefore, a long term perspective may be a cascade of a classical system followed by a NN. The first can provide a training set of reliably classified MUAPs, which can be used for learning of the NN; the NN can then be employed to take the difficult decisions. Similarly, we presently implement a Fuzzy Classifier to resolve these ambiguous EMG segments. But all the work reported in the 80s already used a heuristic logical decision network for this problem, so these new AI approaches cannot be expected to increase performance significantly.

Costs and Performance

Previously, a calculation of how many instructions of what type are required to classify a single MUAP was done commonly and the result related to the performance. Under this viewpoint, the parametric method using the simple features from Fig. 14 shows some advantage since the computational costs are very low in comparison to the nonparametric approach. This may be the reason for that in most EMG machines (usually PC based) still this method is implemented as a standard. But the 95% correct classification rate applies for isolated MUAPs only. This is sufficient for the peripheral EMG; to obtain the shape

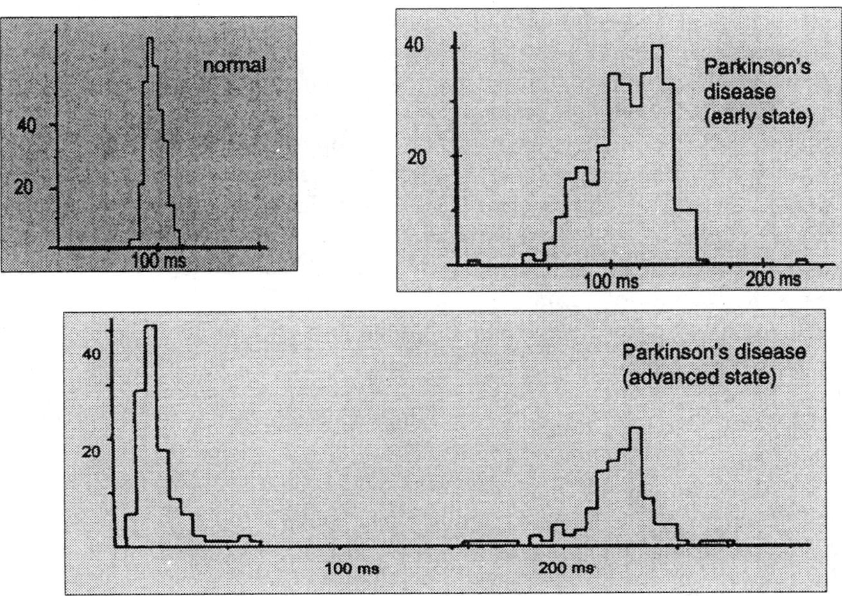

Figure 15. Interval histograms from discharges of a single MU.

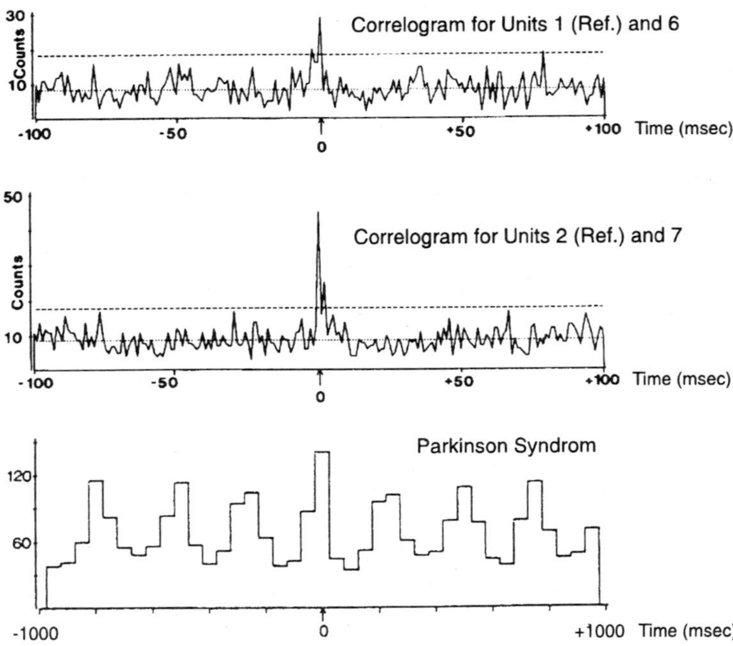

Figure 16. Crosscorrelograms (for method, see Andreassen, 1977).

parameters all reliably classified MUAPs of each MU are averaged, and the resulting curve analyzed. Also the basic examinations in central EMG does also not require a full stream classification: firing rates are derived from some segments which are perfectly done by the classificator. Only research work, especially investigations on the synchronization of MUs, and EMG analysis at higher contraction levels demands for more perfect systems. Presently, nonparametric systems achieves up to 95% correct classification including superimpositions in EMG recordings comprising up to four MUs at low to moderate force development. The only system which keeps high performance when coming up to higher contraction levels is the system of the DeLuca group (DeLuca *et al.*, 1982) due to the usage of a special 4-channel needle electrode; but seemingly this special needle inhibits spreading of this method to other labs. The computational load of those sophisticated methods requires upgrade from PC to workstations or special PC coprocessor boards to obtain results in reasonable time.

Finally, some methods increase their performance by including interval statistics in their decision schema (e.g., Andreassen, 1977). In central motor disorders, however, when we observe the central drive from the brain through the EMG, this discharge schedule is often severely disturbed, thus the timing information cannot further contribute.

A NOTE ON APPLICATION OF NEEDLE EMG

Ten years ago, there was great interest in the central EMG (and sophisticated EMG decomposition) since - as a window to the brain - it apparently offers an access to more detailed analysis of central motor disorders. Interval analysis of MUAP trains of single MUs provides a clear distinction between healthy subjects and patients as shown in Fig. 15. Also, synchronization of different MU (Fig. 16) attracts attention, since it provokes tremor and should not occur in healthy subjects but it does (Wolf and Dengler, 1987). In particular,

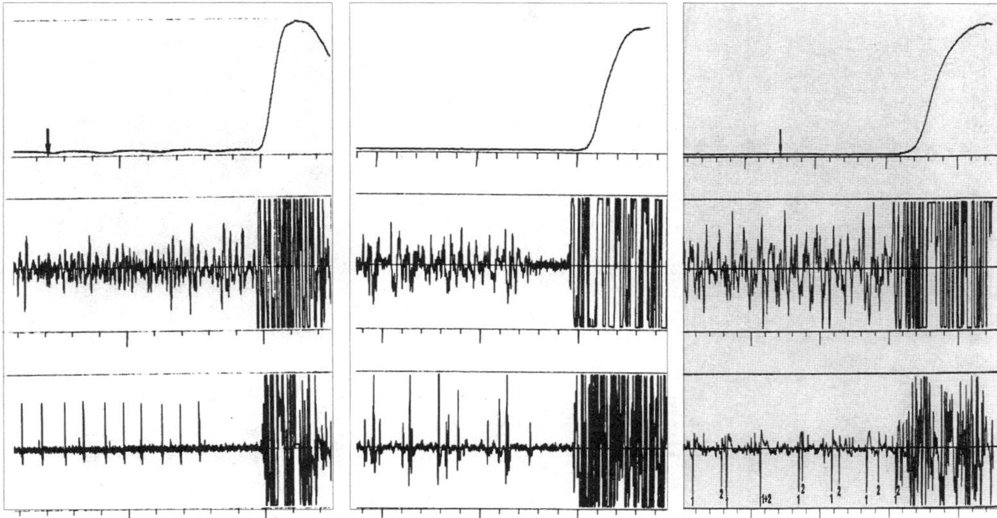

Figure 17. Force (top), surface EMG and needle EMG during a ballistic motor response. The three columns show typical silent period data from three different trials.

hereditary diseases like Chorea urged for a microscopic look to the central drive of MUs aiming at an early diagnosis for risk persons (children of patients). For example, we started a long term follow-up study in those people in 1982 where we could expect first results in 2002. But meanwhile, genetic technology came across and put much of these efforts to

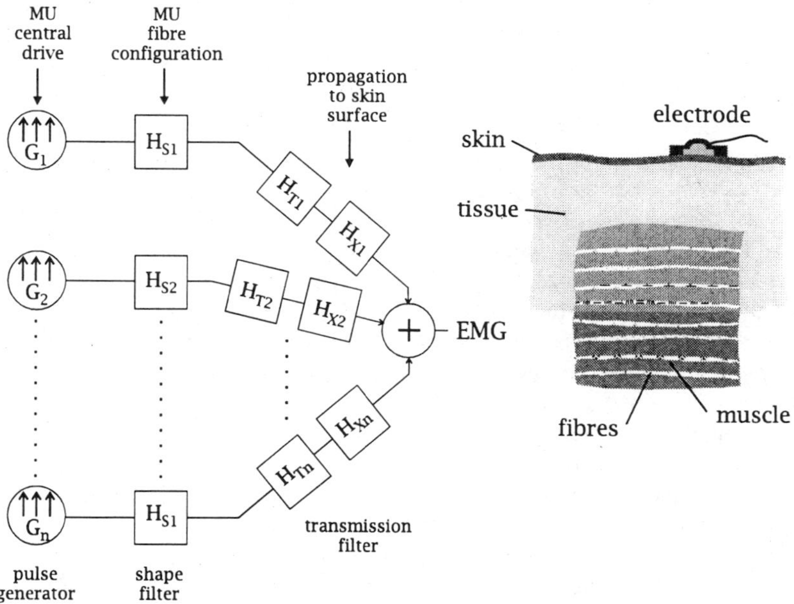

Figure 18. Signal model of surface EMG.

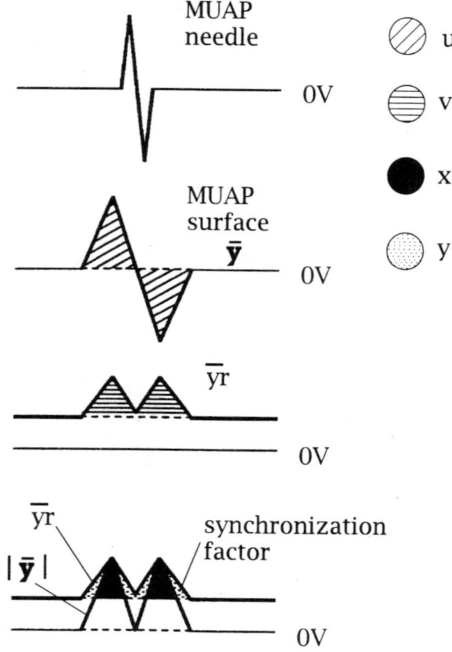

Figure 19. Synchronization factor.

obsolescence. So central EMG is re-orientated to basic research, and represents an important tool for investigating the neuromuscular system.

Certainly, the surface EMG will provide all global information about the muscle, and question is why do we need this tedious decomposition of needle EMG. Results of a study performed on the so-called "premovement silent period" is an example for an answer. Yabe (1976) and several other authors reported that the EMG activity of an already slightly contracted muscle is reduced for 10 - 50 ms just before onset a voluntary ballistic contraction. But Hummelsheim and Hefter (1991) could not reproduce the effect in their lab and came to the conclusion that it is an artefact due to experimental situation. Usage of needle EMG and decomposition techniques in our lab provides an explanation (Fig. 17): in some trials no decrease of surface EMG amplitude can be observed, but the MU picked up by the needle stops firing before onset, and, vice versa, the surface EMG shows some reduction but the single MU on the needle continues firing (Wierzbicka *et al.*, 1993). This finding was recently confirmed by Moritani and Shibata (1994). Certainly, we could not yet explain the mechanism, but at least the microscopic view by the needle EMG provides an explanation for contradicting results obtained through the macroscopic view of surface EMG.

Another story where surface EMG and needle EMG supplements each other is synchronization of MUs. Even if two needle electrodes are inserted in the muscle a maximum of 8-10 different MU can be observed - which is about 1 percent or less even in muscles with a coarse structure like biceps. Therefore, the general significance of results derived from 5-10 MU data may be critically discussed (Buchthal method for diagnosis requires 20 MUAPs for decision); but synchronization, in particular, describes cooperation of MUs in time and, therefore, the recruitment and discharge pattern of a representative number of MUs from a single record is required, but cannot be obtained. Milner-Brown *et al.* (1975) proposed a method based on simultaneously recorded surface EMG x_S and needle EMG x_N to

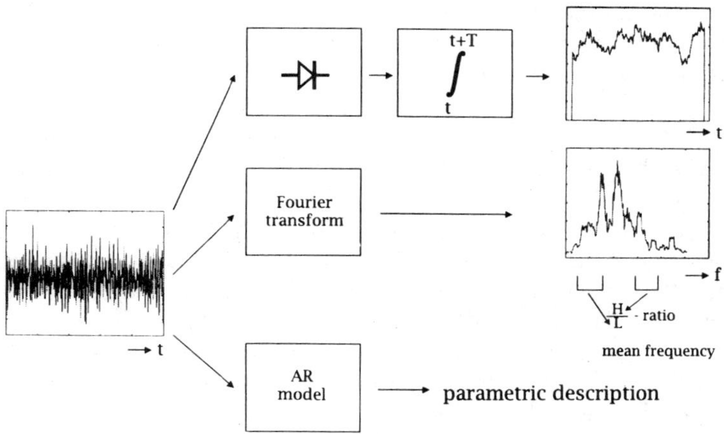

Figure 20. Common evaluation of surface EMG.

determine the synchronization level. Spike triggered averaging of the surface EMG for each MU j, j = 1 .... M, will provide

$$\overline{y}_j(t) = \sum_{n=1}^{N} x_S(t + t_{d_n}), \qquad \overline{yr}_j(t) = \sum_{n=1}^{N} |x_S(t + t_{d_n})| \tag{5}$$

where t_d are the discharge points of MU j. $\overline{y}_j$ represents the MUAP of MU_j as recorded by the surface electrode (area u in Fig. 18), $\overline{yr}_j$ is the same but determined from the rectified EMG (area v). Therefore, $\overline{yr}_j$ contains a bias from the other MUAPs because they do not cancel mutually like in $\overline{y}_j$; synchronized but not strictly time locked components lead to a broadening of the peaks, so that $\overline{yr}_j - |\overline{y}_j|$ (area y) indicates the global synchronization strength.

THE ALTERNATIVE - SURFACE EMG

From subject's and patient's view noninvasive surface EMG recording is preferred. As shown in Fig. 18, the signal model from Fig. 11 must be extended by another additional filter given by the transmission from the muscle through several layers of subcutaneous tissue and skin to the electrode surface, which reduces the EMG to a bandwidth between 10 and 500 Hz. Also, some crosstalk from neighboring muscles is picked up, the more the deeper the muscle investigated. This limits the application of surface electrodes.

But surface EMG provides some global information (Fig. 20); the moving average of the rectified EMG correlates to the force which the muscle develops (e.g., Baratta *et al.*, 1993). The frequency content of the EMG will be influenced by pathological MU firing characteristics (e.g., Lindström and Petersen, 1983) as well as by fatigue (e.g., Merletti *et al.*, 1990, 1991). The signal characteristic of the surface EMG is stochastic due to the contribution of a large amount of fibers which generate APs of constant shape but stochastic intervals. Therefore, description of surface EMG by an AR-model could be done successfully (e.g., Inbar and Noujaim, 1994; Kiryu *et al.*, 1994), but has not yet found broad application in the clinical field. Actually, signal processing requirements in surface EMG evaluation are commonly very low with some exceptions only.

In some patients, the use of needle electrode is not recommended like in hemophiliac people. Therefore, efforts were invested to extract MUAPs from surface EMG and to decompose it, as a substitute for the central (needle) EMG. For this purpose, multi electrode arrays are used, which allows to sharpen the receiver characteristic (e.g., Reucher *et al.*, 1987). First reports sounds promising, but this technique still needs elaboration.

CONCLUSIONS

The progress in computerized EMG-evaluation which was achieved over the last 15 years was really significant, which was partly due to the emerging computer performance, but mostly due to application of modern digital signal processing tools. In particular, decomposition of needle EMG into MUAP trains of single MUs reached a level which is near saturation of performance. Further improvements may be accomplished by a knowledge based expert system or fuzzy expert system which includes the heuristic experience of the specialists for classification of the few uncertain AP. Commercial EMG machines lag far behind the research, since common clinical EMG examinations do not require those sophisticated routines. However, all these decomposition methods are mainly suited for application with low to moderate contraction levels, which represents a clear restriction. This can be overcome not by larger computer systems or further developments of programs but by new electrode systems picking up more information from the muscle site. However, this requires to establish a new set of reference values for the new technique, which inhibits such a development. But nevertheless, there are still a lot of problems in sensorimotor research, which can be investigated by looking to the brain through the small window, which the present EMG technique supplemented by advanced signal processing tools provides.

ACKNOWLEDGMENT

The author is indebted to J. Dochtermann for doing the drawings and S. Schreiber for typing some parts of the manuscript.

REFERENCES

Adrian, E.D., and Bronk, D.W., 1929, The discharge of impulses in motor nerve fibers: II. The frequency of discharge in reflex and voluntary contraction, *J. Physiol. (Lond)* 67:119-151.

Andreassen, S., 1977, Interval pattern of single motor units, Thesis, Technical University of Denmark, Lyngby.

Andreassen, S., 1983, Computerized analysis of motor unit firing, In: Desmedt, J.E., ed., *Computer-Aided Electromyography*, Karger, Basel, Progr. Clin. Neurophysiol. 10:150-163.

Appel, U., and vonBrandt A., 1983, Adaptive Sequential Segmentation of Piecewise Stationary Time Series, *Information Sciences* 29:27-56.

Baratta, R.V., Solomonow, M., Best, R., and D'Ambrosia, R., 1993, Isotonic length/force models of nine different skeletal muscles, *Med. & Biol. Eng. & Comput.* 31:449-458.

Bergmans, J., 1971, Computer assisted on line measurement of motor unit potential parameters in human electromyography, *Electromyography* 2:161-181.

Buchthal, F., Guld, C., and Rosenfalck, P., 1954a, Action potential parameters in normal human muscle and their dependence on physical variables, *Acta Physiol. Scand.* 32:200-218.

Buchthal, F., Pinelli, P., and Rosenfalck, P., 1954b, Action potential parameters in normal human muscle and their physiological determinants, *Acta Physiol. Scand.* 32:219-229.

Gerber, A., Studer, R., deFigueiredo, R.J.P. and Moschytz, F., 1984, A new framework and computer program for quantitative EMG signal analysis, *IEEE Trans. Biomed. Eng.* BME31:857-863.

Guiheneuc, P., Calamel, J., Doncarli, D., Gitton, D., Michel, C., (1983) Automatic detection and pattern recognition of single motor unit potentials in needle EMG. In: Desmedt, J.E., ed., *Computer-Aided Electromyography*, Karger: Basel, pp. 73-127.

Guiheneuc, P., 1985, Computer pattern recognition of motor unit potentials. In: Struppler, A., and Weindl, A., eds., *Electromyography and Evoked Potentials*, Springer: Berlin, pp. 114-121.

Inbar, G.F., and Noujaim, A., 1984, On surface EMG spectral characterization and its application to diagnostic classification. *IEEE Trans. Biomed. Eng.* BME 31:597-604.

Kiryu, T., De Luca, C.J., and Saitoh, A., 1994, AR modeling of myoelectric interference signals during a ramp contraction, *IEEE Trans. Biomed. Eng.* BME41:1031-1038.

Kopec, J., and Hausmanowa-Petrusewicz, I., 1969, Histogram of muscle potentials recorded automatically with the aid of the averaging computer "ANOPS", *Electromyography* 4:371-381.

LeFever, R.S., and DeLuca, C., 1982, A procedure for decomposing the myoelectric signal into constituent action potentials - Part I: Technique, theory and implementation. *IEEE Trans. Biomed. Eng.* BME29:149-157.

Lee, R.G., and White, D., 1973, Computer analysis of motor unit action potentials in routine clinical electromyography. In: Desmedt, J.E., ed., *New Developments in Electromyography and Clinical Neurophysiology*, Karger: Basel, pp. 454-461.

McGill, K.C., and Dorfman, L.F., 1985, Automatic EMG decomposition inbrachial biceps, *Electroenceph. Clin. Neurophysiol.* 61:453-461.

Merletti, R., Knaflitz, M., De Luca, C.J., 1990, Myoelectric manifestations of fatigue in voluntary and electrically elicited contractions, *J. Applied Physiol.* 69:1810-1820.

Merletti, R., Lo Conte, L.R., and Orizio, C., 1991, Indices of muscle fatigue, *J Electromyography and Kinesiology* 1:20-23.

Milner-Brown, H.S., Stein, R.B., and Lee, R.G., 1975, Synchronization of human motor units: Possible roles of exercise and supraspinal reflexes, *Electroneceph. Clin. Neurophysiol.* 38:245-254.

Moritani, T., and Shibata, M., 1994, Premovement electromyographic silent period and alpha-motoneuron excitability, *J Electromyography and Kinesiology* 4:27-36.

Nandedkar, S.D., and Sanders, D., 1984, Special-purpose orthonormal basis functions - application to motor unit action potentials, *IEEE Trans. Biomed. Eng.* BME31:374-377.

Paiss O., and Inbar, G.F., 1987, Autoregressive modeling of surface EMG and its spectrum with application to fatigue, *IEEE Trans. Biomed. Eng.* BME34:761-770.

Pattichis, M., and Pattichis, C.S., 1993, Fast wavelet transform in motor unit action potential analysis, *Proc. 15th Annual Int. Conf., IEEE Eng. in Med. and Biology*, pp. 1225-1226.

Pattichis, C.S., 1992, Artificial neural networks in clincal electromyography, *Ph.D. Thesis*, Queen Mary and Westfield College, University of London, U.K.

Prochazka, V.J., Conrad, B., and Sindermannn, F., 1972, A neuroelectric signal recognition system, *Electroenceph. Clin. Neurophysiol.* 32:95-97.

Prochazka, V.J., Conrad, B., and Sindermann, F., 1973, Computerized single-unit interval analysis and its clinical application, In: Desmedt, J.E., ed., *New Developments in Electromyography and Clinical Neurophysiology*, Karger: Basel, Vol. 2, pp. 462-468.

Schizas, C.N., Pattichis, C.S., Shofield, I.S., Fewcett, P.R., and Middleton, L.T., 1990, Artificial neural nets in computer-aided macro MUAP classification, *IEEE EMBS* 9:31-38.

Stalberg E., and Antoni L., 1983, Computer-aided EMG analysis. In: Desmedt, J.E., ed., *Computer-Aided Electromyography*, Karger: Basel, pp. 186-234.

Stalberg, E., and Stalberg, S., 1989, The use of small computers in the EMG lab. In: Desmedt, J.E., ed., *Computer-Aided Electromyography and Expert System*, Elsevier Science Publ.: Amsterdam, pp. 1-32.

Stashuk D., and deBruin, H., 1988, Automatic decomposition of selective needle-detected myoelectric signals, *IEEE Trans. Biomed. Eng.* BME35:1-10.

Studer, R.M., 1984, Computergestützte Analyse der Signale motorischer Einheiten im Elektromyogramm, *Dissertation*, ETH Zürich.

Studer, R.M., deFigueiredo, R.J.P., and Moschytz, G.S., 1984, An algorithm for sequential signal estimation and system identification for EMG signals, *IEEE Trans. Biomed. Eng.* BME31:285-295.

Usui, S., and Amidor, I.J., 1982, Digital low-pass differentiation for biological signal processing, *IEEE Trans. Biomed. Eng.* 29: 686-693.

Wierzbicka, M.M., Wolf, W., Staude, G., Konstanzer, A., and Dengler, R., 1993, Inhibition of EMG activity in isometrically loaded agonist muscle preceding a rapid contraction, *Electromyogr. Clin. Neurophysiol.* 33:271-278.

Wolf, W., and Dengler, R., 1982, Detection and classification of single motor unit potentials in EMG recordings - a preliminary report. In: *Proc. 6th ICPR, IEEE* IEEE Computer Soc. Press: Los Angeles, Vol. 82, pp. 1195.

Wolf, W., and Dengler, R., 1985, Computerized analysis of single motor unit potentials for the diagnosis of central motor disorders, *Biomedizinische Technik, Erg.-Bd.* 30:8-9.

Wolf, W., and Dengler, R., 1987, Automatic Sorting and Analysis of Multiunit EMG Recordings. In: Struppler, A., and Weindl, A., eds., *Clinical Aspects of Sensory Motor Integration*, Springer: Berlin, pp. 80-85.

Wolf, W., Dengler, R., and Appel, U., 1985, Segmentation and classification in computerized EMG analysis. In: Hamza, M.H., ed., *Applied Signal Processing*, Acta Press: Anaheim, pp. 246-250.

Wolf, W., Dengler, R., and Appel, U., 1987, Computerized analysis of single motor unit activity, In: *Proc. 9th Ann. Conf. IEEE-BME*, IEEE Press: Piscataway, pp. 345-347.

Wolf, W., 1991, Digital signal conditioning, In: Weitkunat, R., ed., *Digital Biosignal Processing*, Elsevier Science Publ.: Amsterdam, pp. 81-126.

Yabe, K., 1976, Premotion silent in rapid voluntary movement, *J. Applied Physiol.* 41:470-473.

24

MULTI-CHANNEL EMG PROCESSING [*]

Edward A. Clancy, William R. Murray, and Neville Hogan

Liberty Mutual Research Center for Safety and Health
Hopkinton, Massachusetts
Department of Mechanical Engineering
University of Washington, Seattle, Washington
Department of Mechanical Engineering and Department of Brain and
    Cognitive Sciences
Massachusetts Institute of Technology, Cambridge, Massachusetts

ABSTRACT

A review of some recent progress in improving the fidelity of electromyogram-based estimates of muscle mechanical activity is presented. Initially, theoretical techniques and experimental investigations into electromyogram amplitude estimation are described. A stochastic functional model of the surface electromyogram, which incorporates single channel optimization (temporal whitening) and spatial combination of multiple channels, is described. Compared to standard electromyogram amplitude estimators, experimental investigations confirm that the model-based estimators improve the amplitude estimate by as much as a factor of 3–4. These experimental investigations also suggest the need for a new functional electromyogram model which includes an additive noise source. Sensitivities of the temporal whitening and spatial combination algorithms are reported.

Because muscles are not commonly used in isolation, it is necessary to model both agonist and antagonist muscles about a joint in order to relate the electromyogram to joint torque and stiffness. A simple model of the elbow joint, which incorporates co-contraction, is reviewed. The model predicts that co-contraction often leads to antagonist muscle activation levels that are significantly *larger* than the corresponding level of agonist activation, that is, muscles supporting the limb against gravity will not work as hard as their antagonist counterparts in the presence of significant co-contraction. Experimental observations confirm this prediction. This same agonist/antagonist muscle model then is used to relate the surface electromyogram to joint torque. Simulation studies that investigated the co-contraction model demonstrate that robust joint torque estimation could be accomplished in the presence of muscular co-contraction. In an experimental trial, the higher fidelity

---

[*] Portions reprinted with permission from; (5,6,7,8,20,21,22), ©IEEE 1990, 1991, 1994, 1995, 1988, 1988, 1988, respectively, and; (9), ©Liberty Mutual Insurance Company, 1995.

electromyogram amplitude estimators produced higher fidelity electromyogram-to-torque processors.

INTRODUCTION

Muscle electrical activity (EMG) is a readily obtainable quantitative measure of muscle activity. Because EMG is a direct consequence of neural activation of a muscle, it has many potential uses in clinical diagnosis, neuromuscular research and engineering applications. The latter include the possibility of using EMG as part of a command, communication and control interface between humans and machines. One notable example that has reached the stage of widespread commercial dissemination uses EMG to provide a means for an amputee to operate a powered artificial limb. Yet despite several decades of effort, using EMG to obtain timely and accurate estimates of mechanical variables such as muscle force or mechanical impedance remains a challenging problem. This paper reviews some recent progress in improving the fidelity of EMG-based estimates of muscle mechanical activity.

This review will focus on techniques for processing and interpreting EMG derived from electrodes on the skin surface, where EMG is obtained noninvasively. A simple phenomenological model of the surface EMG is presented and used to motivate different strategies for improving EMG processing, including weighted combinations of simultaneous recordings from multiple sites and shaping the frequency spectrum (whitening) of each signal. A combination of theoretical, simulation and experimental results is used to show that these approaches do not distort the relation between EMG and the mechanical variables of interest, yet can yield substantial improvements in processor performance.

These approaches appear to demand detailed estimation of the surface EMG characteristics that may vary from subject to subject and from use to use, including the frequency spectrum of each signal, the spatial correlation between signals and the relation of these parameters to mechanical quantities such as joint torque and stiffness. This paper presents experimental results regarding the influence of these factors on EMG processor performance.

One of the deepest problems in using EMG is that muscles are rarely, if ever, active in isolation. This paper examines the consequences of simultaneous antagonist muscle co-activation and shows that known characteristics of muscle biomechanics result in highly counter-intuitive patterns of EMG: *the more electrically active of two antagonist muscles often generates the smaller mechanical force.* Even with high-fidelity EMG processing, failure to account for this factor can introduce grossly inaccurate estimates of muscle mechanical action.

Because there are no noninvasive methods of measuring muscle force directly, identifying the relation between EMG and the mechanical action of the corresponding muscle appears to require isolated contraction of that muscle, which is confounded by the prevalence of muscle co-activation. However, this review shows that identifying the relation between the EMG of multiple muscles and their net action on the skeleton is a much simpler prospect: simultaneous antagonist muscle co-activation need not be eliminated. Thus, despite its apparent complexity, a multi-channel approach to EMG processing can yield both superior EMG amplitude estimation performance and more reliable EMG-to-torque calibration than corresponding single channel approaches.

CHARACTERIZING THE SURFACE ELECTROMYOGRAM

If a bipolar electrode is applied to the skin surface above the muscle, then electrical activity of the muscle can be observed. This surface activity is a complex spatial-temporal summation of the individual muscle fiber action potentials (APs). These APs spread through the body tissue to be recorded at the body surface. Several authors (2,3,10,11,23,27) have developed models of the surface EMG which draw upon the underlying physiology. In general, these models begin by describing the observed electrical activity due to a single motor unit (MU). Each muscle fiber within the MU makes a contribution to the surface potential which is a function of, at least, the spatial arrangement of the fiber, the time onset of the fiber AP, the shape of the fiber AP, the distance from the fiber to the electrode, the filtering effects of tissue, the geometric arrangement of the electrodes, and the electrode recording apparatus. The motor unit action potential (MUAP) has been modeled as the response of the MU to a single neural excitation. Repeated stimulation of an MU produces a series of these MUAPs to form a motor unit action potential train (MUAPT). Typically, many MUs are active during a normal contraction. Trains of impulses traverse the various nerve fibers and excite all muscle fibers within their respective MUs. Each impulse train elicits a characteristic MUAPT response. The surface electrode senses the composite activity of those MUs within its recording field. The recorded EMG, therefore, is modeled as a complex interference pattern of the electrical activity of many MUs.

By considering the EMG as a sum of many independent probabilistic events—the summed contribution of many independent MUs, each MU being excited by a sequence of independently timed nerve impulses—the law of large numbers can be used to describe the EMG as a signal with a Gaussian distributed amplitude. Roesler (26), studying the peak amplitudes of the EMG, found that the probability of deviation from a Gaussian distribution was less than 0.001, using a Chi-square test. Parker *et al.* (23), studying the EMG as a stochastic process, found the first order probability density function (PDF) of the EMG to be well modeled by the Gaussian PDF. The standard deviation of the PDF was a function of the contraction level. (Standard AC coupling of the EMG signal provides a zero mean—the

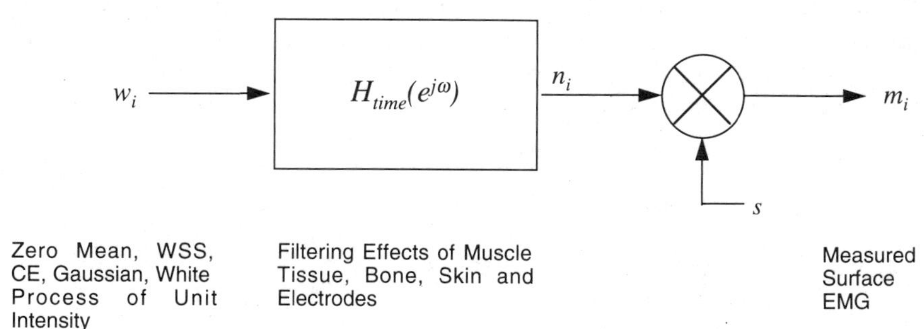

Figure 1. Functional model of a single channel of EMG. (Reprinted from (7) with permission, © 1994 IEEE.)

only other parameter required to characterize a Gaussian random variable.) Further, at a fixed contraction level, the EMG could be described as a random process; Kwatny *et al.* (17) showed that the constant force electromyogram could be considered as a wide-sense stationary, correlation-ergodic, random process, if viewed over a short (1.23 second) time interval. Hogan and Mann (13,14) utilized this probabilistic approach to present the functional mathematical model of EMG shown in Fig. 1.

MODELING THE SURFACE ELECTROMYOGRAM

Single Channel Electromyogram

The simple single channel EMG model of Hogan and Mann (13), (with minor additions due to Clancy and Hogan (7)) applies to the case of isometric, isotonic non-fatiguing contraction. The model is based upon two fundamental assumptions: 1) that the EMG amplitude, defined as the standard deviation of the random process, can be identified from the EMG waveform, and 2) for the case of non-fatiguing, constant-force, isometric muscle contraction, that the EMG amplitude has a constant value. Thus, the discretely sampled EMG waveform is modeled as being formed from the multiplication of a unit intensity, zero mean, wide-sense stationary (WSS), band-limited, correlation-ergodic (CE), jointly Gaussian process and a control signal. The control signal represents the EMG amplitude, s_i. It is assumed that the random process is independent of the EMG amplitude.

The band-limited Gaussian process is modeled as being formed from a zero mean, wide-sense stationary, white, correlation-ergodic, jointly Gaussian process of unit intensity passed through a linear time-invariant (LTI) shaping filter which is stable, causal, and whose inverse exists and is stable and causal. Only the magnitude response of the shaping filter is essential for spectral characterization of the EMG. The stability and causality of the shaping filter and its inverse assure that both filters are realizable. Since the input to the shaping filter is wide-sense stationary and the filter is LTI, the filter output is also wide-sense stationary. Additionally, the LTI assumption guarantees that a jointly Gaussian input to the filter will yield a jointly Gaussian output. For the case of constant force, non-fatiguing contractions, the EMG amplitude simplifies to the constant s, and the shaping filter accounts for *all* of the time dependence in the EMG waveform. This single site model is shown in Fig. 1. Note that the correlation-ergodic assumption is not used in any of the mathematical development, but is used in practice to form power spectral density (PSD) and/or correlation estimates needed in the computation of whitening filters.

If the PSD of the EMG is white (uncorrelated samples), then optimal maximum likelihood estimation of the EMG amplitude s based on the mathematical model is found by computing the root mean square (RMS) of the sampled EMG (7,13,14). The signal-to-noise-ratio (SNR) for this estimator is (13)

$$\text{SNR}_{\hat{s}} = \left[\frac{2\Gamma^2\left(\frac{N+1}{2}\right)}{N\Gamma^2\left(\frac{N}{2}\right) - 2\Gamma^2\left(\frac{N+1}{2}\right)} \right]^{1/2} = \left[\frac{N}{2} \left\{ \frac{\Gamma\left(\frac{N}{2}\right)}{\Gamma\left(\frac{N+1}{2}\right)} \right\}^2 - 1 \right]^{-1/2} \approx \sqrt{2N}$$

where N is the number of samples in the smoothing window, Γ is the gamma function, and "$\hat{s}$" is the EMG amplitude estimate. When successive samples of the EMG are correlated (non-white PSD), a solution to this discrete-time estimation problem can be found by first "whitening" the data with the filter $H_{time}^{-1}(e^{j\omega})$. Because $H_{time}(e^{j\omega})$ is constrained to be LTI, invertibly stable and invertibly causal, $H_{time}^{-1}(e^{j\omega})$ must exist and be stable, causal and LTI. The output of this filter is a white Gaussian process of intensity s. Further, since the filtering operation is invertible, optimal estimation of s from the whitened sequence is equivalent to optimal estimation of s from the original samples. Thus, the optimal estimate is completed as derived in the analysis of the previous case. Intuitively, the whitening filter orthogonalizes the data samples, allowing the detection algorithm to operate on each output sample

$$m_i \longrightarrow \boxed{H_{time}^{-1}\left(e^{j\omega}\right)} \xrightarrow{\hat{v}_i} \boxed{\left[\frac{1}{N}\sum_{k=0}^{N-1}\hat{v}_{i-k}^2\right]^{1/2}} \longrightarrow \hat{s}_i$$

Figure 2. Optimal single channel EMG processor. (Reprinted from (7) with permission, ©1994 IEEE.)

independently. The resultant estimator is shown in Fig. 2. If perfect whitening is achieved, the SNR performance is specified as above. Note that if correlated EMG samples are *not* whitened prior to amplitude estimation, Hogan and Mann (14) suggest the use of the above SNR performance formula with an effective number of samples (degrees of freedom) $N_{effective} = 2\,B_s T$, where B_s is the statistical bandwidth of the signal (in Hertz) and T is the smoothing window duration (in seconds).

Multi-Channel Electromyogram

When multiple bipolar electrodes are placed over a muscle, several correlated EMGs are recorded. In order to form a useful model from this description, a third fundamental assumption must be added to the two assumptions listed above: 3) the mutually uncorrelated information from each recording site should be weighted equally in the determination of the EMG amplitude estimate. The EMG amplitude can thus be more formally defined as the

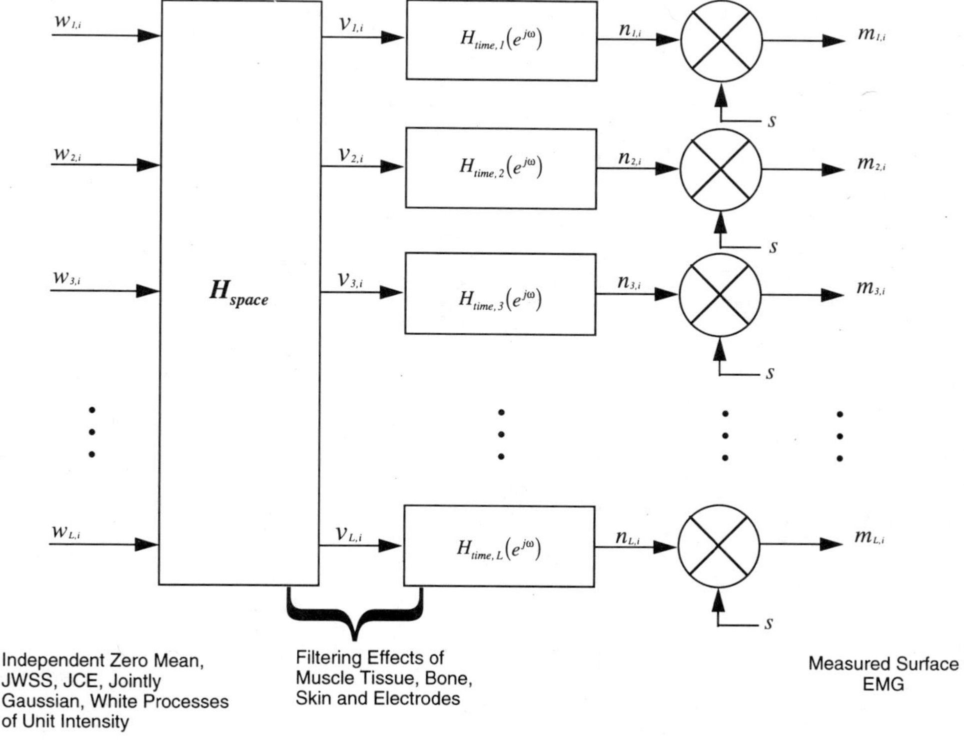

Figure 3. Functional model of multiple channels of EMG. (Reprinted from (8) with permission, © 1995 IEEE.)

common standard deviation of the equally weighted mutually uncorrelated information from all recording sites.

Figure 3 shows a phenomenological model of multiple surface EMG waveforms which embodies the above description and assumptions. Spatial and temporal correlation are both accounted for in this model. L independent, zero mean, jointly wide-sense stationary (JWSS), white, jointly correlation-ergodic (JCE), jointly Gaussian processes of unit intensity are passed through an L-input, L-output, LTI shaping filter, H_{space}, which is stable, causal, and whose inverse exists and is stable and causal. This multi-dimensional shaping filter is restricted to account *only* for the spatial dependence between sites, including differences in signal strength. Such a restriction requires that outputs of the multi-dimensional filter can be based only on knowledge of the present inputs. Any use of past inputs would imply a contribution to the temporal correlation in the EMG. Hence, the multi-dimensional filter has no dynamics and can be represented as a linear transformation. The L outputs from the multi-dimensional shaping filter are passed through a bank of LTI shaping filters to form L dependent, zero mean, JWSS, non-white, JCE, jointly Gaussian processes. The bank of shaping filters accounts for *all* of the time dependence in the EMG. These shaping filters are stable, causal, and have an inverse which is stable and causal. Therefore, the multiple channel EMG amplitude estimation problem is formulated as estimating the EMG amplitude (common standard deviation) from L zero mean, JWSS, JCE, jointly Gaussian processes.

If the PSD of each surface EMG waveform is white (uncorrelated temporally), and the multiple channels are both mutually uncorrelated spatially and of equal intensity, then all N samples over a particular time period from each of L sites could be grouped into one data set. For such a set, sequential causal optimal maximum likelihood estimates of the EMG amplitude, $\hat{s}_i$, are (8,13)

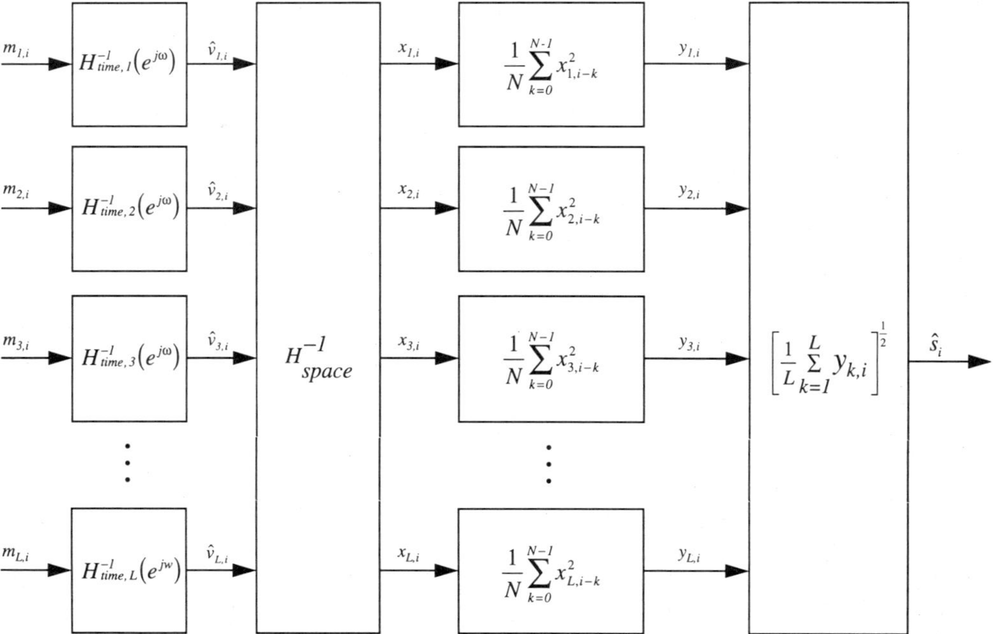

Figure 4. Optimal multi-channel EMG processor. (Reprinted from (8) with permission, © 1995 IEEE.)

$$\hat{s}_i = \left[\frac{1}{N \cdot L} \sum_{n=0}^{N-1} \sum_{l=1}^{L} m_{l,i-n}^2 \right]^{1/2}$$

where i is the time index, l is the site index and $m_{l,i}$ are the surface EMG waveform samples.

For the case in which the surface EMG waveforms are spatially and temporally correlated, consider the following two filter operations. First, filter each respective site by the temporal whitening filter $H_{time,l}^{-1}(e^{j\omega})$. Second, filter the whitened sequences via the L-input, L-output linear transformation H_{space}^{-1}. This spatial uncorrelation filter can be derived from the eigenvalues and eigenvectors of the inter-site covariance matrix of the whitened data. Again, this stable, causal, linear transformation exists and no information is lost due to the filter. The L output processes are mutually uncorrelated, white, Gaussian processes all of equal intensity. Thus, the optimal estimate of $\hat{s}_i$ is the spatial-temporal moving-average

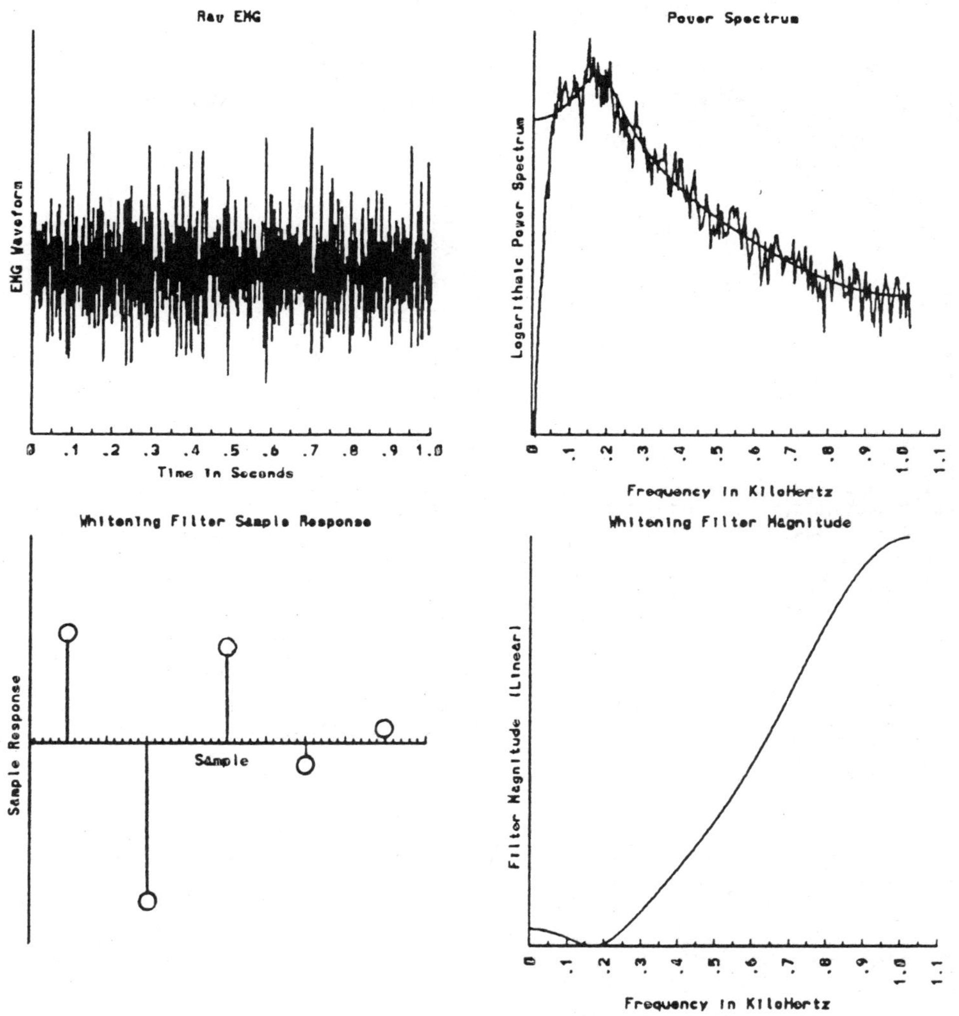

Figure 5. Design of a whitening filter. Smooth spectra use autoregressive technique, jagged power spectrum uses FFT technique. (Reprinted from (7) with permission, © 1994 IEEE.)

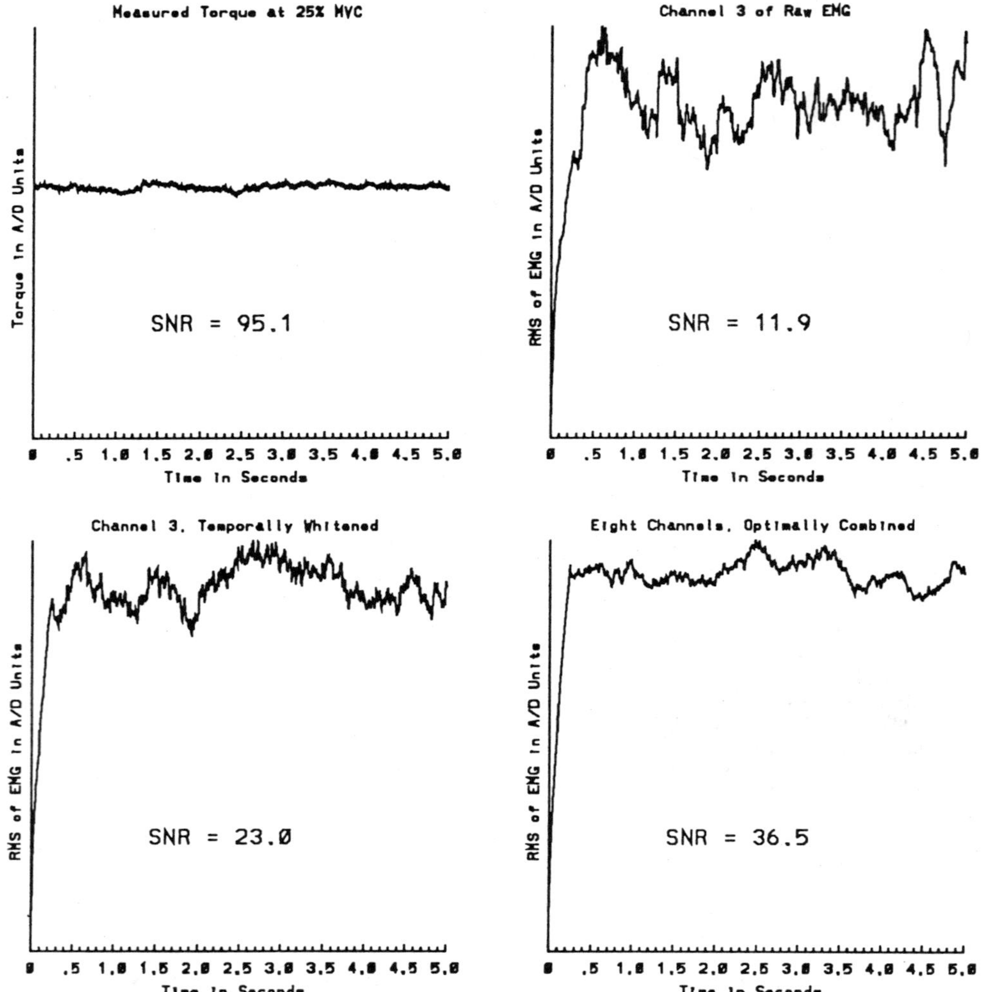

Figure 6. EMG amplitude estimates. (Reprinted from (5) with permission, © 1990 IEEE.)

root mean square filter of the temporally whitened, spatially uncorrelated samples. Figure 4 shows a discrete-time implementation of this estimator. Intuitively, the inverse filters orthogonalize the data samples, allowing the detection algorithm to operate on each output sample independently.

EXPERIMENTAL PERFORMANCE OF ELECTROMYOGRAM AMPLITUDE ESTIMATORS

Experimental Methods

In order to investigate these EMG models, an experimental trial was performed using five subjects. Up to eight channels of biceps/triceps EMGs, and joint torque, were recorded (at a sampling rate of 2048 Hz) while subjects performed isometric, isotonic, non-fatiguing elbow contractions (4,7,8). For single channel whitening, it was assumed that each temporal shaping filter in the EMG model could be approximated as an autoregressive (AR) filter. In order to identify a whitening filter from a data record, the AR PSD coefficients of a data record were estimated. These same coefficients specify a moving average whitening filter—the inverse of the shaping filter. For multiple channel analysis, the eigenvalues and eigenvectors of the sample covariance matrix of the whitened data (evaluated at a reference value of the EMG amplitude) were used to determine the spatial uncorrelation filter. All amplitude estimators used a 245ms duration moving time window. SNRs were computed as the square root of the ratio of the squared amplitude estimate sample mean divided by the amplitude estimate sample variance (4,8,13).

Single Channel EMG Amplitude Estimation Results

Figure 5 illustrates how a whitening filter was derived from the EMG. Figure 6 shows the amplitude estimate formed from a raw and whitened EMG. Note that if perfect whitening could be achieved and the Gaussian EMG model accurately described the surface EMG, then SNR performance for whitened data would depend only on the length of the smoothing window, N. Since $N=500$ (245ms), the ideal whitened SNR performance would be 31.6.

Initially, the *order of the whitening filter* was investigated. Filter orders of 0 (no whitening), 1, 2, 3, 4, 9, 14, 19 and 24 were evaluated. Up to ninth-order, there was a statistically significant performance improvement in the SNR with each additional order ($p<0.008$). Compared to the fourth-order results, all higher order results were significantly different ($p<0.008$), but the strength of the difference in SNR performance was small (≤ 0.2). All fourth-order and above investigations improved the SNR from 10.7 ± 3.3 (no whitening) to 17.4 ± 6.1.

For fourth-order and above filters, the investigations were repeated *varying the length of data used for calibration*. No statistically significant difference in SNR performance between using 5 versus 20 seconds of data to calibrate the whitening filters was found ($p>0.45$). Compared to the five second calibration length, a statistically significant drop in performance occurs for calibration lengths less than 3 seconds ($p<0.0029$). The strength of this drop is small—with a 62.5ms calibration length, the drop in SNR performance is 0.7 ($<5\%$).

The *effect of sampling rate* was studied next. All data were decimated by a factor of two, providing an effective sampling frequency of 1024 Hz. Whitening significantly improved the average SNR performance ($p<0.000001$) from an unwhitened value of 10.7 ± 3.3 to a whitened value of 13.5 ± 4.6. Thus, whitening at a sampling rate of 2048 Hz apparently was able to recover considerable information between 512 and 1024 Hz.

From the above results, it was concluded that fourth-order whitening filters, calibrated from five seconds of data, performed as well as any of the other whitening filters evaluated. These selections were used in all further analysis. All of the previous investigations restricted the whitening filters to be calibrated from recordings during an identical level of contraction. This restriction was then relaxed. *A single trial per subject was chosen and whitening filters designed for each electrode-amplifier*. These whitening filters were then applied to data from their corresponding electrode-amplifier for all remaining trials of that subject. Calibrating the whitening filters from a trial corresponding to 50 or 75% maximum voluntary contraction (MVC) improved the SNR performance from 10.8 ± 3.4 to 17.6 ± 6.0. This improvement degraded slightly when calibration was from a trial corresponding to 25% MVC, and degraded markedly when calibration was from a trial corresponding to 10% MVC.

The next comparison studied the feasibility of applying one *composite whitening filter*, derived from one trial from a subject, to all electrode-amplifiers of all trials for that subject. For the five subjects, calibration from high torque trials improved the SNR performance from 10.7±3.3 to 16.7±5.7. Performance dipped when calibration was from a trial corresponding to 10% MVC. Although these improvements were not as good as the previous set, it is clear that a single composite whitening filter per subject provided considerable SNR improvement.

The last comparison of whitening filters investigated the feasibility of a *universal whitening filter*. That is, one filter which could be applied to all electrode-amplifiers at all contraction levels for all subjects. A universal whitening filter was derived from the data of one subject, then applied to all recordings of all trials of the remaining subjects. Based on the previous results, the universal whitening filters were all designed from high contraction trials (50% MVC). Average SNR improvements varied from 39% to 55%, yielding average SNRs ranging from 14.7±5.4 to 17.4±5.5.

A New Surface EMG Model with Additive Noise

Initial evaluation of whitening filters calibrated with a 50% MVC trial, then applied to data recorded during contractions at less than 10% MVC provided poor results. At low EMG amplitude, additive background noise (not included in the previously described EMG models) seemed to dominate the output of the whitening filters. The relative contribution of background noise to the EMG amplitude estimate seemed to increase progressively as the EMG amplitude decreased. Similar difficulties have been noted by Kaiser and Peterson (16).

An EMG waveform model including additive noise was proposed (4). Unfortunately, formal solutions to the model were not readily apparent. Thus, an ad hoc "adaptive whitening" solution was implemented. The goal of the adaptive filter was to incorporate, in a simple strategy, an EMG amplitude estimator which would simultaneously preserve the benefits of temporal whitening for high contraction *and* the benefits of unwhitened estimation of low contraction. Results using this adaptive amplitude estimator are described later in this paper.

Multi-channel EMG Amplitude Estimation Results

Initially, the *number of EMG channels* participating in the EMG amplitude estimate was investigated. The spatial uncorrelation filter for a particular trial was calibrated from a separate trial from the same subject from the corresponding channels during an identical level of contraction. Two, four, six, seven and eight channel estimators were evaluated. Each increment in the number of EMG channels provided a statistically significant increase in SNR performance (p < 0.002 for all paired comparisons). On one subject, eight channels were acquired, from which optimal combination yielded an SNR of 35.0±13.4. This result improves SNR performance by a factor of 3–4. Increasing the number of electrode-amplifiers beyond eight may continue to provide a statistically significant improvement in SNR performance, however, the strength of this improvement is expected to be small.

The investigations were repeated *varying the length of data used for calibration* . No statistical difference in SNR performance between using 5 and 20 second calibration lengths was found for any number of sites (p > 0.36 for all paired comparisons). Compared to the five second calibration length, *none* of the shorter calibration lengths (15.625ms to 4s) significantly altered the SNR performance (0.09 < p < 0.85 for the various paired comparisons).

The next set of comparisons studied the feasibility of *calibrating a single spatial uncorrelation filter* from one trial from a subject, and applying that spatial uncorrelation filter to all trials for that subject. In addition, estimators were also formed in which no spatial uncorrelation was performed. Rather, the data from each channel were normalized (based on calibration from the data which determine the temporal whitening filters) and then smoothed. This filter was called the equal variance combiner. In general, all of these multiple channel combination filters provided considerable performance improvement. With four, six or eight channels, the equal variance combiner performed progressively poorer, on average, than the other filters ($p < 0.002$ for all paired comparisons). All of these above differences, however, were small in strength. Thus, the simpler equal variance combiner, which uses no spatial uncorrelation filter, is justified for most applications.

When possible, the electrode-amplifiers contributing to multi-channel estimators where chosen from non-adjacent locations on the muscle group for the above investigations. The next examination *contrasted non-adjacent selection with adjacent selection*. Adjacent channels were found to be more correlated than non-adjacent channels. The SNR performance of the adjacent and non-adjacent channels was compared for three different combination filters. The results show little difference in performance between adjacent/ non-adjacent channel selection for two and four channel amplitude estimators. Statistical tests suggest that the poorer performance of the equalized variance combiner may be significant ($p < 0.011$ for both paired comparisons). However, even if these differences are significant, their strengths are weak. Hence, it appears that electrode pairs with inter-channel correlation coefficients as high as 0.5-0.6 perform about as well as more weakly correlated (correlation coefficients of 0.15-0.4) channels.

EFFECTS OF ANTAGONIST CO-CONTRACTION

Given a high-fidelity estimate of EMG amplitude, there remains the problem of relating this information to relevant mechanical consequences of muscle activity, such as joint torque and stiffness. Muscles pull, and to generate the positive and negative torques that are needed to move the skeleton, they are arranged in opposing groups about the joints. Although reciprocal activation of opposing muscles is supported by neural circuits in the spinal cord, simultaneous activation of antagonist muscle groups is commonplace. Because the muscle force induced by neural activation varies with muscle length, muscle exhibits behavior loosely analogous to a spring with variable stiffness, and antagonist muscle co-activation may be a strategy by which the biological system controls how it interacts with its environment (12). Antagonist co-activation and the dependence of muscle force on length, shortening speed and other mechanical variables must be considered to interpret EMG properly. In the following, a highly simplified model of muscle mechanics and corroborating experimental observations are presented to illustrate the scope of the problem: *the more electrically active of two antagonist muscles often generates the smaller mechanical force* (19-22).

Simplified Model of Muscle and Joint Mechanics

The model describes the relation between levels of flexor and extensor activity required to maintain various postures of the elbow and forearm over the full range of voluntary muscle activation. Only the quasi-static aspects of maintaining limb posture are modeled; the dynamics of the arm and hand and the excitation/contraction dynamics of the muscles are ignored. To maintain a posture of the limb with a constant level of muscular activity, neural inputs to the muscles may be assumed constant. The stiffness about the elbow is assumed to be due to the *apparent* stiffness of active muscle, whether due to intrinsic muscle properties or due to the action of neural feedback.

The force-length characteristic of skeletal muscle may be modeled by the bilinear functions shown in Fig. 7(b). This form is a local approximation to *any* analytical function relating muscle force to length and neural activation, obtained by expanding the function as a Taylor series and neglecting terms of order higher than the first product of differentials of length and neural activation (19). At a fixed length, this muscle model yields the linear relationship

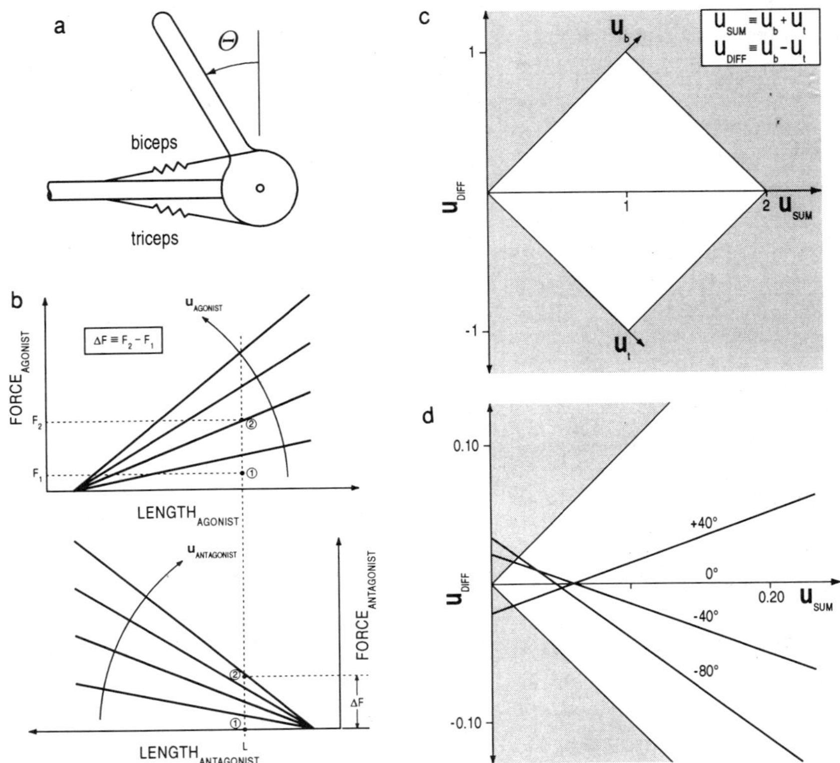

Figure 7. (a) Simple model of the biomechanics of the elbow. (b) Force-length characteristics for an antagonistically-coupled pair of muscles. (c) Two alternative representations of neural input. (d) Analytical predictions for the quasi-static maintenance of posture. (Adapted from (21) with permission, © 1988 IEEE.)

between stiffness and force that is experimentally observed in animal preparations (18) and intact humans (28,29). At a fixed level of neural activation, this model yields a linear force-length characteristic whereas the observed behavior in animal preparations is nonlinear (15,25). Though highly simplified, this model represents essential aspects of *in vivo* muscle behavior over the normal physiological range—both muscle force and the efficiency (or gain) with which neural activation produces muscle force increase with muscle length.

A single mechanical degree of freedom model of antagonist muscle action about the elbow is shown in Fig. 7(a). The upper arm is horizontal, the elbow is supported, and the forearm and hand are free to rotate in an upright vertical plane about the elbow. The flexor and extensor muscle groups are each modeled as a single muscle. The forearm and hand are modeled as a rigid body rotating about a fixed axis. The model is symmetrical about the upright position ($\theta = 0°$), where it is assumed that the flexor and extensor: have equal length, each generate joint stiffness K_{QS} at maximum activation, and each generate active torque τ_0 at maximum activation. The moment arms through which the muscle forces act about the joint vary with elbow angle (1,28), but for simplicity this angle dependence will be neglected.

Using the assumed bilinear muscle model, both the actively generated torque, $\tau_0 u$, and joint stiffness, $K_{QS} u$, are proportional to neural activation, where u represents the relative strength of neural activation of the muscle, that is, $0 \le u \le 1$. In this study, the EMG amplitude scaled by the MVC EMG amplitude is assumed to be a measure of this relative neural activation. Taking elbow torques to be positive in the direction of increasing θ yields the following model of the isometric torques produced by opposing bilinear "muscles" acting about the elbow:

$$\tau_b = \left(\tau_0 - K_{QS}\theta\right)u_b \quad \text{and} \quad \tau_t = -\left(\tau_0 - K_{QS}\theta\right)u_t$$

where subscripts b and t refer to the biceps (flexor) and triceps (extensor), respectively. The net isometric torque produced by the muscles is:

$$\tau_{net} = \tau_b - \tau_t = \tau_0\left(u_b - u_t\right) - K_{QS}\theta\left(u_b + u_t\right)$$

This expression for the net isometric torque shows that it is possible to achieve independent control of the muscle-generated stiffness and torque produced about the joint, regardless of joint angle, by independently controlling the sum and difference of the neural inputs, respectively.

This simple model may be used to identify how the levels of agonist and antagonist muscles activity must be related if opposing muscles are co-activated while posture is maintained. The net isometric torque generated by the muscles must offset the torque due to gravity. Thus,

$$\tau_0\left(u_b - u_t\right) = K_{QS}\theta_P\left(u_b + u_t\right) - Mgl\,\sin\theta_P$$

where M is the mass of the forearm and hand, g is the acceleration due to gravity, l is the distance from the elbow axis to the forearm/hand mass center, and the subscript P indicates θ is a fixed posture to be maintained. It is useful to define the following alternative set of neural inputs for the model:

$$u_{sum} \equiv u_b + u_t \quad \text{and} \quad u_{diff} \equiv u_b - u_t$$

This new set of neural inputs is constrained such that $0 \leq u_{sum} \leq 2$, and $-1 \leq u_{diff} \leq 1$. The relation between these two sets of neural inputs is shown in Fig. 7(c), where the unshaded area is the region of allowable activation—the region for which the inequality constraints on muscle activation are satisfied. Rearranging the equilibrium relationship yields an expression for the difference of muscle activation levels, u_{diff}, required to maintain equilibrium of the forearm and hand at the specified posture θ_P:

$$u_{diff} = \frac{K_{QS}\,\theta_P}{\tau_0}u_{sum} - \frac{Mgl}{\tau_0}\sin\theta_P$$

where u_{sum} determines the muscular stiffness about the joint and remains an independent input to the system.

Model Predictions and Experimental Observations

This model of the biomechanics of the elbow yields two testable predictions, depicted in Fig. 7(d):

1. If posture of the elbow and forearm is maintained at varying degrees of co-contraction, the muscular activity of the agonist and antagonist will be linearly related. The *equilibrium line*, the locus of all possible (u_{sum}, u_{diff}) pairs for which a specific equilibrium posture is maintained, will be a straight line.

2. These equilibrium lines are counter-intuitive—the sign of u_{diff} indicates that activation of the antagonist is larger than activation of the agonist—for all but the lowest levels of muscular activity.

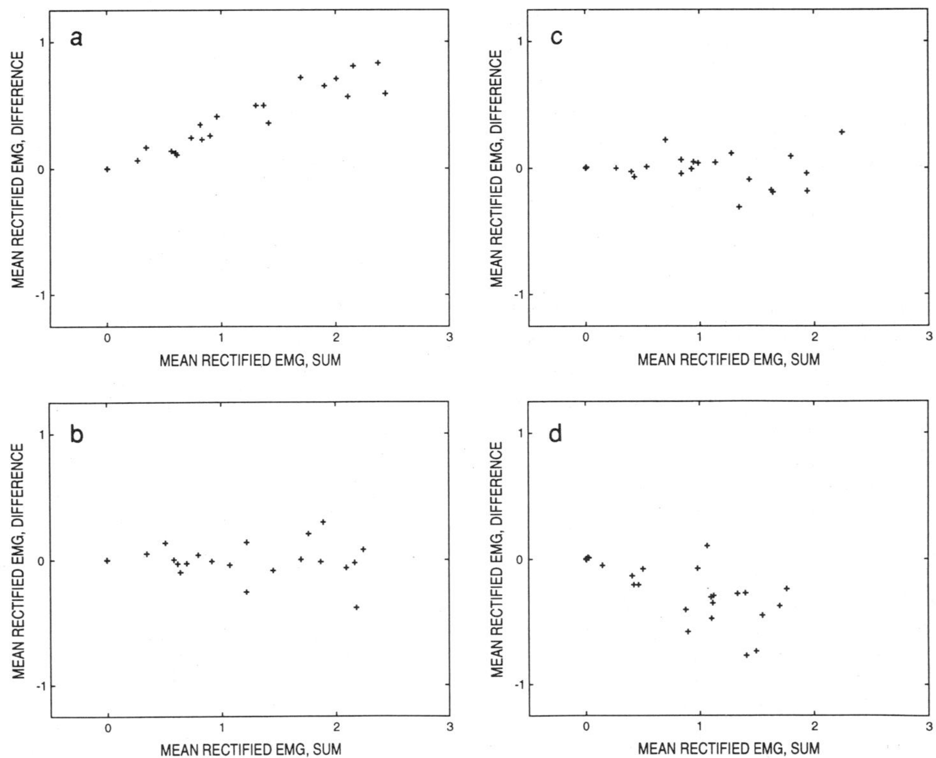

Figure 8. Data from deliberate co-contraction experiments. (a) $\theta_P = +40°$. (b) $\theta_P = 0°$. (c) $\theta_P = -40°$. (d) $\theta_P = -80°$. (Adapted from (21) with permission, © 1988 IEEE.)

Due to the symmetry of the model about the upright posture, the predicted equilibrium line for $\theta_P = 0°$ has a slope of zero degrees in Fig. 7(d). The counter-intuitive nature of the predicted equilibrium lines for the off-upright postures can be explained with the aid of Fig. 7(b). Consider a posture of the model in which the forearm is halfway between upright and fully extended, and assume that this posture is achieved with both muscles operating at the length designated **L** in Fig. 7(b). For maintenance of this posture in the presence of gravity, the agonist and antagonist are the biceps and triceps, respectively. With no co-contraction, only a small amount of force, $\mathbf{F}_1$, is required from the agonist to offset the effect of gravity on the limb. Increasing the neural activation of the agonist to about 50% generates a change in force, $\Delta\mathbf{F}$, which must be matched by co-contraction of the antagonist if the limb is to remain stationary. Because the antagonist is considerably shorter than the agonist at this posture, the generation of a matching $\Delta\mathbf{F}$ requires a much larger change in the neural input to the antagonist, from zero to almost twice that of the agonist—an increase in neural activation that brings the antagonist close to saturation. In other words, shorter muscles are less efficient force producers and require more activation to exert a given force.

Figure 8 shows experimental observations of mean rectified EMG amplitudes recorded from biceps and triceps muscles using bipolar surface electrode arrays (19). The elbow was supported comfortably and the antagonist muscles were deliberately co-activated while the forearm and hand were held in fixed, vertical-plane postures at +40°, 0°, -40° and -80° away from upright. The difference of EMG amplitudes (biceps EMG amplitude minus triceps EMG amplitude) is plotted against the sum of EMG amplitudes. The experimentally observed equilibrium lines are straight and confirm the counter-intuitive model predictions for postures at $\theta_P = 40°$ and $\theta_P = -80°$; the sign of the EMG amplitude difference shows that the antagonist is more active than the agonist for all but the lowest levels of muscular activity. Consideration of the moment arms through which muscle force is converted to elbow torque explains the lack of significant slope for the $\theta_P = 40°$ case (19). Conclusively, maintaining static equilibrium for postures of the forearm away from upright, while simultaneously contracting the agonist and antagonist muscles about the elbow, may require the antagonist to work harder than the agonist.

CALIBRATING THE EMG-TO-TORQUE RELATION

An EMG-to-Torque Model Incorporating Agonist-Antagonist Co-Contraction

EMG amplitude has historically been used to estimate the torque exerted about a joint. Calibration of an EMG-to-torque relation has generally required an assumption of isolated contraction of the muscle of interest—a condition which may not be possible due to co-contraction. Multiple channel EMG can be used to resolve this paradox. Consider again the model shown in Fig. 7(a). Biceps- and triceps-generated torques are related to net isometric torque about the joint as:[*] $\tau_{net} = \tau_b - \tau_t$. Let the biceps and triceps EMG amplitudes (s_b, s_t) be related to τ_b and τ_t as the polynomials:

$$\tau_b = f_{b,1} \cdot s_b + f_{b,2} \cdot s_b^2 + f_{b,3} \cdot s_b^3 + \cdots \quad \text{and} \quad \tau_t = f_{t,1} \cdot s_t + f_{t,2} \cdot s_t^2 + f_{t,3} \cdot s_t^3 + \cdots$$

where $f_{i,j}$ are fit parameters. With these assumptions, the error between the estimated and measured joint torque at each instant in time was: $error = \tau_b - \tau_t - \tau_{net}$. For a sequence of measurements at various levels of biceps-triceps contraction, linear least squares techniques were used to minimize the mean square error (MSE) with respect to the fit parameters.

EMG-to-Torque Simulations

Prior to acquiring experimental data, it was important to determine what tasks a subject should perform, that is, it was important to determine what experimental data to actually acquire. To this end, EMG amplitudes were simulated (representing possible experimental tasks) and the conditioning of the least squares problem evaluated. The results suggested that two contractions, one biceps EMG dominant and the other triceps EMG dominant, should be concatenated to form an experimental task. *Co-contraction should be minimized, but need not be eliminated.*

A second simulation study investigated three aspects of the identification technique— identification accuracy, identification repeatability and the influence of overfitting the model data. Biceps and triceps EMG waveforms were simulated from known EMG amplitudes multiplied by Gaussian noise. Measured joint torque was simulated from the known EMG amplitudes and known polynomial (first through third order) EMG amplitude-to-torque relationships. EMG amplitude estimates were formed from the simulated EMG waveforms using a moving average (245ms window) RMS filter. This filter produced amplitude estimates with an SNR of 31.6. The MSE was typically less than 2% of the combined

---

[*] Torques due to gravity have been ignored, since the subsequently described experimental trial oriented the arm in the plane perpendicular to the direction of gravity.

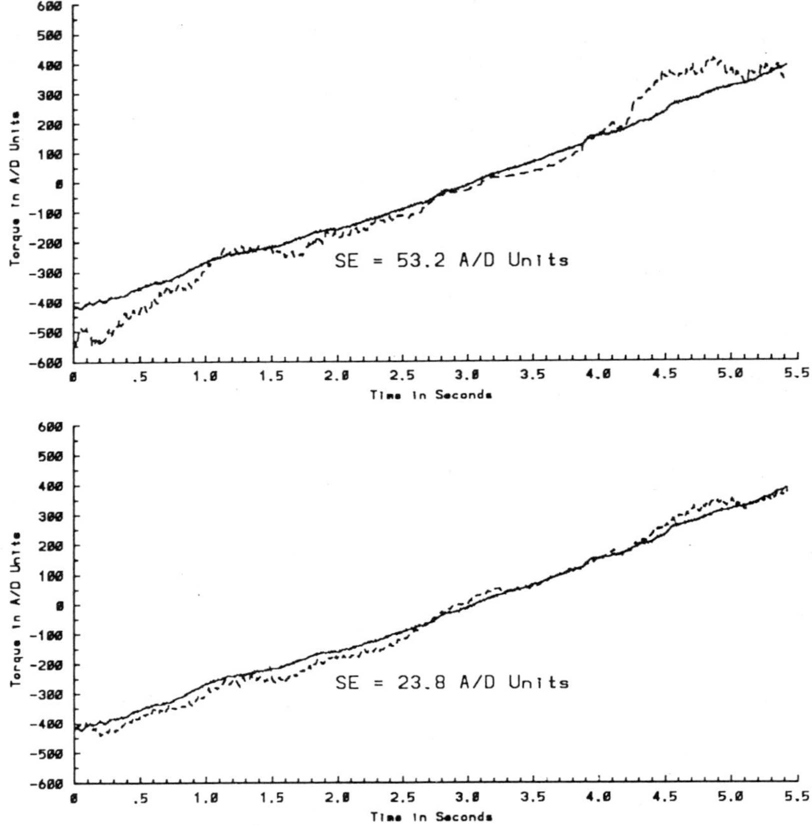

Figure 9. Prediction of elbow torque from the EMG. Solid lines are measured torque, dash lines are predicted torque. Top shows results using standard single-channel EMG amplitude estimation, bottom shows results using adaptive multi-channel EMG amplitude estimation. (Reprinted from (6) with permission, © 1991 IEEE.)

flexion/extension torque range. Overfitting the data by one or two orders only slightly degraded performance. Trial-to-trial repeatability of the identified EMG amplitude-to-torque relationship was influenced by the trial-to-trial repeatability of the EMG amplitude.

Experimental Methods

In order to investigate this EMG-to-torque system identification technique, an experimental trial was performed (4,6). Five EMG channels applied to each of the biceps and triceps muscles, and joint torque, were monitored. Three subjects performed isometric, quasi-isotonic, non-fatiguing elbow contractions which spanned the range from 50% MVC elbow extension to 50% MVC elbow flexion. Six EMG amplitude estimators were constructed: 1) single channel unwhitened, 2) single channel whitened, 3) single channel adaptive whitened, 4) multiple channel unwhitened, 5) multiple channel whitened, and 6) multiple channel adaptive whitened. (The adaptive whitener compensated for additive background EMG noise when the EMG amplitude was low.) All EMG amplitude data were scaled to span the range from zero to one. Least squares fitting of the EMG amplitudes (biceps and triceps) to polynomial basis functions was accomplished with the singular value decomposition method of Press *et al.* (24). EMG-to-torque prediction performance was evaluated by calibrating an EMG-to-torque relationship from one experimental trial, and then applying that relationship to all other experimental trials for that subject.

Results: Prediction of Joint Torque

Figure 9 displays an example of EMG-to-torque results. Each multiple channel predictor performed better than the respective single channel predictor. The adaptive processors (single/multiple), however, had a lower prediction standard error (SE) than the unwhitened or whitened processors. The adaptive multiple channel predictor had an SE approximately 90% of the unwhitened multiple channel predictor. Although this improvement was small, it demonstrated the potential of adaptive whitening. As the model order (number of biceps/triceps coefficients in the linear regression model) increased, the prediction SE attained a minimum at third order, and remained near this value out to sixth order. These results follow the same trends as the simulation results. The best prediction performance was provided by the third order multiple channel adaptive processor, which exhibited 67% of the SE of the unwhitened single channel predictor. For each respective subject, this best predictor had an average SE approximately 3% of the torque range for that subject.

CONCLUSION

Theoretical modeling, simulation studies and experimental investigations which describe recent advances in EMG processing have been reviewed. Optimized, multi-channel surface EMG models have led to improved methods for estimating the EMG amplitude. Experimental investigation has demonstrated that these advanced processing strategies provide large gains in the fidelity of the EMG amplitude estimate. Once an estimate of the EMG amplitude is obtained, it provides a tool to examine the mechanical activity of muscle. Such a tool is particularly useful since muscles are rarely, if ever, active in isolation. One counter-intuitive consequence of muscular co-contraction is that the more electrically active of two antagonist muscles often generates the smaller mechanical torque. For the static maintenance of elbow joint posture, this notion was predicted by a simple joint model, then observed experimentally. Next, simulation studies demonstrated that, with the use of multi-channel EMG, muscular co-contraction need not be eliminated in order to calibrate an EMG-to-torque relation. In addition, experimental studies demonstrated that higher fidelity EMG amplitude estimates produced higher fidelity EMG-to-torque processors. Thus, despite its apparent complexity, a multi-channel approach to EMG processing can yield both superior EMG amplitude estimation performance and more reliable EMG-to-torque calibration than corresponding single channel approaches.

ACKNOWLEDGMENTS

This work was supported by NIH grant AR40029; NIDRR grants H133E80024 and GOO8535121; and the Whitaker Foundation. Experiments were performed at the Eric P. and Evelyn E. Newman Laboratory for Biomechanics and Human Rehabilitation at the Massachusetts Institute of Technology.

REFERENCES

1. Amis, A. A., Dowson, D., and Wright ,V., (1979), Muscle strengths and musculoskeletal geometry of the upper limb, *Engng. Med.* 8(1):41-8.

2. Basmajian, J. V., and DeLuca, C. J. (1985), *Muscles Alive: Their Functions Revealed by Electromyography.* Baltimore, MD: Williams & Wilkins.

3. Brody, G., Scott, R. N., and Balasubramanian, R. A (1974), a model for myoelectric signal generation, *Med. Biol. Eng.* Jan: 29-41.

4. Clancy, E. A. (1991), Stochastic modeling of the relationship between the surface electromyogram and muscle torque, *Ph.D. dissertation,* Dept. Elect. Eng. Comp. Sci., M.I.T., Cambridge, MA.

5. Clancy, E. A., and Hogan, N. (1990), EMG amplitude estimation from temporally whitened, spatially uncorrelated multiple channel EMG, *Ann. Int. Conf. IEEE Eng. Med. Bio. Soc.* 12: 0453-0454.

6. Clancy, E. A., Hogan, N. (1991), Estimation of joint torque from the surface EMG, *Ann. Int. Conf. IEEE Eng. Med. Bio. Soc.* 13: 0877-0878.

7. Clancy, E.A., and Hogan, N. (1994), Single site electromyograph amplitude estimation, *IEEE Trans. Biomed. Eng.* 41:159-167.

8. Clancy, E.A., and Hogan, N. (1995), Multiple site electromyograph amplitude estimation, *IEEE Trans. Biomed. Eng.* 42:203-211.

9. Clancy, E. A., Murray, W. R., and Hogan, N. (1995), A review of some recent progress in processing the surface myoelectric signal. *Technical Memorandum 95-6* (unpublished), Liberty Mutual Research Center for Safety and Health, Hopkinton, MA.

10. DeLuca, C. J. (1979), Physiology and mathematics of myoelectric signals, *IEEE Trans. Biomed. Eng.* 26: 313-25.

11. DeLuca, C. J., and Van Dyk, E. J. (1975), Derivation of some parameters of myoelectric signals recorded during sustained constant force isometric contractions, *Biophys. J.* 15:1167–1180.

12. Hogan, N. (1984), Adaptive control of mechanical impedance by coactivation of antagonist muscles, *IEEE Trans. Automat. Cont.* 16:681-690.

13. Hogan, N., and Mann, R. W. (1980), Myoelectric signal processing: Optimal estimation applied to electromyography—Part I: Derivation of the optimal myoprocessor, *IEEE Trans. Biomed. Eng.* 27:382-95.

14. Hogan, N., and Mann, R. W. (1980), Myoelectric signal processing: Optimal estimation applied to electromyography—Part II: Experimental demonstration of optimal myoprocessor performance, *IEEE Trans. Biomed. Eng.* 27:396-410.

15. Joyce, G. C., Rack, P. M. H., and Westbury, D. R. (1969), The mechanical properties of cat soleus muscle during controlled lengthening and shortening movements, *J. Physiol.* 204:461-474.

16. Kaiser, E., and Petersen, I. (1974), Adaptive filter for EMG control signals. In: *The Control of Upper-Extremity Prostheses and Orthoses.* Springfield, IL: Charles C. Thomas, pp. 54-57.

17. Kwatny, E., Thomas, D. H., and Kwatny, H. G. (1970), An application of signal processing techniques to the study of myoelectric signals, *IEEE Trans. Biomed. Eng.* 17:303-313.

18. Morgan, D. L. (1977), Separation of active and passive components of short-range stiffness of muscle, *Am. J. Physiol.* 232(1):C45-C49.

19. Murray, W. R. (1988), Essential factors in modeling the modulation of impedance about the human elbow. *Ph.D. dissertation,* Dept. Mech. Eng., M.I.T., Cambridge, MA.

20. Murray, W. R. (1988), Maintenance of elbow equilibrium through co-contraction. *Proc. Fourteenth Ann. Northeast Bioeng. Conf. pp.* 29-32.

21. Murray, W. R. (1988), Modeling elbow equilibrium in the presence of co-contraction. *Proc. Fourteenth Ann. Northeast Bioeng. Conf.* pp. 190-193.

22. Murray, W. R., and Hogan, N. (1988), Co-contraction of antagonist muscles: Predictions and observations. *Ann. Int. Conf. IEEE Eng. Med. Biol. Soc.* 10: 1926-1927.

23. Parker, P. A., Stuller. J. A, and Scott, R. N. (1977), Signal processing for the multistate myoelectric channel. *Proc. IEEE* 65(5): 662-674.

24. Press, W. H., Flannery, B. P., Teukolsky, S. A., and Vetterling, W. T. (1988), *Numerical Recipes in C: The Art of Scientific Computing.* New York: Cambridge University Press, pp. 528-539.

25. Rack, P. M. H., and Westbury, D.R. (1969), The effects of length and stimulous rate on tension in the isometric cat soleus muscle. *J. Phys.* 204: 443-460.

26. Roesler, H. (1974), Statistical analysis and evaluation of myoelectric signals for proportional control. In: *The Control of Upper-Extremity Prostheses and Orthoses.* Springfield, IL: Charles C. Thomas, pp. 44-53.

27. Shwedyk, E., Balasubramanian, R., and Scott, R. N. (1977), A nonstationary model for the electromyogram, *IEEE Trans. Biomed. Eng.* 24:417-24.

28. Wilkie, D. R. (1950), The relationship between force and velocity in human muscles, *J. Physiol.* 110:249-280.

29. Zahalak, G. I., Heyman. S. J. (1979). A quantitative evaluation of the frequency-response characteristics of active human skeletal muscle *in vivo, J. Biomech. Eng.* 101:28-37.

25

ESTIMATION OF HUMAN ELBOW JOINT MECHANICAL TRANSFER FUNCTION DURING STEADY STATE AND DURING CYCLICAL MOVEMENTS

Gideon F. Inbar

Department of Electrical Engineering
Technion-IIT
Haifa 32000, Israel

ABSTRACT

Signal processing techniques can be beneficially used in studies of system model identification and model parameter estimation, and as powerful tools in elucidating the structure and function of partially known biological systems. In the present study, pseudorandom band limited white noise mechanical perturbations are used in order to investigate the torque to angular displacement transfer function (TF) of the human elbow joint.

Very high coherence was obtained for the steady state mechanical TF measurements. The mechanical parameters were also estimated during cyclic movements at different cycle times and different loads. A one or two zeroes and two poles ARMA model gave a clear best fit for more than 90% of the data with the model parameters changing with the operating point and with the level of muscle force. Thus, a linear second order model with changing parameters appears to describe well the single degree of freedom (DOF) elbow joint. The model stiffness increased with movement acceleration and inertial load, i.e., with joint torque. Increased stiffness, and with it, damping at the cyclic target points, is a desirable condition to achieve zero velocity at these points. During the high velocity low acceleration portion of the cycle the stiffness could be lower than under static conditions. Models that estimate equilibrium trajectories based on high joint stiffness during movement must therefore be reassessed.

INTRODUCTION

Mathematical model characterization of biological systems is an essential tool in formalizing our knowledge and assumptions about the structure and function of the system, and our ability to investigate and validate existing and new theories. This characterization

Advances in Processing and Pattern Analysis of Biological Signals, Edited by Isak Gath and Gideon F. Inbar
Plenum Press, New York, 1996

requires experimental model identification and parameter estimation of the model equations. In applying this approach, one can write the system equations in their continuous or discrete form, characterize the system about an operating point for linear small signal studies or use large signal nonlinear models, characterize the system in the time or frequency domains, etc. (e.g., Porat, 1995; Van den Bosch & Van der Klauw, 1994). Since the introduction of the model identification work on large signal nonlinear closed loop muscle model (Inbar *et al.*, 1970), most of the parameter estimation techniques have been used in the study of the neuromuscular control system (NMCS). In the present paper, some of the issues associated with biological system characterization are outlined through the application of signal processing and parameter estimation schemes to the mechanical TF of a single DOF human elbow joint peripheral NMCS.

ELBOW JOINT DYNAMICS

Knowledge about joint dynamic parameters, how they change with movement and what parameters, if any, are independently controlled by the central nervous system (CNS) is very important to the understanding of movement control. Inbar and Yafe (1976) demonstrated that adaptation to changes in loading conditions may be carried out by controlling the joint mechanical parameters, the control signal, or both. Modulation of the mechanical stiffness parameter of joint dynamics is an essential feature of the λ and α models of joint control (Feldman, 1986; McIntyre & Bizzi, 1993), the virtual equilibrium trajectory of motor control (Flash, 1987), and to the general understanding of manipulation (Hogan, 1985).

The mechanical stiffness of muscles and joints has therefore been studied extensively (Hoffer & Andreassen, 1981; Agarwal & Gottlieb, 1977; Hunter & Kearney, 1982; Yosef & Inbar, 1986). Past studies, however, were all carried out under steady state conditions and impedance modulation during movement was extrapolated from these results (Flash, 1987).

Recently, techniques have been suggested for the estimation of time varying biological systems (MacNeil *et al.*, 1992). Bennet *et al.* (1992) suggested tracking the mechanical parameters of the human joint during movement based on a small signal linearized model of the joint. This method is applied in the present study to track the elbow joint parameters at different cycle times, i.e., different velocities and with different loads. First, however, the small signal linear model is investigated and its order is determined during steady state and during movement. The parameters of this model are then tracked during movement.

THE EXPERIMENTAL SET-UP

The subject was seated on a chair with his elbow joint centered over the shaft of a motor and his arm strapped to a handle attached to the motor shaft. Movement was restricted to the sagittal plane—no gravitational forces involved—in the range of $\pm30°$, about a center position of $90°$ between the upper arm and forearm. The motor (printed circuit DC) could be used to apply external torque perturbations to the joint and, in addition, to apply a bias load, preloading the joint. The angular position was measured with a servopotentiometer and could be displayed to the subject on a TV monitor. The torque was measured by calibrating the input voltage to the power amplifier driving the motor. The linearity and coherence between the input voltage signal and the torque were measured and calculated in the range of amplitudes and frequencies used in the present study, and were found to be better than 99%.

All recorded channels, EMG for each muscle, pseudo-random binary sequences (PRBS) input voltage signal, and angular position, were sampled at 500 Hz each. The PRBS

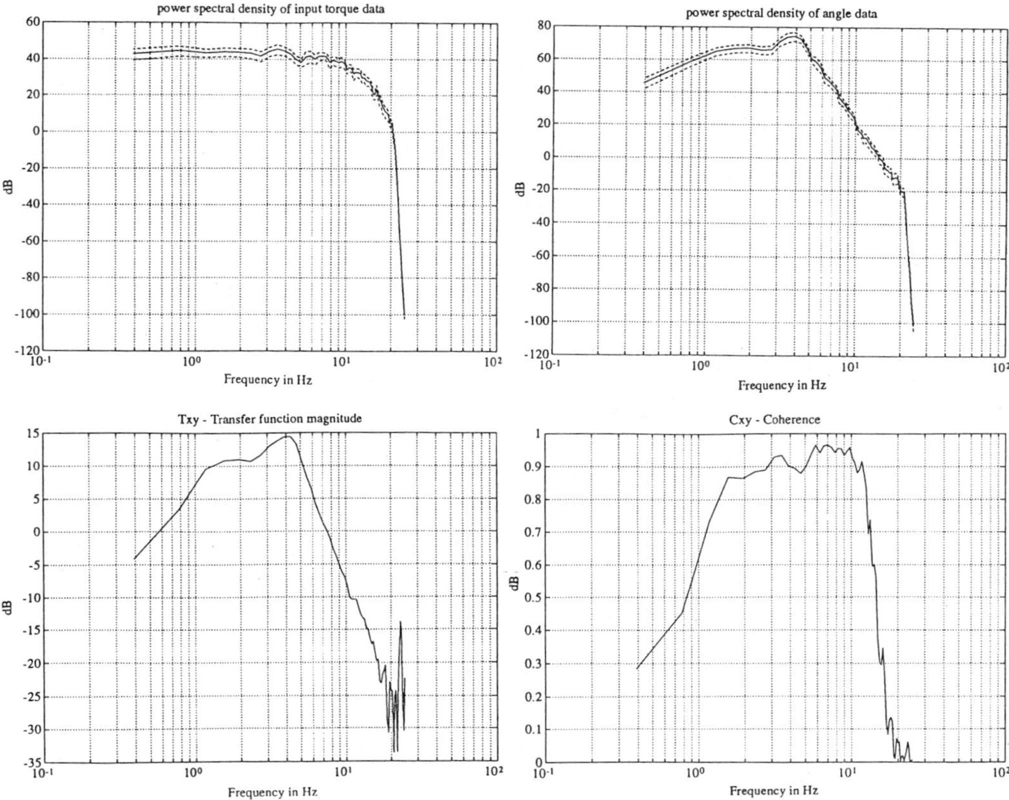

Figure 1. The band limited to 15 Hz PRBS torque input spectrum is shown with the angular position output spectrum in the two upper plates. The transfer function for these signals and their coherence function are shown in the bottom plates.

perturbation input signal was band limited between zero and 35 Hz in the EMG studies and 0.5 - 15 Hz in the impedance measurements (Fig. 1). The EMG measurements are not discussed in the present work.

The Linear Small Signal Joint Dynamics Model

The neuromuscular mechanical joint system is a highly complicated distributive system that has however, been shown to behave like a lumped second order system (McGruer *et al.*, 1968; Agarwal & Gottlieb, 1977; Hunter & Kearney, 1982; Yoseph & Inbar, 1985). The frequency domain behavior of such a joint to small signal mechanical perturbation is shown in Fig. 1. The transfer function (TF) shows the shape of a second order system with a 20 dB/dec slope at the high frequency end, dictated by the joint's inertia. The phase shift (not shown here) is 180° between 1.0 to 12.0 Hz. The coherence between the torque and angular position is above 0.9 at the spectral range of interest, and justifies the small signal linearity assumed. The change in resonance frequency, damping and stiffness, under steady state CNS control signal, with the level of muscular force as determined by the imposed preload level is seen in Fig. 2. The second order linear TF has the following form, with at least K, the stiffness, and B, the viscosity of the joint, being a function of the force operating point. Note, it is theoretically possible that the inertia may also change because of wideband

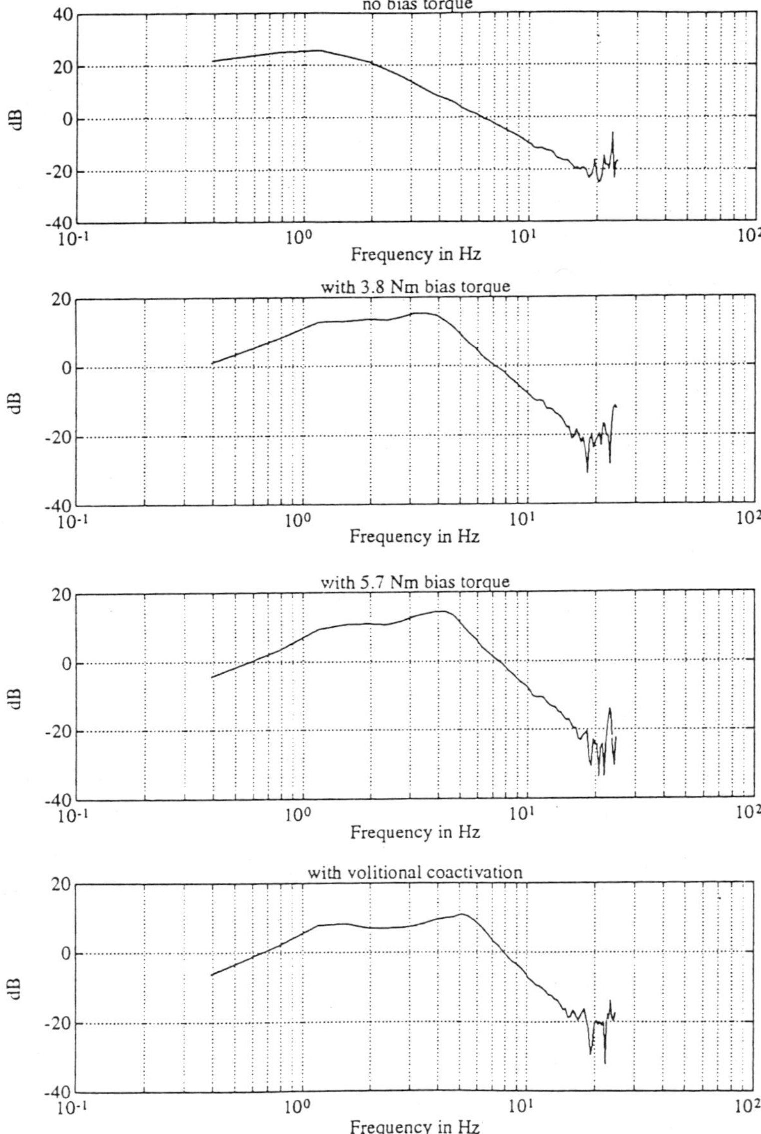

Figure 2. The effect on the joint TF, natural frequency, damping and gain is shown as a function of an increased bias preload, from zero preload on top via 3.8 Nm, 5.7 Nm and muscles co-contraction at the bottom.

acceleration feedback component from proprioceptors and/or spinal cord signal processing properties (Yoseph & Inbar, 1985):

$$H(s) + \frac{\theta(s)}{T(s)} = \frac{1}{JS^2 + BS + K} = \frac{\omega_o^2/K}{S^2 + s\xi\omega_o S + \omega_o^2} \tag{1}$$

where $\theta(s)$ is the Laplace transform of the angular position, $T(s)$ is the transformed torque input, ω_o is the natural resonance frequency and ξ is the damping factor of the joint system.

In order to better establish the order of the system (see Results section), and because all the signals are sampled digitally it is natural to characterize the system by its difference equations in the Z domain. Equation (1) can be transformed to the Z domain using zero order hold (ZOH) approximation:

$$H(z) = \frac{z-1}{z} \cdot Z\left\{L^{-1}\left[\frac{H(s)}{s}\right]\right\}$$

(2)

where Z is the z transform and L^{-1} is the inverse Laplace transform. The transformation result is:

$$H(z) = \left[b_1 z^{-1} + b_2 z^{-2}\right] / \left[1 + a_1 z^{-1} + a_2 z^{-2}\right]$$

(3)

The autoregressive moving average (ARMA) model of this system is:

$$\Delta\theta(t) + a_1(t)\Delta\theta(t-h)\ a_2(t)\Delta\theta(t-2h)\ -$$

$$-b_1(t)\Delta T(t-h) - b_2(t)\Delta T(t-2h) = 0$$

(4)

where h is the sampling interval.

There is a relationship in Eq. (4) between the parameters a_i and b_i, and the parameters of the continuous time model K, B and J. Here we write this relationship only for the stiffness K (Bennet *et al.*, 1992):

$$K(t) = \frac{1 + a_1(t) + a_2(t)}{b_1(t) + b_2(t)}$$

(5)

It is important to note that the order of Eq. (4) can be tested by allowing it to be

$$H(z) = \left[\sum_{k=0}^{q} b_k z^{-k}\right] / \left[1 + \sum_{k=1}^{p} a_k z^{-k}\right]$$

(6)

where p is the number of poles of the AR model order, and q is the number of zeroes of the MA model order. Assuming a model order of two in Eq. (4) is equivalent to a two-poles AR model. In the Results section, it is shown that indeed, according to some criteria, the AR model is two; however, the number of zeroes, q, can be either one or two. Higher order p and q do not significantly improve the fit of the model to the experimental data.

Data Processing

The subject maintained a fixed angular position against a given preload level. Four levels were tested, zero, 3.8 Nm, 5.7 Nm, and at an unknown high level of coactivation. A 10:1 decimation was used and the DC and trends were removed. Data was collected over 54 sec, prefiltered 1-200 Hz and segmented to 0.5 sec segments. Segment length could be changed in order to obtain better frequency resolution. The best resolution tested was at 0.39 Hz. The frequency spectrum and the ARMA parameters were calculated using 50% segment overlap to yield about 200 segments. The input spectrum (Fig. 1) is shown to roll off at about

15 Hz; however, experiments to determine the ARMA model order were run with 0.5 to 35 Hz input spectrum. Model order was determined using two criteria: the Akaike final prediction error (FPE) and the Akaike information criterion (AIC). In addition, the variance of the Z plane poles and the whiteness of the prediction error were investigated (Paiss & Inbar, 1987). For p and q model order, N, the number of sample points, and, the variance of the prediction error, the model order criteria are:

$$FPE(p,q) = [N + (p+q+1)] \ / \ [N - [p+q+1]]\sigma_e^2 \qquad (7)$$

The optimal model order for this criterion is obtained for minimum FPE as a function of both p and q. If minimum is not achieved, the number of samples in the numerator, N, can be multiplied by p<1. The AIC (Paiss & Inbar, 1987) is obtained in a similar way and the optimal model order can be displayed as a minimum in a two dimensional AIC surface.

Experimental Results of Model Order

Three healthy students (one male, two females) were tested to determine model order. For each subject, 25 different combinations of p and q, i.e., up to ARMA model (5,5) were run at each preload level and at different angular positions. Table 1 summarizes the results for the male subject at different levels of input torque perturbation, T_{in}, in arbitrary units. For a noise-free linear system the fraction θ_{out}/T_{in} should stay constant under all load and position conditions. Fatigue sometimes caused subjects to move their joint independent of the input signal. In addition, the level of angular perturbations influenced this value. The model order results can be summarized as follows:

1. The AIC and FPE give different optimal results with the FPE most of the time yielding a (2,2) model order; when it did not, file numbers ds005, ds006, the percent error from the (2,2) result, i.e., $\Delta\%$ was smaller than 6%. The AIC results, when different from the (2,2) model, were smaller than 15%. On some files the AIC did not give an optimal value in the (5,5) range.
2. The female subjects had higher values for the θ_{out}/T_{in} fraction and more often, the optimum model had higher values for p and q. The coherence for their files fell, quite often, to low levels below 4-5 Hz.

Table 1. Optimal model order according to the FPE and AIC criteria at different levels of input perturbation, Tin and preload for subject S.V.

					AIC		FPE	
File	T_{in}	θ_{out}	θ_{out}/T_{in}	Preload	p,q	$\Delta\%$ (2,2)	p,q	$\Delta\%$ (2,2)
ds001	5.7	6.5	1.14	0	2,2	0	2,2	0
ds002	5.7	7.9	1.39	0	3,3	1	2,2	0
ds003	6	8.7	1.45	0	2,2	0	2,2	0
ds004	9	13.5	1.5	0	4,2	7	2,2	0
ds005	12	15	1.25	0	4,2	15	4,4	2
ds006	24	41	1.7	0	1,2	5	1,1	6
ds007	6	7.5	1.25	3	4,2	9	2,2	0
ds008	6	7.5	1.25	3	4,2	3	2,2	0

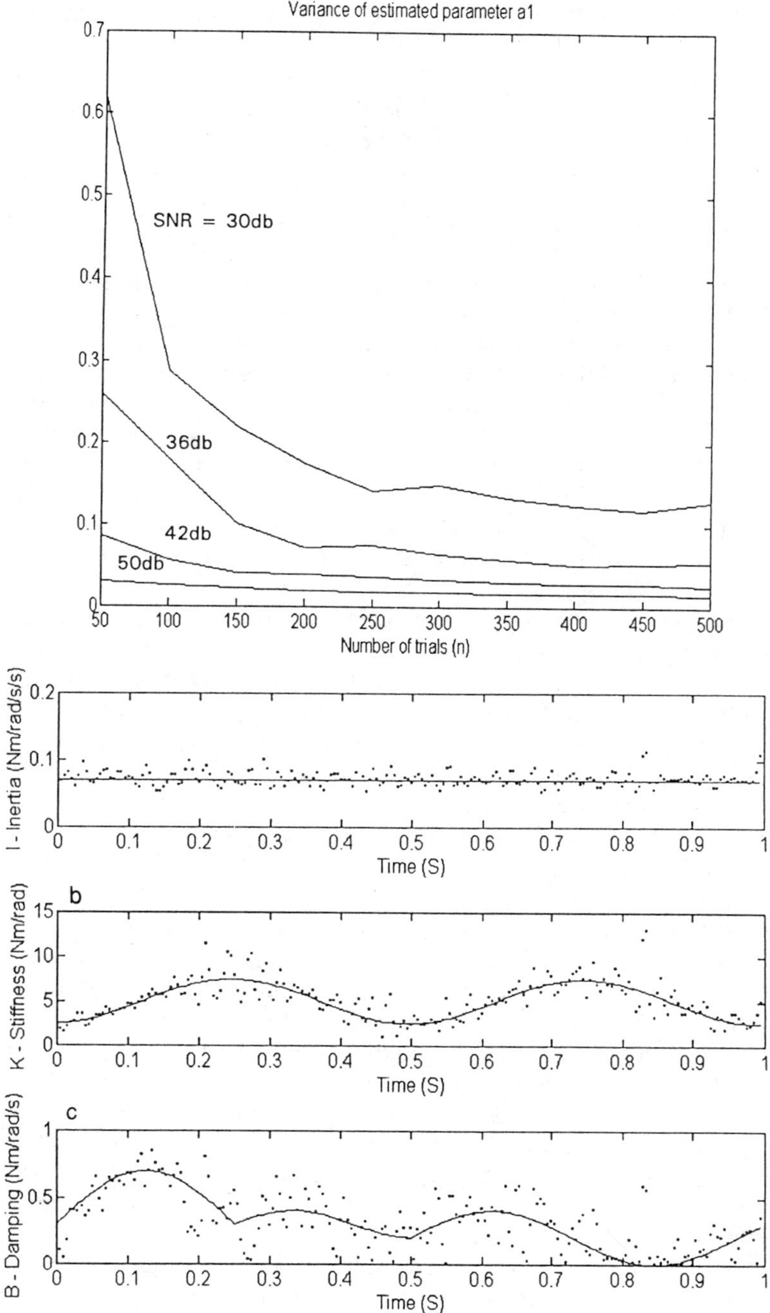

Figure 3. Simulation results of the identification and parameters conversion procedures of a known time varying second order system with output noise. Dots - the identification results of K, B, and J; solid lines, the known value of the parameters. Also shown are the effects of the added noise on the variance of the estimated parameter a_1 as a function of the number of cycles used.

3. The higher values of the perturbation amplitudes, files ds005, ds006, gave higher order models. This may be due to the linear range being exceeded, as was indicated by lower coherences, especially at the higher end of the spectrum.
4. The prediction error-spectrum was estimated for the (2,2) models and for the cases of higher model orders. For the (2,2) results, the prediction error spectrum was already white.
5. All of the above results indicate that the p=2 and q=2, i.e., a two-pole second order system with two zeroes, gives the best small signal fit to the data.

Tracking Model Parameters during Movement

A method to track the parameter values of a model during movement has been suggested (Bennet, 1990, Bennet *et al.*, 1992). The estimation procedure is based on the ability to repeat the same movement n times while the movement is perturbed with pseudo random mechanical perturbations. The perturbations are not synchronized and are uncorrelated with the movement and thus, one can obtain for a particular point in the movement $\theta(t_i)$, n small signal values of $\Delta\theta$ at t_i, with the ΔT that caused it. Theoretically, for a second order system not too many movement cycles are required. However, since the system is noisy and measurement noise is included in the above procedure, *two to three hundred* repetitions of the movement are required in order to reduce the standard deviation of the parameter estimates below 5%. Sinusoidal movements were used and they were synchronized at their maximum velocity.

In the experiments, the subject attempted to maintain constant sinusoidal movement cycle time despite the random perturbations that were continuously applied during the movements. The cycles were aligned to maximum velocity, averaged, and each movement cycle was aligned with the average. This procedure was repeated once more with no further improvement by further repetition, as alignment is done in steps of sampling time h. The final average was subtracted from each movement cycle with the signal difference left assumed to be due to the band limited mechanical perturbations applied by the motor to the joint. Figure 4 shows the results of the above procedure. Movements that were different in their cycle time by more than ±60 msec were not aligned and were discarded (Weissman, 1995).

To study the whole identification and conversion from the discrete sampled data to continuous time, Eq. (5), a simulation procedure was run on a known second order system with time varying parameters (Weissman, 1995). The input perturbation signal was identical to the one used in the experiment, and the large $\theta(t)$ movements were 1.0 Hz sinusoidal movements. The model was identical to Eq. (4) with the parameters a_1, a_2. b_1, b_2 changing in a prescribed way as a function of time in the cycle. The conversion to the continuous time yielded the parameters K, B and J, shown as continuous lines in Fig. 3. The output of the model torque was sampled at 200 Hz and various levels of white band limited Gaussian noise added at the output. Identification results from one example with a 42 dB SNR, where 300 cycles were used, are shown in Fig. 3, 200 dots per cycle. Values of the parameters for each dot in time were estimated from 300 samples, aligned in time from the 300 movements.

Figure 3 shows that a running mean of the parameter values, achieved via low pass filtration, is closely related to the known parameter values marked by the continuous line. On the other hand, it is observed that the variance of the estimates, i.e., the spread of the points about the mean (see for example the value of the estimated inertia J), is considerable. The effect of the number of cycles on the variance of the estimated parameter a_1, for up to 500 cycles, is shown in Fig. 3. The nonmonotonic nature of the curves is due to the different noise segments added at each cycle—identical noise generates a monotonically decreasing

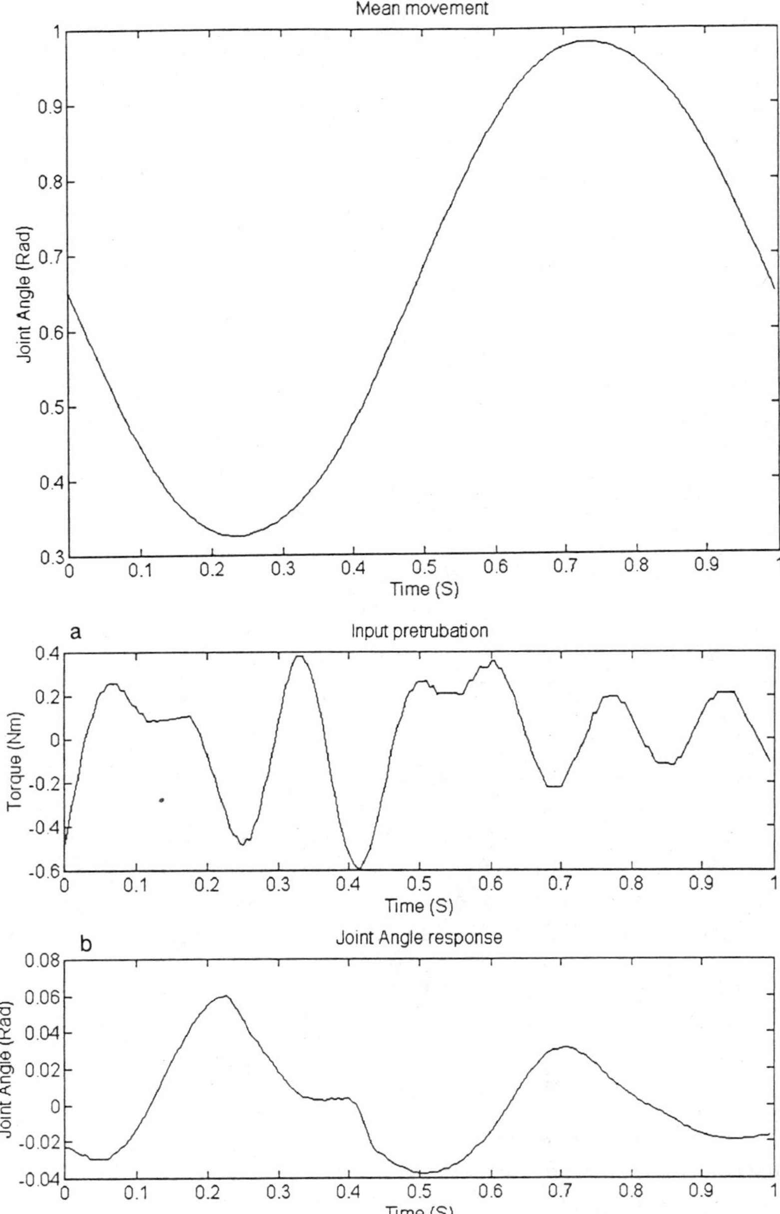

Figure 4. The averaged 1.0 Hz cycle performed by subject S.V. shown at the top frame. a: the band limited random torque input; b: the angular position output.

function. The value of the variance goes down asymptotically in all three cases, with the number of cycles N→∞. This value, which is due to the variance assumed for a_1, is seen for the case of SNR = 50 dB. Figure 3 also shows that little improvement in the reduction of the variance is achieved by increasing the number of cycles above 200. Finally, the influence of the SNR on the variance is shown, where the 42 dB SNR corresponds to the experimental results seen in Fig. 5.

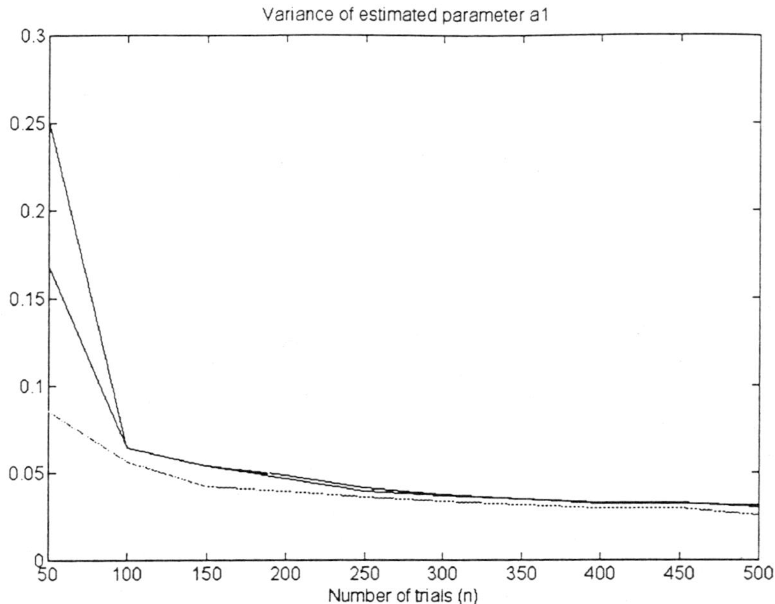

Figure 5. Variance of the discrete time variable a_1 as a function of the number of trials. The two upper curves are for subjects S.V. and O.H., and the thinner lowest line from the simulations (Fig. 3), with SNR = 42 dB.

The model order was then investigated, using the same criteria, AIC and FPE, as before. The results obtained from the male subject, for each of the cycles tested, are shown in Table 2. The number of cycles required to reduce the standard deviation of the estimated parameter a_1 below 5%, is shown in Fig. 5 both for the simulated case of 42 dB signal-to-noise ratio (SNR) and for the experimental results of the two female subjects.

The experimental results were all processed for an ARMA model of second order with two zeroes and at least 150 cycles were used for every subject's condition of cycle time—0.8, 1.0 and 1.2 sec—and loading conditions—0.0245, 0.033, and 0.058 Nm/rad/sec^2. The results are presented in Figs. 6-9.

Analysis of Experimental Results

The problem of model order for the neuromuscular joint is complicated by noise in the system and in the measurements, by nonlinearities, and by variability among subjects, etc. Some of these issues were examined in the present work. The coherence between the applied voltage and torque under various angular velocities, angular displacement and arm loading conditions, was examined and found essentially to be one between 0.05 and 40 Hz.

Table 2. Optimum model order during cyclic movements

Criteria	No. Cycles - (2,2)	No. Cycles - (2,1)	No. Cycles - Other Orders
AIC	156	184	24
FPE	75	278	11

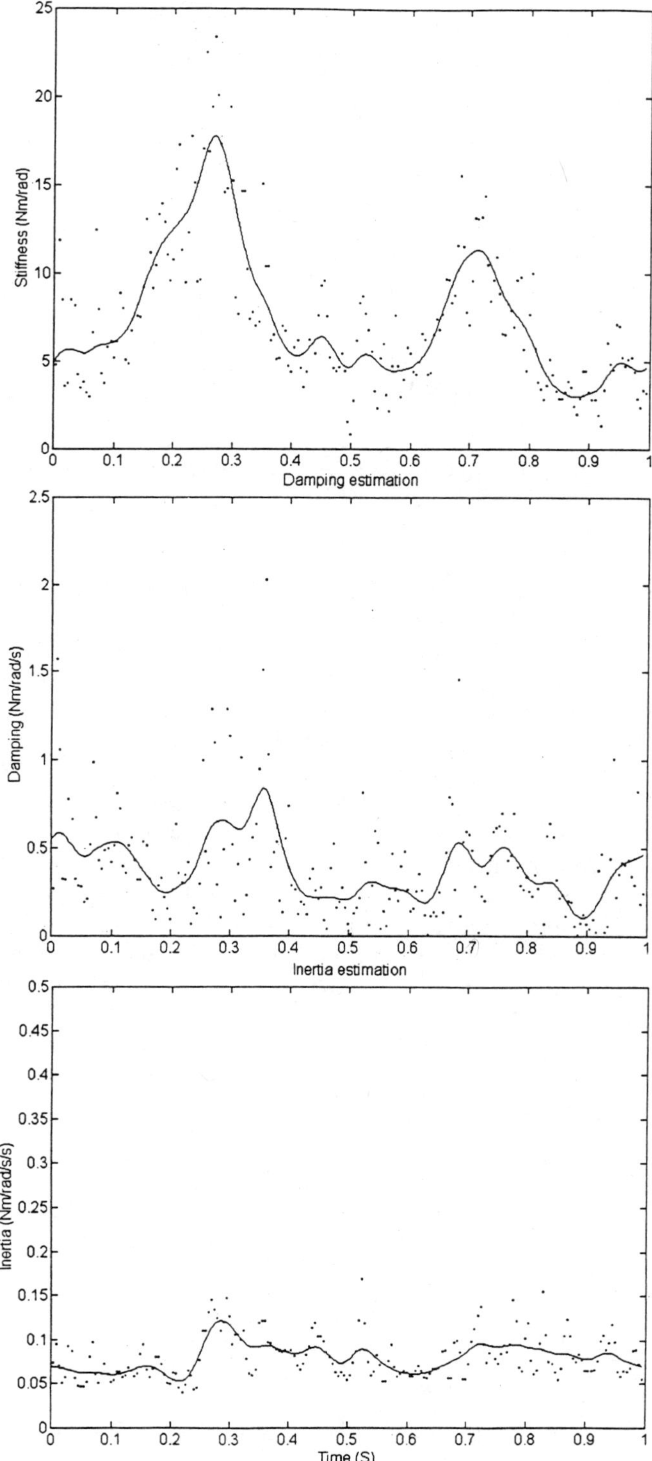

Figure 6. K, B and J for the 1.0 Hz cyclical experiments. Results are from 500 cycles, i.e., 500 input-output values per a shown dot. Solid line - results filtered with a 10 Hz second order Butterworth filter.

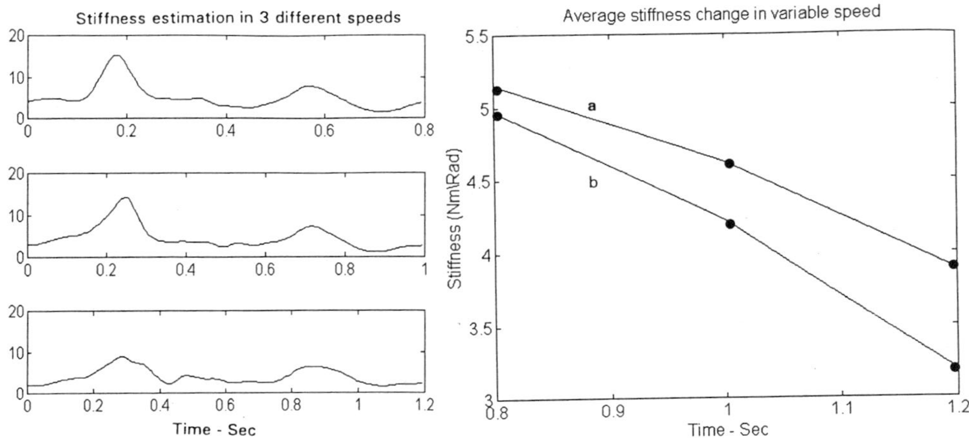

Figure 7. The estimated 10 Hz filtered stiffness parameter K, and its average value for the cycle, both as a function of time and cycle time, respectively.

The small signal linear joint dynamics model was derived from data with coherences above 0.8 and a second order model gave best fit for most of the experimental data both during steady state and during movement (Tables 1 and 2). The conclusion, therefore, in regard to stiffness variations, may hold only in the mean, and exceptions may be common. However, they should be considered in light of their impact on theories of movement control.

In Fig. 6, the J, B and K parameters are seen for 500 cycles of 1.0 Hz movements. The dots show the discrete 200 estimated values of the parameters; the solid line shows the low pass filtered results, filtered with a 10.0 Hz second order Butterworth filter. The stiffness K has two maxima at the target points, where the angular acceleration and torque are maximal and the velocity is zero. K has a minimal value where the acceleration and torque are minimal, at mid-cycle where the velocity is maximal. It is clear that the joint torque is the result of the difference in level of muscular contraction between the biceps and triceps. It is therefore expected that by coactivation, the overall level of stiffness will go up, while the general profile of K(t) will stay the same. Such coactivation can be expected when a subject is asked

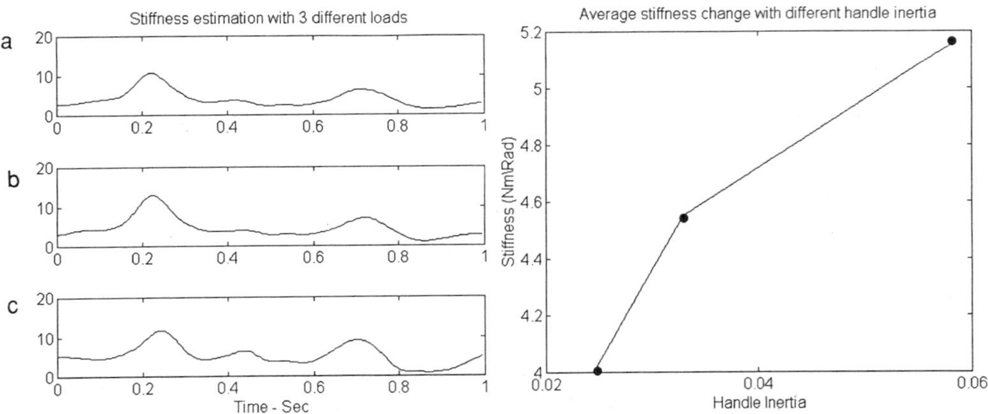

Figure 8. The estimated 10 Hz filtered stiffness parameter K as a function of load, for three inertial loads, 0.0245, 0.033, and 0.048 Nm/rad/sec^2. The average stiffness K for the cycle as a function of load is also shown.

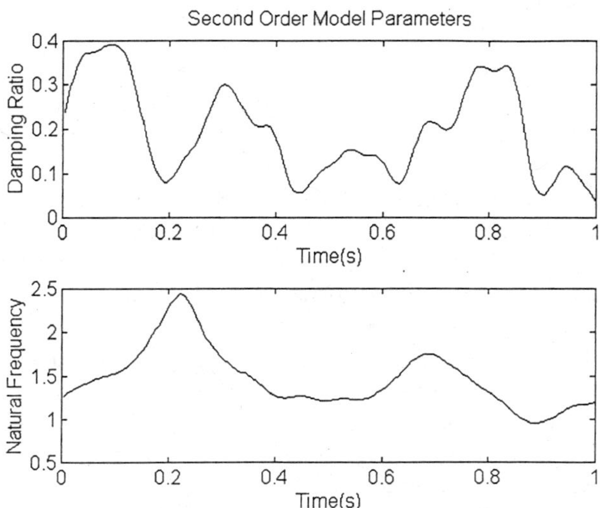

Figure 9. The second order continuous time parameter, ξ and ω_0 as a function of time in the cycle.

to perform the movement with an unknown load for the first time (Inbar & Yafe, 1976). The coactivation level goes down with learning during the repetitions of few movements.

Consistent asymmetry is observed in the stiffness variations, similar to the one seen here in Fig. 5, where one peak is bigger than the other. It is presumed that this asymmetry is due to the asymmetry in the joint muscles, in the flexion phase by the biceps versus the extension phase by the triceps.

The filtered viscous damping parameter values show an increase in phase with the acceleration, similar to K. However it is much less significant and the spread in values is greater. The estimated value of the moment of inertia, J, shows no correlation with the movement parameters and a considerable spread both in its filtered and nonfiltered values. The coefficient of variation for this parameter is 16%. The filtered value of J, as well as J itself, is supposed to stay constant throughout the experiment. Changes in J with movement may take place due to consistent changes in the center of rotation with movement or due to wide band feedback effects from the proprioceptors on spinal cord mechanisms (Joseph & Inbar, 1986). The spread in the estimated J values is due to the noisy measurement and the identification and parameter conversion schemes, as can be seen from the simulation results (Fig. 3).

The changes in the filtered value of K, as a function of movement velocity and with loads is shown in Figs. 7 and 8. The strong and correlated dependency of the stiffness K on the level of joint torque is strengthened by these results. The asymmetry between flexion and extension in the K profile with movement is maintained at different cycle times and with different loads.

The natural frequency of the system, and the damping coefficient of the second order model of the joint, $\xi = B/2(KJ)^{1/2}$, are shown in Fig. 9 for the 1.0 Hz cyclic movement experiment. The natural frequency is consistently highest at the high acceleration points where movement direction is changed. The damping coefficient shows no consistent relationship to movement. This inconsistency is due to the inconsistent relationship and large variance obtained for the estimated parameter B.

DISCUSSION

The NMCS, like other biological systems, is a nonlinear system with some of the noise generated by the CNS through the long reflex loops in response to the perturbations. A large number of experiments, or sufficiently long ones that allow the use of a large number of segments, can overcome some of the estimation errors and large variances in the estimated parameter values. This has been shown for EMGs where the use of 100 segments in spectral estimates elucidated physiological frequencies not detected by a lower number of segments (Paiss & Inbar, 1987). The same effect on the reduction in the variance of the value in the estimated parameters is seen here in the parameters of the elbow joint, both in the steady state experiments and in the cyclical movements.

The NMCS is nonlinear in respect to the angular position and to torque. One can attempt to model these nonlinearities for the whole dynamic input range (Inbar *et al.*, 1970). Alternatively, a small signal dynamic model can be used (McGruer *et al.*, 1968) and its parameters tracked through large changes in angular position (Bennet, 1990). *It has been shown in the present work that a small signal lumped parameter second order model best represents the mechanical TF of the human elbow joint.* The model is a closed loop model that includes the mechanical and neural dynamic properties of the muscles, the proprioceptors, all connective tissue, and the joint. It was further shown that when higher order models gave better fit to the data, this was the result of either extending the input amplitudes perturbations into the nonlinear range, or due to spurious noise introduced by the subjects as a result of fatigue or other unknown factors.

Different results were recently obtained (NacNeil *et al.*, 1992) for the order of an ankle joint system during transitional movements using an impulse response method. However, these results were obtained for different types of movements and a different joint than the ones used in the present study. It should be noted that for other conditions and inputs the small signal linear model may be of a different order, like the third order model obtained for external neuronal stimulation (Allin & Inbar, 1986).

The dynamic variation in the joint TF during cyclical movements was estimated using the method proposed by Bennet (1990). This method and the conversion of the model parameters from their discrete form, Eq. (5), to the continuous form, Eq. (1), introduce large variances in the estimated parameters of the simulated noisy system (Fig. 3). The 10 Hz filtered values of K, B and J gave the general profile of the mean parameter values during the cyclical movement. The most pronounced feature of these results is the consistent changes in stiffness, K, with net joint torque, as seen in Figs. 6-8. The joint was most compliant in mid-cycle and most stiff at the direction changing points where the acceleration is maximal. The average stiffness was found to be in proportion to the net torque (Figs. 7 and 8), contrary to the results of Bennet *et al.* (1992).

The value of stiffness K during the movement could not be compared to its value during steady state postural value, since only net torque was measured in both cases. However, the value of K at mid-cycle point, where the velocity was at its maximum and acceleration at its minimum, was so low—with one female subject it was consistently close to zero—that it is clearly lower at this point than its postural value. Thus, models of motor control based on high stiffness during movement relative to their postural value must be reconsidered. It is noted here that similar results have been obtained (Aaron & Inbar, 1995) for the static stiffness profile of K as a function of angular position, load and velocity, and for cyclical movements using the λ model as suggested by Latash (1993). With the same technique, the static stiffness K was estimated during reaching movements and found to change considerably during the movement with one or two peaks depending on the load.

In accordance with the values obtained for the parameters, the natural frequency and damping factors were calculated (Fig. 9). It is seen that the joint is underdamped throughout the movement with especially low values at mid-cycle, where its average value is close to 0.1. The natural frequency of the joint is less than 3.0 Hz with all subjects tested—less than 2.5 Hz in Fig. 9—throughout the movement. The higher damping at the direction changing point where the velocity is zero is a desirable feature for a manipulendum, as has already been observed when attempting to perform a reaching movement with zero velocity at the target point (van Sonderen & van der Gon, 1990).

Signal processing was applied in the present work to characterize and identify a time varying model of the human elbow joint mechanical TF. It was shown through simulations and experimental results that it is crucial to use large data sets if the variances of the estimated model parameters are to stay within reasonable bounds, e.g., below 5%. It was further shown that data collected in experiments with human subjects can be contaminated with volitionally and reflexively driven noise not related to the input driving signal, e.g., spectral output below 1.0-2.0 Hz. Thus signal processing techniques can be very instrumental in the characterization of biological systems in their normal mode of physiological behavior and for clinical monitoring and classification.

ACKNOWLEDGMENT

This work has been supported by the Fund for the Promotion of Research at the Technion.

REFERENCES

Aaron, A., and Inbar, G., 1995, Elbow joint movement control: An experimental and theoretical study, *The French Israeli Symposium on Robotics*, May 22-23, Herzelia, Israel. Submitted for publication.

Allin J., and Inbar, G., 1986, FNS parameter selection and upper limb characterization, *IEEE Trans on Biomed. Eng.* BME-33(9):809-817.

Agarwal, G. C., and Gottlieb, G. L., 1977, Compliance of human ankle joint, *J. Biomechan. Eng.* 99:166-170.

Bennet, D. J., 1990, The control of human arm movement: models and mechanical constraints, *Ph.D. Thesis*, MIT, Dept. of Brain and Cognitive Sci.

Bennet, D. J., Hollerbach, J. M., Xu, Y., and Hunter, I. W., 1992, Time-varying stiffness of human elbow joint during cyclic voluntary movement, *Exp. Brain Res.* 88:433-442.

Feldman, A. G., 1986, Once more on the equilibrium-point hypothesis (λ model) for motor control, *J. Motor Behavior* 18(1):17-54.

Flash, T., 1987, The control of hand equilibrium trajectories in multi-joint arm movements, *Biol. Cybernetics* 57:257-274.

Hoffer, J. A., and Andreassen, S., 1981, Regulation of soleus muscle stiffness in premammillary cats: intrinsic and reflex components, *J. Neurophysiol.* 45:267-285.

Hogan, N., 1985, Impedance control, An approach to manipulation. Part 1 - Theory; Part 2 - Implementation; Part 3 - Application, *J. Dynamic System Measurement and Control* 107:1-23.

Hunter, I. W., and Kearney, R. E., 1982, Dynamics of human ankle stiffness: Variation with mean ankle torque, *J. Biomechan.* 15(10):747-752.

Inbar, G.F., Baskin, R., and Hsia, S., 1970, Parameter identification analysis of muscle dynamics. *Mathematical Biosciences* 7:61-79.

Inbar, G. F., and Yafe, A., 1976, Parameter and signal adaptation in the stretch reflex loop, *Prog. in Brain Res.* 44:317-337.

Latash, M., 1993, *Control of Human Movement*, Human Kinetics Publishers, Illinois.

MacNeil, J. B., Kearney, R. E., and Hunter, E. W., 1992, Identification of time varying biological systems from ensemble data, *IEEE Trans. on Biomed. Eng.* 39(12):1213-1224.

McIntyre, J., and Bizzi, E., 1993, Servo hypotheses for the biological control of movement, *J. Motor Behavior* 25(3):193-202.

McRuer, D. T., Magdeleno, R. E., and G. P. Moore, 1968, A neuromuscular actuation system model. *IEEE Trans. on Man-Machine Systems* MMS-9(3):61-71.

Paiss, O., and Inbar, G.F., 1987, AR model representation of surface EMG and its application to fatigue measurements. *IEEE Trans. on Biomed. Eng.* 34:761-770.

Porat, B., 1994, *Digital Processing of Random Signals - Theory & Models*, Prentice-Hall Inc., Englewood Cliffs, NJ.

van den Bosch, P. P. J., and van der Klauw, A. C., 1994, *Modeling Identification and Simulation of Dynamical Systems*, CRC Press, Boca Raton.

van Sonderen, J.F., and van der Gon, J.J.D., 1990, A simulation study of a programmed fast two-joint arm movements: Responses to single- and double-step target displacemtns. *Biol. Cybern.* 63:35-44.

Weissman, S., 1995, Identification of elbow joint time varying parameters, *MSc. Dissertation*, Dept. of Electrical Eng., Technion, Haifa.

Yosef, V. A., and Inbar, G. F., 1986, Parameter estimation of the mechanical impedance of the forearm of the human-operator using Gaussian torque input, March, *EE Pub. No. 582*, Technion-Israel Institute of Technology, Haifa.

26

CHARACTERIZING AND MODELING HUMAN ARM MOVEMENTS: INSIGHTS INTO MOTOR ORGANIZATION

Tamar Flash, Irina Gurevich,[1] and Ealan Henis

Department of Applied Mathematics and Computer Science
The Weizmann Institute of Science
Rehovot, Israel
[1] Department of Industrial Engineering and Management
Ben-Gurion University of the Negev
84105 Beer Sheva, Israel

ABSTRACT

The generation of human goal-directed multi-joint arm movements requires the central nervous system (CNS) to deal with complicated motion planning and control problems. These problems include the selection and planning of specific motions for the hand among the large number of possible ones, the transformation of the desired hand trajectory plans into appropriate joint rotations, and the generation of appropriate muscle forces and joint torques in order to execute the desired motion plans. In this article we will review several recent behavioral and modeling studies of human arm trajectory formation. These studies were aimed at characterizing and modeling human multi-joint arm movements and at investigating how the motor system copes with these complicated motion planning and control problems. The topics to be discussed will include hand trajectory planning during reaching and drawing movements, arm trajectory modification in response to unexpected changes in target location and motor adaptation to unexpected force perturbations introduced by means of external elastic loads. Then, the implications of our findings with respect to motor organization and possible neural correlates for the motion planning and motor execution and adaptation schemes suggested here will be discussed.

INTRODUCTION

Motor control investigations belong to the general class of studies that deal with inverse problems whereby we wish to infer from the measured output what is the nature of the internal commands that give rise to this output. The aim of such studies is to gain insights into the principles of operation and organization of the motor system.

Advances in Processing and Pattern Analysis of Biological Signals, Edited by Isak Gath and Gideon F. Inbar
Plenum Press, New York, 1996

The generation of an even simple motor action requires of the motor system to solve complicated motion planning and control problems. The observed motor output emerges as a result of complicated neural activation patterns occurring within many neural structures. Moreover, it is often quite difficult to establish to what extent the features of the observed output reflect explicit motor control strategies or are merely byproducts of the mechanical properties of the controlled plant. There are no general research methodologies for dealing with such problems. Nonetheless, here we will discuss several examples that illustrate how the processing of the observed movement and force outputs can be combined with mathematical modeling techniques in order to infer some general principles about motor organization.

A review of our earlier studies of human arm trajectory formation was presented in a chapter book entitled "The organization of human arm trajectory control" (Flash, 1990). Here we will present several recent additional studies of reaching and drawing movements in which some of those earlier studies were extended and several new problems and issues were addressed. In studying human multi-joint arm movements the basic approach involves recording and analysis of either two-dimensional arm movements performed in the horizontal or vertical planes or less constrained three-dimensional movements. Hence, in our original series of studies, unconstrained two-joint point-to-point movements, performed in the horizontal plane were measured. The recorded movements were then kinematically analyzed and several motor invariants identified by examining movements performed under different speed, amplitude and workspace locations, were identified (see for example Morasso, 1981; Hollerbach & Flash, 1982; Flash & Hogan, 1985). In our own work this kinematic analysis was combined with the development of mathematical models based on optimization theory in order to infer the coordinate systems used and the rules according to which multi-joint arm trajectories are being planned (Flash & Hogan, 1985, 1995). Here we will review applications of optimization theory to studies of reaching, handwriting and drawing movements. Moreover, recent attempts to unify between different approaches to the modeling of drawing movements will be described. These attempts and the analysis of 2D and 3D hand trajectories have stressed the importance of the possibility of identifying within human arm movements, certain elementary building blocks or units of action from which more complicated movements are constructed. This was also one of the major notions that emerged from a recent series of studies involving the analysis of human arm trajectories when the locations of visual targets toward which the arm was reaching were unexpectedly modified. In these studies, mathematical modeling and statistical analysis techniques were combined to infer possible arm trajectory modification strategies. Moreover, mathematical modeling approaches were used to infer the possible spatial locations of the internal representations of visual targets and the time-dependent variations that these locations undergo in the process of arm trajectory planning.

Finally, the paper will discuss the topic of motor execution and possible control strategies for the realization of the desired motion plans. One of the major controversies in motor control pertains to the overall organization of the motor system- namely whether it is valid to assume that there exist certain levels of representation that mainly deal with trajectory planning while others are dedicated to motor execution. To address this issue, here we will review several recent studies of motor adaptation. In particular, we will review our own studies which were aimed at investigating what control strategies might possibly underlie human adaptation to external loads (Gurevich, 1993; Gurevich & Flash, 1995a,b). In these studies, human two-joint reaching movements were measured before and following the unexpected introduction of elastic loads. The arm stiffness field under different load conditions was also measured. Mathematical models of possible load adaptation strategies based on the equilibrium trajectory control hypothesis (Flash, 1987) were developed. Furthermore, the validity of alternative optimization models based on different objective functions was examined. These objective functions were formulated either in terms of dynamic or kinematic variables depending on the extent to which arm trajectory planning is assumed to depend on arm and/or load dynamics. The implications of these findings with respect to neural representations of motor plans will then be discussed.

HAND TRAJECTORY PLANNING STUDIES

Traditionally, motor control studies have focused on anatomical and physiological questions and on single-joint movements. Over the last decade, however, it has become increasingly clear, that to gain deeper insight into motor control, the problems underlying the generation of multi-joint movements and the "computations" performed by the brain in the solution of these problems must also be investigated. Since motor behavior is fundamentally multi-dimensional, movements might be alternatively represented in muscle, joint or end-effector spaces. Hence, a fundamental question in motor control research is in what space(s) or coordinate frame(s) does the brain represent movement. Another fundamental question is what rules or principles govern the selection of specific movement trajectories among the infinite number of possible ones.

Experimental observations of human unconstrained point-to-point reaching movements have indicated that these movements are characterized by straight hand paths and symmetric bell-shaped velocity profiles that tend to remain invariant, despite variations in movement direction, speed, and initial and final positions (Morasso, 1981; Hollerbach & Flash, 1982). Given that the common invariant features of those movements were only evident in the trajectories of the hand through external space and not in the movements of individual limb segments, these results have provided strong indication that planning takes place in terms of hand trajectories rather than in terms of joint rotations. In curved movements, although the hand paths appeared smooth, movement curvature was not uniform and the hand trajectories typically displayed two or more curvature maxima. The hand velocity profiles also had two or more peaks and the minima between adjacent peaks corresponded to the maxima in curvature (Abend *et al.*, 1982; Flash & Hogan, 1985).

To account for those experimental observations, a model based on optimization theory was suggested. Optimization theory provides a convenient way to formulate a coarse-grained model of the underlying neural computations, without requiring specific details of the way those computations are carried out (Flash & Hogan, 1985). Generally speaking, this application of optimization theory consists of defining an objective function that quantifies what is it that the motor system attempts to optimize in human movement and the tools of variational calculus are then applied to identify the specific behavior that achieves such optimum. Hence, in the case of reaching and curved trajectories the kinematic features of the movements were accounted for by assuming that a major goal of motor coordination is to achieve the smoothest motion of the hand in terms of extra-corporeal spatial coordinates. Following an idea originally suggested by Hogan (1984) in the single-joint case, the smoothness of motion of a multi-joint arm was quantified in terms of the integral of squared jerk (rate of change of acceleration) of the hand in space as follows:

$$C_J = \frac{1}{2} \int_0^{t_f} \left(\left(\frac{d^3 x}{dt^3} \right)^2 + \left(\frac{d^3 y}{dt^3} \right)^2 \right) dt \tag{1}$$

where $\vec{r}(t) = (x(t), y(t))$ is the vector of hand position and t_f is the movement duration. It was then suggested that among all possible hand trajectories that join some specified initial and final positions, the motor system selects to generate those trajectories that minimize the above cost function.

Based on this definition the actual movements had to be worked out, their details depending on the conditions assumed at the beginning and end of the movements. The above model has been initially derived for single-joint movements (Hogan, 1984), but most natural movements are multi-dimensional, i.e., they involve simultaneous rotations about several joints. In generalizing, however, the maximum smoothness principle to the multi-joint case, a crucial question is what coordinates should be used in order to define the measure of smoothness. For example, the jerk may relate to the position coordinates of the hand in space, or to the angular coordinates of the different joints (e.g., shoulder and elbow). Since, as discussed above, experimental studies of two-joint arm movements have indicated that arm movements are planned in terms of hand trajectories, the coordinates used to express hand position and consequently jerk are those in which movement planning was assumed to occur, (i.e. the Cartesian coordinates of the hand, $x(t)$ and $y(t)$). If the movement is defined to begin and end at rest (with zero initial and final velocities and accelerations), then it is possible to show that the movements predicted by the model will have the following characteristics:

- The trajectories of the hand follow straight hand paths.
- The velocity profile for moving along that path is smooth and unimodal.
- The shape of the trajectory is invariant under translation, rotation, amplitude and speed scaling.

It was then shown that these predictions agree quite well with the experimental observations. Several studies indeed have reported that during point-to-point motions the path of the hand through space is essentially straight independently of end-point locations (Morasso, 1981; Flash & Hogan, 1985), thus satisfying the prediction that the trajectory is invariant under translation and rotation. It is also true for large and small movements and for movements at different speeds (Flash & Hogan, 1985), thus satisfying the predictions that the trajectories are invariant under amplitude and speed scaling.

To account for the kinematic features of curved movements again the maximum smoothness model was applied assuming that curved motions are generated by specifying a small number of accuracy ("via") points along the trajectory. The hand is then required to pass through these points on its way between the initial and final positions. The time at which the hand passes through these points, however, does not need to be a priori specified, nor does the velocity of passing through these points. These are predictions of the model. As before, limb motion was expressed in terms of hand coordinates. Taking the simplest case of one accuracy point between the initial and final positions, the theory yielded explicit mathematical expressions for the description of curved motions which can be verbally expressed as follows:

- The hand path exhibits a single curvature peak.
- The hand velocity profile exhibits two peaks with a trough between them.
- The depth of this trough is proportional to the deviation of the via point from the straight line joining the initial and final positions.
- The peak in curvature temporally coincides with the valley in tangential velocity.
- The shape of the trajectory is invariant under translation, rotation, amplitude and time scaling.
- The durations of the motions from the initial position to the via point and from the via point to the final position are almost always equal, except for cases in which the via point is located very close to either one of the movement end-points. This behavior will be referred to as the isochrony principle (Vivinai & Flash, 1995).

Careful comparisons of the predicted to observed movements have shown that the minimum-jerk model is quite successful in accounting for the relatively simple curved and obstacle-avoidance movements (Flash & Hogan, 1985), as well as for more complicated handwriting and drawing movements (see below and in Edelman & Flash, 1987). The main tenet of the model was that arm motions are planned in terms of hand trajectories in extra-personal space.

MODELING HANDWRITING AND DRAWING MOVEMENTS

Various applications of optimization theory to the modeling of single-joint and multi-joint straight and simple curved movements were reviewed recently (Flash & Hogan, 1995). Here, we will discuss possible internal representations of more complicated curved and drawing movements. The analysis of simple point-to-point movements has indicated that at least at more central levels, arm movements are represented in terms of spatial trajectories of the hand in extra-personal coordinates. Moreover, it was shown that the kinematic features of straight and relatively simple curved movements are captured by the minimum-jerk model. Another basic question of interest in motor control research is whether there exist a repertoire of basic motor building blocks from which more complex movements are constructed and what rules govern the construction of complex behaviors from such basic alphabet. These questions, naturally arise in the study of relatively complicated motor actions or sequential tasks such as handwriting and drawing movements. Hence it was argued that the motor system need not internally represent all possible letters or figural forms but may use, instead, a limited set of basic motion primitives or strokes which can then be concatenated to form more complicated movements (see Lacquaniti, 1989). Such motion segments have been identified by detecting abrupt changes or discontinuities in certain movement parameters, such as curvature, tangential velocity or in the coefficient relating these two variables to each other, as in the two-thirds power law (Viviani & Cenzato, 1985). The latter law refers to the observation that in naturally executed curved and drawing movements the angular velocity decreases with increasing curvature and is proportional to the two-thirds power of the latter (Lacquaniti, 1989), as follows:

$$A = KC^{2/3} \tag{2}$$

where A is the angular velocity and C is the movement curvature. The gain factor of this relationship K was found to be piece-wise constant and to be determined by the linear extent of each individual segment. These observations were interpreted to suggest that in spite of the apparent continuity of hand movements they are, in fact, intrinsically discontinuous and are constructed of individual segments or strokes. Moreover, it was argued that the hand velocity during any movement need not be explicitly coded but might be automatically derived from the coupling between speed and curvature as expressed by the two-thirds power law (Lacquaniti, 1989).

In our above discussion, two different alternative models of the generation of curved and drawing movements have been presented. According to the two-thirds power law the internal representation of drawing movements consists of the figural forms or spatial attributes of simple strokes such as straight lines, elliptical segments, etc. These internal representations are then transformed into the actual movement trajectories needed to trace these figural forms according to the law of motion as expressed by the two thirds power law. Alternatively, according to the minimum-jerk model or other optimization based models, even the trajectories for simple figural forms are represented by a series of accuracy or via-points and both the hand path and the time-histories of position are prescribed by the

                                                                    T. Flash et al.

generation rules according to the minimum-jerk model. Motivated by the desire to examine whether these two alternative descriptions, may in fact emerge from similar organizing principles, recently Vivinai and Flash (1995) examined whether the minimum-jerk model

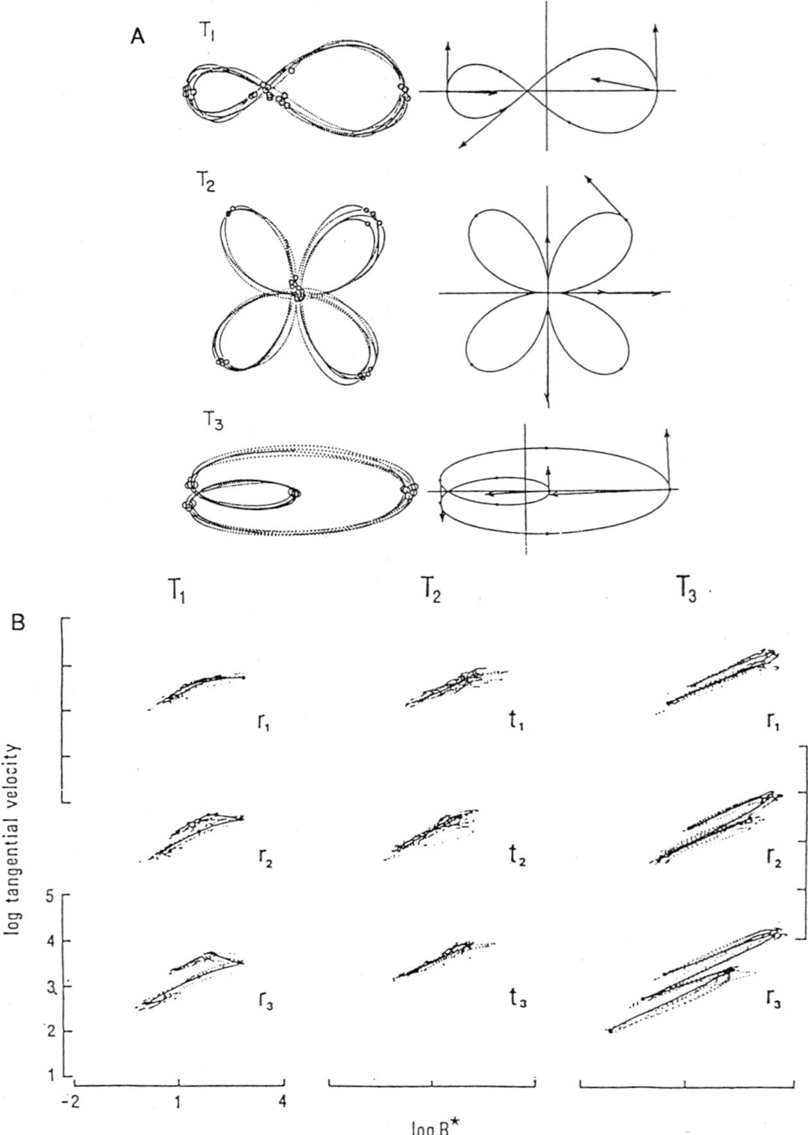

Figure 1. A. The minimum-jerk model versus the two-thirds power law. Comparisons between recorded hand paths ... and the paths predicted by the minimum-jerk model _ for drawing movements of a lemniscate (T_1), a cloverleaf (T_2) and an oblate limacon (T_3). B. The relations between hand speed and the variable $R = R\,(1 + \alpha R)$ where R is the radius of curvature (R is equivalent to $1/C$ where C is the curvature) for the recorded (dotted lines) and model predicted (solid lines) trajectories and $\alpha = 0.05$. The symbols r_1, r_2 and r_3 mark three different ratios between the linear extents of the large and small loops of the T_2 and T_3 templates and t_1, t_2 and t_3 mark three different tempos of generation of the T_3 template. Reproduced by permission from Viviani & Flash (1995).

can successfully account for both the piece-wise segmentation of the movements, as captured by the piece-wise constancy of the gain factor as well as for the aforementioned isochrony principle. To this end, hand trajectories measured during the generation of several figural forms., i.e., a limacon, a clove-leaf, and the figure eight (a leminiscate) (see Fig. 1A) were examined. In applying the minimum-jerk model to such drawing movements, a less constrained version of the minimum-jerk model than the one previously used in (Flash & Hogan, 1985) was required. To satisfy the conditions present during repetitive drawing movements, non-zero accelerations and velocities at the movement end-points had to be considered. Moreover, it was assumed that the trajectories for those figural forms are not traced as a whole but rather as a sequence of two symmetric half cycle segments with one via point per segment. To model those movement segments the minimum-jerk cost function was then augmented by either one of the two following alternative sets of via-point constraints: position constraints, position and velocity constraints, and in (Yalov, 1991) position and movement direction (slope) constraints at the via-points. The difference between using explicit velocity constraints or slope constraints is that in the latter case only the ratio between the x and y velocity components and not their actual values at the via-points are explicitly specified. For all three model versions the time of passage through the via point was not derived from the data but was computed by the model. This, therefore, allowed us to examine whether the predicted trajectories satisfy the isochrony principle. The parameters critical for each model version were extracted from the data and were substituted in the corresponding expressions to derive the form of the optimal trajectories. The match between the predicted and measured movements for each figural form was assessed based on the calculation of the sum of squares of position and velocity errors, the values of the time of passage through the via-point and the relationship between the trajectory velocity and curvature. This analysis has shown that for obvious reasons the model using explicit velocity constraints gives the best fit to the data but the model with the slope constraints (Yalov, 1991) also gave satisfactory results. Moreover, the predicted trajectories were found to satisfy the power law with similar relationship between hand curvature and velocity as was observed for the measured movements (Fig. 1B), and were also found to obey the isochrony principle similarly to the recorded movements. Further mathematical analysis has shown that general fifth-order polynomials (minimum-jerk trajectories) with arbitrary coefficients do not automatically satisfy the power law relationship between velocity and curvature (Yalov, 1991). Nonetheless, the minimum-jerk model combined with the end-point positions, velocities and accelerations as those present during the recorded drawing movements, yielded trajectories with similar spatial and temporal characteristics to the measured ones.

Thus, taken together, the above studies have shown that the minimum-jerk model and the empirically derived power law are consistent with each other. It should be emphasized, however, that the implications of these two models are different. According to the two-thirds power law, movement path is internally represented in its entirety a priori, and the control system uses this law of motion to prescribe the velocity of motion. On the other hand, the minimum-jerk model enables to determine both the hand path and law of motion based on the positions and movement directions at several critical points. Hence optimization based models offer more parsimonious descriptions of complicated trajectories based on a small number of position and/or velocity values at a few via points (for a model of handwriting based on similar ideas see Edelman & Flash (1987)). Those models, however, make no assumptions concerning the existence of intermediate levels of representation of the instantaneous details of the trajectories. Moreover, while in the two-thirds power law an individual stroke is defined according to whether its gain factor remains the same for the entire segment, in the minimum-jerk model a motion segment represents a motor unit over which hand jerk is globally minimized. Is it at all necessary to entertain a notion of intermediate movement representations which are isomorphically related to the actual movements, as implied by the two-thirds power law? Several lines of evidence support the existence of such intermediate representations (Vivinai & Flash, 1995). They include the observation that the length of a motion segment influences, although to a limited extent, the duration required for its generation. It seems, therefore, implausible to assume that information about the movement length is available to the system while its form is not explicitly represented. Secondly, recent neurophysiological data suggest that an accurate representation of the real movement is present at the cortical level by means of the population vector which represents both movements speed and its instantaneous direction (Schwartz, 1993). Finally, the behavior observed during continuous target tracking movements also supports such a detailed intermediate movement representation.

HUMAN ARM TRAJECTORY MODIFICATION STUDIES

Beyond the need to generate movements toward static objects, the nervous system must also correct or modify ongoing movements and control dynamic tasks. Thus, in a recent series of studies the mechanisms subserving arm trajectory modification were investigated in human subjects, using the double-step target displacement paradigm (Henis and Flash, 1991, 1995). In this paradigm, the subject is presented with a visual target and is instructed to move his/her arm toward the target. In some of the trials, however, the target might be suddenly displaced following an inter-stimulus-interval (ISI) to a new location either during the reaction or movement time (Flash & Henis, 1991) and the arm trajectory must be accordingly modified. In considering the problem of rapid modification of an ongoing motion plan, one's first guess is that upon the change in target location the initial motion plan is aborted and the motor system plans a new movement toward the new target location (this will be referred to here as the "abort-replan" scheme). However, mathematical analysis of experimentally recorded human arm movements has demonstrated that the kinematic features of the measured movements were better accounted for by an alternative trajectory modification scheme (Flash & Henis, 1991). According to this scheme, the initial trajectory plan is neither aborted nor modified following the target switch but is instead vectorially summed with a new unconstrained point-to-point trajectory plan for moving between the first and second target locations. The second added trajectory plan was also shown to have similar kinematic characteristics to those of unconstrained point-to-point trajectories, i.e., to follow a kinematic profile which is adequately described by the minimum-jerk model. Thus, the modified trajectories appear to result from the vectorial superposition of two time-shifted point-to-point motions. The first unit is expressed as follows:

$$x(t) = x_0 + (x_1 - x_0) \cdot \left[6\left(\frac{t}{T_1}\right)^5 - 15\left(\frac{t}{T_1}\right)^4 + 10\left(\frac{t}{T_1}\right)^3 \right] \tag{3}$$

where T_1 is the movement duration and x_0 and x_1 are the initial hand position and first target location, respectively.

The vectorially added time-shifted trajectory unit is expressed as follows:

$$x(t) = (x_2 - x_1) \cdot \left[6\left(\frac{t-s}{T_2}\right)^5 - 15\left(\frac{t-s}{T_2}\right)^4 + 10\left(\frac{t-s}{T_2}\right)^3 \right] \tag{4}$$

where x_2 is the second target location, s is the time-shift between the two trajectory units,

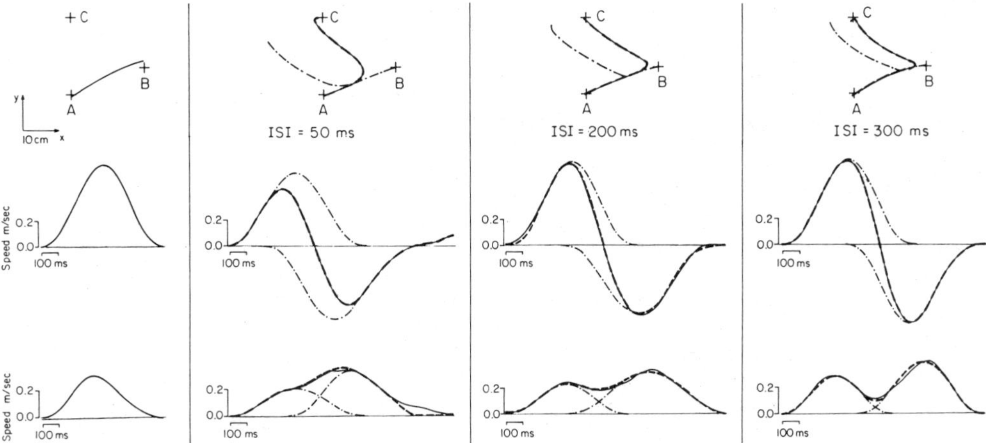

Figure 2. The Superposition scheme. Comparisons between measured and simulated non-averaged modified trajectories for different ISIs. The simulated trajectories were derived based on the superposition scheme. Shown are the hand paths (top rows) and the x and y velocity components (lower rows, V_x and V_y, respectively) of the measured (solid lines), simulated (dashed lines) and of the two superimposed trajectory units (alternating dots and dashes). Reproduced by permission from Flash and Henis (1991).

and T_2 is the second unit duration. The added trajectory unit has zero initial and final velocities and accelerations as well as a zero initial position. Thus, the vectorial summation of the added motion plan with the initial one guarantees continuity in hand position and its first and second derivatives.

The success of the superposition model in capturing the kinematic features of the modified movements is illustrated in Fig. 2 where the measured movements were compared to the simulated ones for the non-averaged movements. The abort-replan scheme was also mathematically modeled and arm movements based on this scheme were predicted and mathematically compared to the measured ones. From a computational perspective, comparing between the two alternative trajectory modification schemes, the superposition scheme is computationally much simpler than the abort-replan scheme, since no information about the expected hand location at the switching time is required in this scheme. Moreover, the superposition scheme was found to be significantly more successful in accounting for the kinematic features of the measured modified movements than the alternative abort-replan scheme. Hence taken together, these findings have suggested that arm trajectory modification may involve parallel planning and superposition of elementary units of action and that there exist a basic repertoire of elementary movements from which more complex movements are constructed.

More recently, arm trajectory modification was further investigated for relatively short ISIs, i.e., ISIs ranging between 10-200 msec (Henis & Flash, 1995). Under such conditions, a high percentage of the movements were found to be initially directed in between the first and second target locations (*averaged* trajectories). The initial direction of movement was found to depend on D: the time difference between the presentation of the second stimulus and movement onset (see Fig. 3). In attempting to account for the kinematic features of the averaged trajectories, the abort-replan and the superposition schemes were again used, and the performance of the superposition scheme was found to surpass that of the abort-replan scheme. This time, however, the modified trajectories were shown to result from the vectorial summation of the two following elemental motions: one for moving between the initial hand position and an intermediate location, and a second one for moving between the intermediate location and the final target. Moreover, it was hypothesized that due to the quick

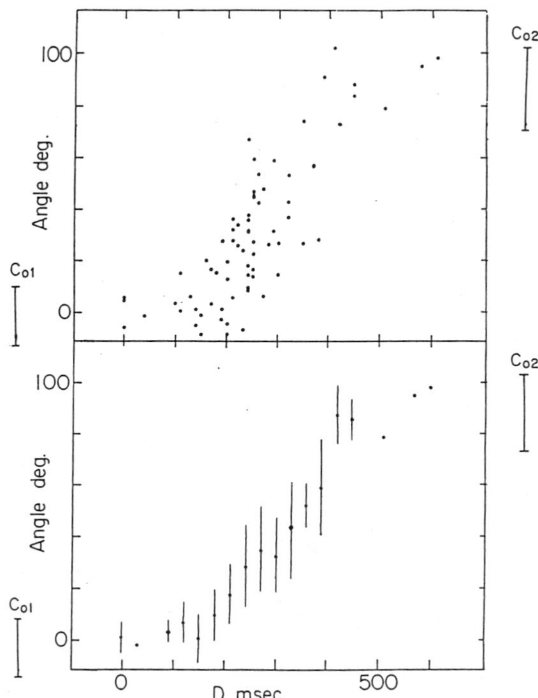

Figure 3. Initial movement direction of the modified trajectories as a function of $D = RT_1 - ISI$, i.e., the duration of the time interval between the second target presentation (i.e., change in target location) and movement initiation. Top panel: raw data. Lower panel: Means and standard deviations of initial directions for movements with similar Ds within 15 msec intervals. Reproduced by permission from Henis & Flash (1995).

displacement of the stimulus, the *internally specified* intermediate goal might be influenced by both stimuli. Hence, its location was hypothesized to be different from that of the first stimulus. Mathematical analysis was further performed to infer the intermediate (i.e., internally represented) target locations. This analysis has shown that for increasing values of the D parameter, these inferred locations gradually shift from the first toward the second target locations along a path that curved toward the initial hand position (Fig. 4). These locations showed also a strong resemblance to the intermediate locations of saccadic eye movements generated in a similar double-step paradigm (Aslin & Shea, 1987). The match between the inferred end-points of the first motion units as derived from our study of arm trajectory modification, and those measured for saccadic movements, may suggest underlying similarities between internally perceived target locations in oculo-motor and manual motor control which may simplify the computations and coordinate transformations involved in visual-motor integration.

In the aforementioned studies the superposition scheme was found to better account for the data than an alternative model involving the abort of the rest of the initial motion plan and the replanning of a new trajectory heading toward the final target. By contrast, Hoff

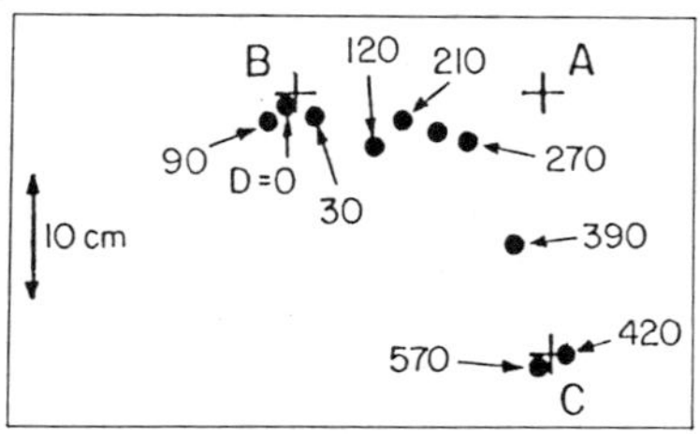

Figure 4. Shifting of the locations of the intermediate targets representing the end-points of the first trajectory units (inferred based on the superposition scheme applied to *average* movements) with increasing values of $D = RT1 - ISI$. Reproduced by permission from Henis and Flash, (1995). A, B and C denote, respectively, the initial hand position and the first and second target locations.

and Arbib (1993), have suggested a trajectory generation and modification scheme which is based on the use of an internal feedback control law. Hence, as the movement proceeds, the system continuously re-evaluates its progress based on incoming albeit delayed information and uses this information to generate minimum-jerk like reaching movements. To deal with external perturbations or to incorporate online information, a scheme involving the generalization of the minimum-jerk model to non-zero initial boundary conditions and compensations (after some delay) for the perturbations was developed. This scheme, however, is mathematically equivalent to the abort-replan scheme and as discussed above, was less successful than the superposition scheme in accounting for the measured data.

MOTOR EXECUTION AND ADAPTATION SCHEMES

To execute any desired motion plan, appropriate joint torques and muscle activation patterns must be generated. One possible way that this can be done is by first transforming the desired hand trajectory plans into appropriate joint rotations by solving the inverse kinematics problem. Then, the necessary joint torques can be derived by solving the "inverse dynamics" problem. Given the complicated dynamic interactions that exist between the moving skeletal segments during multi-joint movements and the need to distribute the necessary joint torques among a highly redundant set of muscles, it was hypothesized that the motor system must have developed alternative means and control schemes for motor execution that do not involve explicit joint torque and muscle force computations. One such scheme is the so called equilibrium trajectory model (Feldman, 1986; Hogan, 1985; Flash, 1987; Bizzi *et al.*, 1992). According to this model, the viscoelastic properties of muscles might play an important role in allowing the motor system to bypass the need for explicit torque computations. Thus, movements are generated by gradually shifting the limb equilibrium position along the desired motions while the equilibrium positions are internally coded by specifying the appropriate neural activities to sets of agonist and antagonist spring-like muscles. Alternatively, the recent progress in neural network research, has led to a renewed interest in the possibility that while the motor system may not solve from scratch the inverse kinematics and dynamics computations each time a new movement is about to be generated, successful solutions to these problems are embedded into the synaptic connections between the elements of cortical and sub-cortical networks subserving various motor functions (e.g, Alexander *et al.* (1992)).

According to the "hierarchical approach" to human arm trajectory control expressed above and elsewhere (e.g., Flash 1990), the motor system is hierarchically organized and desired behavioral goals are gradually transformed into actual movements following a step-by-step transformation whereby higher motor levels specify the desired motion plans while lower levels are concerned with their execution. Recently, an alternative view was suggested (Uno *et al.*, 1989). Thus, while it was again postulated that the motor commands are directly calculated from the goal of the movement, represented by some performance index, by contrast to the minimum-jerk model, Uno *et al.* (1989) have postulated that the underlying criterion for movement selection involves the optimization of the rate of change of joint torques. This latter objective of performance is critically dependent on the dynamics of the musculoskeletal system. Thus, while according to the minimum-jerk model, kinematic motion plans are selected independently of the movement dynamics or external load conditions, according to this alternative model, substantially different hand trajectories are expected to be generated for movements performed in different regions of the workspace, or under different load conditions. These two alternative classes of models would also have substantially different predictions with respect to the characteristics of the movement output following load adaptation.

MOTOR ADAPTATION STUDIES

The question of what is optimized in arm movements can be investigated from motor learning and adaptation perspectives. Investigating motor adaptation to elastic loads, Uno *et al.* (1989) have concluded that the behavior in the presence of the load is different from the one seen in the unloaded case. Completely different results, however, were obtained in another recent study (Gurevich, 1993; Gurevich & Flash, 1995). In this study, static elastic loads were unexpectedly introduced during human reaching toward visual targets. Thus, while in the first few trials following load application the movements were found to be misdirected and to miss the final target, following a relatively small number of practice trials (5-7), the loaded movements tended to converge toward the ones seen in the unperturbed case, i.e., to follow straight hand paths with symmetric velocity profiles (Flash & Gurevich, 1992; Gurevich, 1994). These observations have therefore indicated that human arm trajectories tend to obey the same kinematic plan independently of the external force conditions, thus supporting the idea that the desired behavior is independent of movement dynamics. Similar conclusions were arrived at in another recent study (Shadmehr & Mussa-Ivaldi, 1993) in which velocity-dependent force fields were used to perturb the motion and the perturbed trajectories performed in the presence of the new force fields were again found to converge toward the ones seen in the unloaded case. In a third related study, Wolpert *et al.* (1993) have used altered visual feedback conditions that caused an increase in the perceived curvature of aiming movements achieved through computer manipulations. This has led to significant corrective adaptation in the curvature of the actually produced movements; the hand movements became curved, thereby reducing the visually perceived curvature. These results have again suggested that arm trajectories are planned in extrinsic visual space and are incompatible with the assumptions of the minimum-torque change model or similar models that assume that movement generation is based on the optimization of cost functions which depend on movement dynamics and/or are expressed in terms of intrinsic coordinates.

What is it that the CNS learns during skill acquisition and when adapting to new external conditions? To account for the kinematic characteristics of the movements following load adaptation, an adaptation scheme based on the equilibrium trajectory model was suggested. This scheme involves the modification of both the arm impedance (stiffness and viscosity) and equilibrium trajectory (Gurevich, 1993; Gurevich & Flash, 1995a,b). In the section below the proposed model for load adaptation is described at greater length and its implications with respect to motor organization and the underlying neural computations are discussed.

Human Adaptation to Elastic Loads: Behavioral and Modeling Studies

In our studies (Gurevich & Flash, 1995a,b), unloaded as well as loaded movements, generated in the presence of elastic loads, were performed and recorded using a two-degrees-of-freedom planar manipulandum. A servo control system was used to control the endpoint stiffness and viscosity exerted at the tip of the manipulandum and hence the dynamic properties of the external loads. All the movements were performed in the absence of visual feedback from the moving arm. The subjects held the handle of the manipulandum and were instructed to move it under the visual targets upon their illumination. Each block of trials was composed of 45 - 50 point-to-point movements consisting of unloaded movements and of movements performed against elastic loads. During a single block of trials, movements with fixed start and end positions were performed and recorded. Each block of trials began with 10 - 15 control movements after which the load was unexpectedly introduced. After the first loaded movement, 10 - 15 more movements, performed against the load, were recorded.

Then the load was unexpectedly removed and the subject performed 5 - 10 movements without the load. To simulate arm trajectories before and following the application of the load, the arm stiffness field was measured either in the presence or absence of external bias forces. Stiffness measurements were conducted using a two-degrees-of-freedom planar manipulandum according to the procedure described in Mussa-Ivaldi *et al.* (1985).

The kinematic analysis of the hand trajectories recorded in these experiments has demonstrated that goal-directed arm movements performed immediately following the introduction of the external elastic loads showed typical deviations from the straight lines connecting the initial hand positions and the final target locations and were characterized by large end-point errors. Typical examples of first loaded (FL) and released (RLS) trajectories are presented in Fig. 5 together with the trajectories of all unloaded (UL) and loaded skilled (LSK) movements performed within the same block of trials. The end points for the UL, LSK, FL and RLS movements, marked by the filled circle, empty triangle and filled triangle, respectively, are also shown in these figures.

Regarding the loaded movements following practice, it was found that the compensation for external loads was remarkably fast and efficient. Thus, even though no instruction was given to the subject to attempt and restore the original kinematic form of the movement, already after five to seven trials following the introduction of the load, the movement trajectories converged toward roughly straight hand paths with bell-shaped speed profiles similarly to those observed in unconstrained planar horizontal reaching movements. The

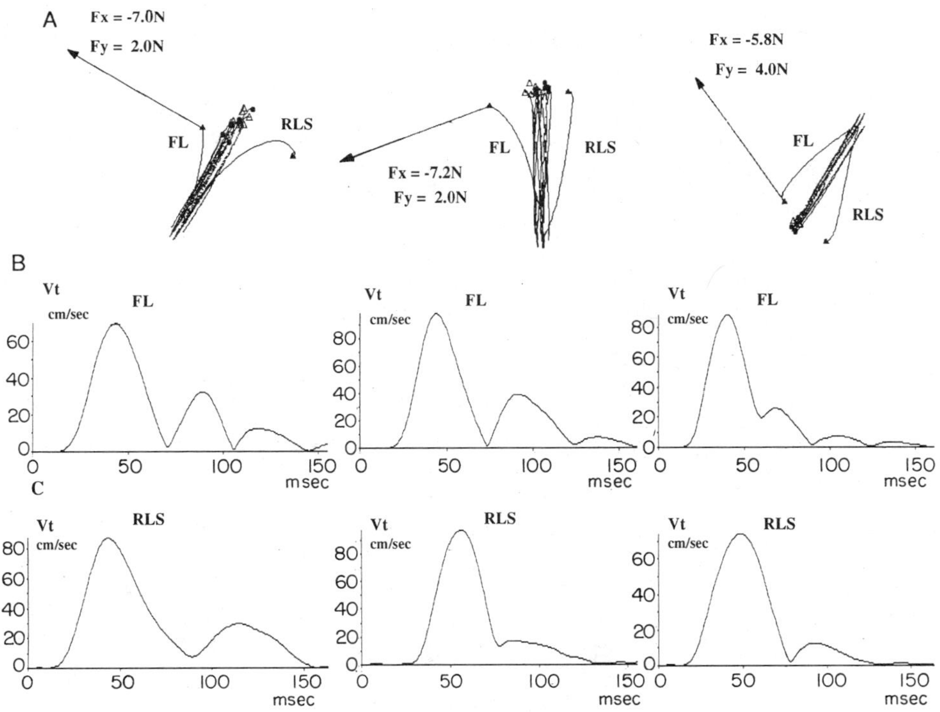

Figure 5. A: Hand paths and end-point locations of UL, LSK, FL and RLS movements, where the end-points of the UL movements are marked by the filled circles and those of the LSK movements by the triangles. The end-points of the FL and RLS movements are marked by filled triangles. The direction of the load is indicated by the arrow. C: Velocity profiles of the FL movements. C: Velocity profiles of the RLS movements.

quick adaptation to external loads is illustrated by the values of the correlation indices of the loaded trajectories with respect to the unloaded movements (calculated as in Edelman and Flash, 1987) which are displayed in Fig. 6 in a successive order versus the number of movement trial within one block of movements. Notice the relatively low values of the correlation indices for the first loaded movement (FL) with respect to those of the unloaded movements and then the gradual increase in their magnitudes toward values similar to those obtained for the unloaded movements.

Beyond the results of the kinematic analysis described above, we were interested in investigating what mechanisms might possibly subserve the tuning of motor commands in order for the movements following practice to have similar kinematic characteristics to those observed for the unloaded movements. Thus, a control scheme which extends the concept of the *Equilibrium Trajectory Control* hypothesis to motor adaptation to external loads was developed (Gurevich, 1993; Gurevich & Flash, 1995).

According to the *Equilibrium Trajectory Control* hypothesis (Flash, 1987), multi-joint arm movements result from the gradual shifting of the hand equilibrium position along a desired trajectory which in the case of point-to-point movements in the horizontal plane was shown to be a straight line connecting the end-points of the movement. In earlier papers it was assumed (Hogan, 1985; Flash 1987) that, if for the sake of simplicity, the different muscle groups that act on the arm are modelled as linear springs in series with viscous damping elements, the resulting joint torques underlying movement generation can be shown to depend on the instantaneous differences between the actual and equilibrium joint positions and on the instantaneous joint velocities as follows:

$$\vec{N} = R(\vec{\Theta} - \vec{\Theta}_0) - B\dot{\vec{\Theta}} \tag{5}$$

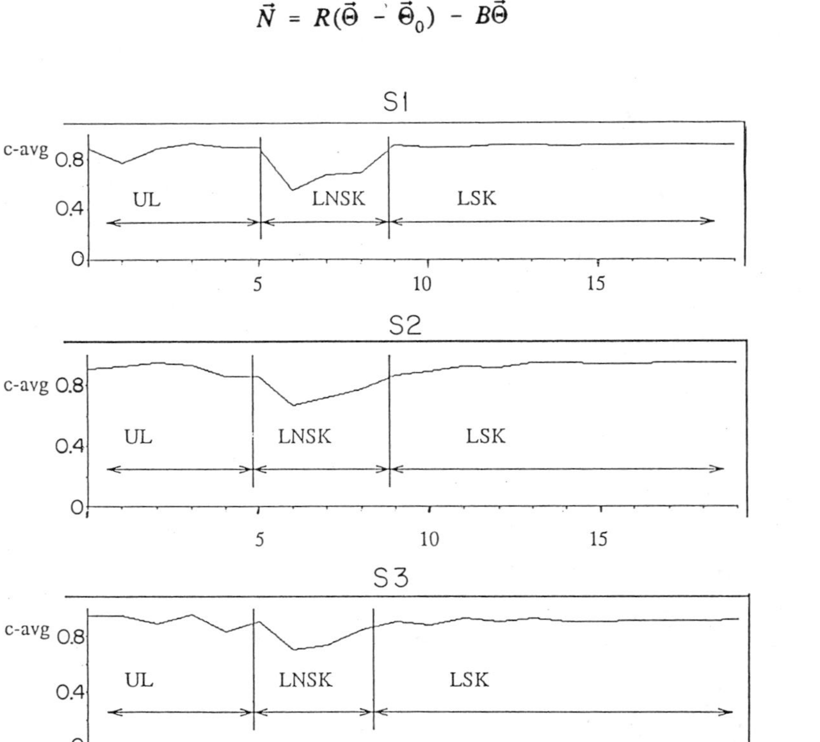

Figure 6. Average correlation indices versus the trial number for three different subjects (S1, S2 and S3).

where $\vec{\Theta}$ and $\vec{\Theta}_0$ are, respectively, the vectors of actual and equilibrium joint angles, R and B are, respectively, the joint stiffness and viscosity matrices, and $\dot{\Theta}$ is the vector of joint velocities.

On the other hand, in order to generate a movement in the multi-joint case, the resultant joint torques required to drive the arm along any given trajectory must obey the following expression (Hollerbach & Flash, 1992):

$$\vec{N} = I(\vec{\Theta})\,\ddot{\vec{\Theta}} + C(\vec{\Theta},\,\dot{\vec{\Theta}})\,\dot{\vec{\Theta}} \tag{6}$$

where $\vec{N}$ is the vector of shoulder and elbow joint torques, $I\,\vec{\Theta}$ is the inertia matrix, $C(\vec{\Theta}\vec{\Theta}$ is a matrix specifying the centrifugal and Coriolis interaction torques, $\vec{\Theta}$ is a vector describing the shoulder and elbow joint angles, $\dot{\Theta}$ is a vector of joint velocities and $\ddot{\Theta}$ is a vector of joint accelerations. In the unloaded case the torques acting at the different joints are actively generated by the muscles acting about these joints. Thus, in the two-joint case, when an external torque is applied to the subject's hand, the total torque driving the arm along the measured movement can be expressed as:

$$\vec{N}^L + \vec{T}^L = I(\vec{\Theta}^L)\ddot{\vec{\Theta}}^L + C(\sigma^L,\vec{\Theta}^L)\dot{\sigma}^L \tag{7}$$

where $\vec{\Theta}^L$, $\dot{\vec{\Theta}}^L$ and $\ddot{\vec{\Theta}}^L$ describe, respectively, the joint angular position, velocity and acceleration vectors of the loaded movement, $\vec{T}^L$ is the vector of joint torques exerted by the external load and $\vec{N}^L$ represents the vector of joint torques generated by the muscles in the loaded case.

Our experimental observations have indicated that following several practice trials in the presence of the load, the adapted loaded movements tended to converge toward trajectories having very similar kinematic forms to those seen in the unloaded case both in terms of hand paths and velocity profiles. For this to occur, the torques generated by the muscles in the unloaded and loaded cases, $\vec{N}$ and $\vec{N}^L$ respectively, should satisfy the following relationship:

$$\vec{N} = \vec{N}^L + \vec{T}^L \tag{8}$$

In order to satisfy this relationship and based on the equilibrium trajectory control model, a load adaptation scheme was developed that successfully accounted for the behavior observed in our experimental studies. In developing the proposed adaptation scheme, the main assumption being made was that load adaptation during arm movements involves the modification of both the arm stiffness field and the equilibrium trajectory. To test the validity of the proposed scheme, a computer simulation model implementing this scheme for a two-joint arm was developed. In developing the model several assumptions were made. In particular, the muscles acting on the arm were modeled as non-linear viscoelastic elements with tunable rest-lengths (Feldman, 1986), whereby the total torque, generated by a muscle or by a force generating element was expressed as follows:

$$N = A(\epsilon^{\alpha r(\theta-\theta_0)-b\dot{\theta}} - 1)\quad, \tag{9}$$

where r is the muscle moment arm, θ is the actual joint angle, θ_0 is the equilibrium joint angle corresponding to the angle at which the muscle exerts zero force, $\dot{\Theta}$ is the joint angular velocity, and b is a viscosity parameter. Hence, the joint stiffness contributed by each such force generating element in the unloaded case was assumed to be:

$$R = \frac{dN}{d\theta} = \alpha r(N + A) \tag{10}$$

Three different muscle groups, namely, the single-joint shoulder and elbow and the two-joint muscles were considered in our simulations. The motor commands to those muscle groups were assumed to be associated with the setting of the muscle group's rest angle. Finally, the control of movement was assumed to involve a gradual shift of the equilibrium state of the arm, achieved by modifying the rest angles of the different muscle groups.

Considering the above assumptions, the motor commands to each muscle group required to oppose the external load were assumed to be associated with a shift λ of the instantaneous equilibrium position of the unloaded movement θ_0 to a new equilibrium position. Hence the joint torque N^L generated by any muscle group in the loaded case was expressed as follows:

$$N^L = A(\epsilon^{\alpha r(\theta - \theta_0^l - b\dot\theta)} - 1), \text{ where } \theta_0^l = \theta_0 + \lambda \tag{11}$$

and the corresponding stiffness in the loaded case was therefore:

$$R^L = \frac{dN^L}{d\theta} = \alpha r(N + A)\beta = R\beta \quad , \tag{12}$$

where N and R are expressed according to (9) and (10), and β is defined as:

$$\beta = \epsilon^{-\alpha r \Lambda} \tag{13}$$

To achieve load adaptation, condition (8) must be satisfied. Hence, using a linear approximation of the muscle torque rest-length relationship (Eq. 9), the following relationship between the loaded and unloaded instantaneous joint equilibrium positions should hold:

$$\vec{\Theta}_0^L \approx \vec{\Theta}_0 + \frac{\vec{T}^L}{R^L} \tag{14}$$

Where $\vec{\Theta}_0^L = (\theta_{01}^L, \theta_{02}^L)^T$ is the vector of instantaneous shoulder and elbow equilibrium positions in the loaded case, $\vec{\Theta}_0$ is the corresponding vector in the unloaded case, $\vec{T}_0$ is the vector of externally applied joint torques and R^L is the arm joint stiffness matrix in the loaded case. Based on the above considerations it was therefore suggested (Gurevich, 1993; Gurevich & Flash 1995a,b) that in adapting to the presence of external loads, the instantaneous joint equilibrium position vector $\vec{\Theta}_0$ must be shifted by $\vec{\Delta}$ where:

$$\vec{\Delta} = \vec{\Theta}_0^L - \vec{\Theta}_0 \approx \frac{\vec{T}^L}{R^L} \tag{15}$$

Thus, according to the proposed adaptation scheme the *adapted* equilibrium trajectory for the LSK movement corresponds to a vectorial sum of the original equilibrium trajectory, generated in order to drive the arm along the desired straight hand path, and a shift of the equilibrium position generated in order to compensate for the presence of an external load.

With respect to the arm stiffness field that accompanies the introduction of external loads, recently we have measured the static arm stiffness field both in the loaded and unloaded cases (Flash & Gurevich, 1995). A mathematical model was then developed describing the variations of the stiffness field with external loads. A full description of this model can be found elsewhere (Gurevich, 1993; Flash & Gurevich, 1995). Nonetheless, for our purposes here it is important to mention that the model has successfully accounted for the measured stiffness field under different experimental conditions (i.e., different external bias forces and measuring positions). The model was based on the assumption that the

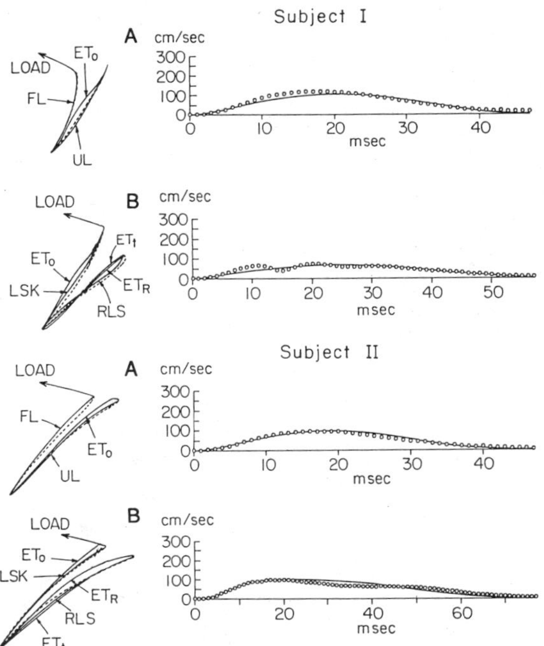

Figure 7. Simulations of UL, FL, LSK and RLS movements performed in two different directions by one subject. The actual and equilibrium trajectories are represented by solid lines and the simulated actual trajectories - by dashed lines. The direction of the load is indicated by the arrow. *From left to right* are drawn the actual and simulated hand paths, equilibrium trajectory paths and velocity profiles of the actual and simulated movements. **A:** Simulations of UL and FL movements; ET_0 - is the derived equilibrium trajectory of the unloaded movement; the actual and simulated velocity profiles of the UL movement. **B:** The actual and simulated trajectories of the LSK and RLS movements; ET_t is the total equilibrium trajectory of the LSK movement derived according to the summation scheme; ET_r is the ET of the LSK and RLS movements derived from the RLS movements; the actual and simulated velocity profiles of the LSK movement.

relationship between the torque generated by each individual muscle group and the joint angle can be expressed by eq. 9. Moreover, our experimental observations have demonstrated that the polar orientation of the arm stiffness field, observed in the unloaded case (Flash & Mussa-Ivaldi, 1990), is maintained in the presence of external loads for all measuring positions and for different magnitudes and directions of the external load (Gurevich, 1993; Flash & Gurevich, 1995). Hence, based on those observations and our aforementioned assumptions, all four elements of the loaded joint stiffness matrix were modelled as sums of two different terms. One term describes the stiffness of the corresponding stiffness element in the unloaded case while the second term shows a linear dependency on the magnitudes of the shoulder and elbow joint torques, T_1 and T_2, respectively, which are exerted by the external force acting on the subject's hand. Thus, all four elements of the joint stiffness matrix in the loaded case, R_{ij}^L ($i = 1...4$), were described as follows:

$$R_{ij}^L = CR_{ij}^0 + a_{ij}T_1 + b_{ij}T_2, \tag{16}$$

where C is a constant gain factor, R_{ij}^0 is the corresponding joint stiffness element during the unloaded case and a_{ij} and b_{ij} are coefficients whose values depend on the arm configuration and on the muscle parameters. It should be emphasized that in our model the same gain factor was used for all four elements of the stiffness matrix hence guaranteeing that both the shape and orientation of the unloaded static stiffness ellipses will remain invariant for any given postural location (see Mussa-Ivaldi *et al.*, 1985). Moreover, according to the model the values of all three different terms R_{ij}^0, a_{ij}, and b_{ij}, remain fixed for any given static postural position, although the value of C, representing the amount of co-contraction, may change among trials even under the same external bias force conditions.

Thus, considering the adaptation strategy described above and our model of the loaded arm stiffness field (see Eq. (16) above), we have proposed that compensation and

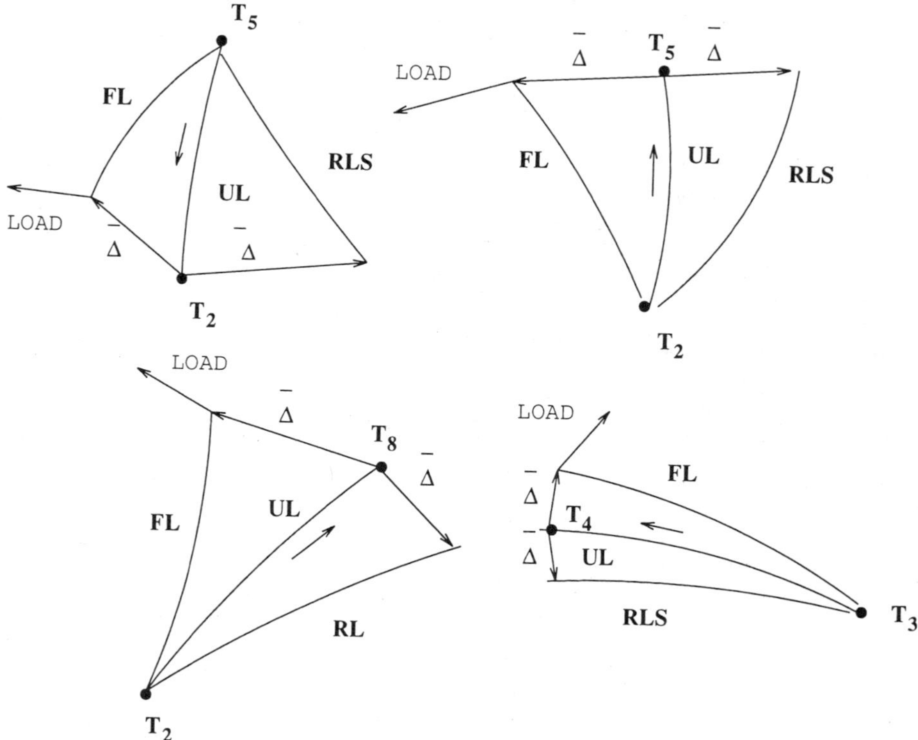

Figure 8. Typical examples of paths of UL, FL and RLS movements. Each panel represents the UL, FL and RLS movements performed within the same block of trials. As can be seen, the trajectories of RLS movements are close to being mirror images of those of the corresponding FL movements with respect to the corresponding UL movements. The vectors represent shifts of the end-positions of the FL and RLS movements with respect to those of the UL movements.

adaptation to external loads reflect the summation of movement-related and load-related components of both the arm equilibrium trajectory and stiffness field.

The proposed summation scheme was implemented in computer simulations, using the measured values of the static stiffness field and the aforementioned model (for details of the simulation procedure see Gurevich, 1993 and Gurevich & Flash, 1995b). The success of our simulations in predicting the kinematic profiles of the UL, FL, LSK and RLS movements based on the above scheme was then examined by comparing the measured to the corresponding computer simulated movements. As illustrated by the typical results presented in Fig. 7, our model was quite successful in accounting for the kinematic properties of all these four movement types. Shown also are the equilibrium trajectories inferred for the UL, LSK and RLS movements, illustrating that following practice in the presence of the load, the equilibrium trajectory for the LSK movements is shifted by $\vec{\Delta}$ as compared to the equilibrium trajectory derived for the unloaded movement.

Thus, taken together the success of our model in simulating the various movement types (UL, FL, LSK, RLS) may suggest that motor adaptation is achieved by appropriately adjusting the equilibrium trajectory and the arm impedance parameters in order for the actual movement to follow the desired kinematic plan in spite of variations in external loads. In this work the equilibrium trajectory control scheme was successfully generalized to deal with the behavior observed immediately after the application of the load and following practice. By contrast, our analysis has shown that the alternative minimum torque change model (Uno *et al.*, 1989) cannot successfully account for the behavior observed following load adaptation since this model assumes that the objective of performance of skilled movements does depend on movement dynamics. Hence, our results provide further support for the idea that arm trajectory generation processes are hierarchically organized whereby movements having the same kinematic forms are generated by appropriately modifying the motor commands so as to adjust to different external loads and/or dynamic conditions.

Our results further showed that, typically within the same block of trials, the trajectories of the RLS movements are almost mirror images of those of the FL movements (see Fig. 8). We have shown that the directions of the new equilibrium trajectories after practice nearly coincide with those of the RLS movements. Therefore, we suggest that instead of deriving explicitly a load-related equilibrium trajectory from the time-sequences of joint torques needed to compensate for the external loads, the motor system may learn based on the end-point errors the amount by which the direction of the equilibrium trajectory should be rotated, in order for these errors to be corrected. Thus, it might not be necessary for the motor system to compute a time-sequence of a load-related equilibrium trajectory, but, as shown in Fig. 8, only an appropriate shift of the equilibrium trajectory. The above strategy may provide effective means for motor adaptation to new external conditions.

The proposed strategy is consistent with recent neurophysiological studies by (Kalaska *et al.*, 1990) indicating that the activity of motor cortical cells, measured during movements performed against external loads, can be described as a summation of phasic activities responsible for driving the arm along the desired trajectory and tonic activities related to arm posture and generated in order to oppose external loads. Those observations support the idea that movement dynamics is somehow encoded by the activity of motor cortical cells. On the other hand, the activity of most of the cells recorded in area 5 was found to be only weakly affected by the load, while showing a strong dependence on movement direction (Kalaska *et al.*, 1983, 1990). This therefore has indicated that the activities of area 5 neurons encode movement kinematics. Hence those results are in accordance with our ideas concerning the hierarchical organization of the motor system, whereby higher levels deal with the planning of movement kinematics in terms of spatial trajectories of the hand while lower levels are concerned with motor execution. Here it is interesting to mention the results reported by Georgopoulos *et al.* (1992) that have indicated that the neural activities in the motor cortex are invariant under directional changes in bias forces, thus disagreeing with the findings of Kalaska (see also Wise 1993).

SUMMARY

In this chapter several aspects of the generation of multi-joint arm movements have been discussed. The first topic dealt with was that of arm trajectory planning during reaching and drawing movements. Evidence for the notion that arm movements are planned in terms of hand trajectories in extra-personal space and the minimum-jerk model that postulates smoothness maximization were reviewed. Then, the extension of this model to curved movements and the attempt to unify between alternative models of drawing movements, namely the minimum-jerk model and the two-thirds power law were discussed. In particular, we reviewed the evidence that the minimum-jerk model can successfully account for the piece-wise segmentation of the movements according to the relation between velocity and curvature as expressed by the two-thirds power law and for the isochrony principle that refers to the durations of these different segments. The idea that complex movements are composed of elementary building blocks received additional support from human arm trajectory modification studies where we have shown that the superposition of such elementary trajectory units can successfully account for the kinematic properties of arm trajectories modified in response to unexpected target switches. Finally, various possible motor execution schemes were discussed and the results from a study of motor adoptation to externally applied elastic loads were presented. In particular, our results have demonstrated that following a few practice trials the loaded movements tended to converge toward straight hand paths with bell-shaped velocity profiles which are characteristic of arm movements recorded in the unloaded case. These results are consistent with the idea that arm generation processes are hierarchically planned and that it is possible to distinguish between control levels that are involved in planning desired kinematic plans and other levels that take care of motor execution.

ACKNOWLEDGMENTS

This research was partially supported by grants 85-00395 and 88-00141 from the United States - Israel Binational Science Foundation (BSF), Jerusalem, Israel and by the McDonnel-Pew Program in Cognitive Neuroscience.

REFERENCES

Abend, W., Bizzi, E., and Morasso, P., 1982, Human arm trajectory formation, *Brain* 105:331-348.

Alexander, G. E., DeLong, M. R., and Crutcher, M. D., 1992, Do cortical and basal ganglia motor areas use "motor programs" to control movement, *Behavioral Brain Sci.* 15:644-655.

Aslin, R. N., and Shea, S. L., 1987, The amplitude and angle of saccades to double-step target displacements, *Vision Res.* 27:1925-1942.

Bizzi, E., Hogan, N., Mussa-Ivaldi, F. A., and Giszter, S., 1992, Does the nervous system use equilibrium point control to guide single and multiple-joint movements, *Behavioral Brain Sci.* 15:603-613.

Edelman, S., and Flash, T., 1987, A model of handwriting, *Biol. Cybern.* 57:25-36.

Feldman, A. G., 1986, Once more on the equilibrium-point hypothesis (λ model) for motor control, *J. Motor Behavior* 18:17-54.

Flash, T., and Hogan, N., 1985, The coordination of arm movements: an experimentally confirmed mathematical model, *J. Neurosci.* 7:1688-1703.

Flash T., 1987, The control of hand equilibrium trajectories in multi-joint arm movements, *Biol. Cybern.* 57:257-274.

Flash, T., 1990, The organization of human arm trajectory control, in *Multiple Muscle Systems: Biomechanics and Movement Organization* (Winters, J. & Woo, S., eds.), Springer-Verlag, pp. 282-301.

Flash, T., and Mussa-Ivaldi, F., 1990, Human arm stiffness characteristics during the maintenance of posture, *Exp. Brain Res.*, 82:315-326.

Flash, T., and Henis, E., 1991, Arm trajectory modification during reaching towards visual targets, *J. Cognitive Neurosci.* 3:220-230.

Flash, T., and Gurevich, I., 1992, Human motor adaptation to external loads, *Ann. Int. Conf. of the IEEE Eng. in Med., and Biol. Soc.* 13:885-886

Flash, T., and Hogan, N., 1995, Optimization principles in motor control, in *The Handbook of Brain Theory and Neural Networks* (Arbib, M. A., ed.), MIT Press, Cambridge, MA. pp. 682-685.

Flash, T. and Gurevich, I., 1995, Arm postural stiffness field variations with external load. Submitted.

Georgopoulos, P., Ashe, J., Smyrnis, N., and Taira, M., 1992, The motor cortex and the coding of force, *Science*, 1692-1695.

Gurevich, I., 1993, Learning and adaptation in human multi-joint motor behavior, *Ph. D. Thesis*, Dept. of Applied Math., and Computer Science, Weizmann Inst. of Science, Israel.

Gurevich, I., and Flash, T., 1995a, Motor adaptation to external loads in planar two-joint arm movement, part 1. Submitted.

Gurevich, I., and Flash, T., 1995b, Motor adaptation to external loads in planar two-joint arm movement, part 2. Submitted.

Henis, E., and Flash, T., 1995, Mechanisms underlying the generation of averaged modified trajectories, *Biol. Cybern.* 72:407-419.

Hoff, B., and Arbib, M. A., 1993, Models of trajectory formation and temporal interaction of reach and grasp, *J. Motor Behavior* 25:175-192.

Hogan, N., 1984, An organizing principle for a class of voluntary movements, *J. Neurosci.* 4:2745-2754.

Hogan, N., 1985, The mechanics of multi-joint posture and movement, *Biol. Cybern.* 52:315-331.

Hollerbach, J. M., 1982, Computers, brains and the control of movement, *Trends in Neurosci.* 5:189-192.

Hollerbach, J. M., and Flash, T., 1982, Dynamic interactions between limb segments during planar arm movement, *Biol. Cybern.* 44:67-77.

Kalaska, J. F., Cohen, D. A., Prud'homme, M., and Hyde, M. L., 1990, Parietal area 5 neuronal activity encodes movement kinematics, not movement dynamics, *Exp. Brain Res.* 80:351-364.

Kalaska, J. F., Kaminiti, R. and Georgopoulos, A. P., 1983, Cortical mechanisms related to the direction of two-dimensional arm movements: relations in parietal area 5 and comparison with motor cortex, *Exp. Brain Res.* 51:247-260.

Lacquaniti, F., 1989, Central representations of human limb movement as revealed by studies of drawing and handwriting, Trends in Neurosci. 12:287-291.

Morasso, P., 1981, Spatial control of arm movements, *Exp. Brain Res.* 42:223-227.

Mussa-Ivaldi, F. A., Hogan, N., and Bizzi, E., 1985, Neural, mechanical and geometric factors subserving arm posture in humans, *J. Neurosci.* 5:2732-2743.

Shadmehr, R., and Mussa-Ivaldi, F. A., 1994, Adaptive representation of dynamics during learning of a motor task. *J. Neurosci.* 5:3208-3224.

Schwartz, A. B., 1993, Motor control activity during movement population representation during sinusoid tracing, *J. Neurophys.*

Uno, Y., Kawato, M., and Suzuki, R., 1989, Formation and control of optimal trajectory in human multijoint arm movement-minimum torque-change model, *Biol. Cybern.* 61:89-101.

Viviani, P., and Flash, T., 1995, Minimum-jerk, two-thirds power law and isochrony: converging approaches to the study of movement planning, *J. Exp. Psychol. : Human Perception and Performance* 21:32-53.

Viviani, P., and Cenzato, M., 1985, Segmentation and coupling in complex movements, *J. Exp. Psychol. : Human Perception and Performance* 13:62-78.

Wise, S. P., 1993, Monkey motor cortex: movements, muscles, motoneurons and metrics, *Trends in Neurosci.* 16:46-49.

Wolpert, D. M., Ghahramani, Z., and Jordan, M. Z., 1993, On the role of extrinsic coordinates in arm trajectory planning: evidence from an adaptation study, *Computational Cognitive Science Tech. Report 9308*, MIT, Cambridge, MA.

Yalov, S., 1991. Computational models of hand trojectory planning in drawing movements. M. Sc Thesis, Dept. of Appl. Math and Computer Sci., Weizmann Inst. of Science, Israel.

27

PROCESSING AND PATTERN ANALYSIS OF HANDWRITING MOVEMENTS

Ehud Bar-On and Anna Tolmacheva

Technion IIT & BarOn Technologies Ltd.
Gutwirth Park, Technion City, Haifa 32000, Israel

ABSTRACT

A new model for hand movement during writing that assumes three dimensional movement "of the hand", (as opposed to two dimensional x-y pentip movement) has been proposed. Based upon this model, BarOn technologies has constructed a pen that includes built in motion sensors that provide accurate measurements of the three orthogonal accelerations of hand movement during writing. An approach to pattern recognition and identification problem solutions based on hierarchical similarity method is proposed. The results of an experiment in which the aforementioned pattern recognition and identification approach was examined using data collected with the "acceleration" pen are described.

INTRODUCTION

Research activity in handwriting recognition started during the late 1950's, following the invention of electronic tablets that accurately capture x-y coordinates of pen-tip position. On-line handwriting recognition, (in contrast to off-line recognition, e.g., Optical Character Recognition), is based on digitized information of the x-y coordinates recorded over time. One model for hand movement during writing, as captured in the x-y coordinates, is suggested by Hollerbach (1). According to the Hollerbach model, handwriting can be viewed as generated by two coupled oscillators (for horizontal and vertical pen-tip movements) and a constant horizontal movement. In Bar-On and Rumelhart (2) the primitives of hand movement motoric patterns during handwriting were investigated. The main finding of this study was that pen-strokes are specific to an individual writer, and characterize the writer's unique motor control mechanism. The dynamic data from many thousands of handwritten characters, produced by many writers, had been segmented into pen-strokes and subjected to cluster analysis. Cluster analysis of pen-strokes vectors was performed using a variety of clustering techniques, and yielded very consistent results. It was found that despite the fact that hand movement during writing can take any shape or form, a particular writer employs a very limited repertoire of hand movements. The clustering revealed a limited set of twelve to fourteen clusters of pen strokes categorizing more than 95% of total pen strokes. When this study was extended to Japanese characters (Kanji), where characters are composed of many short straight lines pen strokes, the Hollerbach model seemed less appropriate.

Since it is unreasonable to accept that human being of the different cultures posses distinct motor control mechanisms, it was assumed that a more general model is needed. Therefore, a new model that assumes three dimensional movement "of the hand", (as opposed to two dimensional x-y pentip movement) has been proposed. To this end, BarOn technologies (3) constructed a pen that includes a built in motion sensors that provide accurate measurements of the three orthogonal accelerations, translations, and rotations of the hand movement during writing. It was found that hand movement during writing, which is an automated behavior, is more consistent than pen-tip tracing on an electronic tablet surface. The practical benefit of this invention (that has been patented) is better handwriting recognition and verification, and an improved input device for man-machine communication.

PROCESSING AND MODELLING OF HANDWRITING MOVEMENT

In any behavioral model of handwriting, behavior is influenced both by the general properties of handwriting and by properties which are specific to an individual writer. It was found that a considerable part of the variance in the movement sequences can be attributed to individual differences. Many researchers have noted that handwriting style is so distinctive that writers can be recognized according to their handwriting. This is common knowledge and therefore signatures are recognized as a unique identifier of a specific writer. There are however, biomechanic constraints on the hand that must be considered. Some general principles have been suggested as governing the handwriting control mechanism. For example, Flash and Hogan (4), proposed that humans tend to write in a way that minimizes jerk (the third time derivative of the position signal). As we will see, there are alternative hypotheses about the type of constraints imposed by the biomechanics of handwriting.

Despite the fact that the governing principles of handwriting might be universal, each writer has his own unique variation. Differences between individuals were more pronounced in the unwritten strokes (the pen movements that do not touch the writing surface) than in the written ones. The friction of the pen with the writing surface diminishes the characteristics of the hand control mechanism which are better revealed when the pen is up.

A connectionist model proposed by BarOn and Rumelhart (2) is more biologically and cognitive plausible than purely mechanical Hollerbach model (1). The connectionist view proposes that stored knowledge-atoms are dynamically assembled into context-sensitive schemata at the time of inference. This is consistent with our conjecture that there is no essential difference between the so-called "cognitive" and "motoric" brain mechanisms. The connectionist schema-model is also consistent with the neural evidence demonstrating, that specialization among different cortical motor areas is related to a certain sequences of movements (not to transformations as proposed by the " motor program " approach). According to our conjecture, preparatory units and movement executing units belong to the same scheme.

2-D VELOCITY MODEL

The "recognition" procedure proves that there is a consistent pattern of hand movement for a specific writer for each character. We start by presenting results from 2-D velocity signal handwriting recognition. One of the key problems in recognizing cursive handwriting is the segmentation problem. Rumelhart (4) has devised a learning algorithm for cursive handwriting recognition which combined word recognition and letter recognition. The letter recognition was based on recognizing PMP (Primitive Motor Patterns), and PMP sequences that made letters. This system learned to recognize cursive script as it was generated by a writer. It involved simultaneously learning to recognize and segment letters from examples of cursive script produced by and recorded from a number of writers. He collected approximately 1000 words from 58 writers. Although Rumelhart's experiment was done for handwriting recognition, there were several things that could be learned from it concerning PMPs and their sequencing during handwriting. Rumelhart's recognition procedure was based on the velocity of the handwriting signals, as calculated from the x-y position of the pen-tip.

In addition to the data from Rumelhart's experiment, several thousand pen strokes of Japanese handwriting were collected (2). Most of the data was collected from hand written Hiragana characters, but some data was collected during writing Kanji (idiographic) Japanese characters. Hiragana characters have the curved shapes of English hand printed characters, but without the ligature of cursive handwriting. Preprocessing of the collected raw handwriting data was made in an effort to extract features that could be used to segment and characterize the "pen-strokes". A pen stroke was defined as a segment of the cursive writing signal, between two consecutive zero crossings of the vertical velocity of pen movement. Each character was segmented into several segments or "pen-strokes" The principle of segmentation and feature extraction was to segment the continuous velocity signals into discrete segments and to represent each segment by a feature vector in the feature space. Once a "pen-stroke" was defined, there were many ways to represent it in a feature space, because on-line character recognition research employs several orthogonal transformations such as a discrete Fourier transform of the curve segments corresponding to the pen-strokes. That is, a pen-stroke can be represented by its Fourier coefficients obtained from its $x(t)$ and $y(t)$ signals. Segmentation and feature extraction methods depend, of course, upon the goal. If the goal is pattern recognition, then the segmentation and feature extraction are geared toward discrimination between the various patterns. Consequently, the first step of our work was to investigate only features that might be explained by the neurobiological control structures (direction of the strokes, their curvature, etc). The segmentation and feature extraction mechanism employed was to develop a model of the underlying handwriting process and to describe the data in terms of model parameters.

The model employed was derived from that of Hollerbach (1) and involved the assumption that the generation process could be described as a constrained modulation of an underlying oscillatory process. A pair of coupled oscillators (in the vertical and horizontal directions) produce letter forms. Modulation of the vertical oscillation controls the letter heights, while modulation of the horizontal oscillation is responsible for control of the corner shapes by changing amplitude or phase shifting. The coupled harmonic oscillators model is just one of the many models that exist. Hollerbach himself cites previous research that factored handwriting into two functional degrees of freedom: horizontally and vertically. These orthogonal movements are ascribed to the wrist and the finger independent movements. If we speak about writing in a notebook (small size letters), we can think about the wrist's horizontal movements and the finger's flexion and extension movements causing vertical motion. In a case when the size of the letters is more than an inch, the arm muscles are also involved. The Hollerbach model is a purely mechanical model, it doesn't deal with neural system control, but some of the parameters might be interpreted in terms of the human neuromechanical system.

The dynamic data from many thousands of handwritten characters, produced by many writers, has been segmented into pen-strokes and subjected to cluster analysis. The basic units of clustering were the pen-strokes, each of which was represented as a point in an n dimensional space. Of the six features that we extracted for each stroke, only three have been used. First, we used only one frequency for the modeling, so the rare strokes that involved higher harmonies were removed. Second, we did not differentiate between Up-strokes and Down-strokes. Up strokes contain higher order harmonies, but we limited our analysis to basic movements, trying to ignore the fluctuation induced by bio-mechanical control mechanisms. Cluster analysis of the pen-stroke vectors by a variety of clustering methods yielded very consistent results. We found that despite the fact that hand movement during writing can take any shape or form, a particular writer employs only a very limited repertoire of hand movements. The clustering revealed a limited set of twelve to fourteen clusters of pen strokes categorizing more than 95% of total pen stroke movement. These pen-strokes are primitive "motoric patterns", of which handwriting is composed. Furthermore, these motoric patterns reveal an hierarchical structure. Since these primitive patterns are arranged in an hierarchical structure, the actual control mechanism resembles more a "tree" structure more than "look-up" table.

When this study was extended to Japanese characters (Kanji), where characters are composed of many short straight lines pen strokes, the Hollerbach model seemed less appropriate. However, we have shown that they can also be modeled by sinusoidal velocity curves, although not by continuous oscillations as was assumed by Hollerbach. Therefore, a new model has been proposed which assumes three dimensional movement, when the modeling is of the hand movement and not the x-y movements of the tip. To this end, BarOn technologies (3) constructed a pen that includes motion built in motion sensors that provide accurate measurements of the hand during writing. The practical benefit from this invention is better handwriting recognition and improved an input device for man-machine communication.

3-D ACCELERATIONS PATTERN ANALYSIS

The techniques (7, 8) used to solve pattern recognition and identification problems can be grouped into two general classes: the decision-theoretic approach, where the recognition (identification) is based on partitioning the pattern feature space, and the syntactic approach, where the structural pattern models are defined as the union of subpatterns via various ways of composition, and recognition (identification) is based on comparing the model of new patterns against the models, defined in training stage. In both cases the problem cannot be solved without defining a similarity procedure for pattern comparison.

The proposed approach is based on a hierarchical comparison procedure. At each level of comparison a definite membership function for fuzzy comparison procedure is chosen from a set of predefined membership functions. This function is used in order to define a pattern model subset similar to the recognized or identified pattern. On the lower level of the hierarchy only the subset of representatives from the previous level is considered. This method provides flexibility in usage, and is adaptable to different types of writing patterns and users.

Recognition Scheme

During the first stage, define pattern structure, as union of subpatterns and find the structural similarity between the patterns. In the system of handwritten symbol, word, and picture recognition based on acceleration pen signals, subpattern definition is done according to signal levels (8). The acceleration pen captures $x(t)$, $y(t)$, and $z(t)$ accelerations and pen position status (up or down). Segmentation for the acceleration pen is done according to the points t where the pen is changing position. To have the minimal value equal to zero, the sampled $x(t)$, $y(t)$, and $z(t)$ acceleration signals are normalized and after that divided to k levels from minimal until the maximal level and each level is assigned an identity label *0*,

1, 2,..., k-1. After such labeling, we will have the pen stroke representations, as sequences of labels. Before signal segments A^1 and A^2 are compared, interpolation of the signal segments is performed in order to create the segments with equal lengths. The structural similarity function $S(A^1, A^2)$ is defined for all labelled signals segment pairs in the following way:

$$S(A^1, A^2) = \frac{\sum\limits_{i=1}^{l}(a_i^1 - a_i^2)}{l*(k-1)} \tag{1}$$

where l is the maximal length of two segments and k is the number of levels in segments

$$A(a_1^1, a_2^1, ..., a_l^1), A(a_1^2, a_2^2, ..., a_l^2) \tag{1a}$$

The similarity function $S^i(X, Z)$ between the i-th segment of the symbol Z set $\{Z^j\}$ and the i-th segment of the symbol X according to all acceleration signals x, y, z is equal to:

$$S^i(X,Z) = \max_{j}(S_x^i(Z^j) + S_y^i(Z^j) + S_z^i(Z^j)) \tag{1b}$$

where Z^j is symbol Z, which represents the training set, and X is the recognized symbol.

During the second stage, extract main features of patterns which may be useful for pattern recognition, and analyze feature behavior. Individual features may be ordered in the set of pattern classes $P = \{p_1, p_2, p_3,..., p_m\}$ with respect to their ability to divide set P. The function *delta(p_i, p_j)* may be used for this estimation. The average *(mean(p_i))* and deviation *(dev(p_i))* of feature F for all the representatives of the p_i pattern class have to be defined. We assume that

$$mean(p_i) <= mean(p_j), \; if \; i < j. \tag{2}$$

Definition 1.
delta = (mean(p_i) + dev(p_i)) - (mean(p_j) - dev(p_j))
if *(delta < 0)* then *delta(p_i, p_j) = 0*
else *delta(p_i, p_j) = delta*
deltaFinal(F) = (delta(p_1, p_2) + delta(p_2, p_3) + ...
+ delta(p_{m-1}, p_m)) / (mean(p_m)(m-1))*
Definition 2.
Feature F is better than feature $F1$ in the set of patterns P
if *deltaFinal(F) < deltaFinal(F1)*
if delta*(p_1, p_m) > 0* then the feature F is useless in the set of P divisions.

During the third stage, use the basic similarity function for all the feature similarity definitions. The basic similarity function B(s, R) characterizes the similarity between parameter s « [a, b] and the sample of parameters R with known mean M, standard deviation D and range of definition [a, b] and it is defined in the following way:

$$B(s,R) = \begin{cases} 1, \\ \text{if } s \in [M-D, M+D] \\ 1 - (M-D-s)/(b-a-2D), \\ \text{if}((s<(M-D))\&(b-a \neq 2D)) \\ 1 - (s-M-D)/(b-a-2D)), \\ \text{if}((s>(M+D)) \& (b-a \neq 2D)) \end{cases} \tag{3}$$

If in the training set there is only one representative, or an approach to pattern recognition without feature statistic generalization is used, then $D = 0$. In these cases triangular similarity function instead of a trapezoidal similarity function in fuzzy reasoning process is applied.

PATTERN RECOGNITION AND IDENTIFICATION

Algorithm for Recognition

The recognition algorithm is based on the defined structural similarity function $S(A^1, A^2)$ and the basic similarity function $B(s, R)$. It is assumed, that pattern structure is the most informative attribute, and that features are ordered according to their ability to divide the set of patterns. For pattern recognition realization do the following:

Store all the templates $T_1, T_2,... T_n$. Define the useful features $F1, F2,..., Fk$ and the sequential similarity functions $R_1, R_2,..., R_k$ based upon $B(s, R)$ function.

Compute and store tables with entries

Table_0(i, j) = S (T_i, T_j) (Structural similarity function)
Table_1(i, j) = R_1(T_i, T_j) (First feature similarity function)
Table_2(i, j) = R_2(T_i, T_j) (Second feature similarity function)

.

.

.

Table_k(i, j) = R_k(T_i, T_j)(k-th feature similarity function)

Compute and store thresholds using similarity functions values on different templates.

$t_0(i_0) = Min(1 - S(T_i, T_j))$
$t_1(i_1) = Min(1 - R_1(T_i, T_j))$
$t_2(i_1) = Min(1 - R_2(T_i, T_j))$

.

.

.

$t_k(i_k) = Min(1 - R_k(T_i, T_j))$

For each incoming input pattern I, perform the following steps.
Step 0("Qualify" candidates according to structural similarity function definition)

Mark all templates as "qualified" candidates.
for all j { All "qualified" templates except i_0 },
    if $S(I, T_{i0}) - S(I, T_j) > t_0$,
        "Disqualify" template T_j
    end if
end for
Step 1("Qualify" candidates according to the feature similarity function R_1 definition) Use all the templates marked as "qualified" for the structural similarity function S.

    for all j { All "qualified" templates except i_1 },
      if $R_1(I, T_{i1}) - R_1(I, T_j) > t_1$,
        "Disqualify" template T_j
    end if
    end for

.
.
.

Step k("Qualify" candidates according to the feature similarity function R_k definition) Use all the templates marked as "qualified" for the similarity function R_{k-1}

    for all j { All "qualified" templates except i_k },

        if $R_k(I, T_{ik}) - R_k(I, T_j) > t_k$,

            "Disqualify" template T_j

        end if

    end for

If the number of "qualified" templates is greater than one, "disqualify" all templates whose main similarity function $R(I) = (S + R_1 + ... + R_k)/(k+1)$ is less than maximum one in the last step.

Algorithm for Identification

Store all the templates T_1, T_2,..., T_n. Take the set of incoming templates I_1, I_2, ..., I_n and compute the main similarity function $R(I_1)$, $R(I_2)$, ..., $R(I_n)$ using the algorithm for recognition.

Find $T = max\ R(I_j)$, $t = min\ R(I_j)$, $m = Mean\ R(I_j)$, $D = Deviation\ R(I_j)$

Define an interval of uncertain range $U = (t - D, t + D)$

For each input pattern I and number k determine if I belongs to the template class T_k. Perform the following steps:

Compute main similarity function $R(I)$, using algorithm for recognition.

if $R(I, T_k) > t + D$

    mark T_k as "qualified"

end if

if $R(I, T_k) < t - D$

    mark T_k as "unqualified"

end if

if $t-D < R(I, T_k) < t + D$

    mark T_k as "uncertain"

end if

CONCLUSION

The recognition software system was implemented in C++ and runs on IBM PC. Two preliminary experiments were conducted to test the recognition system. Samples of hand-printed characters from 5 writers were collected. For each writer prototypes were created for an alphabet of upper and lowercase letters. Test samples were obtained from each writer in order to test the system. Data for this experiment were collected using an acceleration pen (resolution 12 bit, X-Y-Z accelerometers and pressure sensor, sampling rate 100 Hz). The results of the experiment, in which the writer wrote 5 sets of symbols during the training and recognition stages are : for writer 1 Recognition Rate (RR) equalled to 99%, for writer 2 RR equalled to 93%, for writer 3 RR equalled to 97%, and for writer 4 RR was equalled to 97%. Writer 4 wrote symbols during the recognition stage twice: the first time on the same day as the training stage, the second time - after week. RR for him after week equalled to 95%. The identification algorithm was checked for two problem solutions :signature verification and word signal to symbol signal division.

REFERENCES

1. Hollerbach, J. M., 1981, An Oscillation Theory of Handwriting. *Biol. Cybernetics* 39:139-156.
2. BarOn, E., and Rumelhart, D. E., 1992, What the Human Brain Tells the Human Hand. A Behavioral Perspective. *Technical Report*, Stanford University.
3. PCT/US92/08703. 1992, Apparatus for Reading Handwriting.
4. Flash, T., and Hogan, N., 1985, The Coordination of Arm Movements: an Experimentally Confirmed Mathematical Model. *J.Neurosci.*, 5:1688-1703.
5. Rumelhart, D. E., 1992, Segmentating and Recognizing Online Cursive Handwriting. *Technical Report*, PDP Lab., Stanford University.
6. Castro, J. L., 1995, Fuzzy Logic Controllers Are Universal Approximators. *IEEE Transactions on Systems, Man, and Cybernetics* 25(4):629-635.
7. Avi-Itzak, H., and Thanh D., 1994, Lossless Acceleration for Correlation-based Nearest-neighbor Pattern Recognition. *2ND IAPR Int. Conf. on Pattern Recognition. Jerusalem, Israel, October 9-13.*
8. BarOn, E., Tolmacheva, A., 1994, A Development System for Intelligent Tutors. *Technical Report Part 1. User - Friendly Interface Development.* Technion.

INDEX